Springer Handbook of Enzymes Volume 16

Springer
Berlin
Heidelberg
New York
Hong Kong
London
Milan
Paris
Tokyo

Dietmar Schomburg and
Ida Schomburg (Eds.)

Springer Handbook of Enzymes

Volume 16

Class 1 · Oxidoreductases I

EC 1.1.1.1–1.1.1.50

coedited by Antje Chang

Second Edition

Springer

Professor Dietmar Schomburg
e-mail: d.schomburg@uni-koeln.de

Dr. Ida Schomburg
e-mail: i.schomburg@uni-koeln.de

Dr. Antje Chang
e-mail: a.chang@uni-koeln.de

University to Cologne
Institute for Biochemistry
Zülpicher Strasse 47
50674 Cologne
Germany

ISBN 3-540-20596-9 2nd Edition Springer-Verlag Berlin Heidelberg New York

The first edition was published as Volume 9 (ISBN 3-540-59077-3) of the "Enzyme Handbook".

Library of Congress Cataloging-in-Publication Data

Springer handbook of enzymes /
Dietmar Schomburg and Ida Schomburg (ed.) –
Berlin ; Heidelberg ; New York ; Barcelona ; Hong Kong ;
London ; Milan ; Paris ; Tokyo : Springer

Vol. 16 Class 1, oxidoreductases I: EC 1.1.1.1–1.1.1.50 / coed. by Antje Chang. – 2. ed. – 2004
ISBN 3-540-20596-9

Springer-Verlag Berlin Heidelberg New York
a member of BertelsmannSpringer Science+Business Media GmbH

springeronline.com

Cover design: Erich Kirchner, Heidelberg
Typesetting: medio Technologies AG, Berlin

2/3020xv-5 4 3 2 1 0 – printed on acid-free paper

Attention all Users of the "Springer Handbook of Enzymes"

Information on this handbook can be found on the internet at http://www.springer.de/enzymes/

A complete list of all enzyme entries either as an alphabetical Name Index or as the EC-Number Index is available at the above mentioned URL. You can download and print them free of charge.

A complete list of all synonyms (> 25,000 entries) used for the enzymes is available in print form (ISBN 3-540-41830-X).

Save 15 %

We recommend a standing order for the series to ensure you automatically receive all volumes and all supplements and save 15 % on the list price.

Preface

Today, as the full information about the genome is becoming available for a rapidly increasing number of organisms and transcriptome and proteome analyses are beginning to provide us with a much wider image of protein regulation and function, it is obvious that there are limitations to our ability to access functional data for the gene products – the proteins and, in particular, for enzymes. Those data are inherently very difficult to collect, interpret and standardize as they are widely distributed among journals from different fields and are often subject to experimental conditions. Nevertheless a systematic collection is essential for our interpretation of genome information and more so for applications of this knowledge in the fields of medicine, agriculture, etc. Progress on enzyme immobilisation, enzyme production, enzyme inhibition, coenzyme regeneration and enzyme engineering has opened up fascinating new fields for the potential application of enzymes in a wide range of different areas.

The development of the enzyme data information system BRENDA was started in 1987 at the German National Research Centre for Biotechnology in Braunschweig (GBF) and is now continuing at the University at Cologne, Institute of Biochemistry. The present book "Springer Handbook of Enzymes" represents the printed version of this data bank. The information system has been developed into a full metabolic database.

The enzymes in this Handbook are arranged according to the Enzyme Commission list of enzymes. Some 3,700 "different" enzymes are covered. Frequently enzymes with very different properties are included under the same EC- number. Although we intend to give a representative overview on the characteristics and variability of each enzyme, the Handbook is not a compendium. The reader will have to go to the primary literature for more detailed information. Naturally it is not possible to cover all the numerous literature references for each enzyme (for some enzymes up to 40,000) if the data representation is to be concise as is intended.

It should be mentioned here that the data have been extracted from the literature and critically evaluated by qualified scientists. On the other hand, the original authors' nomenclature for enzyme forms and subunits is retained. In order to keep the tables concise, redundant information is avoided as far as possible (e.g. if K_m values are measured in the presence of an obvious cosubstrate, only the name of the cosubstrate is given in parentheses as a commentary without reference to its specific role).

The authors are grateful to the following biologists and chemists for invaluable help in the compilation of data: Cornelia Munaretto and Dr. Antje Chang.

Cologne
Winter 2003

Dietmar Schomburg, Ida Schomburg

List of Abbreviations

A	adenine
Ac	acetyl
ADP	adenosine 5'-diphosphate
Ala	alanine
All	allose
Alt	altrose
AMP	adenosine 5'-monophosphate
Ara	arabinose
Arg	arginine
Asn	asparagine
Asp	aspartic acid
ATP	adenosine 5'-triphosphate
Bicine	N,N'-bis(2-hydroxyethyl)glycine
C	cytosine
cal	calorie
CDP	cytidine 5'-diphosphate
CDTA	trans-1,2-diaminocyclohexane-N,N,N,N-tetraacetic acid
CMP	cytidine 5'-monophosphate
CoA	coenzyme A
CTP	cytidine 5'-triphosphate
Cys	cystcinc
d	deoxy-
D-	(and L-) prefixes indicating configuration
DFP	diisopropyl fluorophosphate
DNA	deoxyribonucleic acid
DPN	diphosphopyridinium nucleotide (now NAD)
DTNB	5,5'-dithiobis(2-nitrobenzoate)
DTT	dithiothreitol (i.e. Cleland's reagent)
EC	number of enzyme in Enzyme Commission's system
E. coli	Escherichia coli
EDTA	ethylene diaminetetraacetate
EGTA	ethylene glycol bis(-aminoethyl ether) tetraacetate
ER	endoplasmic reticulum
Et	ethyl
EXAFS	extended X-ray absorption fine structure
FAD	flavin-adenine dinucleotide
FMN	flavin mononucleotide (riboflavin 5'-monophosphate)
Fru	fructose
Fuc	fucose
G	guanine
Gal	galactose

GDP	guanosine 5'-diphosphate
Glc	glucose
GlcN	glucosamine
GlcNAc	N-acetylglucosamine
Gln	glutamine
Glu	glutamic acid
Gly	glycine
GMP	guanosine 5'-monophosphate
GSH	glutathione
GSSG	oxidized glutathione
GTP	guanosine 5'-triphosphate
Gul	gulose
h	hour
H4	tetrahydro
HEPES	4-(2-hydroxyethyl)-1-piperazineethane sulfonic acid
His	histidine
HPLC	high performance liquid chromatography
Hyl	hydroxylysine
Hyp	hydroxyproline
IAA	iodoacetamide
Ig	immunoglobulin
Ile	isoleucine
Ido	idose
IDP	inosine 5'-diphosphate
IMP	inosine 5'-monophosphate
ITP	inosine 5'-triphosphate
K_m	Michaelis constant
L-	(and D-) prefixes indicating configuration
Leu	leucine
Lys	lysine
Lyx	lyxose
M	mol/l
mM	millimol/l
m-	*meta-*
Man	mannose
MES	2-(N-morpholino)ethane sulfonate
Met	methionine
min	minute
MOPS	3-(N-morpholino)propane sulfonate
Mur	muramic acid
MW	molecular weight
NAD^+	nicotinamide-adenine dinucleotide
NADH	reduced NAD
$NADP^+$	NAD phosphate
NADPH	reduced NADP
NAD(P)H	indicates either NADH or NADPH
NBS	N-bromosuccinimide

NDP	nucleoside 5'-diphosphate
NEM	N-ethylmaleimide
Neu	neuraminic acid
NMN	nicotinamide mononucleotide
NMP	nucleoside 5'-monophosphate
NTP	nucleoside 5'-triphosphate
o-	*ortho-*
Orn	ornithine
p-	*para-*
PBS	phosphate-buffered saline
PCMB	*p*-chloromercuribenzoate
PEP	phosphoenolpyruvate
pH	$-\log 10[H^+]$
Ph	phenyl
Phe	phenylalanine
PHMB	*p*-hydroxymercuribenzoate
PIXE	proton-induced X-ray emission
PMSF	phenylmethane-sulfonylfluoride
p-NPP	*p*-nitrophenyl phosphate
Pro	proline
Q_{10}	factor for the change in reaction rate for a 10 °C temperature increase
Rha	rhamnose
Rib	ribose
RNA	ribonucleic acid
mRNA	messenger RNA
rRNA	ribosomal RNA
tRNA	transfer RNA
Sar	N-methylglycine (sarcosine)
SDS-PAGE	sodium dodecyl sulfate polyacrylamide gel electrophoresis
Ser	serine
T	thymine
t_H	time for half-completion of reaction
Tal	talose
TDP	thymidine 5'-diphosphate
TEA	triethanolamine
Thr	threonine
TLCK	N^{α}-*p*-tosyl-L-lysine chloromethyl ketone
T_m	melting temperature
TMP	thymidine 5'-monophosphate
Tos-	tosyl-(*p*-toluenesulfonyl-)
TPN	triphosphopyridinium nucleotide (now $NADP^+$)
Tris	tris(hydroxymethyl)-aminomethane
Trp	tryptophan
TTP	thymidine 5'-triphosphate
Tyr	tyrosine
U	uridine
U/mg	μmol/(mg*min)

UDP	uridine 5'-diphosphate
UMP	uridine 5'-monophosphate
UTP	uridine 5'-triphosphate
Val	valine
Xaa	symbol for an amino acid of unknown constitution in peptide formula
XAS	X-ray absorption spectroscopy
Xyl	xylose

Index of Recommended Enzyme Names

Description of Data Fields

All information except the nomenclature of the enzymes (which is based on the recommendations of the Nomenclature Committee of IUBMB (International Union of Biochemistry and Molecular Biology) and IUPAC (International Union of Pure and Applied Chemistry) is extracted from original literature (or reviews for very well characterized enzymes). The quality and reliability of the data depends on the method of determination, and for older literature on the techniques available at that time. This is especially true for the fields Molecular Weight and Subunits.

The general structure of the fields is: **Information – Organism – Commentary – Literature**

The information can be found in the form of numerical values (temperature, pH, K_m etc.) or as text (cofactors, inhibitors etc.).

Sometimes data are classified as Additional Information. Here you may find data that cannot be recalculated to the units required for a field or also general information being valid for all values. For example, for Inhibitors, Additional Information may contain a list of compounds that are not inhibitory.

The detailed structure and contents of each field is described below. If one of these fields is missing for a particular enzyme, this means that for this field, no data are available.

1 Nomenclature

EC number

The number is as given by the IUBMB, classes of enzymes and subclasses defined according to the reaction catalyzed.

Systematic name

This is the name as given by the IUBMB/IUPAC Nomenclature Committee

Recommended name

This is the name as given by the IUBMB/IUPAC Nomenclature Committee

Synonyms

Synonyms which are found in other databases or in the literature, abbreviations, names of commercially available products. If identical names are frequently used for different enzymes, these will be mentioned here, cross references are given. If another EC number has been included in this entry, it is mentioned here.

CAS registry number

The majority of enzymes have a single chemical abstract (CAS) number. Some have no number at all, some have two or more numbers. Sometimes two enzymes share a common number. When this occurs, it is mentioned in the commentary.

2 Source Organism

For listing organisms their systematic name is preferred. If these are not mentioned in the literature, the names from the respective literature are used. For example if an enzyme from yeast is described without being specified further, yeast will be the entry. This field defines the code numbers for the organisms in which the enzyme with the respective EC number is found. These code numbers (form <_>) are displayed together with each entry in all fields of Brenda where organism-specific information is given.

3 Reaction and Specificity

Catalyzed reaction

The reaction as defined by the IUBMB. The commentary gives information on the mechanism, the stereochemistry, or on thermodynamic data of the reaction.

Reaction type

According to the enzyme class a type can be attributed. These can be oxidation, reduction, elimination, addition, or a name (e.g. Knorr reaction)

Natural substrates and products

These are substrates and products which are metabolized in vivo. A natural substrate is only given if it is mentioned in the literature. The commentary gives information on the pathways for which this enzyme is important. If the enzyme is induced by a specific compound or growth conditions, this will be included in the commentary. In *Additional information* you will find comments on the metabolic role, sometimes only assumptions can be found in the references or the natural substrates are unknown.
In the listings, each natural substrate (indicated by a bold **S**) is followed by its respective product (indicated by a bold **P**). Products are given with organisms and references included only if the respective authors were able to demonstrate the formation of the specific product. If only the disappearance of the substrate was observed, the product is included without organisms of references. In cases with unclear product formation only a ? as a dummy is given.

Substrates and products

All natural or synthetic substrates are listed (not in stoichiometric quantities) The commentary gives information on the reversibility of the reaction, on

isomers accepted as substrates it and compares the efficiency of substrates. If a specific substrate is accepted by only one of several isozymes, this will be stated here.
The field *Additional Information* summarizes compounds that are not accepted as substrates or general comments which are valid for all substrates.
In the listings, each substrate (indicated by a bold **S**) is followed by its respective product (indicated by a bold **P**). Products are given with organisms and references included if the respective authors demonstrated the formation of the specific product. If only the disappearance of the substrate was observed, the product will be included without organisms or references. In cases with unclear product formation only a ? as a dummy is given.

Inhibitors

Compounds found to be inhibitory are listed. The commentary may explain experimental conditions, the concentration yielding a specific degree of inhibition or the inhibition constant. If a substance is activating at a specific concentration but inhibiting at a higher or lower value, the commentary will explain this.

Cofactors, prosthetic groups

This field contains cofactors which participate in the reaction but are not bound to the enzyme, and prosthetic groups being tightly bound. The commentary explains the function or, if known, the stereochemistry, or whether the cofactor can be replaced by a similar compound with higher or lower efficiency.

Activating Compounds

This field lists compounds with a positive effect on the activity. The enzyme may be inactive in the absence of certain compounds or may require activating molecules like sulfhydryl compounds, chelating agents, or lipids. If a substance is activating at a specific concentration but inhibiting at a higher or lower value, the commentary will explain this.

Metals, ions

This field lists all metals or ions that have activating effects. The commentary explains the role each of the cited metal has, being either bound e.g. as Fe-S centers or being required in solution. If an ion plays a dual role, activating at a certain concentration but inhibiting at a higher or lower concentration, this will be given in the commentary.

Turnover number (min^{-1})

The k_{cat} is given in the unit min^{-1}. The commentary lists the names of the substrates, sometimes with information on the reaction conditions or the type of reaction if the enzyme is capable of catalyzing different reactions with a single substrate. For cases where it is impossible to give the turnover number in the defined unit (e.g., substrates without a defined molecular weight, or an undefined amount of protein) this is summarized in Additional Information.

Specific activity (U/mg)

The unit is micromole/minute/milligram of protein. The commentary may contain information on specific assay conditions or if another than the natural substrate was used in the assay. Entries in Additional Information are included if the units of the activity are missing in the literature or are not calculable to the obligatory unit. Information on literature with a detailed description of the assay method may also be found.

K_m-Value (mM)

The unit is mM. Each value is connected to a substrate name. The commentary gives, if available, information on specific reaction condition, isozymes or presence of activators. The references for values which cannot be expressed in mM (e.g. for macromolecular, not precisely defined substrates) are given in Additional Information. In this field we also cite literature with detailed kinetic analyses.

pH-Optimum

The value is given to one decimal place. The commentary may contain information on specific assay conditions, such as temperature, presence of activators or if this optimum is valid for only one of several isozymes. If the enzyme has a second optimum, this will be mentioned here.

pH-Range

Mostly given as a range e.g. 4.0–7.0 with an added commentary explaining the activity in this range. Sometimes, not a range but a single value indicating the upper or lower limit of enzyme activity is given. In this case, the commentary is obligatory.

Temperature optimum (°C)

Sometimes, if no temperature optimum is found in the literature, the temperature of the assay is given instead. This is always mentioned in the commentary.

Temperature range (°C)

This is the range over which the enzyme is active. The commentary may give the percentage of activity at the outer limits. Also commentaries on specific assay conditions, additives etc.

4 Enzyme Structure

Molecular weight

This field gives the molecular weight of the holoenzyme. For monomeric enzymes it is identical to the value given for subunits. As the accuracy depends on the method of determination this is given in the commentary if provided in the literature. Some enzymes are only active as multienzyme complexes for which the names and/or EC numbers of all participating enzymes are given in the commentary.

Subunits

The tertiary structure of the active species is described. The enzyme can be active as a monomer a dimer, trimer and so on. The stoichiometry of subunit composition is given. Some enzymes can be active in more than one state of complexation with differing effectivities. The analytical method is included.

Posttranslational modifications

The main entries in this field may be proteolytic modification, or side-chain modification, or no modification. The commentary will give details of the modifications e.g.:

- proteolytic modification <1> (<1>, propeptide Name) [1];
- side-chain modification <2> (<2>, N-glycosylated, 12% mannose) [2];
- no modification [3]

5 Isolation / Preparation / Mutation / Application

Source / tissue

For multicellular organisms, the tissue used for isolation of the enzyme or the tissue in which the enzyme is present is given. Cell-lines may also be a source of enzymes.

Localization

The subcellular localization is described. Typical entries are: cytoplasm, nucleus, extracellular, membrane.

Purification

The field consists of an organism and a reference. Only references with a detailed description of the purification procedure are cited.

Renaturation

Commentary on denaturant or renaturation procedure.

Crystallization

The literature is cited which describes the procedure of crystallization, or the X-ray structure.

Cloning

Lists of organisms and references, sometimes a commentary about expression or gene structure.

Engineering

The properties of modified proteins are described.

Application

Actual or possible applications in the fields of pharmacology, medicine, synthesis, analysis, agriculture, nutrition are described.

6 Stability

pH-Stability

This field can either give a range in which the enzyme is stable or a single value. In the latter case the commentary is obligatory and explains the conditions and stability at this value.

Temperature stability

This field can either give a range in which the enzyme is stable or a single value. In the latter case the commentary is obligatory and explains the conditions and stability at this value.

Oxidation stability

Stability in the presence of oxidizing agents, e.g. O_2, H_2O_2, especially important for enzymes which are only active under anaerobic conditions.

Organic solvent stability

The stability in the presence of organic solvents is described.

General stability information

This field summarizes general information on stability, e.g., increased stability of immobilized enzymes, stabilization by SH-reagents, detergents, glycerol or albumins etc.

Storage stability

Storage conditions and reported stability or loss of activity during storage.

References

Authors, Title, Journal, Volume, Pages, Year.

Alcohol dehydrogenase 1.1.1.1

1 Nomenclature

EC number

1.1.1.1

Systematic name

alcohol:NAD$^+$ oxidoreductase

Recommended name

alcohol dehydrogenase

Synonyms

40 kDa allergen [Swissprot]
ADH
ADH-A2 [Swissprot]
ADH-B2 [Swissprot]
ADH-C2 [Swissprot]
ADH-HT [Swissprot]
ADH3 [Swissprot]
Alcohol dehydrogenase-B2 [Swissprot]
FALDH [Swissprot]
FDH [Swissprot]
GSH-FDH [Swissprot]
Gastric alcohol dehydrogenase [Swissprot]
Glutathione-dependent formaldehyde dehydrogenase [Swissprot]
NAD-dependent alcohol dehydrogenase
NAD-specific aromatic alcohol dehydrogenase
NADH-alcohol dehydrogenase
NADH-aldehyde dehydrogenase
Octanol dehydrogenase [Swissprot]
Retinol dehydrogenase [Swissprot]
alcohol dehydrogenase (NAD)
aldehyde reductase
aliphatic alcohol dehydrogenase
dehydrogenase, alcohol
ethanol dehydrogenase
primary alcohol dehydrogenase
yeast alcohol dehydrogenase

CAS registry number

9031-72-5

2 Source Organism

<1> *Cricetulus griseus* [1]
<2> *Ctenopharyngodon idellus* [2]
<3> *Kluyveromyces marxianus* [3]
<4> *Saccharomyces cerevisiae* (isoenzyme I, II and III [91]; alcohol dehydrogenase IV [87]) [3, 87, 90, 91, 94]
<5> *Vicia faba* [4]
<6> *Suncus murinus* [5]
<7> *Oryza sativa* [6, 80]
<8> *Najas marina* [7]
<9> *Drosophila melanogaster* (alleloenzymes AdhS and AdhF [43]; alleloenzymes AdhF' and Adhus [58]; alleloenzyme AdhS [62]) [8, 43, 58, 62, 63, 64, 65]
<10> *Drosophila simulans* [8, 63, 64]
<11> *Drosophila virilis* [8, 64]
<12> *Drosophila funebris* [8]
<13> *Drosophila imigrans* [8]
<14> *Drosophila lebanonensis* [8, 38, 61]
<15> *Mus musculus* (class III enzyme [9]; isoenzyme A2, B2 and C2 [48]) [9, 48]
<16> *Homo sapiens* (class I isoenzymes: α,γ_2, $\gamma_2\gamma_2$, α,γ_1, α,β_1, β_1,γ_2, γ_1,γ_1, β_1,γ_1, β_1,β_1 [13]; class II isoenzyme: pi-ADH [14]; class III isoenzyme [11, 16]; isoenzymes: $\beta_2\beta_2$, $\beta_2\beta_1$, $\alpha\beta_2$ and $\beta_2\gamma_1$ [15]; isoenzyme chi-ADH [16]; isoenzyme pi [23]; isoenzyme α,α, β_1,β_1, γ_1,γ_1 and γ_2,γ_2 [17]; anodic enzyme form [18,21]; isoenzymes β_1,β_1, β_2,β_2 and β_3,β_3 [20]; chi-ADH [21]; ADH Indianapolis form 1 and 2 and 3 [22]; class IV enzyme, σ-ADH [53]; μ-alcolhol dehydrogenase [96]) [10, 11, 12, 13, 14, 15, 16, 17, 18, 19, 20, 21, 22, 23, 53, 96]
<17> *Rattus norvegicus* (class IV enzyme [50]; isoenzyme 1, 2, 3 and 4 [51]) [10, 49, 50, 51]
<18> *Saimiri sciureus* (pyrazole-insensitive isoenzyme and pyrazole-sensitive isoenzyme [24]) [24]
<19> *Macaca mulatta* (cationic pyrazole-sensitive isoenzyme, anodic pyrazole-sensitive isoenzyme and cathionic pyrazole-in sensitive isoenzyme [25]) [25]
<20> *Equus caballus* (SS isoenzyme [28,41,42]; isoenzymes EE, ES, and SS [35]) [26, 27, 28, 29, 30, 31, 32, 33, 34, 35, 36, 37, 39, 40, 41, 42, 54, 55, 86, 92, 93, 98, 102]
<21> *Gallus gallus* [44]
<22> *Oryctolagus cuniculus* [45, 46]
<23> *Mesocricetus auratus* (3 major isoenzymes: TT-ADH, AA-ADH and BB-ADH [95] [95]) [47, 95]
<24> *Uromastix hardwickii* [52]
<25> *Sulfolobus solfataricus* [56, 66, 70]
<26> *Naja naja* [57]

<27> *Salmo gairdnerii* [59]
<28> *Coturnix coturnix* (japonica. ADH-1, ADH-2, ADH-3 and ADH-An [60]) [60]
<29> *Zymomonas mobilis* (enzyme form ADH-I and ADH-II [67]; ADH I and ADH II [68]) [67, 68]
<30> *Chlamydomonas moewusii* [69]
<31> *Glycine max* (4 isoenzymes [71]) [71, 80]
<32> *Petroselinum crispum* [72]
<33> *Camellia sinensis* [73]
<34> *Brassica napus* [74]
<35> *Zea mays* (3 non-allelic alcohol dehydrogenase isoenzymes [75]; wild-type enzyme Adh1-1S and mutant enzyme Adh1-1S1108 [97]) [75, 76, 80, 97]
<36> *Triticum turgidum* (3 isoenzymes [77]) [77]
<37> *Triticum monococcum* [78]
<38> *Hordeum vulgare* (3 isoenzymes [79]) [79]
<39> *Avena sativa* [80]
<40> *Tricticum aestivum* [80]
<41> *Pisum sativum* [80]
<42> *Citrullus lanatus* [80]
<43> *Aspergillus nidulans* [81]
<44> *Candida sp.* (N-16 [82]) [82]
<45> *Astasia longa* [83]
<46> *Sporotrichum pulverulentum* [84]
<47> *Candida guilliermondii* (strain B10-05 [85]) [85]
<48> *yeast* [86, 88, 89]
<49> *Schizosaccharomyces pombe* [91]
<50> *Desulfovibrio gigas* [99]
<51> *Entamoeba histolytica* [100]
<52> *Alligator mississippiensis* [101]

3 Reaction and Specificity

Catalyzed reaction

an alcohol + NAD^+ = an aldehyde or ketone + NADH (<16>, ordered bi bi mechanism with cofactor adding first to form a binary enzyme complex [23]; <9, 20>, ordered bi-bi mechanism [31, 43]; <20>, isoenzyme EE and SS: ordered bi bi mechanism [35]; <14, 30, 34>, Theorell-Chance mechanism [38, 69, 74]; <20>, kinetic mechanism is random for ethanol oxidation and compulsory ordered for acetaldehyde reduction [41]; <9, 10>, rapid equilibrium random mechanism [63]; <47>, oxidizes ethanol in an ordered bi-bi mechanism with NAD^+ as the first substrate fixed [85]; <4>, compulsory-order mechanism with the rate-limiting step being the dissociation of the product enzyme-NAD^+ complex [90]; <4, 49>, mechanism is predominantly ordered with ethanol, but partially random with butanol [91])

Reaction type

oxidation
redox reaction
reduction

Natural substrates and products

S ethanol + NAD^+ <15, 16, 20> (<16>, enzyme is responsible for the major ethanol oxidation capacity in the body. The products acetaldehyde and NADH are responsible for the most of the toxic effects and metabolic disturbances produced by ethanol ingestion [10]; <16>, the anodic enzyme form may contribute significantly to alcohol elimination in man, particularly at high concentrations when the other enzyme species are saturated [18]; <16>, χ-ADH plays an important role in the metabolism of long chain alcohols and aldehydes [21]; <15>, role of the major liver isoenzyme A2 in ethanol metabolism [48]; <16>, the enzyme plays a significant role in first-pass metabolism of ethanol in human [96]; <20>, plays an important role in the metabolism of ethanol [102]) (Reversibility: ? <15, 16, 20> [10, 18, 21, 48, 96, 102]) [10, 18, 21, 48, 96, 102]

P acetaldehyde + NADH <16> [10]

S Additional information <4, 14, 23, 27, 29, 44> (<14>, the enzyme oxidizes alcohols to aldehydes or ketones both for detoxification and metabolic purposes [38]; <23>, involvement in the development of male hamsters reproductive system [47]; <27>, enzyme may be involved in the metabolism of dietary wax esters in salmonid fish [59]; <29>, key enzyme in ethanol production [68]; <44>, one constitutive enzyme, ADH-MI and one inducible enzyme, ADH-MII [82]; <4>, constitutive enzyme [94]) [38, 47, 59, 68, 82, 94]

P ?

Substrates and products

S (E)-hex-2-en-1-ol + NAD^+ <15, 33> (<33>, 43.2% of the activity with ethanol [73]) (Reversibility: ? <15, 33> [48, 73]) [48, 73]

P (E)-hex-2-en-1-one + NADH

S (Z)-hex-2-en-1-ol + NAD^+ <33> (<33>, 9.7% of the activity with ethanol [73]) (Reversibility: ? <33> [73]) [73]

P (Z)-hex-2-en-1-one + NADH

S 1,3-propanediol + NAD^+ <50> (<50>, 7% of the activity with ethanol [99]) (Reversibility: ? <50> [99]) [99]

P ? + NADH

S 11-cis-retinal + NAD^+ <20> (Reversibility: ? <20> [29]) [29]

P 11-cis-retinol + NADH

S 11-cis-retinol + NAD^+ <20> (Reversibility: ? <20> [93]) [93]

P 11-cis-retinal + NADH <20> [93]

S 12-hydroxydodecanoate + NAD^+ <16, 17, 23, 28> (<23>, oxidized at pH 10, not oxidized at pH 7.5 [47]) (Reversibility: ? <16, 17, 23, 28> [11, 14, 16, 49, 53, 60, 95, 96]) [11, 14, 16, 49, 53, 60, 95, 96]

P 12-oxododecanoic acid + NADH

S 13-cis-retinal + NAD^+ <20> (Reversibility: ? <20> [29]) [29]

P 13-cis-retinol + NADH

S 13-cis-retinol + NAD^+ <20> (Reversibility: ? <20> [93]) [93]

P 13-cis-retinal + NADH <20> [93]

S 16-hydroxyhexadecanoate + NAD^+ <16> (Reversibility: ? <16> [13, 14]) [13, 14]

P 16-oxohexadecanoic acid + NADH

S 17β-hydroxyetiocholan-3-one + NAD^+ <16> (Reversibility: ? <16> [16]) [16]

P ethiocholan-3,17-dione + NADH

S 2,3-pentanedione + NADH <32> (<32>, 5.5% of the activity with acetaldehyde [72]) (Reversibility: ? <32> [72]) [72]

P ? + NADH

S 2-butene-1-ol + NAD^+ <17, 35> (Reversibility: ? <17, 35> [49, 75]) [49, 75]

P ? + NADH

S 2-deoxy-D-ribose + NAD^+ <16> (Reversibility: ? <16> [14]) [14]

P ? + NADH

S 2-ethylhexan-1-ol + NAD^+ <46> (<46>, weak activity [84]) (Reversibility: ? <46> [84]) [84]

P 2-ethylhexanal + NADH

S 2-methylbutan-1-ol + NAD^+ <14> (Reversibility: ? <14> [61]) [61]

P ? + NADH

S 2-methylpropan-1-ol + NAD^+ <14, 35, 36, 37, 46> (<35>, weak activity [75]) (Reversibility: ? <14, 35, 36, 37, 46> [61, 75, 84, 97]) [61, 75, 77, 78, 84, 97]

P 2-methyl-propan-1-one + NADH

S 2-pentanone + NADH <20> (Reversibility: r <20> [31]) [31]

P pentan-2-ol + NAD^+ <20> [31]

S 3-butene-1-ol + NAD^+ <35> (Reversibility: ? <35> [75]) [75]

P ? + NADH

S 3-methyl-1-butanol + NAD^+ <20> (Reversibility: ? <20> [92]) [92]

P ? + NADH

S 3-methylbutan-2-ol + NAD^+ <14> (<14>, R-(-)-3-methylbutan-2-ol and S-(+)-3-methylbutan-2-ol [61]) (Reversibility: ? <14> [61]) [61]

P 3-methylbutan-2-one + NADH

S 3-methylbutan-2-one + NADH <25> (Reversibility: ? <25> [66]) [66]

P 3-methylbutan-2-ol + NAD^+

S 3-methylbutanol + NAD^+ <36, 37> (Reversibility: ? <36, 37> [77, 78]) [77, 78]

P ? + NADH

S 3-methylcyclohexanol + NAD^+ <25> (Reversibility: r <25> [66]) [66]

P 3-methylcyclohexanone + NADH

S 3-methylcyclohexanone + NADH <25> (Reversibility: r <25> [66]) [66]

P 3-methylcyclohexanol + NAD^+

S 3-oxo-5β-androstan-17β-ol + NADH <17, 20> (Reversibility: ? <17, 20> [28, 42, 49]) [28, 42, 49]

P 3β,17β-dihydroxy-5β-androstane + NAD^+

S 3-pentanone + NADH <20> (Reversibility: r <20> [31]) [31]
P pentan-3-ol + NAD$^+$ <20> [31]
S 3-phenyl-1-propanol + NAD$^+$ <16> (Reversibility: ? <16> [13]) [13]
P 3-phenyl-1-propanone + NADH
S 3-phenyl-2-propen-1-ol + NAD$^+$ <35> (Reversibility: ? <35> [75]) [75]
P ? + NADH
S 3-pyridylcarbinol + NAD$^+$ <16> (Reversibility: ? <16> [18]) [18]
P pyridine-3-carbaldehyde + NADH
S 3β,12α-dihydroxy-5β-cholanoic acid + NAD$^+$ <17> (Reversibility: ? <17> [51]) [51]
P ? + NADH
S 3β,17β-dihydroxy-5β-androstane + NAD$^+$ <20> (Reversibility: ? <20> [28]) [28]
P 5β-androstan-3,17dione + NADH
S 3β,7α,12α-trihydroxy-5β-cholanoic acid + NAD$^+$ <17> (Reversibility: ? <17> [51]) [51]
P ? + NADH
S 3β,7α-dihydroxy-5β-cholanoic acid + NAD$^+$ <17> (Reversibility: ? <17> [51]) [51]
P ? + NADH
S 3β-hxdroxy-5α-cholanoate + NAD$^+$ <20> (Reversibility: ? <20> [28]) [28]
P 3-oxo-5α-cholanoate + NADH
S 3β-hydroxy-5β-androstan-17-one + NAD$^+$ <20, 23> (<23>, catalyzed by isoenzyme BB-ADH, no activity with isoenzyme AA-ADH and TT-ADH [95]) (Reversibility: ? <20, 23> [28, 95]) [28, 95]
P 5β-androstan-3,17-dione + NADH
S 3β-hydroxy-5β-cholanoate + NAD$^+$ <20> (Reversibility: ? <20> [28]) [28]
P 3-oxo-5β-cholanoate + NADH
S 4-methyl-1-pentanol + NAD$^+$ <14> (Reversibility: ? <14> [61]) [61]
P 4-methyl-1-pentanal + NADH
S 5α-androstan-17β-ol-3-one + NADH <17> (Reversibility: ? <17> [51]) [51]
P 3β,17β-dihydroxy-5α-androstan + NADH
S 5β-pregnan-21-ol-3,20-dione hemisuccinate + NADH <20> (Reversibility: ? <20> [42]) [42]
P 5β-pregnan-3,20,21-trione hemisuccinate + NADH
S 9-cis-retinol + NAD$^+$ <20> (Reversibility: ? <20> [93]) [93]
P 9-cis-retinal + NADH <20> [93]
S D-glucitol + NAD$^+$ <36, 37> (Reversibility: ? <36, 37> [77, 78]) [77, 78]
P ? + NADH
S R-(+)-1-phenylethanol + NAD$^+$ <14> [61]
P 1-phenylethanone + NADH
S R-(+)-cis-verbenol + NAD$^+$ <14> (Reversibility: ? <14> [61]) [61]
P ? + NADH
S R-(+)-trans-bicyclo(2.2.1)-heptanol + NAD$^+$ <14> (<14>, R-(+)-trans-bicyclo(2.2.1)-heptanol and S-(-)-trans-bicyclo(2.2.1)-heptanol [61]) (Reversibility: ? <14> [61]) [61]

P R-(+)-trans-bicyclo(2.2.1)-heptanal + NADH
S S-(-)-1-phenylethanol + NAD^+ <14> [61]
P 1-phenylethanone + NADH
S S-(-)-cis-bicyclo(2.2.1)-heptanol + NAD^+ <14> (Reversibility: ? <14> [61]) [61]
P S-(-)-cis-bicyclo(2.2.1)-heptanal + NADH
S acetaldehyde + NADH <4, 5, 7, 9, 15, 17, 19, 20, 29, 43, 44, 46, 50> (<5>, reduced at the highest rate of all aldehydes tested [4]; <15>, reduced by isoenzyme A_2 and C2, no activity with isoenzyme B2 [48]; <50>, the reduction of acetaldehyde is 4.9fold faster than the oxidation of ethanol [99]) (Reversibility: r <4, 7, 9, 17, 19, 20, 29, 43, 44, 46, 50> [6, 25, 28, 41, 42, 43, 51, 68, 81, 84, 90, 99]) [4, 6, 25, 28, 41, 42, 43, 48, 49, 51, 68, 81, 82, 84, 90, 99]
P ethanol + NAD^+ <4, 5, 7, 9, 19, 20, 44, 46> [4, 6, 25, 28, 41, 42, 43, 82, 84, 90]
S acetone + NADH <20, 25> (Reversibility: r <20, 25> [31, 66]) [31, 66]
P isopropanol + NAD^+ <20, 25> [31, 66]
S all-trans-retinal + NAD^+ <20> (Reversibility: ? <20> [93]) [93]
P all-trans-retinol + NADH
S all-trans-retinol + NAD^+ <16, 20, 23> (Reversibility: ? <16, 20, 23> [47, 53, 93, 95]) [47, 53, 93, 95]
P all-trans-retinal + NADH <20> [93]
S allyl alcohol + NAD^+ <5, 29, 35, 45, 47> (<4>, no activity [87]) (Reversibility: ? <5, 29, 35, 45, 47> [4, 67, 68, 75, 83, 85, 97]) [4, 67, 68, 75, 83, 85, 97]
P prop-2-en-1-al + NADH
S anisaldehyde + NADH <25> (Reversibility: ? <25> [66, 70]) [66, 70]
P anisic alcohol + NADH
S benzaldehyde + NADH <15> (<15>, reduced by isoenzyme A2 and C2, no activity with isoenzyme B2 [48]) (Reversibility: r <15> [48]) [48]
P benzyl alcohol + NAD^+
S benzyl alcohol + NAD^+ <15, 16, 17, 20, 23, 25, 28> (<15>, oxidized by isoenzyme A2 and C2 no activity with isoenzyme B2 [48]; <17>, oxidation with isoenzyme ADH-1 and ADH-3, no activity with isoenzyme ADH-2 [49]; <35>, no activity [75]) (Reversibility: ? <15, 16, 17, 20, 23, 25, 28> [13, 14, 16, 42, 47, 48, 49, 60, 66, 70, 96]) [13, 14, 16, 42, 47, 48, 49, 60, 66, 70, 96]
P benzaldehyde + NADH
S butan-2-ol + NAD^+ <9, 10, 11, 14, 16, 20, 23, 25, 27, 35, 45, 46, 47> (<20>, (R)-2-butanol and (S)-2-butanol [31]; <35>, weak activity [75]; <35>, no activity [97]) (Reversibility: ? <9, 10, 11, 14, 16, 20, 23, 25, 27, 35, 45, 46, 47> [20, 31, 47, 53, 59, 61, 64, 66, 75, 83, 84, 85, 92, 95]) [20, 31, 47, 53, 59, 61, 64, 66, 75, 83, 84, 85, 92, 95]
P butan-2-one + NADH <20> [31]
S butan-2-one + NADH <20, 25> (Reversibility: r <20, 25> [31, 66]) [31, 66]
P butan-2-ol + NAD^+ <20> [31]

S butanol + NAD^+ <4, 9, 10, 11, 14, 16, 17, 20, 22, 23, 25, 27, 28, 29, 35, 36, 37, 45, 47, 52> (<29>, oxidized by enzyme form ADH-I, no activity with enzyme form ADH-II [67]; <29>, activity with ADH I, no activity with ADH II [68]; <4>, weak [87]; <17>, pH 10.0: oxidized by ADH-1 and ADH-3, no activity with isoenzyme ADH-2 [49]) (Reversibility: ? <4, 9, 10, 11, 14, 16, 17, 20, 22, 23, 25, 27, 28, 29, 35, 36, 37, 45, 47, 52> [16, 18, 20, 24, 42, 45, 47, 49, 53, 59, 60, 61, 64, 66, 68, 75, 77, 78, 83, 85, 90, 95, 96, 97, 99, 101]) [16, 18, 20, 24, 42, 45, 47, 49, 53, 59, 60, 61, 64, 66, 68, 75, 77, 78, 83, 85, 90, 95, 96, 97, 99, 101]
P butyraldehyde + NADH
S butyraldehyde + NADH <20, 29, 50> (<29>, activity with enzyme form ADH I, no activity with enzyme form ADH II [68]) (Reversibility: r <20, 29, 50> [42, 68, 99]) [42, 68, 99]
P n-butanol + NAD^+ <29> [68]
S cinnamyl alcohol + NAD^+ <20, 45, 47> (Reversibility: ? <20, 45, 47> [83, 85, 92]) [83, 85, 92]
P cinnamaldehyde + NADH
S cis-4-methylcyclohexanol + NAD^+ <14> (Reversibility: ? <14> [61]) [61]
P 4-methylcyclohexanone + NADH
S cyclohexanol + NAD^+ <14, 15, 16, 17, 19, 23, 28, 36, 37, 52> (<19>, reaction is catalyzed by the cathodic pyrazole-sensitive enzyme, no activity with the cathodic pyrazole-insensitive enzyme and by the anodic pyrazole-insensitive enzyme [25]; <15>, oxidized by isoenzyme A2, no activity with isoenzymes B2 and C2 [48]; <35>, no activity [75]) (Reversibility: ? <14, 15, 16, 17, 19, 23, 28, 36, 37, 52> [13, 14, 25, 47, 48, 49, 60, 61, 77, 78, 95, 101]) [13, 14, 25, 47, 48, 49, 60, 61, 77, 78, 95, 101]
P cyclohexanone + NADH
S cyclohexanone + NADH <20> (Reversibility: ? <20> [28, 42]) [28, 42]
P cyclohexanol + NAD^+
S cyclopentanol + NAD^+ <36, 37> (Reversibility: ? <36, 37> [77, 78]) [77, 78]
P ? + NADH
S cyclopetanone + NADH <25> (Reversibility: ? <25> [66]) [66]
P cyclopentanol + NADH
S digitose + NAD^+ <16> (Reversibility: ? <16> [16]) [16]
P ? + NADH
S ethanol + 3-benzoylpyridine-adenine dinucleotide <20> (<20>, rapid equilibrium bi bi mechanism [33]) [33]
P acetaldehyde + ?
S ethanol + NAD^+ <2, 4, 7, 9, 10, 11, 15, 16, 17, 18, 19, 20, 22-24, 27, 28, 29, 30-37, 43, 44, 45, 46, 50, 52> (<15>, no activity with isoenzyme B2, oxidized by isoenzyme A2 and C2 [48]; <16>, class III isoenzyme chi-ADH oxidizes ethanol very poorly [16]; <16>, no oxidation with class III enzyme [11]; <17>, isoenzyme ADH-1 and ADH-3, no activity with isoenzyme ADH-2 [49]; <44>, the reduction of acetaldehyde of ADH-MII is about 7times higher than that of the oxidation of ethanol [82]; <50>, the reduction of acetaldehyde is 4.9fold faster than the oxidation of ethanol

[99]) (Reversibility: r <4, 7, 9, 17, 19, 20, 29, 43, 44, 45, 46, 50> [6, 25, 28, 41, 42, 43, 51, 68, 81, 82, 83, 84, 90, 99]; ? <2, 10, 11, 15, 16, 18, 22, 23, 24, 27, 28, 30, 31, 32, 33, 34, 35, 36, 52> [2, 10, 12-14, 16-18, 20, 24, 35, 45, 47, 48, 49, 52, 53, 57, 59, 60, 64, 65, 67, 69, 71-78, 87, 92, 95, 96, 97, 101, 102]) [2, 6, 10, 12-14, 16, 17, 18, 20, 24, 25, 28, 35, 41-43, 45, 47-49, 51-53, 57, 59, 60, 64, 65, 67-69, 71-78, 81-84, 87, 90, 92, 95-97, 99, 101, 102]

P acetaldehyde + NADH <4, 7, 17, 19, 20, 32, 44, 46, 50> [6, 25, 28, 41, 42, 49, 72, 82, 84, 90, 99]

S ethylene glycol + NAD^+ <15, 16, 47> (<16>, class III isoenzyme χ-ADH shows no activity [16]; <15>, oxidized by isoenzyme A2, no activity with isoenzyme B2 and C2 [48]; <28>, no activity [60]) (Reversibility: ? <15, 16, 47> [13, 14, 48, 85]) [13, 14, 48, 85]

P ? + NADH

S formaldehyde + NAD^+ <15, 29, 50> (<15>, reduced by isoenzyme A2, no activity with isoenzyme B2 and C2 [48]; <29>, activity with ADH I, no activity with ADH II [68]) (Reversibility: r <15> [48]; ? <29, 50> [68, 99]) [48, 68, 99]

P methanol + NADH

S furfural + NADH <29> (<29>, activity with enzyme form ADH I, no activity with enzyme form ADH II [68]) (Reversibility: r <29> [68]) [68]

P furfuryl alcohol + NADH <29> [68]

S furfuryl alcohol + NAD^+ <16, 29> (<29>, activity with ADH I, no activity with ADH II [68]) (Reversibility: r <29> [68]; ? <16> [16]) [16, 68]

P furfural + NADH <29> [68]

S heptan-2-ol + NAD^+ <14> (Reversibility: ? <14> [61]) [61]

P heptan-2-one + NADH

S heptan-4-ol + NAD^+ <14> (Reversibility: ? <14> [61]) [61]

P heptan-4-one + NADH

S hexadecanol + NAD^+ <27> (<27>, very low activity [59]) (Reversibility: ? <27> [59]) [59]

P hexadecanal + NADH

S hexan-1-ol + NAD^+ <14, 23, 35, 36, 37, 47> (Reversibility: ? <14, 23, 35, 36, 37, 47> [47, 61, 75, 77, 78, 85, 97]) [47, 61, 75, 77, 78, 85, 97]

P n-hexanal + NADH

S hexan-2-ol + NAD^+ <14> (Reversibility: ? <14> [61]) [61]

P 2-hexanone + NADH

S hexanol + NAD^+ <15, 16, 20, 23, 29, 46, 52> (<29>, activity with ADH I, no activity with ADH II [68]) (Reversibility: ? <15, 16, 20, 23, 29, 46, 52> [21, 42, 48, 53, 84, 92, 101]) [20, 21, 42, 48, 53, 68, 84, 92, 95, 101]

P n-hexanal + NADH <20> [42]

S isobutyl alcohol + NAD^+ <16> (Reversibility: ? <16> [20]) [20]

P ? + NADH

S isopentenyl alcohol + NAD^+ <16> (Reversibility: ? <16> [20]) [20]

P isopentanone + NADH

S *m*-nitrobenzaldehyde + NAD^+ <17> [49]

P *m*-nitrobenzyl alcohol + NADH <17> [49]

S methanol + NAD$^+$ <15, 16, 17, 18, 19, 22, 15, 27, 44, 52> (<16>, class I isoenzymes [13]; <16>, class III isoenzyme χ-ADH shows no activity [16]; <16>, oxidized by isoenzyme β_1,β_1 [20]; <16>, activity detected with class II isoenzyme pi-ADH [14]; <18>, reaction is catalyzed by the pyrazole-sensitive enzyme, no activity with the pyrazole-insensitive enzyme [24]; <19>, reaction is catalyzed by the cathodic pyrazole-sensitive enzyme, no activity by the cathodic pyrazole-insensitive enzyme and by the anodic pyrazole-insensitive enzyme [25]; <22>, no activity at pH 7.5, slight activity at pH 10.8 [45]; <15>, weak activity with isoenzyme A2, no activity with isoenzyme B2 and C2 [48]; <17>, oxidized with ADH-3, no activity with ADH-1 and ADH-2 [49]; <44>, reaction is extremly weak [82]; <16>, anodic enzyme form shows no activity [18]; <16>, no activity with class III enzyme [11]; <16>, no activity with isoenzyme β_3,β_3 [20]; <16, 28, 29, 35, 46, 47>, no activity [21, 60, 67, 68, 75, 84, 85]) (Reversibility: r <15> [48]; ? <16, 17, 18, 19, 22, 15, 27, 44, 52> [12, 13, 20, 24, 25, 45, 48, 49, 59, 66, 82, 101]) [12, 13, 20, 24, 25, 45, 48, 49, 59, 66, 82, 101]

P formaldehyde + NADH

S methylcrotonyl alcohol + NAD$^+$ <23> (Reversibility: ? <23> [95]) [95]

P methylcrotonaldehyde + NADH

S methylglyoxal + NADH <32> (<32>, 11.9% of the activity with acetaldehyde [72]) (Reversibility: ? <32> [72]) [72]

P ? + NADH

S n-pentylaldehyde + NAD$^+$ <29> (Reversibility: r <29> [68]) [68]

P n-pentanol + NADH

S octan-1-ol + NAD$^+$ <15, 16, 17, 19, 28, 29, 35, 46, 47, 52> (<29>, activity with ADH I, no activity with ADH II [68]; <46>, weak activity [84]) (Reversibility: ? <15, 16, 17, 19, 28, 29, 35, 46, 47, 52> [11, 13, 14, 21, 25, 48, 49, 60, 68, 75, 84, 85, 101]) [11, 13, 14, 21, 25, 48, 49, 60, 68, 75, 84, 85, 101]

P n-octanal + NADH

S octan-2-ol + NAD$^+$ <14, 16, 20, 46> (<20>, (R)-2-octanol and (S)-2-octanol [31]; <46>, weak activity [84]) (Reversibility: ? <14, 16, 20, 46> [16, 31, 61, 84]) [16, 31, 61, 84]

P octan-2-one + NADH

S octanal + NADH <16, 17, 28> (Reversibility: ? <16, 17, 28> [16, 49, 60]) [16, 49, 60]

P octanol + NAD$^+$

S *p*-nitrobenzaldehyde + NADH <15> (Reversibility: ? <15> [48]) [48]

P *p*-nitrobenzyl alcohol + NAD$^+$

S *p*-nitrophenyl octanoate + H_2O <20> (<20>, acid-assisted nucleophilic catalysis involving the ammonium ion of Lys and the thiolate of Cys in the acyl-oxygen cleavage [39]) (Reversibility: ? <20> [39]) [39]

P *p*-nitrophenol + octanoate <20> [39]

S pentan-2-ol + NAD$^+$ <14, 20, 25, 46, 47> (<20>, (R)-2-pentanol and (S)-2-pentanol [31]) (Reversibility: r <20> [31]; ? <14, 25, 46, 47> [61, 66, 84, 85]) [31, 61, 66, 84, 85]

P 2-pentanone + NADH <20> [31]

S pentan-3-ol + NAD$^+$ <20, 25> (Reversibility: r <20> [31]; ? <25> [66]) [31, 66]
P 3-pentanone + NADH <20> [31]
S pentanol + NAD$^+$ <15, 16, 18, 19, 20, 46> (<15>, oxidized by isoenzyme A2 and B2, no activity with isoenzyme C2 [48]) (Reversibility: ? <15, 16, 18, 19, 20, 46> [14, 16, 18, 20, 24, 25, 48, 52, 53, 57, 84, 92, 96]) [14, 16, 18, 20, 24, 25, 48, 52, 53, 57, 84, 92, 96]
P pentanone + NADH
S pentanol + NAD$^+$ <14, 16, 17, 22, 28, 29, 30, 31, 35, 36, 37, 47> (<29>, activity with ADH I, no activity with ADH II [68]) (Reversibility: ? <14, 16, 17, 22, 28, 29, 30, 31, 35, 36, 37, 47> [11, 45, 49, 60, 61, 68, 69, 71, 75, 77, 78, 85]) [11, 45, 49, 60, 61, 68, 69, 71, 75, 77, 78, 85]
P n-pentanal + NADH
S phenylalaninol + NAD$^+$ <16> (Reversibility: ? <16> [16]) [16]
P ? + NADH
S phenylethanol + NAD$^+$ <14> (<14>, R-(+)-phenylethanol and S-(-)-phenylethanol [61]) (Reversibility: ? <14> [61]) [61]
P phenylacetaldehyde + NADH
S prop-2-en-1-ol + NAD$^+$ <33> (<33>, 140.6% of the activity with ethanol [73]) (Reversibility: ? <33> [73]) [73]
P ? + NADH
S propan-1,2-diol + NAD$^+$ <30, 50> (Reversibility: ? <30, 50> [69, 99]) [69, 99]
P ? + NADH
S propan-2-ol + NAD$^+$ <4, 9, 10, 11, 14, 16, 20, 22, 25, 35, 43, 46, 47> (<50>, 3% of the activity with ethanol [99]; <29>, very low activity [67]; <4>, no activity [87]) (Reversibility: ? <4, 9, 10, 11, 14, 16, 20, 22, 25, 29, 35, 43, 46, 47> [14, 43, 45, 61, 64, 65, 66, 81, 84, 85, 90, 92, 97, 99]) [14, 43, 45, 61, 64, 65, 66, 81, 84, 85, 90, 92, 97, 99]
P acetone + NADH <20> [31, 42]
S propanol + NADH <4, 9, 14, 16, 22, 25, 27, 29, 31, 35, 36, 37, 45, 46, 47, 50> (Reversibility: ? <4, 9, 14, 16, 22, 25, 27, 31, 35, 36, 37, 45, 46, 47, 50> [20, 42, 45, 53, 59, 61, 65, 66, 67, 68, 71, 75, 77, 78, 83, 84, 85, 87, 90, 92, 96, 97, 99]) [20, 42, 45, 53, 59, 61, 65, 66, 67, 68, 71, 75, 77, 78, 83, 84, 85, 87, 90, 92, 96, 97, 99]
P propionaldehyde + NADH <29> [68]
S propionaldehyde + NADH <20, 29, 32, 50> (<32>, 94% of the activity with acetaldehyde [72]) (Reversibility: ? <20, 29, 32, 50> [42, 68, 72, 99]) [42, 68, 72, 99]
P propanol + NAD$^+$
S trans-4-(N,N-dimethylamino)-cinnamaldehyde + NADH <16> (Reversibility: ? <16> [19]) [19]
P trans-4-(N,N-dimethylamino)-cinnamyl alcohol + NAD$^+$
S trans-4-methylcyclohexanol + NAD$^+$ <14> (Reversibility: ? <14> [61]) [61]
P 4-methylcyclohexanone + NADH
S tryptophol + NAD$^+$ <16> (Reversibility: ? <16> [14]) [14]

P ? + NADH
S vanillyl alcohol + NAD^+ <16> (Reversibility: ? <16> [14, 16]) [14, 16]
P vanillin + NADH

Inhibitors

1,10-phenanthroline <2, 15, 16, 18, 19, 22, 23, 29, 30, 35> (<23>, mixed type inhibition [47]; <15>, inhibition of isoenzyme A2 and C2, no inhibition of isoenzyme B2 [48]; <29>, 0.2 mM, strong inhibition [68]) [2, 14, 21, 24, 25, 45, 47, 48, 68, 69, 75, 95]
1,2-dithioglycerol <20> (<20>, competitive [98]) [98]
1,2-ethanedithiol <20> (<20>, competitive [98]) [98]
1,3-propanedithiol <20> (<20>, competitive [98]) [98]
1,4-butanedithiol <20> (<20>, competitive [98]) [98]
1,4-dithioerythritol <20> (<20>, competitive [98]) [98]
1,4-dithiothreitol <20> (<20>, competitive [98]) [98]
1-hydroxypyridine-2-thione <20> (<20>, competitive [98]) [98]
1-thio-1-phenylmethane <20> (<20>, competitive [98]) [98]
1-thioacetamide <20> (<20>, competitive [98]) [98]
1-thioacetate <20> (<20>, competitive [98]) [98]
1-thiobutane <20> (<20>, competitive [98]) [98]
1-thioethane <20> (<20>, competitive [98]) [98]
1-thioglycerol <20> (<20>, competitive [98]) [98]
1-thiopropane <20> (<20>, competitive [98]) [98]
1-thiosorbitol <20> (<20>, competitive [98]) [98]
12-hydroxydodecanoate <23> [47]
2,2'-bipyridine <16, 19> [14, 25]
2,2'-dipyridyl <22, 50> [45, 99]
2,2,2-trifluoroethanol <4, 14> (<14>, competitive towards ethanol [61]) [61, 94]
2-chloroethanol <14> (<14>, competitive towards ethanol [61]) [61]
2-fluoroethanol <4> [94]
2-mercapto-1-methylimidazole <20> (<20>, competitive [98]) [98]
2-mercaptobenzimidazole <20> (<20>, competitive [98]) [98]
2-mercaptobenzothiazole <20> (<20>, competitive [98]) [98]
2-mercaptoethanol <20, 37> (<20>, competitive [98]) [78, 98]
2-mercaptoimidazole <20> (<20>, competitive [98]) [98]
2-phenylethanethiol <20> (<20>, competitive [98]) [98]
2-pyridylethanethiol <20> (<20>, competitive [98]) [98]
2-thioacetate <20> (<20>, competitive [98]) [98]
2-thiobutane <20> (<20>, competitive [98]) [98]
2-thiopropane <20> (<20>, competitive [98]) [98]
2-thiopyridine <20> (<20>, competitive [98]) [98]
2-thiopyrimidine <20> (<20>, competitive [98]) [98]
3-mercapto-1,2,4-triazole <20> (<20>, competitive [98]) [98]
3-thiopropionate <20> (<20>, competitive [98]) [98]
4-bromopyrazole <16> [23]
4-cyanopyrazole <16> [23]

4-iodopyrazole <16> [23]
4-methoxypyrazole <4, 16, 17, 22, 23> (<16>, 1 mM, 28% inhibition [23]; <22>, class I ADHs: migrate towards cathode on starch gel and are very sensitive to 4-methylpyrazole inhibition, class II ADH: migrates slowly towards anode and is less sensitive to 4-methylpyrazole, class II ADH: migrates rapidly towards anode and is insensitive to 4-methylpyrazole [46]; <23>, competitive [47]; <17>, 0.1-10 mM, ADH-2 is practically insensitive, ADH-3 is very sensitive [49]; <17>, competitive inhibitor of all four isoenzymes [51]) [23, 45, 46, 47, 49, 51, 53, 91]
4-methylpyrazole <2, 16, 18, 19, 23> (<16>, class III enzyme is completely insensitive to inhibition [11,16]; <16>, poor inhibitor, class II isoenzyme [14]; <16>, no inhibition by 12 mM [21]; <16>, 1 mM, 31% inhibition [23]; <23>, isoenzyme AA-ADH, BB-ADH and TT-ADH [95]; <16>, competitive against ethanol [96]) [2, 14, 23, 24, 25, 95, 96]
4-nitropyrazole <16> [23]
4-octylpyrazole <16> [12]
4-pentylpyrazole <16> [12, 23]
4-propylpyrazole <16> [23]
5-β-D-ribofuranosylnicotinamide adenine dinucleotide <20> (<20>, potent and specific inhibitor [55]) [55]
6-thioguanine <20> (<20>, competitive [98]) [98]
6-thioguanosine <20> (<20>, competitive [98]) [98]
8-amino-6-methoxyquinoline <16> [12]
8-hydroxyquinoline 5-sulfonic acid <16, 18> [21, 24]
Ag^+ <22> [45]
$B_4O_7^{2-}$ <35> [76]
Cibacron blue <14> (<14>, competitive towards NAD^+ [61]) [61]
Co^{2+} <50> [99]
Cu^{2+} <22> (<35>, no effect [75]) [45]
Cys <20> (<20>, competitive [98]) [98]
EDTA <16, 18, 19, 25, 35, 50> (<25>, 15 mM, 85% inhibition [66]) [14, 24, 25, 66, 68, 76, 99]
H_2O_2 <20> (<20>, inactivation is caused by oxidation of its functional Cys residues, coenzyme protects from inactivation [40]) [40]
Hg^{2+} <22> (<35>, no effect [75]) [45]
Mg^{2+} <35> [76]
Mn^{2+} <50> [99]
NADH <9, 14, 47> (<9,14>, competitive towards NAD^+ [61, 63]; <47>, linear competitive inhibition [85]) [43, 61, 63, 85]
$NADP^+$ <16> [21]
NEM <22, 35, 50> [45, 75, 99]
NO <20> (<20>, Cys residues contained within the zinc/thiolate active center may be primary sites of NO interaction [102]) [102]
NaCN <4> [94]
NaN_3 <30> [69]
Ni^{2+} <50> [99]
PCMB <22, 29, 35, 47, 50> (<35>, no effect [76]) [45, 68, 75, 85, 99]

PMSF <35> [75]
S-2-chloro-3-(imidazol-5-yl)propionate <20> (<20>, inactivation at pH 8.2, R-2-chloro-3-(imidazol-5-yl)propionate has no effect [29]) [29]
Zn^{2+} <22, 35, 44, 50> (<44>, 0.1 mM, 20% decrease of activity [82]) [45, 76, 82, 99]
acetaldehyde <47> (<47>, linear noncompetitive inhibition [85]) [85]
acetamide <4> [94]
acetone <9> (<9>, product inhibition [43]) [43]
biochanin A <23> (<23>, inhibition of isoenzyme BB-ADH, no inhibition of isoenzyme AA-ADH and TT-ADH [95]) [95]
butyramide <4> [91]
captopril <20> (<20>, competitive [98]) [98]
coenzyme A <20> (<20>, competitive [98]) [98]
cysteamine <20> (<20>, competitive [98]) [98]
daidzein <23> (<23>, inhibition of isoenzyme BB-ADH, no inhibition of isoenzyme AA-ADH and TT-ADH [95]) [95]
diethyldithiocarbamate <20> (<20>, competitive [98]) [98]
dipicolinic acid <16, 18> [21, 24]
disulfiram <20> (<20>, competitive [98]) [98]
genistein <23> (<23>, inhibition of isoenzyme BB-ADH, no inhibition of isoenzyme AA-ADH and TT-ADH [95]) [95]
glutathione <20, 22, 37> (<20>, competitive [98]) [45, 78, 98]
heptafluorobutanol <4> [91]
imidazole <14> (<14>, pure competitive inhibition towards ethanol and (R)-(+)-phenylethanol [61]) [61]
iodoacetamide <30, 35> [69, 75, 76]
iodoacetic acid <22, 47> [45, 85]
isoburyramide <14> (<14>, competitive towards ethanol and butan-2-ol [61]) [61]
octanoic acid <20> (<20>, competitive inhibition of the hydrolysis of *p*-nitrophenyl octanoate [39]) [39]
p-chloromercuribenzene sulfonate <50> [99]
p-hydroxymercuribenzoate <35> [75]
p-nitrophenol <20> (<20>, noncompetitive inhibition of the hydrolysis of *p*-nitrophenyl octanoate [39]) [39]
penicillamine <20> (<20>, competitive [98]) [98]
pyrazole <4, 14, 15, 16, 17, 18, 19, 22, 28, 31> (<16>, 0.05 mM, 50% inhibition [10]; <17>, 0.05 mM, complete inhibition [10]; <16>, no inhibition at 1.0 mM [23]; <18,19>, organism has a pyrazole-sensitive isoenzyme and a pyrazole-insensitive enzyme [24,25]; <14>, competitive [38]; <15>, inhibition of isoenzyme A2 and C2. Isoenzyme B2 is insensitive to pyrazole inhibition with trans-2-hexen-1-ol as substrate [48]; <17>, 0.1-10 mM, ADH-2 is practically insensitive, ADH-3 is very sensitive [49]; <28>, pyrazole-sensitive enzyme forms ADH-1, ADH-2, ADH-3 and the pyrazole-insensitive form ADH-An [60]; <14>, competitive towards ethanol [61]) [10, 12, 24, 25, 38, 45, 48, 49, 60, 61, 71, 94]

pyridoxal 5'-phosphate <5> (<5>, inactivates by modifying its ε-amino group, NAD$^+$ protects [4]) [4]
sulfonic acid <16> [21]
tert-butyl hydroperoxide <48> (<48>, irreversible, inactivation is associated with -SH group oxidation [86]) [86]
testosterone <23> (<23>, inhibition of isoenzyme BB-ADH, no inhibition of isoenzyme AA-ADH and TT-ADH [95]) [95]
thiophenol <20> (<20>, competitive [98]) [98]
thiourea <20> (<20>, competitive [98]) [98]
trifluoroethanol <4> [91]
Additional information <4, 20, 39> (<20>, design of inhibitors [31]; <7, 31, 35, 39-42>, inactivator from rice seedlings [80]; <4>, use of competitive dead-end inhibitors to determine the chemical mechanism of action of yeast alcohol dehydrogenase [94]; <20>, thiols and dithiols have their main effect, at the enzyme-NAD$^+$-thiol complex level, by competition with alcohol for the catalytic zinc [98]) [31, 80, 94, 98]

Cofactors/prosthetic groups

3-benzoylpyridine-adenine dinucleotide <20> (<20>, can be used as coenzyme [33]) [33]
NAD$^+$ <1-52> (<20>, kinetics of coenzyme binding in the pH-range 10-12 [26]; <20>, NAD$^+$ acts as an activator which induces an active form of the enzyme [34]; <20>, preferred substrate [42]; <9>, NAD$^+$-plus-acetone-induced conversion [62]) [1-102]
NADH <1-52> (<20>, kinetics of coenzyme binding in the pH range 10-12 [26]; <20>, preferred coenzyme [42]) [1-102]
NADP$^+$ <20, 50> (<20>, also utilized as coenzyme by the SS isoenzyme [28]; <50>, at 1.5% of the activity with NAD$^+$ [99]; <22, 23, 30, 43>, no activity as coenzyme [45, 47, 69, 81]) [28, 99]
NADPH <20, 50> (<20>, also utilized as coenzyme by the SS isoenzyme [28]; <43>, no activity as coenzyme [81]; <50>, at 1.6% of the activity with NADH [99]) [28, 99]
Additional information <14, 20, 48> (<14>, crystallographic study of the coenzyme binding mode [38]; <20>, kinetics of native and modified enzyme with coenzyme analogues [54]; <48>, kinetic study on the binding of monomeric and polymeric derivatives of NAD$^+$ [88]) [38, 54, 88]

Activating compounds

EDTA <44> (<44>, enhances activity [82]) [82]
tert-butyl hydroperoxide <20> (<20>, stimulation up to 100 mM [86]) [86]
Additional information <20> (<20>, isonicotinimidylation and methylation increases activity [35]) [35]

Metals, ions

Fe^{2+} <44> (<44>, enhances activity [82]) [82]
Mg^{2+} <44> (<44>, enhances activity [82]) [82]
Zn^{2+} <4> (<4>, activates [87]) [87]
copper <19> (<19>, contains 0.5 mol of copper per mol of enzyme [25]) [25]

zinc <15, 16, 17, 18, 19, 20, 22, 23, 25, 29, 30, 35, 43, 50> (<16>, zinc containing enzyme [12]; <16>, from $\beta_1\gamma_1$ and $\gamma_1\gamma_1$ isoenzyme the active-site zinc is extracted much more slowly than from $\beta_1\beta_1$ isoenzyme. Characterization of active-site-specific zinc-depleted and reconstituted cobalt(II) alcohol dehydrogenase [19]; <16>, 3.6-4.2 gatom of zinc per mol [21]; <19>, contains 3.7-4.2 mol of zinc per mol of enzyme [25]; <16>, contains 3.7 gatom of zinc per mol of enzyme [23]; <16,22>, contains 3.8 mol of Zn per mol of protein [16,45]; <18>, contains 3.9 gatom of zinc per mol of enzyme [24]; <20>, substitution of catalytic and /or noncatalytic zinc ions by cobaltous ions [36]; <20>, contains 4 zinc atoms per molecule [42]; <15>, isoenzyme A2 contains 2.7 mol of zinc per mol of enzyme, isoenzyme b2 contains 1.9 mol of zinc per mol of enzyme, isoenzyme C2 contains 3.2 mol of zinc per mol of enzyme. A and C subunits each contain two atoms of zinc, with at least one being involved catalytically, the b subunit probably contains a single non-catalytic zinc atom [48]; <17>, ADH-1 contains 3.9 mol of zinc per mol of subunit, ADH-2 contains 4.2 mol of zinc per mol of subunit [49]; <25>, 2 mM required for optimal activity [66]; <29>, enzyme form ADH I and ADH II contain one zinc atom per subunit [67]; <30>, contains 4 zinc atoms per dimer [69]; <35>, required for activity, tightly bound within the enzyme [75]; <43>, contains 1 zinc atom per subunit [81]; <23>, isoenzyme AA-ADH contains 4.3 zinc atoms per dimeric molecule, isoenzyme BB-ADH contains 3.7 zinc atoms per dimeric molecule, isoenzyme AA-ADH contains 4.1 zinc atoms per dimeric molecule [95]; <16>, enzyme contains 7.59 zinc atoms per molecule [96]; <50>, contains 1.2 Zn atom per subunit [99]) [12, 16, 19, 21, 23, 24, 25, 36, 42, 45, 48, 49, 66, 67, 69, 70, 75, 81, 95, 96, 99]

Turnover number (min^{-1})

1.08 <20> (13-cis-retinal, <20>, reduction with NADH [93]) [93]
1.08 <20> (13-cis-retinol, <20>, oxidation with NAD^+ [93]) [93]
1.26 <20> (11-cis-retinal, <20>, reduction with NADH [93]) [93]
1.32 <20> (11-cis-retinol, <20>, oxidation with NAD^+ [93]) [93]
1.32 <20> (all-trans-retinal, <20>, oxidation with NAD^+ [93]) [93]
1.38 <20> (9-cis-retinal, <20>, reduction with NADH [93]) [93]
1.5 <20> (9-cis-retinol, <20>, oxidation with NAD^+ [93]) [93]
3 <52> (methanol) [101]
3.18 <20> ((R)-2-octanol) [31]
4.28 <20> (2-pentanone) [31]
4.8 <20> (3-methyl-1-butanol, <20>, mutant enzyme W54L [92]) [92]
5 <18> (butanol, <18>, pyrazole-insensitive enzyme, pH 7.5 [24]) [24]
5 <18> (ethanol, <18>, pyrazole-insensitive enzyme, pH 7.5 [24]) [24]
6.12 <16> (methanol) [12]
7 <18> (methanol, <18>, pyrazole-sensitive enzyme, pH 7.5 [24]) [24]
7.2 <20> (2-butanone) [31]
9 <52> (octanol) [101]
10 <52> (cyclohexanol) [101]
10 <52> (hexanol) [101]
11 <16> (1-pentanol) [11]

13 <18> (methanol, <18>, pyrazole-sensitive enzyme, pH 10.0 [24]) [24]
14.4 <20> (*p*-nitrophenyl octanoate) [39]
14.7 <16> (pentanol) [16]
16.2 <20> (3-methyl-1-butanol, <20>, wild-type enzyme Adh 1 [92]) [92]
17 <18> (pentanol, <18>, pyrazole-sensitive enzyme, pH 7.5 [24]) [24]
18.6 <20> (butan-2-ol, <20>, mutant enzyme W54L [92]) [92]
19 <25> (acetone) [66]
19.8 <20> (acetone) [31]
20 <16> (12-hydroxydodecanoic acid) [11]
20 <18> (pentanol, <18>, pyrazole-sensitive enzyme, pH 10.0 [24]) [24]
21 <23> (all-trans retinol, <23>, isoenzyme AA-ADH [95]) [95]
21 <18> (butanol, <18>, pyrazole-sensitive enzyme, pH 7.5 [24]) [24]
21.6 <20> (3-pentanone) [31]
23 <17> (cyclohexanol, <17>, isoenzyme ADH-2, pH 10.0 [49]) [49]
24 <16> (all-trans-retinol) [53]
24 <18> (ethanol, <18>, pyrazole-sensitive enzyme, pH 7.5 [24]) [24]
27 <18> (butanol, <18>, pyrazole-sensitive enzyme and pyrazole-insensitive enzyme, pH 10.0 [24]) [24]
27 <18> (ethanol, <18>, pyrazole-insensitive enzyme, pH 10.0 [24]) [24]
28 <52> (ethanol) [101]
28 <16> (vanillyl alcohol) [16]
34 <18> (ethanol, <18>, pyrazole-sensitive enzyme [24]) [24]
34.8 <20> (isopropanol) [31]
35 <16> (cyclohexanol) [14]
40 <16> (pentanol, <16>, pH 7.5, anodic enzyme form [18]) [18]
41 <16> (3-pyridylcarbinol, <16>, pH 7.5, anodic enzyme form [18]) [18]
42 <16> (butanol, <16>, pH 7.5, anodic enzyme form [18]) [18]
42 <16> (ethanol, <16>, pH 7.5, anodic enzyme form [18]) [18]
43 <16> (NAD^+, <16>, pH 7.5, anodic enzyme form [18]) [18]
45 <16> (2-propanol) [14]
45 <28> (cyclohexanol, <28>, enzyme form ADH-2 [60]) [60]
45 <16> (ethylene glycol) [14]
46.8 <20> (benzyl alcohol) [42]
50 <17> (1-butanol, <17>, isoenzyme ADH-3, pH 10.0 [49]) [49]
52.2 <20> ((S)-2-pentanol) [31]
54 <20> (5β-pregnan-21-ol-3,20-dione hemisuccinate) [42]
56.4 <20> (butyl alcohol) [42]
59 <25> (anisaldehyde) [66]
59.4 <20> (hexyl alcohol) [42]
60 <20> ((S)-2-butanol) [31]
60 <17> (1-octanol, <17>, isoenzyme ADH-3, pH 10.0 [49]) [49]
60 <17> (benzyl alcohol, <17>, isoenzyme ADH-3, pH 10.0 [49]) [49]
60 <17> (ethanol, <17>, isoenzyme ADH-3, pH 10.0 [49]) [49]
60.6 <20> ((R)-2-pentanol) [31]
61 <25> (cyclopentanone) [66]
62 <23> (12-hydroxydodecanoate, <23>, isoenzyme TT-ADH [95]) [95]
66 <52> (butanol) [101]

69.9 <20> (propyl alcohol) [42]
70 <17> (1-pentanol, <17>, isoenzyme ADH-3, pH 10.0 [49]) [49]
73 <25> (3-methylbutan-2-one) [66]
76 <17> (12-hydroxydodecanoate, <17>, isoenzyme ADH-3, pH 10.0 [49]) [49]
76.2 <20> (ethanol) [42]
78 <23> (all-trans retinol, <23>, isoenzyme TT-ADH [95]) [47, 95]
82 <23> (benzyl alcohol) [47]
87 <28> (12-hydroxydodecanoate, <28>, enzyme form ADH-3 [60]) [60]
90 <17> (cyclohexanol, <17>, isoenzyme ADH-3, pH 10.0 [49]) [49]
91.2 <20> ((S)-2-octanol) [31]
93 <20> (all-trans-retinol, <20>, oxidation with NAD^+ [93]) [93]
100 <17> (2-buten-1-ol, <17>, isoenzyme ADH-3, pH 10.0 [49]) [49]
101 <25> (benzyl alcohol) [66]
105 <17> (cyclohexanol, <17>, isoenzyme ADH-1, pH 10.0 [49]) [49]
105 <23> (cyclohexanol, <23>, isoenzyme BB-ADH [95]) [95]
110 <16> (tryptophol) [14]
111 <28> (octan-1-ol, <28>, enzyme form ADH-3 [60]) [60]
113 <25> (ethanol) [66]
114 <23> (all-trans retinol, <23>, isoenzyme BB-ADH [95]) [95]
116 <25> (butan-2-one) [66]
118 <25> (butan-1-ol) [66]
120 <20> ((R)-2-butanol) [31]
122 <23> (cyclohexanol) [47]
123 <20> (3-oxo-5β-androstan-17β-ol) [42]
130 <16> ((S)-2-butanol) [53]
130 <25> (3-methylcyclohexanone) [66]
132 <23> (12-hydroxydecanoate, <23>, isoenzyme BB-ADH [95]) [95]
135.6 <20> (3-pentanol) [31]
140 <17> (1-octanol, <17>, isoenzyme ADH-2, pH 10.0 [49]) [49]
141 <23> (n-hexanol, <23>, isoenzyme BB-ADH [95]) [95]
144 <23> (cyclohexanol, <23>, isoenzyme AA-ADH [95]) [95]
146 <23> (12-hydroxydodecanoate) [47]
150 <20> (all-trans-retinal, <20>, reduction with NADH [93]) [93]
150 <16> (ethanol, per active site [12]) [12]
151 <28> (benzyl alcohol, <28>, enzyme form ADH-3 [60]) [60]
155 <23> (n-butanol, <23>, isoenzymee BB-ADH [95]) [95]
158 <23> (methylcrotonyl alcohol, <23>, isoenzyme BB-ADH [95]) [95]
166 <23> (12-hydroxydecanoate, <23>, isoenzyme AA-ADH [95]) [95]
168 <23> (n-hexanol, <23>, isoenzyme AA-ADH [95]) [95]
170 <16> (1-octanol) [11]
170 <16> (2-deoxy-D-ribose) [14]
170 <23> (ethanol, <23>, isoenzyme AA-ADH [95]) [95]
175 <28> (pentan-1-ol, <28>, enzyme form ADH-3 [60]) [60]
175 <25> (pentan-3-ol) [66]
177 <16> (12-hydroxydodecanoate) [53]
180 <23> (methylcrotonyl alcohol, <23>, isoenzyme AA-ADH [95]) [95]

182 <23> (1-hexanol) [47]
182 <16> (12-hydroxydodecanoate) [16]
186 <25> (methanol) [66]
200 <18> (acetaldehyde, <18>, pyrazole-insensitive enzyme, pH 7.0 [24]) [24]
200 <23> (n-butanol, <23>, isoenzyme AA-ADH [95]) [95]
200 <16> (octanal) [16]
203 <23> (ethanol, <23>, isoenzyme BB-ADH [95]) [95]
212 <17> (1-pentanol, <17>, isoenzyme ADH-2, pH 10.0 [49]) [49]
213 <25> (butan-2-ol) [66]
213 <28> (cyclohexanol, <28>, enzyme form ADH-3 [60]) [60]
229 <25> (pentan2-ol) [66]
230 <17> (12-hydroxydodecanoate, <17>, isoenzyme ADH-12, pH 10.0 [49]) [49]
230 <16> (ethanol, <16>, isoenzyme γ_1,γ_1 [13]) [13]
240 <16> (12-hydroxydodecanoate) [14]
240 <16> (ethanol, <16>, isoenzyme α,γ_1 [13]) [13]
242 <25> (propan-1-ol) [66]
245 <23> (n-hexanol, <23>, isoenzyme TT-ADH [95]) [95]
260 <16> (octanol, <16>, isoenzyme α,γ_1 [13]) [13]
280 <28> (NAD^+, <28>, enzyme form ADH-2 [60]) [60]
280 <16> (benzyl alcohol, <16>, isoenzyme α,γ_1 [13]) [13]
285 <23> (2-butanol) [47]
295 <28> (ethanol, <28>, enzyme form ADH-2 [60]) [60]
348 <20> (propan-2-ol, <20>, mutant enzyme W54L [92]) [92]
350 <17> (2-buten-1-ol, <17>, isoenzyme ADH-2, pH 10.0 [49]) [49]
354 <20> (cyclohexanone) [42]
370 <16> (16-hydroxyhexadecanoate) [14]
371 <28> (NAD^+, <28>, enzyme form ADH-3 [60]) [60]
398 <28> (ethanol, <28>, enzyme form ADH-3 [60]) [60]
402 <20> (4-methyl-1-pentanol, <20>, wild-type enzyme Adh 1 [92]) [92]
415 <25> (propan-2-ol) [66]
431 <23> (n-butanol, <23>, isoenzyme TT-ADH [95]) [95]
433 <28> (butan-1-ol, <28>, enzyme form ADH-3 [60]) [60]
440 <16> (octanol) [16]
450 <23> (1-butanol) [47]
450 <16> (3-phenyl-1-propanol) [14]
463 <25> (3-methyl-cyclohexanol) [66]
465 <23> (NAD^+, <23>, oxidation of ethanol [47]) [47]
470 <16> (ethanol) [14]
480 <23> (ethanol) [47]
480 <16> (pentanol) [14]
500 <16> (octanol) [14]
510 <23> (cyclohexanol, <23>, isoenzyme TT-ADH [95]) [95]
515 <23> (methylcrotonyl alcohol, <23>, isoenzyme TT-ADH [95]) [95]
520 <16> (vanillyl alcohol) [14]
533 <28> (NADH, <28>, enzyme form ADH-3 [60]) [60]

550 <16> (benzyl alcohol) [14]
575 <23> (ethanol, <23>, isoenzyme TT-ADH [95]) [95]
610 <16> (ethanol, <16>, pH 10.0, anodic enzyme form [18]) [18]
714 <20> (butan-2-ol, <20>, wild-type enzyme Adh 1 [92]) [92]
960 <16> (pentanol) [53]
1000 <18> (acetaldehyde, <18>, pyrazole-sensitive enzyme, pH 7.0 [24]) [24]
1020 <28> (octanal, <28>, enzyme form ADH-3 [60]) [60]
1030 <16> (propanol) [53]
1104 <20> (pentanol, <20>, mutant enzyme W54L [92]) [92]
1116 <20> (4-methyl-1-pentanol, <20>, mutant enzyme W54L [92]) [92]
1170 <16> (hexanol) [53]
1200 <17> (12-hydroxydodecanoate, <17>, isoenzyme ADH-1, pH 10.0 [49]) [49]
1434 <20> (hexanol, <20>, mutant enzyme W54L [92]) [92]
1722 <20> (propanol, <20>, mutant enzyme W54L [92]) [92]
1770 <20> (benzaldehyde) [42]
1776 <20> (hexanol, <20>, wild-type enzyme Adh 1 [92]) [92]
1840 <16> (ethanol) [53]
1908 <20> (acetaldehyde) [42]
1962 <20> (butanol, <20>, mutant enzyme W54L [92]) [92]
2090 <16> (butanol) [53]
2124 <20> (butyraldehyde) [42]
2514 <20> (propionaldehyde) [42]
2930 <17> (1-butanol, <17>, isoenzyme ADH-1, pH 10.0 [49]) [49]
2930 <17> (1-pentanol, <17>, isoenzyme ADH-1, pH 10.0 [49]) [49]
3186 <20> (propan-2-ol, <20>, wild-type enzyme Adh 1 [92]) [92]
3372 <20> (pentanol, <20>, wild-type enzyme Adh 1 [92]) [92]
3396 <20> (hexaldehyde) [42]
3650 <17> (1-octanol, <17>, isoenzyme ADH-1, pH 10.0 [49]) [49]
3760 <17> (ethanol, <17>, isoenzyme ADH-1, pH 10.0 [49]) [49]
5380 <17> (benzyl alcohol, <17>, isoenzyme ADH-1, pH 10.0 [49]) [49]
5880 <20> (cinnamyl alcohol, <20>, mutant enzyme W54L [92]) [92]
6120 <20> (ethanol, <20>, mutant enzyme W54L [92]) [92]
7080 <20> (butanol, <20>, wild-type enzyme Adh 1 [92]) [92]
7380 <20> (ethanol, <20>, wild-type enzyme Adh 1 [92]) [92]
7980 <20> (cinnamyl alcohol, <20>, wild-type enzyme Adh 1 [92]) [92]
12800 <17> (2-buten-1-ol, <17>, isoenzyme ADH-1, pH 10.0 [49]) [49]
18480 <20> (ethanol, <20>, wild-type enzyme Adh 1 [92]) [92]
Additional information <9, 10, 16, 20> (<16>, turnover-numbers for the class I isoenzymes with the substrates ethanol, methanol, ethylene glycol, benzyl alcohol, octanol, cyclohexanol and 16-hydroxyhexadecanoic acid [13]; <20>, K_m-values of active-site Co(II)substituted enzyme [31]; <9, 10>, kinetics of ethanol oxidation [63]) [13, 28, 31, 63]

Specific activity (U/mg)

0.1 <17> (<17>, isoenzyme 4 [49]) [49]
0.12 <17> (<17>, isoenzyme 1 [49]) [49]

0.2 <17> (<17>, isoenzyme 3 [49]) [49]
0.6 <16> [16]
0.62 <17, 24> (<17>, isoenzyme 2 [49]) [49, 52]
0.65 <16> [21]
1.14 <17> (<17>, ADH-2 [49]) [49]
1.3 <16, 17> (<17>, ADH-3 [49]) [18, 49]
1.36 <15> (<15>, isoenzyme A2 [48]) [48]
1.47 <16> [14]
2 <20> [27]
2.1 <23> (<23>, isoenzyme AA-ADH [95]) [95]
2.5 <23> (<23>, isoenzyme BB-ADH [95]) [95]
2.69 <22> [45]
3.2 <20> (<20>, reduction of 3-oxo-5β-androstan-17β-ol [42]) [42]
3.3 <16> [12]
3.92 <25> [66]
4 <23> [47]
4.2 <20> (<20>, reduction of cyclohexanone [42]) [42]
5.3 <25> [70]
8.08 <15> (<15>, isoenzyme B2 [48]) [48]
17.17 <32> [72]
31.8 <50> [99]
32.5 <17> (<17>, ADH-1 [49]) [49]
49.3 <44> [82]
55 <9> [65]
64 <15> (<15> isoenzyme C2 [48]) [48]
65 <35> (<35>, mutant enzyme Adh1-1S1108 [97]) [97]
70 <35> (<35>, wild-type enzyme Adh1-1S [97]) [97]
79.1 <47> [85]
93 <29> (<29> enzyme form ADHI [68]) [68]
196.3 <46> [84]
470 <29> (<29> enzyme form ADHII [68]) [68]
55900 <37> [78]
Additional information <20, 21, 22, 30, 35, 36, 43> [35, 44, 46, 57, 69, 75, 77, 81]

K_m-Value (mM)

0.0007 <52> (hexanol) [101]
0.0014 <52> (octanol) [101]
0.0017 <17> (NADH, <17>, isoenzyme ADH-2, pH 7.5 [49]) [49]
0.0025 <16> (NADH) [16]
0.0025 <25> (anisaldehyde) [66]
0.0051 <28> (NADH, <28>, reduction of octanal [60]) [60]
0.0064 <16> (NADH, <16>, isoenzyme β_1,β_1 [16]) [16]
0.00694 <22> (NAD^+, <22>, pH 7.5 [45]) [45]
0.007 <16> (NADH, <16>, isoenzyme γ_1,γ_1 [16]) [16]
0.007 <16> (benzyl alcohol) [14]
0.007 <16> (octanol) [14]

0.0074 <16> (NAD^+, <16>, isoenzyme β_2,β_2 [16]; <16>, isoenzyme β_1,β_1 [17]) [16, 17]
0.0079 <16> (NAD^+, <16>, isoenzyme γ_1,γ_1 [16, 17]) [16, 17]
0.008 <16> (cyclohexanol, <16>, isoenzyme α,γ_1 [13]) [13]
0.0086 <46> (acetaldehyde) [84]
0.0087 <16> (NAD^+, <16>, isoenzyme γ_2,γ_2 [16, 17]) [16, 17]
0.009 <16> (all-trans-retinol) [53]
0.011 <14> (NAD^+, <14>, oxidation of ethanol [61]) [61]
0.011 <16> (NADH, <16>, isoenzyme α,α [16]) [16]
0.011 <28> (octan-1-ol, <28>, enzyme form ADH-3 [60]) [60]
0.012 <29> (NADH, <29>, enzyme form ADH II [68]) [68]
0.012 <18> (butanol, <18>, pyrazole-sensitive enzyme, pH 7.5 [24]) [24]
0.012 <20> (hexaldehyde) [42]
0.012 <18> (pentanol, <18>, pyrazole-sensitive enzyme, pH 7.5 [24]) [24]
0.013 <17> (12-hydroxydodecanoate, <17>, isoenzyme ADH-3, pH 10.0 [49]) [49]
0.013 <16> (NAD^+, <16>, isoenzyme α,α [16, 17]) [16, 17]
0.013 <25> (benzyl alcohol) [66]
0.013 <28> (benzyl alcohol, <28>, enzyme form ADH-2 [60]) [60]
0.018 <18> (pentanol, <18>, pyrazole-sensitive enzyme, pH 10.0 [24]) [24]
0.019 <25> (propan-1-ol) [66]
0.02 <23> (12-hydroxydodecanoate, <23>, isoenzyme BB-ADH [95]) [95]
0.02 <50> (NADH, <50>, reduction of acetaldehyde [99]) [99]
0.02 <23> (all-trans-retinol, <23>, isoenzyme TT-ADH [95]) [95]
0.02 <18> (butanol, <18>, pyrazole-sinensitive enzyme, pH 10.0 [24]) [24]
0.02 <23> (n-hexanol, <23>, isoenzyme TT-ADH [95]) [95]
0.021 <25> (2-methylcyclohexanol) [66]
0.021 <17> (5α-androstan-17β-ol-3-one, <17>, isoenzyme 2 [51]) [51]
0.021 <46> (NADH) [84]
0.022 <16> (NAD^+) [12]
0.022 <16> (ethanol, <16>, isoenzyme β_1,β_1 [20]) [20]
0.025 <17> (1-octanol, <17>, isoenzyme ADH-3, pH 10.0 [49]) [49]
0.025 <17> (5α-androstan-17β-ol-3-one, <17>, isoenzyme 2 [51]) [51]
0.025 <16> (NAD^+) [16]
0.027 <17> (5α-androstan-17β-ol-3-one, <17>, isoenzyme 4 [51]) [51]
0.027 <29> (NADH, <29>, enzyme form ADH I [68]) [68]
0.028 <14> (NAD^+, <14>, oxidation of propan-2-ol [61]) [61]
0.03 <23> (all-trans-retinol, <23>, isoenzyme BB-ADH [95]) [95]
0.03 <20> (hexanol) [42]
0.03 <23> (n-butanol, <23>, isoenzyme BB-ADH [95]) [95]
0.031 <20> (3-oxo-5β-androstan-17β-ol) [42]
0.032 <16> (3-phenyl-1-propanol) [14]
0.032 <17> (3β-7α-dihydroxy-5β-cholanoate, <17>, isoenzyme 3 [51]) [51]
0.034 <16> (vanillyl alcohol) [14]
0.036 <25, 47> (NADH) [70, 85]
0.037 <35> (NAD^+, <35>, wild-type enzyme ADH1-1S [97]) [97]
0.038 <35> (NAD^+, <35>, mutant enzyme ADH1-1S1108 [97]) [97]

0.04 <28> (12-hydroxydodecanoate, <28>, enzyme form ADH-3 [60]) [60]
0.04 <17> (5α-androstan-17β-ol-3-one, <17>, isoenzyme 1 [51]) [51]
0.04 <17> (NAD^+, <17>, isoenzyme ADH-2, pH 10.0 and pH 7.5 [49]) [49]
0.04 <23> (cyclohexanol, <23>, isoenzyme BB-ADH [95]) [95]
0.04 <23> (n-hexanol, <23>, isoenzyme BB-ADH [95]) [95]
0.042 <35> (NADH, <35>, mutant enzyme ADH1-1S1108 [97]) [97]
0.043 <28> (octanal, <28>, enzyme form ADH-2 [60]) [60]
0.044 <17> (NAD^+, <17>, isoenzyme 4 [51]) [51]
0.044 <16> (pentanol) [96]
0.046 <20> (benzyl alcohol) [42]
0.047 <16> (12-hydroxydodecanoate) [96]
0.047 <20> (5β-pregnan-21-ol-3,20-dione hemisuccinate) [42]
0.048 <16> (12-hydroxydodecanoate) [53]
0.049 <16> (ethanol, <16>, isoenzyme β_1,β_1 [15, 17]) [15, 17]
0.05 <23> (12-hydroxydodecanoate, <23>, isoenzyme BB-ADH and isoenzyme ADH-TT [95]) [95]
0.05 <17> (NAD^+, <17>, isoenzyme ADH-1, pH 7.5 [49]) [49]
0.05 <23> (all-trans-retinol, <23>, isoenzyme AA-ADH [95]) [95]
0.05 <17> (benzyl alcohol, <17>, isoenzyme ADH-3, pH 10.0 [49]) [49]
0.05 <25> (pentan-2-ol) [66]
0.051 <17> (NADH, <17>, isoenzyme 4 [51]) [51]
0.053 <43> (acetaldehyde) [81]
0.0531 <22> (NAD^+, <22>, pH 10.8 [45]) [45]
0.055 <25> (butan-1-ol) [66]
0.056 <16> (12-hydroxydodecanoic acid) [11]
0.057 <17> (NAD^+, <17>, isoenzyme 1 [51]) [51]
0.059 <4> (NAD^+) [87]
0.059 <20> (all-trans-retinol, <20>, oxidation with NAD^+ [93]) [93]
0.06 <23> (12-hydroxydodecanoate, <23>, isoenzyme AA-ADH [95]) [95]
0.06 <16> (16-hydroxyhexadecanoate) [14]
0.06 <18> (butanol, <18>, pyrazole-insensitive enzyme, pH 7.5 [24]) [24]
0.06 <20> (butyraldehyde) [42]
0.064 <17> (3β-7α-dihydroxy-5β-cholanoate, <17>, isoenzyme 2 [51]) [51]
0.066 <17> (3β-7α-dihydroxy-5β-cholanoate, <17>, isoenzyme 1 [51]) [51]
0.07 <23> (n-butanol, <23>, isoenzyme AA-ADH [95]) [95]
0.071 <17> (3β-7α-dihydroxy-5β-cholanoate, <17>, isoenzyme 4 [51]) [51]
0.071 <17> (NADH, <17>, isoenzyme 2 [51]) [51]
0.073 <29> (NAD^+, <29>, enzyme form ADH I [68]) [68]
0.074 <50> (NAD^+, <50>, oxidation of ethanol [99]) [99]
0.076 <47> (NAD^+) [85]
0.076 <17> (NADH, <17>, isoenzyme 1 [51]) [51]
0.08 <17> (1-pentanol, <17>, isoenzyme ADH-3, pH 10.0 [49]) [49]
0.08 <16> (octanol, <16>, isoenzyme α,γ_1 [13]) [13]
0.085 <35> (NADH, <35>, wild-type enzyme ADH1-1S [97]) [97]
0.085 <16> (acetaldehyde, <16>, isoenzyme β_1,β_1 [15,20]) [15, 20]
0.086 <29> (acetaldehyde, <29>, enzyme form ADH I [68]) [68]
0.09 <25> (anisaldehyde) [70]

0.09 <18> (butanol, <18>, pyrazole-sensitive enzyme, pH 10.0 [24]) [24]
0.09 <16> (pentanol) [14]
0.091 <17> (NADH, <17>, isoenzyme 3 [51]) [51]
0.091 <20> (butanol) [42]
0.099 <17> (NAD^+, <17>, isoenzyme 2 [51]) [51]
0.1 <17> (1-octanol, <17>, isoenzyme ADH-2 [49]) [49]
0.1 <16, 17> (12-hydroxydodecanoate, <17>, isoenzyme ADH-1, pH 10.0 [49]) [16, 49]
0.1 <17, 36, 37> (NAD^+, <17>, isoenzyme ADH-3 [49]) [49, 77, 78]
0.1 <18> (acetaldehyde, <18>, pyrazole-sensitive enzyme, pH 7.0 [24]) [24]
0.105 <16> (NADH, <16>, isoenzyme β_2,β_2 [16]) [16]
0.11 <20> (13-cis-retinal, <20>, oxidation with NAD^+ [93]) [93]
0.11 <29> (NAD^+, <29>, enzyme form ADH II [68]) [68]
0.11 <30> (NADH) [69]
0.115 <17> (NAD^+, <17>, isoenzyme 2 [51]) [51]
0.122 <4> (NADH) [87]
0.123 <17> (5β-androstan-3β-ol-17 one, <17>, isoenzyme 4 [51]) [51]
0.125 <17> (5β-androstan-3β-ol-17 one, <17>, isoenzyme 2 [51]) [51]
0.125 <30> (NAD^+) [69]
0.128 <17> (3β,12α-dihydroxy-5β-cholanoate, <17>, isoenzyme 2 [51]) [51]
0.13 <16> (hexanol) [53]
0.136 <34> (NAD^+, <34>, 23°C [74]) [74]
0.137 <20> (13-cis-retinal, <20>, reduction with NADH [93]) [93]
0.14 <17> (5β-androstan-3β-ol-17 one, <17>, isoenzyme 3 [51]) [51]
0.148 <20> (benzaldehyde) [42]
0.149 <17> (3β,7α,12α-trihydroxy-5β-cholanoate, <17>, isoenzyme I [51]) [51]
0.15 <16> (benzyl alcohol) [96]
0.15 <50> (ethanol) [99]
0.155 <34> (NAD^+, <34>, 27°C [74]) [74]
0.16 <17> (3β,12α-dihydroxy-5β-cholanoate, <17>, isoenzyme 3 [51]) [51]
0.16 <17> (3β,7α,12α-trihydroxy-5β-cholanoate, <17>, isoenzyme 2 [51]) [51]
0.16 <20> (9-cis-retinal, <20>, oxidation with NAD^+ [93]) [93]
0.164 <17> (5β-androstan-3β-ol-17 one, <17>, isoenzyme 1 [51]) [51]
0.17 <17> (1-butanol, <17>, isoenzyme ADH-3, pH 10.0 [49]) [49]
0.17 <46> (NAD^+) [84]
0.18 <20> ((S)-2-octanol) [31]
0.18 <16> (NAD^+, <16>, isoenzyme β_1,β_1 [16]) [16]
0.18 <18> (ethanol, <18>, pyrazole-sensitive enzyme, pH 7.5 [24]) [24, 76]
0.182 <17> (3β,7α,12α-trihydroxy-5β-cholanoate, <17>, isoenzyme 3 [51]) [51]
0.185 <34> (NAD^+, <34>, 33°C [74]) [74]
0.19 <10> (NAD^+) [64]
0.194 <47> (acetaldehyde) [85]
0.195 <44> (NAD^+) [82]
0.2 <20> (11-cis-retinal, <20>, reduction with NADH [93]) [93]

0.2 <11> (NAD^+) [64]
0.2 <17> (NAD^+, <17>, isoenzyme ADH-1, pH 7.5 [49]) [49]
0.2 <28> (pentan-1-ol, <28>, enzyme form ADH-3 [60]) [60]
0.2 <16> (tryptophol) [14]
0.217 <17> (acetaldehyde, <17>, isoenzyme 3 [51]) [51]
0.224 <34> (NAD^+, <34>, 43°C [74]) [74]
0.225 <25> (pentan-3-ol) [66]
0.23 <16> (12-hydroxydodecanoate) [14]
0.23 <28> (NAD^+, <28>, oxidation of ethanol [60]) [60]
0.23 <17> (acetaldehyde, <17>, isoenzyme 2 [51]) [51]
0.24 <16> (acetaldehyde, <16>, isoenzymes β_2,β_2 [15, 20]; <16>, isoenzyme γ_2,γ_2 [15]) [15, 20]
0.24 <16> (butanol) [96]
0.242 <20> (9-cis-retinal, <20>, reduction with NADH [93]) [93]
0.248 <17> (3β,12α-dihydroxy-5β-cholanoate, <17>, isoenzyme 4 [51]) [51]
0.25 <25> ((+)-2-methylcyclohexanone) [66]
0.25 <17> (NAD^+, <17>, isoenzyme ADH-1, pH 10.0 [49]) [49]
0.257 <20> (13-cis-retinol, <20>, oxidation with NAD^+ [93]) [93]
0.26 <9> (NAD^+) [64]
0.26 <20> (all-trans-retinal, <20>, oxidation with NAD^+ [93]) [93]
0.26 <25> (ethanol) [66]
0.26 <23> (n-hexanol, <23>, isoenzyme TT-ADH [95]) [95]
0.276 <17> (acetaldehyde, <17>, isoenzyme 4 [51]) [51]
0.28 <20> (11-cis-retinol, <20>, oxidation with NAD^+ [93]) [93]
0.28 <16> (pentanol) [53]
0.28 <50> (propan-1-ol) [99]
0.3 <20> (9-cis-retinol, <20>, oxidation with NAD^+ [93]) [93]
0.3 <17> (octanal, <17>, isoenzyme ADH-1, pH 7.5 [49]) [49]
0.31 <17> (3β,12α-dihydroxy-5β-cholanoate, <17>, isoenzyme I [51]) [51]
0.31 <17> (acetaldehyde, <17>, isoenzyme 1 [51]) [51]
0.33 <16> (acetaldehyde, <16>, isoenzyme γ_1,γ_1 [15]) [15]
0.34 <16> (NAD^+) [96]
0.34 <20> (all-trans-retinal, <20>, reduction with NADH [93]) [93]
0.35 <17> (2-buten-1-ol, <17>, isoenzyme ADH-3, pH 10.0 [49]) [49]
0.35 <17> (3β,7α-12α-trihydroxy-5β-cholanoate, <17>, isoenzyme 4 [51]) [51]
0.35 <43> (NADH) [81]
0.35 <25> (butan-2-ol) [66]
0.35 <52> (butanol) [101]
0.39 <52> (cyclohexanol) [101]
0.41 <10> (butan-2-ol) [64]
0.45 <16, 22> (ethanol, <22>, pH 10.8 [45]) [12, 45]
0.487 <44> (NAD^+) [82]
0.5 <17> (1-octanol, <17>, isoenzyme ADH-1, pH 7.5 [49]) [49]
0.5 <23> (ethanol, <23>, isoenzyme AA-ADH [95]) [95]
0.5 <25> (propan-2-ol) [66]
0.51 <17> (1-octanol, <17>, isoenzyme ADH-2, pH 10.0 [49]) [49]

0.55 <16> (1-octanol) [11]
0.6 <32> (acetaldehyde) [72]
0.63 <16> (ethanol, <16>, isoenzyme γ_2,γ_2 [15]) [15]
0.7 <27> (ethanol) [59]
0.72 <14> (cyclohexanol) [61]
0.73 <20> ((S)-2-pentanol) [31]
0.76 <23> (cyclohexanol, <23>, isoenzyme AA-ADH [95]) [95]
0.76 <17> (ethanol, <17>, isoenzyme 2 [51]) [51]
0.76 <23> (ethanol, <23>, isoenzyme BB-ADH [95]) [95]
0.79 <16> (butanol) [53]
0.8 <17, 18> (ethanol, <18>, pyrazole-sensitive enzyme, pH 10.0 [24]) [10, 24]
0.8 <16> (octanol) [21]
0.81 <14> (4-methylpentan-1-ol) [61]
0.84 <16> (ethanol, <16>, isoenzyme β_2,β_2 [20]) [20]
0.85 <14> (S-(-)-trans-bicyclo(2.2.1)-heptanol) [61]
0.86 <9> (butan-2-ol) [64]
0.9 <43> (NAD^+, <43>, reduction of ethanol [81]) [81]
0.91 <16> (propanol) [96]
0.93 <10> (propan-2-ol) [64]
0.94 <16> (ethanol, <16>, isoenzyme β_2,β_2 [15]) [15]
0.95 <14> (S-(-)-2-methylbutan-1-ol) [61]
1 <14> (cis-4-methylcyclohexanol) [61]
1 <16> (ethanol, <16>, isoenzyme γ_1,γ_1 [15]) [15]
1.1 <14> (2-propanol) [61]
1.1 <14> (trans-4-methylcyclohexanol) [61]
1.2 <14> (R-(+)-cis-verbenol) [61]
1.2 <16> (octanol) [16]
1.23 <52> (ethanol) [101]
1.24 <20> (propionaldehyde) [42]
1.25 <30> (acetaldehyde) [69]
1.25 <38> (acetaldehyde, <38>, isoenzyme II [79]) [79]
1.3 <29> (acetaldehyde, <29>, enzyme form ADH II [68]) [68]
1.3 <25> (cyclopentanone) [66]
1.3 <43> (ethanol) [81]
1.3 <43> (propan-2-ol) [81]
1.35 <20> ((S)-2-butanol) [31]
1.37 <17> (ethanol, <17>, isoenzyme 1 [51]) [51]
1.39 <16> (propanol) [53]
1.4 <17> (12-hydroxydodecanoate, <17>, isoenzyme ADH-1, pH 10.0 [49]) [49]
1.4 <17> (benzyl alcohol, <17>, isoenzyme ADH-1, pH 10.0 [49]) [49]
1.4 <17> (ethanol, <17>, isoenzyme ADH-3, pH 10.0 and pH 7.5 [49]) [49]
1.4 <17> (*m*-nitrobenzaldehyde, <17>, isoenzyme ADH-1, pH 7.5 [49]) [49]
1.41 <17> (ethanol, <17>, isoenzyme 4 [51]) [51]
1.45 <38> (acetaldehyde, <38>, isoenzyme I [79]) [79]
1.5 <14> (R-(+)-trans-bicyclo(2.2.1)-heptanol) [61]

1.5 <20> (cinnamyl alcohol, <20>, mutant enzyme W54L [92]) [92]
1.5 <16> (ethanol, <16>, isoenzyme α,α [17]) [17]
1.6 <20> (3-pentanol) [31]
1.6 <11> (butan-2-ol) [64]
1.6 <16> (ethanol, <16>, isoenzyme β_1,β_1 [17]) [17]
1.6 <17> (octanol, <17>, isoenzyme ADH-2, pH 7.5 [49]) [49]
1.7 <16, 38> (ethanol, <16>, isoenzyme α,γ_1 [13]; <16>, isoenzyme γ_2,γ_2 [17]; <38>, isoenzyme III [79]) [13, 17, 79]
1.8 <50> (acetaldehyde) [99]
1.8 <16> (ethanol) [10]
1.87 <17> (ethanol, <17>, isoenzyme 3 [51]) [51]
1.87 <9> (propan-2-ol) [64]
1.9 <14> (butan-1-ol) [61]
1.9 <14> (propan-1-ol) [61]
1.92 <20> ((R)-2-octanol) [31]
1.99 <11> (propan-2-ol) [64]
2 <28> (butan-1-ol, <28>, enzyme form ADH-3 [60]) [60]
2.2 <17> (cyclohexanol, <17>, isoenzyme ADH-3, pH 10.0 [49]) [49]
2.3 <23> (12-hydroxydodecanoate, <23>, isoenzyme TT-ADH [95]) [95]
2.4 <16> (octanal) [16]
2.5 <17> (2-buten-1-ol, <17>, isoenzyme ADH-1, pH 10.0 [49]) [49]
2.5 <46> (ethanol) [84]
2.6 <23> (n-butanol, <23>, isoenzyme TT-ADH [95]) [95]
2.7 <14> (2-methylpropan-1-ol) [61]
2.83 <4> (acetaldehyde) [87]
2.9 <20> (propanol) [42]
3 <17> (butanal, <17>, isoenzyme ADH-1, pH 7.5 [49]) [49]
3.01 <10> (butan-1-ol) [64]
3.1 <17> (1-pentanol, <17>, isoenzyme ADH-1, pH 10.0 [49]) [49]
3.2 <16> (ethanol, <16>, isoenzyme γ_1,γ_1 [17]) [17]
3.31 <9> (butan-1-ol) [64]
3.33 <22> (ethanol, <22>, pH 7.5 [45]) [45]
3.33 <10> (propan-2-ol) [64]
3.4 <14> (R-(+)-1-phenylethanol) [61]
3.4 <16> (acetaldehyde, <16>, isoenzyme β_3,β_3 [20]) [20]
3.42 <9> (propan-2-ol) [64]
3.5 <38> (ethanol, <38>, isoenzyme III from leaf [79]) [79]
3.5 <17> (octanal, <17>, isoenzyme ADH-1, pH 7.5 [49]) [49]
3.8 <14> (S-(-)-cis-bicyclo(2.2.1)-heptanol) [61]
3.9 <14> (S-(-)-1-phenylethanol) [61]
4 <18, 20> (ethanol, <18>, pyrazole-insensitive enzyme, pH 7.5 [24]; <20>, wild-type enzyme Adh1 [92]) [24, 92]
4.1 <14> (pentan-1-ol) [61]
4.16 <11> (propan-2-ol) [64]
4.2 <16> (ethanol, <16>, isoenzyme α,α [15]) [15]
4.2 <16> (ethanol, <16>, isoenzyme α,α [17]) [17]
4.3 <16> (acetaldehyde, <16>, isoenzyme α,α [15]) [15]

4.58 <20> (cinnamyl alcohol, <20>, wilde-type enzyme ADh1 [92]) [92]
4.68 <47> (ethanol) [85]
4.8 <29> (ethanol, <29>, enzyme form ADH I [68]) [68]
5 <18> (ethanol, <18>, pyrazole-insensitive enzyme, pH 10.0 [24]) [24]
5 <52> (methanol) [101]
5.2 <28> (cyclohexanol, <28>, enzyme form ADH-3 [60]) [60]
5.22 <10> (ethanol) [64]
6 <20> (acetaldehyde) [28]
6 <20> (acetaldehyde) [42]
6 <14> (ethanol) [61]
6.2 <14> (ethanol) [61]
6.4 <16> (NADH, <16>, isoenzyme β_1,β_1 [20]) [20]
6.44 <11> (ethanol) [64]
6.5 <20> (hexanol, <20>, mutant enzyme W54L [92]) [92]
6.6 <25> (acetone) [66]
6.6 <14> (heptan-4-ol) [61]
6.7 <20> (4-methyl-1-pentanol, <20>, mutant enzyme W54L [92]) [92]
7 <20> (ethanol, <20>, pH 10.0 [28]) [28]
7.05 <7> (acetaldehyde) [6]
7.2 <38> (ethanol, <38>, isoenzyme II from leaf [79]) [79]
7.4 <16> (NAD^+, <16>, isoenzyme β_1,β_1 [20]) [20]
7.4 <14> (hexan-1-ol) [61]
7.5 <20> ((R)-2-butanol) [31]
7.5 <25> (3-methylbutan-2-one) [66]
8.1 <28> (ethanol, <28>, enzyme form ADH-3 [60]) [60]
8.92 <9> (ethanol) [64]
9 <20> (isopropanol) [31]
10.4 <16> (methanol) [12]
11 <16> (vanillyl alcohol) [16]
11.5 <18> (acetaldehyde, <18>, pyrazole-insensitive enzyme, pH 7.0 [24]) [24]
11.8 <20> (pentanol, <20>, mutant enzyme W54L [92]) [92]
12 <36, 37> (ethanol) [77, 78]
12 <18> (methanol, <18>, pyrazole-sensitive enzyme, pH 7.5 [24]) [24]
13.4 <20> (ethanol, <20>, mutant enzyme W54L [92]) [92]
13.6 <20> (propanol, <20>, wilde-type enzyme Adh1 [92]) [92]
13.8 <20> (hexanol, <20>, wilde-type enzyme Adh1 [92]) [92]
14 <25> (butan-2-one) [66]
14 <20> (cyclohexanone) [28]
14 <18> (methanol, <18>, pyrazole-sensitive enzyme, pH 10.0 [24]) [24]
14.1 <20> (cyclohexanone) [42]
14.2 <20> (2-pentanone) [31]
15 <20> (2-butanone) [31]
16 <31> (ethanol) [71]
16 <20> (ethanol, <20>, pH 8.8 [28]) [28]
16.7 <4> (ethanol) [87]
17 <17> (1-butanol, <17>, isoenzyme ADH-1, pH 10.0 [49]) [49]

17 <16> (hexanol) [21]
18 <16> (ethanol) [96]
20 <34> (ethanol, <34>, 23°C [74]) [74]
22 <20, 35> (ethanol, <20>, pH 7.6 [28]; <35>, mutant enzyme ADH1S-1108 [97]) [28, 97]
22.2 <30> (ethanol) [69]
23 <16> (cyclohexanol, <16>, isoenzyme β_1,β_1 [13]) [13]
24 <35> (ethanol, <35>, wild-type enzyme ADH1-1S [97]) [97]
25 <38> (NADH, <38>, isoenzyme III from leaf [79]) [79]
25 <34> (ethanol, <34>, 27°C [74]) [74]
27 <16> (1-pentanol) [11]
27 <29> (ethanol, <29>, enzyme form ADH II [68]) [68]
27 <25> (methanol) [66]
27.5 <38> (ethanol, <38>, isoenzyme I from leaf [79]) [79]
28 <16, 23> (ethanol, <23>, isoenzyme TT-ADH [95]) [53, 95]
29.4 <20> (butanol, <20>, mutant enzyme W54L [92]) [92]
30 <38> (NAD^+, <38>, isoenzyme II from leaf [79]) [79]
30 <34> (ethanol, <34>, 33°C [74]) [74]
31 <38> (NAD^+, <38>, isoenzyme III from leaf [79]) [79]
31.1 <7> (NADH) [6]
31.7 <20> ((R)-2-pentanol) [31]
35 <32> (methylglyoxal) [72]
36 <16> (ethanol, <16>, isoenzyme β_3,β_3 [20]) [20]
40 <27> (NAD^+, <27>, enzyme from liver [59]) [59]
40 <20> (ethanol) [42]
41.3 <34> (ethanol, <34>, 43°C [74]) [74]
42 <20> (ethanol, <20>, pH 7.0 [28]) [28]
43.9 <20> (propanol, <20>, mutant enzyme W54L [92]) [92]
47.3 <7> (ethanol) [6]
50 <16> (ethylene glycol, <16>, isoenzyme α,γ_1 [13]) [13]
50 <50> (formaldehyde) [99]
51.7 <7> (NAD^+) [6]
53.8 <20> (4-methyl-1-pentanol, <20>, wilde-type enzyme ADh1 [92]) [92]
53.9 <20> (butanol, <20>, wilde-type enzyme Adh1 [92]) [92]
56.2 <20> (pentanol, <20>, wilde-type enzyme Adh1 [92]) [92]
60 <17> (2-buten-1-ol, <17>, isoenzyme ADH-2, pH 10.0 [49]) [49]
60 <27> (NAD^+, <27>, enzyme from caecum [59]) [59]
64 <46> (ethanol) [84]
64 <46> (propan-2-ol) [84]
70 <38> (NADH, <38>, isoenzyme II from leaf [79]) [79]
70 <35> (acetaldehyde, <35>, mutant enzyme ADH1-1S1108 [97]) [97]
73 <38> (NAD^+, <38>, isoenzyme I from leaf [79]) [79]
75 <20> (3-pentanone) [31]
78 <17> (1-pentanol, <17>, isoenzyme ADH-2, pH 10.0 [49]) [49]
85 <35> (acetaldehyde, <35>, wild-type enzyme ADH1-1S [97]) [97]
93.3 <20> (butan-2-ol, <20>, wilde-type enzyme Adh1 [92]) [92]
105 <16> (NADH, <16>, isoenzyme β_2,β_2 [20]) [20]

118 <20> (ethanol, <20>, pH 6.1 [28]) [28]
120 <16> ((S)-2-butanol) [53]
120 <16> (ethanol) [14]
121 <38> (NADH, <38>, isoenzyme I from leaf [79]) [79]
135 <20> (acetone) [31]
150 <16> (methanol, <16>, isoenzyme α,γ_1 [13]) [13]
180 <16> (NAD^+, <16>, isoenzyme β_2,β_2 [20]) [20]
193 <20> (propan-2-ol, <20>, mutant enzyme W54L [92]) [92]
210 <16> (cyclohexanol) [14]
220 <17> (cyclohexanol, <17>, isoenzyme ADH-1, pH 10.0 [49]) [49]
230 <17> (1-butanol, <17>, isoenzyme ADH-1, pH 7.5 [49]) [49]
260 <16> (NADH, <16>, isoenzyme β_3,β_3 [20]) [20]
268 <20> (2-propanol, <20>, wilde-type enzyme ADh1 [92]) [92]
290 <16> (ethylene glycol) [14]
310 <16> (2-deoxy-D-ribose) [14]
320 <23> (cyclohexanol, <23>, isoenzyme TT-ADH [95]) [95]
340 <17> (ethanol, <17>, isoenzyme ADH-1, pH 10.0 [49]) [49]
550 <43> (NAD^+, <43>, reduction of propan-2-ol [81]) [81]
560 <16> (2-propanol) [14]
710 <16> (NAD^+, <16>, isoenzyme β_3,β_3 [20]) [20]
1900 <17> (cyclohexanol, <17>, isoenzyme ADH-2, pH 10.0 [49]) [49]
5000 <17> (ethanol, <17>, isoenzyme ADH-1, pH 7.5 [49]) [49]
Additional information <9, 10, 16, 20, 35, 45, 48> (<16>, K_m values for the class I isoenzymes with the substrates ethanol, methanol, ethylene glycol, benzyl alcohol, octanol, cyclohexanol and 16-hydroxyhexadecanoic acid [13]; <20>, K_m-values of active-site Co(II)substituted enzyme [31]; <20>, kinetics of native and modified enzyme with coenzyme analogues [54]; <9, 10>, kinetics of ethanol oxidation [63]; <48>, pH-dependence of K_m-value [89]) [13, 22, 23, 31, 54, 63, 75, 83, 89]

pH-Optimum

4.5 <29> (<29>, reduction of acetaldehyde, enzyme form ADH I [68]) [68]
5.6-6.5 <46> (<46>, oxidation of ethanol [84]) [84]
5.7 <32> (<32>, reduction of acetaldehyde [72]) [72]
5.9 <16> (<16>, acetaldehyde reduction, isoenzyme β_1,β_1 [15]) [15]
6.5 <7, 28, 29> (<7>, reduction of acetaldehyde [6]; <28>, reduction of octanal [60]; <29>, reduction of acetaldehyde, enzyme form ADH II [68]) [6, 60, 68]
6.7 <31> (<31>, reduction of acetaldehyde [71]) [71]
6.8 <30> (<30>, oxidation of NADH [69]) [69]
7 <16, 44> (<16>, isoenzyme β_3,β_3 [20]; <16>, ADH Indianapolis form 1 [22]) [20, 22, 82]
7-7.5 <16> (<16>, and a second optimum at pH 10.0-10.5, ADH Indianapolis form 2 and 3 [22]) [22]
7.4 <16> (<16>, acetaldehyde reduction of isoenzyme β_2,β_2 [15]) [15]

7.5 <5, 10, 25, 50> (<5>, reduction of substrate [4]; <10>, and second optimum at pH 9.9 [64]; <25>, reduction of anisaldehyde [66]; <50>, reduction of acetaldehyde [99]) [4, 64, 66, 99]
8 <9, 11, 35> (<9, 11>, and a second optimum at pH 9.5 [64]; <35>, mutant enzyme ADH1-1S1108 [97]) [64, 97]
8-8.4 <46> (<46>, reduction of aldehyde [84]) [84]
8-8.5 <47> [85]
8.1 <43> (<43>, reduction of propan-2-ol or ethanol [81]) [81]
8.2 <18, 45> (<18>, and a second optimum at pH 10.6, oxidation of ethanol, pyrazole-sensitive enzyme [24]) [24, 83]
8.3 <4> (<4>, alcohol dehydrogenase IV [87]) [87]
8.5 <5, 16, 25, 35> (<5>, oxidation of substrate [4]; <16>, isoenzyme β_2,β_2 [20]; <25>, oxidation of benzyl alcohol [66]; <35>, wild-type enzyme ADH1-1S [97]) [4, 20, 66, 97]
8.5-8.8 <16> (<16>, ethanol oxidation, isoenzyme β_2,β_2, β_2,β_1, α,β_2 and $\beta_2\gamma_1$ [15]) [15]
8.6-8.8 <38> (<38>, reduction of acetaldehyde, isoenzyme I [79]) [79]
8.7 <35> [76]
8.9 <32> (<32>, oxidation of ethanol [72]) [72]
9 <35> (<35>, oxidation of ethanol [97]) [97]
9-10 <23, 31> (<31>, oxidation of acetaldehyde [71]) [47, 71]
9.1 <30> (<30>, reduction of NAD^+ [69]) [69]
9.5 <7, 9, 11, 17, 29, 50> (<7,17>, oxidation of ethanol [6, 51]; <29>, enzyme form ADH-I, oxidation of ethanol [67, 68]; <17>, isoenzyme 1 [51]; <9, 11>, and a second optimum at pH 8.0 [64]; <50>, oxidation of ethanol [99]) [6, 51, 64, 67, 68, 99]
9.9 <10> (<10>, and second optimum at pH 7.5 [64]) [64]
10 <35> [75]
10 <16, 29> (<29>, enzyme form ADH-II, oxidation of ethanol [67, 68]) [67, 96]
10-10.5 <16> (<16>, and a second optimum at pH 10.0-10.5, ADH Indianapolis form 2 and 3 [22]) [22]
10.4 <16> [14]
10.5 <2, 16, 17> (<2>, oxidation of ethanol [2]; <16>, isoenzyme β_1,β_1 [20]) [2, 10, 20]
10.6 <18, 19> (<18>, and a second optimum at pH 8.2, oxidation of ethanol, pyrazole-sensitive enzyme [24]; <19>, pyrazole-sensitive enzyme [25]) [24, 25]
10.7 <17> (<17>, oxidation of ethanol, isoenzyme 2, 3 and 4 [49]) [49]
10.8 <16, 22> (<22>, and a second lower maximum at pH 7.5 [45]) [12, 45]
11 <17, 18, 28> (<18>, oxidation of ethanol, pyrazole-insensitive enzyme [24]; <17>, oxidation of octanol [49]; <28>, ethanol oxidation, enzyme form ADH-2 and ADH-3 [60]) [24, 49, 60]
11.2 <19> (<19>, oxidation of ethanol, pyrazole-sensitive enzyme form [25]) [25]

pH-Range

5.2-6.8 <46> (<46>, pH 5.2: about 35% of maximal activity, pH 6.8: about 55% of maximal activity, oxidation of ethanol [84]) [84]
7.2-9.2 <46> (<46>, pH 7.2: about 50% of maximal activity, pH 9.2: about 60% of maximal activity [84]) [84]
8-10.5 <16> (<16>, pH 8.0: about 40% of maximal activity, pH 10.5: about 85% of maximal activity [96]) [96]
8-12 <16> (<16>, about 30% of maximal activity at pH 8.0 and at pH 12.0 [14]) [14]
8.2-9.5 <4> (<4>, pH 8.2: about 10% of maximal activity, pH 9.5: about 40% of maximal activity [87]) [87]
9-12 <19> (<19>, pH 9.0: about 25% of maximal activity, pH 12.0: about 80% of maximal activity, pyrazole-sensitive enzyme [25]) [25]
9.5-11.5 <28> (<28>, pH 9.5: about 40% of maximal activity, pH 11.5: about 85% of maximal activity, ethanol oxidation enzyme form ADH-3 [60]) [60]
10-11 <19> (<19>, pH 10.0: about 50% of maximal activity, pH 11.0: about 75% of maximal activity, pyrazole-insensitive enzyme [25]) [25]

Temperature optimum (°C)

30 <46> [84]
40 <50> [99]
45-50 <4> [3]
50-55 <47> [85]
70 <44> (<44>, above [82]) [82]

Temperature range (°C)

10-45 <34> (<34>, activity increases from 10°C to 45°C [74]) [74]
10-60 <46> (<46>, 10°C: about 65% of maximal activity, 50°C: about 40% of maximal activity [84]) [84]
45-95 <25> (<25>, continous increase in activity from 40°C to 95°C [66]) [66]

4 Enzyme Structure

Molecular weight

54600 <11> (<11>, gel filtration [64]) [64]
55600 <9> (<9>, gel filtration [64]) [64]
56000 <14> (<14>, gel filtration [61]) [61]
57600 <10> (<10>, gel filtration [64]) [64]
60000-80000 <4> (<4>, gel filtration [87]) [87]
67000 <29> (<29>, enzyme form ADH-II, glycerol density gradient centrifugation [67]) [67]
68000 <32> (<32>, gel filtration [72]) [72]
70000 <25, 35> (<25>, gel filtration, sucrose density gradient centrifugation [66]; <35>, gel filtration [75]) [66, 75]
72000 <22> (<22>, gel filtration [45]) [45]

74000 <25, 31> (<25>, gel filtration, sucrose density gradient centrifugation [70]; <31>, gel filtration sedimentation equilibrium centrifugation [71]) [70, 71]
74500 <22> (<22>, gel filtration [46]) [46]
76000-77000 <19> (<19>, equilibrium sedimentation [25]) [25]
77330 <18> (<18>, equilibrium sedimentation [24]) [24]
78000 <16> (<16>, ultracentrifugation under non-denaturing conditions [23]) [23]
78000-85000 <16> (<16>, amino acid analysis, ultracentrifugation [18]) [18]
79000 <15> (<15>, isoenzyme B2, gel filtration [48]) [48]
79000-84000 <16> (<16>, ultracentrifugal analysis [21]) [21]
80000 <17, 23, 27, 29> (<17, 23, 27>, gel filtration [47, 49, 59, 95]; <29>, enzyme form ADHI, gel filtration [68]) [47, 49, 49, 59, 68, 95]
82700 <16> (<16>, equilibrium sedimentation [16]) [16]
82750 <28> (<28>, enzyme form ADH-3, equilibrium sedimentation [60]) [60]
83000 <15> (<15>, isoenzyme A2, gel filtration [48]) [48]
85000 <15> (<15>, isoenzyme C2, gel filtration [48]) [48]
86000 <7> [6]
89000-91000 <38> (<38>, gel filtration [79]) [79]
94000 <30> (<30>, gel filtration [69]) [69]
95000 <29> (<29>, enzyme form ADHII, gel filtration [68]) [68]
116000 <36, 37> (<36, 37>, gel filtration [77, 78]) [77, 78]
145000 <29> (<29>, enzyme form ADH-I, glycerol density gradient centrifugation [67]) [67]
150000 <33> (<33>, gel filtration [73]) [73]
152000 <47> (<47>, gel filtration [85]) [85]
176000 <46> (<46>, non-denaturing PAGE [84]) [84]
180000 <44> (<44>, gel filtration [82]) [82]
290000 <43> (<43>, gel filtration [81]) [81]
320000 <50> (<50>, gel filtration [99]) [99]

Subunits

? <2, 16, 21, 24, 43> (<2>, x * 28000, SDS-PAGE [2]; <43>, x * 37500, SDS-PAGE [81]; <16, 21, 24>, x * 40000, SDS-PAGE [11, 44, 52]; <16>, x * 42000, SDS-PAGE [14]) [2, 11, 14, 44, 46, 52, 81]
decamer <50> (<50>, 10 * 43000, SDS-PAGE [99]) [99]
dimer <4, 5, 7, 9, 10, 11, 14, 15, 16, 17, 18, 22, 23, 25, 27, 28, 29, 30, 31, 32, 35, 36, 37, 38> (<9, 10, 11>, 2 * 27800, SDS-PAGE [64]; <14>, 2 * 28000, SDS-PAGE [61]; <5>, 2 * 30000 [4]; <29>, 2 * 31100, enzyme form ADH-II, SDS-PAGE [67]; <29>, 2 * 34700, enzyme form ADH-I, SDS-PAGE [67]; <35>, 2 * 35000, SDS-PAGE [75]; <22>, 2 * 36000, SDS-PAGE [45]; <25, 32>, 2 * 37000, SDS-PAGE [66, 70, 71, 72]; <29>, 2 * 38000, enzyme form ADHII, SDS-PAGE [68]; <17>, 2 * 39000, ADH-2, SDS-PAGE [49]; <15>, 2 * 39000, isoenzyme B2, SDS-PAGE [48]; <29>, 2 * 40000, enzyme form ADHI [68]; <4, 16, 18, 23, 27>, 2 * 40000, SDS-PAGE [16, 23, 24, 59, 87, 95]; <17>, 2 * 40000, ADH-3, SDS-PAGE [49]; <23>, 2 * 41000, SDS-PAGE [47]; <16>, 2 * 41000, class III

isoenzyme chi ADH, SDS-PAGE [16]; <22>, 2 * 41700, enzyme form CM-I: a polypeptide chain + C polypeptide chain, enzyme form CM-II: B-chain + C-chain, enzyme form CM III, homodimer of C chains, SDS-PAGE [46]; <28>, 2 * 42000, enzyme form ADH-2 and ADH-3, SDS-PAGE [60]; <16>, 2 * 42000, anodic enzyme form, SDS-PAGE [18]; <30>, 2 * 42000, SDS-PAGE [69]; <17>, 2 * 43000, ADH-1, SDS-PAGE [49]; <15>, 2 * 43000, isoenzyme A2, SDS-PAGE [48]; <38>, 2 * 45000, SDS-PAGE [79]; <7>, 2 * 46000, SDS-PAGE [6]; <15>, 2 * 47000, isoenzyme C2, SDS-PAGE [48]; <36, 37>, 2 * 58000, SDS-PAGE [77,78]) [4, 6, 16, 21, 23, 24, 45, 46, 47, 48, 49, 59, 60, 61, 64, 66, 67, 68, 69, 70, 71, 72, 75, 77, 78, 79, 87, 95]
tetramer <44, 46, 47> (<47>, 4 * 38000, SDS-PAGE [85]; <46>, 4 * 42000, SDS-PAGE [84]; <44>, 4 * 45000, SDS-PAGE [82]) [82, 84, 85]

5 Isolation/Preparation/Mutation/Application

Source/tissue

adrenal gland <17> [49]
brain <17> [49]
cecum <27> [59]
coleoptile <7> [80]
cornea <17> [49]
embryo <31, 36> [71, 77]
epididymis <17> [49]
heart <17> [49]
intestine <17> [49]
kernel <35> [97]
kidney <17> [49]
leaf <32, 38> [72, 79]
liver <1, 2, 6, 15, 16, 17, 18, 19, 20, 21, 22, 23, 24, 27, 28, 52> (<15>, isoenzyme A2 and B2 [48]; <23>, isoenzyme AA-ADH and BB-ADH most abundant in [95]) [1, 2, 5, 10, 12, 13, 14, 15, 16, 17, 18, 19, 20, 21, 22, 23, 24, 25, 26, 27, 28, 29, 30, 31, 32, 33, 34, 35, 36, 37, 39, 40, 41, 42, 44, 45, 46, 48, 49, 51, 52, 54, 55, 57, 59, 60, 86, 92, 93, 95, 98, 101]
lung <17> [49]
muscle <17> [49]
ovary <17> [49]
pancreas <17> [49]
root <38> [79]
scutellum <35> [75]
seed <5, 7, 8, 33, 34, 38> [4, 6, 7, 73, 74, 79]
seedling <35> [76]
shoot <38> [79]
spleen <17> [49]
stomach <15, 16, 17> (<15>, isoenzyme C2 [48]; <16>, stomach mucosa, μ-alcohol dehydrogenase [96]) [48, 49, 50, 53, 96]

testis <16, 17, 23> (<16>, the class III enzyme contributes by far the bulk of the total alcohol dehydrogenase activity [11]; <23>, activity increases during the prepubertal development [47]; isoenzyme TT-ADH is only found in testis [95]) [11, 47, 49, 95]
thymus <17> [49]
trophozoite <51> [100]
uterus <17> [49]

Purification

<1> [1]
<3> [3]
<4> [3, 87]
<6> [5]
<7> [6]
<9> (alleloenzymes AdhF' and Adhus [58]; alleloenzyme Adh71k [63]) [8, 58, 63, 65]
<10> [8, 63]
<11> [8]
<12> [8]
<13> [8]
<14> [8]
<15> (liver isoenzyme A2 and B2 and stomach isoenzyme C2 [48]) [48]
<16> (class I isoenzymes [13]; class II isoenzyme: π-ADH [14]; class III isoenzyme chi-ADH [16]; anodic enzyme form [18]; isoenzyme β_3,β_3 [20]) [12, 13, 14, 15, 16, 18, 20]
<17> (ADH-1, ADH-2 and ADH-3 [49]; isoenzyme 1, 2, 3, and 4 [51]) [49, 51]
<18> [24]
<19> [25]
<20> (SS-isoenzyme [28]) [26, 28, 42, 92]
<21> [44]
<22> (class I isoenzyme [46]) [45, 46]
<23> [47]
<25> [66, 70]
<28> (ADH-2 and ADH-3 [60]) [60]
<29> [67, 68]
<30> (partial [69]) [69]
<31> [71]
<32> [72]
<33> [73]
<35> (wild-type enzyme Adh1-1S and mutant enzyme Adh1-1S1108 [97]) [75, 76, 97]
<36> (3 isoenzymes [77]) [77]
<37> [78]
<43> [81]
<44> (ADH-MII [82]) [82]
<45> (partial [83]) [83]

<46> [84]
<47> [85]
<50> [99]
<52> [101]

Crystallization

<14> [38]
<16> [12]
<20> [30, 34, 35]
<25> (holo-enzyme form and apo-enzyme form [56]) [56]

Cloning

<4> [87, 91]
<15> (class III enzyme [9]) [9]
<16> (class IV enzyme, expression in Escherichia coli [53]) [53]
<49> [91]

Engineering

M294L <4> (<4>, 7-10fold increase in reactivity, V/K_m, with butanol, pentanol and hexanol [91]) [91]
W54L <20> (<20>, less active than the wild-type enzyme with ethanol, 1-propanol and 1-butanol. With 1-pentanol and 1-hexanol the mutant enzyme is a better catalyst than the wild-type enzyme [92]) [92]

6 Stability

pH-Stability

6-9 <35> (<35>, isoenzyme 2 and isoenzyme 3 [75]) [75]
6.5-7 <34> (<34>, most stable at [74]) [74]
7-10.6 <16> (<16>, stable [12]) [12]
7.5 <29> (<29>, 4°C, stable for 2 days [68]) [68]

Temperature stability

-20 <47> (<47>, loss of activity after 24 h [85]) [85]
23 <16> (<16>, unstable at room temperature and above [12]) [12]
40 <7, 47> (<7>, 60 min, stable [6]; <47>, stable below [85]) [6, 85]
50 <25, 35> (<25>, 24 h, 30% loss of activity [66]; <35>, wild-type enzyme Adh1-1S is more stable than mutant enzyme Adh1-1S1108 [97]) [66, 97]
55 <35> (<35>, 60 min, 73% loss of activity of enzyme 1, 10% loss of activity of enzyme 2, 21% loss of activity of enzyme 3 [75]) [75]
60 <25, 34> (<25>, half-life: 20 h [66]; <34>, denaturation above [74]) [66, 74]
65 <25, 29> (<29>, inactivation constant for ADHI: 1.23 per min, inactivation constant for ADH II: 0.13 per min [68]; <25>, pH 8.0, protein concentration 2.5 mg/ml, 16 h, stable. 50% loss of activity after 8 h at protein concentration of 0.5 mg/ml [70]) [68, 70]
70 <25> (<25>, half-life: 5 h [66]) [66]

85 <25> (<25>, pH 8.0, protein concentration 0.5 mg/ml, 3 h, 50% loss of activity [70]) [70]
Additional information <20> (<20>, distinct subunits have different deactivation properties [37]) [37]

Oxidation stability

<50>, pronounced oxygen lability at pH 9, even at pH 7, 2 h under air, 80% loss of activity [99]

General stability information

<16>, 100fold purified enzyme is destroyed by freezing [12]
<20>, the catalytic zinc ions have an important stabilizing effect on the tertiary and quaternary structure of the immobilized enzyme [36]
<22>, dialysis against 50 mM Tris-HCl buffer, stable after 5 h, 3% loss of activity after 1 day, 82% loss of activity after 6 days [45]
<22>, dithiothreitol stabilizes activity at all stages of purification [46]
<29>, enzyme form ADH I is more stable during purification than enzyme form ADH-II [67]

Storage stability

<16>, 4°C, 10 mM HEPES buffer, 1 mM dithioerythritol, pH 7.5, stable for 2 weeks [16]
<16>, 4°C, 5 mM Na phosphate, pH 7.5, the half-life is 24 h. 0.01 mM ethanol effectively stabilizes for several weeks [18]
<16>, 4°C, 5 mM Na-phosphate, pH 7.5, 50% loss of activity after 1 day. Enzyme can be stabilized for up to 2 weeks by storage in buffer containing 10 mM ethanol [23]
<16>, 4°C, pH 7.5, stable for 2-3 weeks [14]
<22>, 4°C, in presence of 1.0 mM dithiothreitol, stable for 10 days [45]
<29>, 4°C, most stable at pH 7.5 during storage for 2 days [68]
<35>, -20°C, isoenzyme 2 is stable in 50% ethylene glycol for at least 3 months, isoenzyme 1 and 3 maintain less than 50% of their original activity after 2 weeks [75]

References

[1] Talbot, B.G.; Qureshi, A.A.; Cohen, R.; Thirion, J.P.: Purification and properties of two distinct groups of ADH isozymes from Chinese hamster liver. Biochem. Genet., **19**, 813-829 (1981)

[2] Fong, W.P.: Isolation and characterization of grass carp (Ctenopharyngodon idellus) liver alcohol dehydrogenase. Comp. Biochem. Physiol. B, **98**, 297-302 (1991)

[3] Pessione, E.; Pergola, L.; Cavaletto, M.; Giunta, C.; Trotta, A.; Vanni, A.: Extraction, purification and characterization of ADH1 from the budding yeast Kluyveromyces marxianus. Ital. J. Biochem., **39**, 71-82 (1990)

[4] Leblova, S.; El Ahmad, M.: Characterization of alcohol dehydrogenase isolated from germinating bean (Vicia faba) seeds. Collect. Czech. Chem. Commun., **54**, 2519-2527 (1989)
[5] Keung, W.M.; Ho, Y.W.; Fong, W.P.; Lee, C.Y.: Isolation and characterization of shrew (Suncus murinus) liver alcohol dehydrogenase. Comp. Biochem. Physiol. B Comp. Biochem., **93**, 169-173 (1989)
[6] Tong, W.F.; Lin, S.W.: Purification and biochemical properties of rice alcohol dehydrogenase. Bot. Bull. Acad. Sin., **29**, 245-253 (1988)
[7] Van Geyt, J.; Jacobs, M.; Triest, L.: Characterization of alcohol dehydrogenase in Najas marina L. Aquat. Bot., **28**, 129-141 (1987)
[8] Vilageliu, L.; Juan, E.; Gonzalez-Duarte, R.: Determination of some biochemical features of alcohol dehydrogenase from Drosophila melanogaster, Drosophila simulans, Drosophila virilis, Drosophila funebris, Drosophila imigrans and drosophila lebanonensis. Comparison of their properties and estimation of the homology of the ADH enzyme of different species. Adv. Genet., Dev., Evol. Drosophila, [Proc. Eur. Drosophila Res. Conf.] (Lakovaara, S., ed.) Plenum N.Y., **7**, 237-250 (1982)
[9] Edenberg, H.J.; Brown, C.J.; Carr, L.G.; Ho, W.H.; Hu, M.W.: Alcohol dehydrogenase gene expression and cloning of the mouse χ-like ADH. Adv. Exp. Med. Biol., **284**, 253-262 (1991)
[10] Herrera, E.; Zorzano, A.; Fresneda, V.: Comparative kinetics of human and rat liver alcohol dehydrogenase. Biochem. Soc. Trans., **11**, 729-730 (1983)
[11] Dafeldecker, W.P.; Vallee, B.L.: Organ-specific human alcohol dehydrogenase: isolation and characterization of isozymes from testis. Biochem. Biophys. Res. Commun., **134**, 1056-1063 (1986)
[12] Woronick, C.L.: Alcohol dehydrogenase from human liver. Methods Enzymol., **41 B**, 369-374 (1975)
[13] Wagner, F.W.; Burger, A.R.; Vallee, B.L.: Kinetic properties of human liver alcohol dehydrogenase: oxidation of alcohols by class I isoenzymes. Biochemistry, **22**, 1857-1863 (1983)
[14] Ditlow, C.C.; Holmquist, B.; Morelock, M.M.; Vallee, B.L.: Physical and enzymatic properties of a class II alcohol dehydrogenase isozyme of human liver: π-ADH. Biochemistry, **23**, 6363-6368 (1984)
[15] Yin, S.-J.; Bosron, W.F.; Magnes, L.J.; Li, T.-K.: Human liver alcohol dehydrogenase: purification and kinetic characterization of the $\beta 2 \beta 2$, $\beta 2 \beta 1$, $\alpha \beta 2$, and $\beta 2 \gamma 1$ Oriental isoenzymes. Biochemistry, **23**, 5847-5853 (1984)
[16] Wagner, F.W.; Pares, X.; Holmquist, B.; Vallee, B.L.: Physical and enzymatic properties of a class III isozyme of human liver alcohol dehydrogenase: chi-ADH. Biochemistry, **23**, 2193-2199 (1984)
[17] Bosron, W.F.; Magnes, L.J.; Li, T.-K.: Kinetic and electrophoretic properties of native and recombined isoenzymes of human liver alcohol dehydrogenase. Biochemistry, **22**, 1852-1857 (1983)
[18] Bosron, W.F.; Li, T.-K.: Isolation and characterization of an anodic form of human liver alcohol dehydrogenase. Biochem. Biophys. Res. Commun., **74**, 85-91 (1977)

[19] Schneider-Bernloehr, H.; Formicka-Kozlowska, G.; Buehler, R.; Wartburg, J.-P.; Zeppezauer, M.: Active-site-specific zinc-depleted and reconstituted cobalt(II) human-liver alcohol dehydrogenase. Preparation, characterization and complexation with NADH and trans-4-(N,N-dimethylamino)-cinnamaldehyde. Eur. J. Biochem., **173**, 275-280 (1988)
[20] Burnell, J.C.; Li, T.-K.; Bosron, W.F.: Purification and steady-state kinetic characterization of human liver β 3 β 3 alcohol dehydrogenase. Biochemistry, **28**, 6810-6815 (1989)
[21] Pares, X.; Vallee, B.L.: New human liver alcohol dehydrogenase forms with unique kinetic characteristics. Biochem. Biophys. Res. Commun., **98**, 122-130 (1981)
[22] Bosron, W.F.; Li, T.-K.; Vallee, B.L.: New molecular forms of human liver alcohol dehydrogenase: isolation and characterization of ADHIndianapolis. Proc. Natl. Acad. Sci. USA, **77**, 5784-5788 (1980)
[23] Bosron, W.F.; Li, T.-K.; Dafeldecker, W.P.; Vallee, B.L.: Human liver π-alcohol dehydrogenase: kinetic and molecular properties. Biochemistry, **18**, 1101-1105 (1979)
[24] Dafeldecker, W.P.; Pares, X.; Vallee, B.L.; Bosron, W.F.; Li, T.-K.: Simian liver alcohol dehydrogenase: isolation and characterization of isoenzymes from Saimiri sciureus. Biochemistry, **20**, 856-861 (1981)
[25] Dafeldecker, W.P.; Meadow, P.E.; Pares, X.; Vallee, B.L.: Simian liver alcohol dehydrogenase: isolation and characterization of isoenzymes from Macaca mulatta. Biochemistry, **20**, 6729-6734 (1981)
[26] Kvassman, J.; Pettersson, G.: Kinetics of coenzyme binding to liver alcohol dehydrogenase in the pH range 10-12. Eur. J. Biochem., **166**, 167-172 (1987)
[27] Andersson, L.; Mosbach, K.: Alcohol dehydrogenase from horse liver by affinity chromatography. Methods Enzymol., **89**, 435-445 (1982)
[28] Pietruszko, R.: Alcohol dehydrogenase from horse liver, steroid-active SS isoenzyme. Methods Enzymol., **89**, 429-435 (1982)
[29] Dahl, K.H.; Eklund, H.; McKinley-McKee, J.S.: Enantioselective affinity labelling of horse liver alcohol dehydrogenase. Correlation of inactivation kinetics with the three-dimensional structure of the enzyme. Biochem. J., **211**, 391-396 (1983)
[30] Ramaswamy, S.: Dynamics in alcohol dehydrogenase elucidated from crystallographic investigations. Enzymol. Mol. Biol. Carbonyl Metab., **7**, 275-284 (1999)
[31] Adolph, H.W.; Maurer, P.; Schneider-Bernloehr, H.; Sartorius, C.; Zeppezauer, M.: Substrate specificity and stereoselectivity of horse liver alcohol dehydrogenase. Kinetic evaluation of binding and activation parameters controlling the catalytic cycles of unbranched, acyclic secondary alcohols and ketones as substrates of the native and active-site-specific Co(II)-substituted enzyme. Eur. J. Biochem., **201**, 615-625 (1991)
[32] Freudenreich, C.; Samama, J.-P.; Biellmann, J.-F.: Design of inhibitors from the three-dimensional structure of alcohol dehydrogenase. Chemical synthesis and enzymatic properties. J. Am. Chem. Soc., **106**, 3344-3353 (1984)

[33] Samama, J.-P.; Hirsch, D.; Goulas, P.; Biellmann, J.-F.: Dependence of the substrate specificity and kinetic mechanism of horse-liver alcohol dehydrogenase on the size of the C-3 pyridinium substituent. 3-Benzoylpyridine-adenine dinucleotide. Eur. J. Biochem., **159**, 375-380 (1986)
[34] Eklund, H.: Coenzyme binding in alcohol dehydrogenase. Biochem. Soc. Trans., **17**, 293-296 (1989)
[35] Dworschack, R.T.; Plapp, B.V.: Kinetics of native and activated isozymes of horse liver alcohol dehydrogenase. Biochemistry, **16**, 111-116 (1977)
[36] Maret, W.; Andersson, I.; Dietrich, H.; Schneider-Bernloehr, H.; Einarsson, R.; Zeppezauer, M.: Site-specific substituted cobalt(II) horse liver alcohol dehydrogenases. Preparation and characterization in solution, crystalline and immobilized state. Eur. J. Biochem., **98**, 501-512 (1979)
[37] Skerker, P.S.; Clark, D.S.: Thermostability of alcohol dehydrogenase: evidence for distinct subunits with different deactivation properties. Biotechnol. Bioeng., **33**, 62-71 (1989)
[38] Benach, J.; Atrian, S.; Gonzalez-Duarte, R.; Ladenstein, R.: The catalytic reaction and inhibition mechanism of Drosophila alcohol dehydrogenase: observation of an enzyme-bound NAD-ketone adduct at 1.4 A resolution by X-ray crystallography. J. Mol. Biol., **289**, 335-355 (1999)
[39] Tsai, C.S.: Multifunctionality of liver alcohol dehydrogenase: kinetic and mechanistic studies of esterase reaction. Arch. Biochem. Biophys., **213**, 635-642 (1982)
[40] Favilla, R.; Cavatorta, P.; Mazzini, A.; Fava, A.: The peroxidatic reaction catalyzed by horse liver alcohol dehydrogenase. 2. Steady-state kinetics and inactivation. Eur. J. Biochem., **104**, 223-227 (1980)
[41] Ryzewski, C.N.; Pietruszko, R.: Kinetic mechanism of horse liver alcohol dehydrogenase SS. Biochemistry, **19**, 4843-4848 (1980)
[42] Ryzewski, C.N.; Pietruszko, R.: Horse liver alcohol dehydrogenase SS: purification and characterization of the homogenous isoenzyme. Arch. Biochem. Biophys., **183**, 73-82 (1977)
[43] Winberg, J.O.; McKinley-McKee, J.S.: Drosophila melanogaster alcohol dehydrogenase: product-inhibition studies. Biochem. J., **301**, 901-909. (1994)
[44] von Bahr-Lindstroem, H.; Andersson, L.; Mosbach, K.; Joernvall, H.: Purification and characterization of chicken liver alcohol dehydrogenase. FEBS Lett., **89**, 293-297 (1978)
[45] Hoshino, T.; Ishigura, I.; Ohta, Y.: Rabbit liver alcohol dehydrogenase: purification and properties. J. Biochem., **97**, 1163-1172 (1985)
[46] Keung, W.M.; Yip, P.K.: Rabbit liver alcohol dehydrogenase: isolation and characterization of class I isozymes. Biochem. Biophys. Res. Commun., **158**, 445-453 (1989)
[47] Keung, W.M.: A genuine organ specific alcohol dehydrogenase from hamster testes: isolation, characterization and developmental changes. Biochem. Biophys. Res. Commun., **156**, 38-45 (1988)
[48] Algar, E.M.; Seeley, T.-L.; Holmes, R.S.: Purification and molecular properties of mouse alcohol dehydrogenase isozymes. Eur. J. Biochem., **137**, 139-147 (1983)

[49] Julia, P.; Farres, J.; Pares, X.: Characterization of three isoenzymes of rat alcohol dehydrogenase. Tissue distribution and physical and enzymatic properties. Eur. J. Biochem., **162**, 179-189 (1987)
[50] Pares, X.; Moreno, A.; Cederlund, E.; Hoeoeg, J.-O.; Joernvall, H.: Class IV mammalian alcohol dehydrogenase. Structural data of the rat stomach enzyme reveal a new class well separated from those already characterized. FEBS Lett., **277**, 115-118 (1990)
[51] Mezey, E.; Potter, J.J.: Separation and partial characterization of multiple forms of rat liver alcohol dehydrogenase. Arch. Biochem. Biophys., **225**, 787-794 (1983)
[52] Hjelmqvist, L.; Shafqqat, J.; Siddiqi, A.R.; Joernvall, H.: Linking of isoenzyme and class variability patterns in the emergence of novel alcohol dehydrogenase functions. Characterization of isozymes in Uromastix hardwickii. Eur. J. Biochem., **236**, 563-570 (1996)
[53] Kedishvili, N.Y.; Bosron, W.F.; Stone, C.L.; Hurley, T.D.; Peggs, C.F.; Thomasson, H.R.; Popov, K.M.; Carr, L.G.; Edenberg, H.J.; Li, T.-K.: Expression and kinetic characterization of recombinant human stomach alcohol dehydrogenase. Active-site amino acid sequence explains substrate specificity copared with liver isozymes. J. Biol. Chem., **270**, 3625-3630 (1995)
[54] Plapp, B.V.; Sogin, D.C.; Dworschack, R.T.; Bohlken, D.P.; Woenckhaus, C.: Kinetics of native and modified liver alcohol dehydrogenase with coenzyme analogues: isomerization of enzyme-nicotinamide adenine dinucleotide complex. Biochemistry, **25**, 5396-5402 (1986)
[55] Li, H.; Hallows, W.H.; Punzi, J.S.; Marquez, V.E.; Carrell, H.L.; Pankiewicz, K.W.; Watanabe, K.A.; Goldstein, B.M.: Crystallographic studies of two alcohol dehydrogenase-bound analogues of thiazole-4-carboxamide adenine dinucleotide (TAD), the active anabolite of the antitumor agent tiazofurin. Biochemistry, **33**, 23-32 (1994)
[56] Pearl, L.H.; Demasi, D.; Hemmings, A.M.; Sica, F.; Mazzarella, L.; Raia, C.A.; D'Auria, S.; Rossi, M.: Crystallization and preliminary X-ray analysis of an NAD(+) dependent alcohol dehydrogenase from the extreme thermophilic archaebacterium Sulfolobus solfataricus. J. Mol. Biol., **229**, 782-784 (1993)
[57] Shafqat, J.; Hjelmqvist, L.; Joernvall, H.: Liver class-I alcohol dehydrogenase isozyme relationships and constant patterns in a variable basic structure. Distinctions from characterization of an ethanol dehydrogenase in cobra, Naja naja. Eur. J. Biochem., **236**, 571-578 (1996)
[58] Retzios, A.; Thatcher, D.R.: Characterization of the AdhF' and Adhus alleloenzymes of Drosophila melanogaster (fruitfly) alcohol dehydrogenase. Biochem. Soc. Trans., **9**, 298-299 (1981)
[59] Bauermeister, A.; Sargent, J.: Purification and properties of an alcohol dehydrogenase from the liver and intestinal caecum of rainbow trout (Salmo gairdnerii). Biochem. Soc. Trans., **6**, 222-224 (1978)
[60] Nussrallah, B.A.; Dam, R.; Wagner, F.W.: Characterization of Coturnix quail liver alcohol dehydrogenase enzymes. Biochemistry, **28**, 6245-6251 (1989)

[61] Winberg, J.-O.; Hovik, R.; McKinley-McKee, J.S.; Juan, E.; Gonzalez-Duarte, R.: Biochemical properties of alcohol dehydrogenase from Drosophila lebanonensis. Biochem. J., **235**, 481-490 (1986)
[62] Winberg, J.-O.; McKinley-McKee, J.S.: Drosophila melanogaster alcohol dehydrogenase. Biochemical properties of the NAD^+-plus-acetone-induced isoenzyme conversion. Biochem. J., **251**, 223-227 (1988)
[63] Heinstra, P.W.H.; Thoerig, G.E.W.; Scharloo, W.; Drenth, W.; Nolte, R.J.M.: Kinetics and thermodynamics of ethanol oxidation catalyzed by genetic variants of the alcohol dehydrogenase from Drosophila melanogaster and D. simulans. Biochim. Biophys. Acta, **967**, 224-233 (1988)
[64] Juan, E.; Gonzalez-Duarte, R.: Determination of some biochemical and structural features of alcohol dehydrogenases from Drosophila simulans and Drosophila virilis. Comparison of their properties with the Drosophila melanogaster Adhs enzyme. Biochem. J., **195**, 61-69 (1981)
[65] Lee, C.-Y.: Alcohol dehydrogenase from Drosophila melanogaster. Methods Enzymol., **89**, 445-450 (1982)
[66] Rella, R.; Raia, C.A.; Pensa, M.; Pisani, F.M.; Gambacorta, A.; de Rosa, M.; Rossi, M.: A novel archaebacterial NAD^+-dependent alcohol dehydrogenase. Purification and properties. Eur. J. Biochem., **167**, 475-479 (1987)
[67] Wills, C.; Kratofil, P.; Londo, D.; Martin, T.: Characterization of the two alcohol dehydrogenases of Zymomonas mobilis. Arch. Biochem. Biophys., **210**, 775-785 (1981)
[68] Kinoshita, S.; Kakizono, T.; Kadota, K.; Das, K.; Taguchi, H.: Purification of two alcohol dehydrogenases from Zymomonas mobilis and their properties. Appl. Microbiol. Biotechnol., **22**, 249-254 (1985)
[69] Grondal, E.J.M.; Betz, A.; Kreuzberg, K.: Partial purification and properties of alcohol dehydrogenase from unicellular green alga Chlamydomonas moewusii. Phytochemistry, **22**, 1695-1699 (1983)
[70] Ammendola, S.; Raia, C.A.; Caruso, C.; Camardella, L.; DÁuria, S.; De Rosa, M.; Rossi, M.: Thermostable NAD(+)-dependent alcohol dehydrogenase from Sulfolobus solfataricus: gene and protein sequence determination and relationship to other alcohol dehydrogenases. Biochemistry, **31**, 12514-12523 (1992)
[71] Tihanyi, K.; Talbot, B.; Brzezinski, R.; Thirion, J.-P.: Purification and characterization of alcohol dehydrogenase from soybean. Phytochemistry, **28**, 1335-1338 (1989)
[72] Liang, Z.Q.; Hayase, F.; Nishimura, T.; Kato, H.: Purification and characterization of NAD-dependent alcohol dehydrogenase and NADH-dependent 2-oxoaldehyde reductase from parsley. Agric. Biol. Chem., **54**, 1717-1719 (1990)
[73] Hatanaka, A.; Harada, T.: Purification and properties of alcohol dehydrogenase from tea seeds. Agric. Biol. Chem., **36**, 2033-2035 (1972)
[74] Stiborova, M.; Leblova, S.: Kinetics of the reaction catalysed by rape alcohol dehydrogenase. Phytochemistry, **18**, 23-24 (1979)
[75] Lai, Y.-K.; Chandlee, J.M.; Scandalios, J.G.: Purification and characterization of three non-allelic alcohol dehydrogenase isoenzymes in maize. Biochim. Biophys. Acta, **706**, 9-18 (1982)

[76] Leblova, S.; Ehlichova, D.: Purification and some properties of alcohol dehydrogenase from maize. Phytochemistry, **11**, 1345-1346 (1972)
[77] Langston, P.J.; Pace, C.N.; Hart, G.E.: Wheat alcohol dehydrogenase isozymes. Purification, characterization, and gene expression. Plant Physiol., **65**, 518-522 (1980)
[78] Langston, P.J.; Hart, G.E.; Pace, C.N.: Purification and partial characterization of alcohol dehydrogenase from wheat. Arch. Biochem. Biophys., **196**, 611-618 (1979)
[79] Mayne, M.G.; Lea, P.J.: Properties of three sets of isoenzymes of alcohol dehydrogenase isolated from barley (Hordeum vulgare). Phytochemistry, **24**, 1433-1438 (1985)
[80] Shimomura, S.; Beevers, H.: Alcohol dehydrogenase inactivator from rice seedlings. Properties and intracellular location. Plant Physiol., **71**, 742-746 (1983)
[81] Creaser, E.H.; Porter, R.L.; Britt, K.A.; Pateman, J.A.; Doy, C.H.: Purification and preliminary characterization of alcohol dehydrogenase from Aspergillus nidulans. Biochem. J., **225**, 449-454 (1985)
[82] Yabe, M.; Shitara, K.; Kawashima, J.; Shinoyama, H.; Ando, A.; Fujii, T.: Purification and properties of an alcohol dehydrogenase isozyme from a methanol-using yeast, Candida sp. N-16. Biosci. Biotechnol. Biochem., **56**, 338-339 (1992)
[83] Morosoli, R.; Begin-Heick, N.: The partial purification and characterization of cytosol alcohol dehydrogenase from Astasia. Biochem. J., **141**, 469-475 (1974)
[84] Rudge, J.; Bickerstaff, G.F.: Purification and properties of an alcohol dehydrogenase from Sporotrichum pulverulentum. Enzyme Microb. Technol., **8**, 120-124 (1986)
[85] Indrati, R.; Ohita, Y.: Purification and properties of alcohol dehydrogenase from a mutant strain of Candida guilliermondii. Can. J. Microbiol., **38**, 953-957 (1992)
[86] Tkachenko, A.G.; Winston, G.W.: Interaction of alcohol dehydrogenase with tert-butylhydroperoxide: stimulation of the horse liver and inhibition of the yeast enzyme. Arch. Biochem. Biophys., **380**, 165-173 (2000)
[87] Drewke, C.; Ciriacy, M.: Overexpression, purification and properties of alcohol dehydrogenase IV from Saccharomyces cerevisiae. Biochim. Biophys. Acta, **950**, 54-60 (1988)
[88] Yamazaki, Y.; Maeda, H.; Satoh, A.; Hiromi, K.: A kinetic study on the binding of monomeric and polymeric derivatives of NAD^+ to yeast alcohol dehydrogenase. J. Biochem., **95**, 109-115 (1984)
[89] Mazid, M.A.; Laidler, K.J.: pH dependence of free and immobilized yeast alcohol dehydrogenase kinetics. Can. J. Microbiol., **60**, 100-107 (1982)
[90] Dickinson, F.M.; Monger, G.P.: A study of the kinetics and mechanism of yeast alcohol dehydrogenase with a variety of substrates. Biochem. J., **131**, 261-270 (1973)
[91] Ganzhorn, A.J.; Green, D.W.; Hershey, A.D.; Gould, R.M.; Plapp, B.V.: Kinetic characterization of yeast alcohol dehydrogenases. Amino acid residue 294 and substrate specificity. J. Biol. Chem., **262**, 3754-3761 (1987)

[92] Weinhold, E.G.; Benner, S.A.: Engineering yeast alcohol dehydrogenase. Replacing Trp54 by Leu broadens substrate specificity. Protein Eng., **8**, 457-461 (1995)

[93] Pocker, Y.; Li, H.: Mechanistic enzymology of liver alcohol dehydrogenase. Kinetic and stereochemical characterization of retinal oxidation and reduction. Enzymol. Mol. Biol. Carbonyl Metab., **6**, 331-338 (1996)

[94] Leskovac, V.; Trivic, S.; Anderson, B.M.: Use of competitive dead-end inhibitors to determine the chemical mechanism of action of yeast alcohol dehydrogenase. Mol. Cell. Biochem., **178**, 219-227 (1998)

[95] Keung, W.-M.: Isolation and characterization of three alcohol dehydrogenase isozymes from Syrian golden hamster. Alcohol. Clin. Exp. Res., **20**, 213-220 (1996)

[96] Yin, S.-J.; Wang, M.-F.; Liao, C.-S.; Chen, C.-M.; Wu, C.-W.: Identification of a human stomach alcohol dehydrogenase with distinctive kinetic properties. Biochem. Int., **22**, 829-835 (1990)

[97] Osterman, J.C.; Chiang, Y.; Markwell, J.: Characterization of mutation-induced changes in the maize (Zea mays L.) ADH1-1S1108 alcohol dehydrogenase. Biochem. Genet., **31**, 497-506 (1993)

[98] Langeland, B.T.; Morris, D.L.; McKinley-McKee, J.S.: Metal binding properties of thiols: complexes with horse liver alcohol dehydrogenase. Comp. Biochem. Physiol. B, **123**, 155-162 (1999)

[99] Hensgens, C.M.H.; Vonck, J.; van Beeumen, J.; Bruggen, E.F.J.; Hansen, T.A.: Purification and characterization of an oxygen-labile, NAD-dependent alcohol dehydrogenase from Desulfovibrio gigas. J. Bacteriol., **175**, 2859-2863 (1993)

[100] Flores, B.M.; Stanley, S.L.; Yong, T.-S.; Ali, M.; Yang, W.; Diedrich, D.L.; Torian, B.E.: Surface localization, regulation, and biologic properties of the 96-kDA alcohol/aldehyde dehydrogenase (EhADH2) of pathogenic Entamoeba histolytica. J. Infect. Dis., **173**, 226-231 (1996)

[101] Persson, B.; Bergman, T.; Keung, W.M.; Waldenstroem, U.; Holmquist, B.; Vallee, B.L.; Joernvall, H.: Basic features of class-I alcohol dehydrogenase variable and constant segments coordinated by inter-class and intra-class variability. Conclusions from characterization of the alligator enzyme. Eur. J. Biochem., **216**, 49-56 (1993)

[102] Gergel, D.; Cederbaum, A.I.: Inhibition of the catalytic activity of alcohol dehydrogenase by nitric oxide is associated with S nitrosylation and the release of zinc. Biochemistry, **35**, 16186-16194 (1996)

Alcohol dehydrogenase (NADP) 1.1.1.2

1 Nomenclature

EC number
1.1.1.2

Systematic name
alcohol:NADP oxidoreductase

Recommended name
alcohol dehydrogenase (NADP)

Synonyms
2° Adh <28> [34]
3-DG-reducing enzyme [Swissprot]
ADH [Swissprot]
ALDR <5> [19]
ALR [Swissprot]
ALR 1
Alcohol dehydrogenase [$NADP^+$] [Swissprot]
Aldehyde reductase [Swissprot]
Aldo-keto reductase family 1 member A1 [Swissprot]
D glucuronate reductase <1> [1]
L-hexonate dehydrogenase <1, 5, 24, 21> [14, 18, 23, 27, 57]
NADP-alcohol dehydrogenase
NADP-aldehyde reductase
NADP-dependent aldehyde reductase
NADP-linked aryl alcohol dehydrogenase <37> [30]
NADPH-aldehyde reductase
NADPH-cytochrome c reductase <2> [25]
NADPH dependent aldehyde reductase
NADPH-linked aldehyde reductase <1> [1]
TBADH <35> [38]
TPN-L-hexonate dehydrogenase <4> [4]
TPNH-linked aldehyde reductase <4> [2]
TPNH-specific aldehyde reductase <4> [2]
alcohol dehydrogenase C <55> [48]
aldehyde reductase (NADPH2)
aldehyde/ketone reductase <23> [15]
aldo-keto reductase <5> [24]
bovine brain aldehyde reductase <1> [2]
daunorubicin reductase <2, 4, 24> [11, 14, 27]
hexogenate dehydrogenase <1, 2> [11, 15]

high-K_m aldehyde reductase
liver alcohol dehydrogenase <7> [5, 9]
low-K_m aldehyde reductase
mevaldate reductase <2> [11]
nonspecific succinic semialdehyde reductase
octopine dehydrogenase <6> [5, 7, 8]
putative iron alcohol dehydrogenase <20> [44]
rabbit kidney aldehyde-ketone reductase <3> [2]
triphosphopyridine nucleotide-linked aldehyde reductase <4> [2]
yeast alcohol dehydrogenase <17> [16]

CAS registry number
9028-12-0

2 Source Organism

<1> *Bos taurus* (bovine [1, 2, 18, 20, 33, 57]; ox [3, 12, 15, 16, 21, 27, 55]) [1-3, 12, 13, 15, 16, 18, 20, 21, 27, 33, 55, 57]
<2> *Rattus norvegicus* (rat [1-3, 7, 9-14, 18-20, 27, 29, 33, 41, 51, 52]; Wistar albino [11, 12]; FBNxKGH, AugxWF, BNLr-1, LEW, KGH, DA [19]) [1-3, 7, 9-14, 18-21, 25, 27, 29, 33, 41, 51, 52]
<3> *Oryctolagus cuniculus.* (rabbit [1, 2, 4, 13, 14, 19, 33]; New Zealand albino, New Zealand red, Mixed breed, Chinchilla, Dutch belted [19]) [1, 2, 4, 13, 14, 19, 33]
<4> *Sus scrofa* (pig [2-5, 7-11, 13-19, 22, 26, 33]) [2-5, 7-11, 13-19, 22, 26, 33, 45, 49, 54]
<5> *Homo sapiens* (human [5, 7, 10, 12-14, 18, 19, 21, 23, 24, 32, 33, 36-38, 45, 52, 53, 56, 58]) [5, 7, 10, 12-14, 18, 19, 21, 23, 24, 32, 33, 36-38, 45, 52, 53, 56, 58]
<6> *Pecten maximus* (mollusc [5, 7, 8]) [5, 7, 8]
<7> *Equus sp.* (horse [5, 9, 38]) [5, 9, 38, 48]
<8> *Phycomyces blakesleeanus* [6]
<9> *Rhizopus oryzae* [6]
<10> *Pseudomonas aeruginosa* [6]
<11> *Penicillium urticae* [6]
<12> *Neurospora crassa* [6, 30]
<13> *Candida tropicalis* [6]
<14> *Mycobacterium tuberculosis* (var. bovis BCG [48]) [6, 48]
<15> *Solanum tuberosum* [6]
<16> *Cucumis melo* [6]
<17> *Saccharomyces cerevisiae* (yeast [6, 7, 14, 16, 28, 34, 35, 38-40, 44]; var. uvarum W34 [40]) [6, 7, 14, 16, 28, 34, 35, 38-40, 44]
<18> *Gallus sp.* (chicken [7, 18, 19, 21]) [7, 18, 19, 21]
<19> *Drosophila melanogaster* (fruit fly [7, 14]) [7, 14, 35, 39]
<20> *Schizosaccharomyces pombe* [44]

<21> *Mus musculus* (mouse [9, 12, 14, 18, 19, 33]; strain B10.A, LP/J, CBAxC57, C57L/J, Nude [19]) [9, 12, 14, 18, 19, 33]
<22> *hamster* [12]
<23> *Cavia porcellus* (guinea pig [14, 15, 19, 20, 24, 33]; Hartley outbred. [19]) [14, 15, 19, 20, 24, 33]
<24> *Ovis aries* (sheep [9, 18, 19, 27, 33]; Suffolk-Hampshire) [9, 18, 19, 27, 33]
<25> *Felis sp.* (cat, mixed breed [19]) [19]
<26> *Canis familiaris* (dog, beagle, mixed breed [19]) [19, 33]
<27> *squirrel monkey* [19]
<28> *Thermoanaerobacter ethanolicus* (strain ATCC 31550 [28]; 39E (A.T.C.C. 33223), formerly Clostridium thermohydrosulfuricum 39E [34]; strain JW200 [34]) [28, 34, 43]
<29> *Zymomonas mobilis* [28, 34, 35, 39, 44]
<30> *Clostridium beijerinckii* (NRRL B592 [34]) [28, 34, 43]
<31> *Clostridium thermohydrosulfuricum* (new name Thermoanaerobacter ethanolicus strain 39 E [34]) [28, 34]
<32> *Escherichia coli* [28, 34, 39, 48]
<33> *Acinetobacter sp.* (strain M-1 [47]) [28, 47]
<34> *Bacillus stearothermophilus* [28]
<35> *Thermoanaerobacter brockii* (formely Thermoanaerobium brockii [38]) [28, 34, 35, 38, 43, 46]
<36> *Trypanosoma cruzi* (Tulahuen strain, Tul 0 stock [30]) [30]
<37> *Crithidia fasciculate* [30]
<38> *Sporobolomyces salmonicolor* (strain AKU 4429 [31]) [31]
<39> *Clostridium thermocellum* [34]
<40> *Clostridium acetobutylicum* (NRRL B643 [34]) [34, 39, 44]
<41> *Clostridium kluyveri* [34]
<42> *Propionibacterium freudenreichii* [34]
<43> *Leuconostoc mesenteroides* [34]
<44> *Vibrio harveyi* [34]
<45> *Thermococcus litoralis* (DSM 5473 [35]) [35, 44]
<46> *Sulfolobus solfataricus* [35, 39, 44]
<47> *Methanogenium thermophilum* [35]
<48> *Methanogenium liminatans* [35]
<49> *Methanobacterium palustre* [35]
<50> *Acetobacter aceti* [35]
<51> *Thermococcus sp.* (strain AN1 [39]) [39, 42]
<52> *Vigna radiata* (mung bean [42]) [42]
<53> *Rana perezi* (frog [43]) [43]
<54> *Thermococcus hydrothermalis* [44]
<55> *Mycobacterium smegmatis* [48]
<56> *monkey* [50]
<57> *Entamoeba histolytica* [43, 53]
<58> *Bacillus methanolicus* [39, 44]
<59> *Acinetobacter calcoaceticus* (strainNCIB8250, strain HO1-N [47]) [40, 47]
<60> *Mycobacterium bovis* (strain BCG [48]) [48]
<61> *Mycobacterium avium* [48]

<62> *Mycobacterium paratuberculosis* [48]
<63> *Mycobacterium leprae* [48]
<64> *Bacillus subtilis* [48]
<65> *Helicobacter pylori* [48]
<66> *Thermococcus zilligii* [44]
<67> *Pyrococcus furiosus* [44]
<68> *Archaeoglobus fulgidus* [44]
<69> *Aquifex aeolicus* [44]

3 Reaction and Specificity

Catalyzed reaction

an alcohol + $NADP^+$ = an aldehyde + NADPH (<1> a zinc protein. Some members of this group oxidize only primary alcohols; others act also on secondary alcohols. May be identical with EC 1.1.1.19 (L-glucuronate reductase), EC 1.1.1.33 [mevaldate reductase (NADPH)] and EC 1.1.1.55 [lactaldehyde reductase (NADPH)]. A-specific with respect to NADH [1])

Natural substrates and products

S (4-hydroxy-3-[3-methyl-3-butene-1-ynyl]benzaldehyde) + NADPH <52> (<52> eutypine, a toxin produced by Eutypa lata, the causal agent of Eutypa dieback in the grapevine Vitis vinifera [42]) (Reversibility: ir <52> [42]) [42]
P (4-hydroxy-3-[3-methyl-3-butene-1-ynyl]benzyl alcohol) + $NADP^+$ <52> (<52> eutypinol, non-toxic metabolite [42]) [42]
S D-glucuronate + NADPH <1-5, 21, 23-27> (Reversibility: r <1-5, 21, 23-27> [2, 4, 9, 11-15, 19, 23, 24, 27, 29, 32, 33, 36, 37, 45]) [2, 4, 9, 11-16, 19, 23, 24, 27, 29, 32, 33, 36, 37, 45]
P L-gulonate + $NADP^+$
S D-glyceraldehyde + NADPH <2-5, 21, 23-27> (Reversibility: r <2-5, 21, 23-27> [2, 4, 5, 8, 9, 11-16, 19, 23, 24, 29, 32, 33, 36, 41]) [2, 4, 5, 8, 9, 11-16, 19, 23, 24, 29, 32, 33, 36, 41]
P glycerol + $NADP^+$
S L-glyceraldehyde + NADPH <2-5, 21, 23-27> (Reversibility: r <2-5, 21, 23-27> [5, 11-13, 15, 16, 19, 23, 29, 24, 32, 33, 36, 41]) [5, 11-13, 15, 16, 19, 23, 24, 29, 32, 33, 36, 41]
P glycerol + $NADP^+$
S L-gulonic acid + $NADP^+$ <1-4> (Reversibility: r <1-4> [2, 4, 9, 11-15]) [2, 4, 9, 11-16]
P D-glucoronic acid + NADPH
S α-diketone + NADPH <4> (Reversibility: ? <4> [2]) [2]
P ? + $NADP^+$
S ethylaceetoacetate + NADPH <2, 18> (Reversibility: ? <2, 18> [20, 21]) [20, 21]
P ethyl L-(+)-3-hydroxybutyrate + $NADP^+$

S lauryl aldehyde + NADPH <18> (<18> isozyme AR3 [21]) (Reversibility: ? <18> [21]) [21]
P lauryl alcohol + NADP$^+$
S succinic semialdehyde + NADPH <1-5> (Reversibility: r <1-5> [9, 12-16, 22, 32, 36, 37, 52]) [9, 12-16, 20, 22, 32, 36, 37, 52]
P γ-hydroxybutyrate + NADP$^+$ <2> [20]

Substrates and products

S (+)-camphorquinoneone + NADPH <4, 5> (Reversibility: ? <4, 5> [2, 5]) [2, 5]
P ? + NADP$^+$
S (-)-camphorquinoneone + NADPH <4> (<4> relative activity 2% [2]) (Reversibility: ? <4> [2]) [2]
P ? + NADP$^+$
S (4-hydroxy-3-[3-methyl-3-butene-1-ynyl]benzaldehyde) + NADPH <52> (<52> eutypine, a toxin produced by Eutypa lata, the causal agent of Eutypa dieback in the grapevine Vitis vinifera [42]) (Reversibility: ir <52> [42]) [42]
P (4-hydroxy-3-[3-methyl-3-butene-1-ynyl]benzyl alcohol) + NADP$^+$ <52> (<52> eutypinol, non-toxic metabolite [42]) [42]
S 1,2-cyclohexanedione + NADPH <18> (Reversibility: ? <18> [21]) [21]
P ? + NADP$^+$
S 1,2-propanediol + NADP$^+$ <28> (Reversibility: ? <28> [28]) [28]
P ? + NADPH
S 1,3-propanediol + NADP$^+$ <28, 45, 68> (<28> primary alcohol dehydrogenase [28]) (Reversibility: ? <28, 45, 68> [28, 35, 44]) [28, 35, 44]
P ? + NADPH
S 1-3-butanediol + NADP$^+$ <28> (Reversibility: ? <28> [28]) [28]
P ? + NADPH
S 1-heptanol + NADP$^+$ <8, 17, 28, 45> (<28> primary alcohol dehydrogenase [28]) (Reversibility: r <8, 17, 28, 45> [6, 28, 35, 40]) [6, 28, 35, 40]
P 1-heptanal + NADPH
S 1-octanol + NADP$^+$ <45, 51, 53> (Reversibility: r <45, 51, 53> [35, 39, 43]) [35, 39, 43]
P octanal + NADPH
S 1-pentanol + NADP$^+$ <8, 28, 36, 45, 51, 53> (<28> primary alcohol dehydrogenase [28]) (Reversibility: r <8, 28, 36, 45, 51, 53> [6, 28, 30, 35, 39, 43]) [6, 28, 30, 35, 39, 43]
P 1-pentanal + NADPH
S 13-cis-retinal + NADPH <53> (Reversibility: ? <53> [43]) [43]
P 13-cis-retinol + NADP$^+$
S 1-hexanol + NADP$^+$ <8, 17, 45, 53> (Reversibility: r <8, 17, 45, 53> [6, 35, 40, 43]) [6, 35, 40, 43]
P 1-hexanal + NADPH
S 2,3-pentanedione + NADPH <4, 28> (<28> secondary alcohol dehydrogenase [28]) (Reversibility: ? <4, 28> [2, 28]) [2, 28]
P ? + NADP$^+$

S 2,4-hexadienal + NADPH <33> (Reversibility: ? <33> [47]) [47]
P 2,4-hexadien-1-ol + $NADP^+$
S 2-butanol + $NADP^+$ <28, 35, 53> (<28> 2° Adh [34]) (Reversibility: r <28, 35, 53> [28, 34, 43, 46]) [28, 34, 43, 46]
P butan-2-one + NADPH
S 2-chloroheptanal + NADPH <4> (Reversibility: r <4> [9]) [9]
P 2-chloroheptanol + $NADP^+$
S 2-decanone + NADPH <33> (Reversibility: ? <33> [47]) [47]
P 2-decanol + $NADP^+$
S 2-heptanol + $NADP^+$ <28> (Reversibility: r <28> [28]) [28]
P heptan-2-one + NADPH
S 2-methoxybenzaldehyde + NADPH <52> (Reversibility: ? <52> [42]) [42]
P 2-methoxybenzyl alcohol + $NADP^+$
S 2-methyl-1-propanol + NADP <28> (Reversibility: ? <28> [28]) [28]
P 2-methylpropanal + NADPH
S 2-methylbutanal + NADPH <17> (Reversibility: r <17> [40]) [40]
P 2-methylbutanol + $NADP^+$
S 2-methylundecanal + NADPH <33> (Reversibility: ? <33> [47]) [47]
P 2-methylundecanol + $NADP^+$
S 2-pentanol + $NADP^+$ <28> (Reversibility: r <28> [28]) [28]
P pentan-2-one + NADPH
S 2-propen-1-ol + $NADP^+$ <8> (Reversibility: r <8> [6]) [6]
P acrylaldehyde + NADPH
S 3,4-dihydroxyphenylacetaldehyde + NADPH <56> (Reversibility: r <56> [50]) [50]
P 3,4-dihydroxyphenylethanol + $NADP^+$
S 3,4-dihydroxyphenylglycolaldehyde + NADPH <2, 56> (<2> low K_m reductase [52]) (Reversibility: r <2, 56> [50, 52]) [50, 52]
P ? + $NADP^+$
S 3-cyanobenzaldehyde + NADPH <4> (Reversibility: ? <4> [2]) [2]
P 3-cyanobenzyl alcohol + $NADP^+$
S 3-deoxyglucosone + NADPH <5> (Reversibility: r <5>) [45]
P ? + $NADP^+$
S 3-heptanone + NADPH <33> (Reversibility: ? <33> [47]) [47]
P 3-heptanol + $NADP^+$
S 3-hexanol + $NADP^+$ <28> (Reversibility: r <28> [28]) [28]
P hexan-3-one + NADPH
S 3-methylbutanal + NADPH <17> (Reversibility: r <17> [40]) [40]
P 3-methylbutanol + $NADP^+$
S 3-octanone + NADPH <33> (Reversibility: ? <33> [47]) [47]
P 3-octanol + $NADP^+$
S 3-pyridinecarboxaldehyde + NADPH <1-5, 18, 23-26, 38> (<1> AR3 [15]) (Reversibility: r <2-5, 23-26> [2, 4, 9, 11-19, 21-24, 27]; ir <38> [31]) [2, 4, 9, 11-19, 21-24, 27, 31]
P 3-pyridinemethanol + $NADP^+$
S 3-pyridinemethanol + $NADP^+$ <4> (Reversibility: r <4> [2]) [2]
P 3-pyridinecarboxaldehyde + NADPH

S 4-benzoylpyridine + ? <3> (Reversibility: r <3> [13]) [13]

P ?

S 4-carboxyacetophenone + NADPH <4> (Reversibility: r <4> [9]) [9]

P ? + $NADP^+$

S 4-carboxyphenylglyoxal + NADPH <4> (Reversibility: r <4> [9]) [9]

P ? + $NADP^+$

S 4-hydroxy-3-methoxybenzyl alcohol + $NADP^+$ <5> (Reversibility: r <5> [5]) [5]

P 4-hydroxy-3-methoxybenzaldehyde + NADPH

S 4-hydroxy-3-methoxyphenethylene glycol + $NADP^+$ <4> (Reversibility: r <4> [3]) [3]

P ? + NADPH

S 4-hydroxybutyrate + $NADP^+$ <5> (Reversibility: r <5> [53]) [53]

P 4-ketobutyrate + NADPH

S 4-methylbenzyl alcohol+ $NADP^+$ <4> (Reversibility: r <4> [2]) [2]

P 4-methylbenzaldehyde + NADPH

S 4-nitroacetophenone + NADPH <4> (Reversibility: ? <4> [2]) [2]

P ? + $NADP^+$

S 4-pyridinecarboxaldehyde + NADPH + <2-4, 18, 52> (Reversibility: r <2-4, 18> [2, 13, 20, 21, 25]; ? <52> [42]) [2, 13, 20, 21, 25, 42]

P 4-pyridinemethanol + $NADP^+$

S 5-hydroxyindoleacetaldehyde + NADPH <56> (Reversibility: ? <56> [50]) [50]

P ? + $NADP^+$

S 5α-dihydrotestosterone + NADPH <18> (Reversibility: ? <18> [21]) [21]

P ? + $NADP^+$

S 9-cis-retinal + NADPH <53> (Reversibility: r <53> [43]) [43]

P 9-cis-retinol + $NADP^+$

S 9-cis-retinol + $NADP^+$ <53> (Reversibility: r <53> [43]) [43]

P 9-cis-retinal + NADPH

S D-arabinose + NADPH <4> (Reversibility: r <4> [2]) [2]

P ? + $NADP^+$

S D-arabinose + NADPH <4> (Reversibility: ? <4> [4]) [4]

P ? + $NADP^+$

S D-arabitol + $NADP^+$ <4> (Reversibility: r <4> [2]) [2]

P ? + NADPH

S D-erythrose + NADPH <4> (Reversibility: ? <4> [2, 4]) [2, 4]

P ? + $NADP^+$

S D-fructose + NADPH <4> (Reversibility: ? <4> [2]) [2]

P ? + $NADP^+$

S D-galactose + NADPH <4> (Reversibility: ? <4> [2]) [2]

P ? + $NADP^+$

S D-galacturonate + NADPH <18> (Reversibility: ? <18> [21]) [21]

P ? + $NADP^+$

S D-glucose + NADPH <4, 5> (<5> placenta aldehyde reductase I [24]) (Reversibility: ? <4, 5> [2, 24]) [2, 24]

P ? + $NADP^+$

S D-glucuronate + NADPH <1-5, 18, 21, 23-27> (<1> AR1 [15]; <5> aldehyde reductase II [24]) (Reversibility: ? <1-5, 18, 21, 23-27> [2, 4, 9, 11-16, 19, 21-24, 27, 29, 32, 33, 36, 37]) [2, 4, 9, 11-16, 19, 21-24, 27, 29, 32, 33, 36, 37]

P L-gulonate + $NADP^+$

S D-glucuronolactone + NADPH <2, 4, 18> (Reversibility: ? <2, 4, 18> [2, 18, 21]) [2, 18, 21]

P ? + $NADP^+$

S D-glyceraldehyde + NADPH <2-5, 18, 21, 23-27, 38> (Reversibility: r <2-5, 18, 21, 23-27> [2, 4, 5, 8, 9, 11-14, 16, 18, 19, 21, 24, 29, 32, 33, 36, 41]; ir <38> [31]) [2, 4, 5, 8, 9, 11-14, 16, 18, 19, 21, 24, 29, 31, 32, 33, 36, 41]

P glycerol + $NADP^+$

S D-lactaldehyde + NADPH <1, 4, 5> (Reversibility: r <1, 4, 5> [1, 2, 5]) [1, 2, 5]

P propane-1,2-diol + $NADP^+$

S D-ribose + NADPH <4, 18> (Reversibility: ? <4, 18> [2, 21]) [2, 21]

P ? + $NADP^+$

S D-sorbitol + $NADP^+$ <4> (Reversibility: r <4> [2]) [2]

P ? + NADPH

S D-xylose + NADPH <1, 2, 4, 18, 26> (<1> AR3 [15]) (Reversibility: ? <1, 2, 4, 26> [2, 11, 15, 21, 33]) [2, 11, 15, 21, 33]

P ? + $NADP^+$

S L-arabitol + $NADP^+$ <4> (Reversibility: r <4> [2]) [2]

P ? + NADPH

S L-arabitol + $NADP^+$ <4> (Reversibility: r <4> [2]) [2]

P ? + NADPH

S L-glyceraldehyde + NADPH <2-5, 18, 21, 23-27> (Reversibility: r <2-5, 18, 21, 23-27> [5, 11-13, 16, 18, 19, 21, 24, 29, 32, 33, 36, 41]) [5, 11-13, 16, 18, 19, 21, 24, 29, 32, 33, 36, 41]

P glycerol + $NADP^+$

S L-gulonic acid + NADPH <1-4> (Reversibility: r <1-4> [2, 4, 9, 11-15]) [2, 4, 9, 11-16]

P D-glucoronic acid + $NADP^+$

S L-sorbose + NADPH <4> (Reversibility: r <4> [2]) [2]

P ? + $NADP^+$

S L-xylose + NADPH <4, 18> (Reversibility: ? <4> [2, 21]) [2, 21]

P ? + $NADP^+$

S Tris + $NADP^+$ <4> (Reversibility: ? <4> [2]) [2]

P ? + NADPH

S acetaldehyde + NADPH <2, 4, 28, 33, 36, 51, 53> (Reversibility: ? <2, 4, 28, 33, 36, 51, 53> [2, 20, 28, 30, 34, 39, 43, 47]) [2, 20, 28, 30, 34, 39, 43, 47]

P ethanol + $NADP^+$

S acetoin + NADPH <18> (<18> isozyme AR3 [21]) (Reversibility: ? <18> [21]) [21]

P butan-2,3-diol + $NADP^+$

S acetone + NADPH <28> (Reversibility: ? <28> [28]) [28]

P 2-propanol + $NADP^+$

S acetyl-CoA + NADPH <28> (Reversibility: ? <28> [34]) [34]
P acetaldehyde + CoA + $NADP^+$
S adonitol + $NADP^+$ <4> (Reversibility: r <4> [2]) [2]
P ? + NADPH
S all-trans-retinal + NADPH <53> (Reversibility: r <53> [43]) [43]
P all-trans-retinol + $NADP^+$
S all-trans-retinol + $NADP^+$ <53> (Reversibility: r <53> [43]) [43]
P all-trans-retinal + NADPH
S allactose + NADPH <4> (Reversibility: ? <4> [2]) [2]
P ? + $NADP^+$
S androstane-3,17-dione + NADPH <18> (Reversibility: ? <18> [21]) [21]
P ? + $NADP^+$
S anisaldehyde + NADPH <4, 33, 36> (Reversibility: r <4, 33, 36> [2, 30, 47]) [2, 30, 47]
P anisyl alcohol + $NADP^+$
S anisyl alcohol + $NADP^+$ <4> (Reversibility: r <4> [2]) [2]
P anisaldehyde + NADPH
S benzalacetone + NADPH <2> (Reversibility: ? <2> [25]) [25]
P ? + $NADP^+$
S benzaldehyde + NADPH <1, 2, 4, 5, 33, 36, 52, 55, 60-65> (<4> relative activity 4% [2]) (Reversibility: r <1, 2, 4, 5, 33, 36, 55, 60-65> [1, 2, 5, 11, 30, 47, 48]; ? <52> [42]) [1, 2, 5, 11, 30, 42, 47, 48]
P benzyl alcohol + $NADP^+$
S benzyl alcohol + $NADP^+$ <4, 36> (Reversibility: r <4, 36> [2, 9, 30]) [2, 9, 30]
P benzaldehyde + NADPH
S butane 2,3-dione + NADPH <4> (Reversibility: r <4> [9]) [9]
P ? + $NADP^+$
S chloral hydrate + NADPH <2> (Reversibility: ? <2> [11]) [11]
P ? + $NADP^+$
S cinnamyl alcohol + $NADP^+$ <55> (Reversibility: r <55> [48]) [48]
P cinnamylaldehyde + NADPH
S coniferaldehyde + NADPH <52> (<52> converted less efficiently than eutypine [42]) (Reversibility: ? <52> [42]) [42]
P coniferyl alcohol + $NADP^+$
S cyclohexanol + $NADP^+$ <1> (Reversibility: r <1> [15]) [15]
P cyclohexanal + NADPH
S cyclohexanone + NADPH <1, 4> (<1> G2 [15]) (Reversibility: ? <1, 4> [2, 15]) [2, 15]
P cyclohexanol + $NADP^+$
S cyclopentanol + $NADP^+$ <35> (Reversibility: r <35> [38]) [38]
P cyclopentanal + NADPH
S daunorubicin + NADPH <2, 4, 5> (Reversibility: ? <2, 4, 5> [9, 20, 32]) [9, 20, 32]
P ? + $NADP^+$
S dehydro-L-ascorbate + NADPH <4> (Reversibility: ? <4> [2]) [2]
P ? + $NADP^+$

S diacetyl + NADPH <4, 18, 38> (Reversibility: ? <4, 18, 38> [2, 21, 31]) [2, 21, 31]
P ? + NADP$^+$
S erythritol + NADP$^+$ <4> (Reversibility: r <4> [2]) [2]
P ? + NADPH
S ethanol + NADP$^+$ <8, 17, 28-36, 39-45, 51, 53> (Reversibility: r <8, 17, 28-36, 39-45, 51, 53> [6, 28, 30, 34, 35, 39, 43]) [6, 28, 30, 34, 35, 39, 43]
P acetaldehyde + NADPH
S ethyl 4-bromo-3-oxobutanoate + NADPH <38> (Reversibility: ? <38> [31]) [31]
P (R)-4-bromo-3-hydroxybutanoic acid ethyl ester + NADP$^+$
S ethyl 4-chloro-3-oxobutanoate + NADPH <38> (Reversibility: ? <38> [31]) [31]
P (R)-4-chloro-3-hydroxybutanoic acid ethyl ester + NADP$^+$
S ethylacetoacetate + NADPH <2, 18> (Reversibility: ? <2, 18> [20, 21]) [20, 21]
P ethyl L-(+)-3-hydroxybutyrate + NADP$^+$
S ethylene glycol + NADP$^+$ <28> (Reversibility: ? <28> [28]) [28]
P ? + NADPH
S formaldehyde + NADPH <8, 28> (<8> methanal [6]) (Reversibility: ? <8, 28> [6, 34]) [6, 34]
P methanol + NADP$^+$
S furaldehyde + NADPH <52> (Reversibility: ? <52> [42]) [42]
P furfuryl alcohol + NADP$^+$
S glycerol + NADP$^+$ <2, 4, 54> (Reversibility: r <2, 4, 54> [2, 18, 44]) [2, 18, 44]
P DL-glyceraldehyde + NADPH
S glycolaldehyde + NADPH <18> (Reversibility: r <18> [21]) [21]
P ethane-1,2-diol + NADP$^+$
S hexadecanol + NADP$^+$ <59> (Reversibility: r <59> [47]) [47]
P hexadecanal + NADPH
S hexanal + NADPH <17, 33, 52, 53> (Reversibility: r <17, 33, 53> [40, 43, 47]; ? <52> [42]) [40, 42, 43, 47]
P hexanol + NADP$^+$
S hydroxyindol-3-ylacetaldehyde + NADPH <4> (Reversibility: r <4> [3]) [3]
P ? + NADP$^+$
S hydroxypyruvate + ? <1> (Reversibility: r <1> [1]) [1]
P ?
S indol-3-ylacetaldehyde + NADPH <4, 18> (Reversibility: r <4, 18> [3, 21]) [3, 21]
P indol-3-ethanol + NADP$^+$
S indolacetaldehyde sodium bisulfate complex + NADPH <4> (Reversibility: ? <4> [2, 9]) [2, 9]
P ? + NADP$^+$
S iodoacetaldehyde + NADPH <1> (Reversibility: r <1> [1]) [1]
P iodoethan-2-ol + NADP$^+$

S iodohexadecanal + NADPH <5> (Reversibility: r <5> [45]) [45]
P iodohexadecanol + $NADP^+$
S isobutanol + $NADP^+$ <45> (Reversibility: r <45> [35]) [35]
P isobutanal
S lauryl aldehyde + NADPH <18> (<18> isozyme AR3 [21]) (Reversibility: ? <18> [21]) [21]
P lauryl alcohol + $NADP^+$
S *m*-chlorobenzaldehyde + NADPH <1> (Reversibility: r <1> [1]) [1]
P *m*-chlorobenzyl alcohol + $NADP^+$
S *m*-fluorobenzaldehyde + NADPH <33, 52> (Reversibility: ? <33, 52> [42, 47]) [42, 47]
P *m*-fluorobenzyl alcohol + $NADP^+$
S *m*-methoxybenzaldehyde + NADPH <1, 52> (Reversibility: r <1> [1]; ? <52> [42]) [1, 42]
P *m*-methoxybenzyl alcohol + $NADP^+$
S *m*-methylbenzaldehyde + NADPH <33, 52> (Reversibility: ? <33, 52> [42, 47]) [42, 47]
P *m*-methylbenzyl alcohol + $NADP^+$
S *m*-nitrobenzaldehyde + NADPH <4, 5, 18, 38, 52, 53> (Reversibility: r <4, 5, 18, 53> [2, 5, 21, 42, 43]; ir <38> [31]; ? <52> [42]) [2, 5, 21, 31, 42, 43]
P *m*-nitrobenzylalcohol + $NADP^+$
S menadione + NADPH <1> (<1> G2 [15]) (Reversibility: ? <1> [15]) [15]
P ? + $NADP^+$
S methanol + $NADP^+$ <54> (Reversibility: r <54> [44]) [44]
P formaldehyde + NADPH
S methylglyoxal + NADPH <2, 4, 5, 38> (Reversibility: r <2, 4, 5, 18> [2, 5, 21, 41, 45]; ? <38> [31]) [2, 5, 21, 31, 41, 45]
P ? + $NADP^+$
S mevaldate + NADPH <5> (Reversibility: ? <5> [32]) [32]
P ? + $NADP^+$
S n-butanol + $NADP^+$ <4, 8, 28, 29, 36, 45, 51, 53> (Reversibility: r <4, 8, 28, 29, 36, 45, 51, 53> [2, 6, 28, 30, 34, 35, 39, 43]) [2, 6, 28, 30, 34, 35, 39, 43]
P butyraldehyde + NADPH
S n-butyraldehyde + NADPH <4, 5, 18, 28, 33, 53, 55> (<4> relative activity 2% [2]) (Reversibility: r <4, 5, 18, 28, 33, 53, 55> [2, 5, 18, 21, 34, 43, 47, 48]) [2, 5, 9, 18, 21, 34, 43, 47, 48]
P n-butanol + $NADP^+$
S n-decanal + NADPH <2, 33, 52> (<52> converted less efficiently than eutypine [42]) (Reversibility: r <2, 33> [20, 47]; ? <52> [42]) [20, 42, 47]
P n-decanol + $NADP^+$
S n-dodecanal + NADPH <2, 33> (Reversibility: r <2, 33> [20, 47]) [20, 47]
P n-dodecanol + $NADP^+$
S n-heptanal + NADPH <4, 33> (Reversibility: r <4, 33> [9, 47]) [9, 47]
P heptanol + $NADP^+$
S n-hexadecanal + NADPH <2> (Reversibility: r <2> [20, 25]) [20, 25]
P hexadecanol + $NADP^+$
S n-octadecanal + NADPH <2> (Reversibility: r <2> [20, 25]) [20, 25]

P n-octadecanol + NADP$^+$
S n-octanal + NADPH <2, 33, 53, 55> (Reversibility: r <2, 33, 53, 55> [20, 43, 47, 48]) [20, 43, 47, 48]
P n-octanol + NADP$^+$
S n-propanol + NADP$^+$ <8, 28, 36, 45, 51> (Reversibility: r <8, 28, 36, 45, 51> [6, 28, 30, 34, 35, 39]) [6, 28, 30, 34, 35, 39]
P propionaldehyde + NADPH
S n-tetradecanal + NADPH <2> (Reversibility: r <2> [20]) [20]
P n-tetradecanol + NADP$^+$
S nonanal + NADPH <33> (Reversibility: ? <33> [47]) [47]
P ? + NADP$^+$
S *o*-fluorobenzaldehyde + NADPH <33, 52> (Reversibility: ? <33, 52> [42, 47]) [42, 47]
P *o*-fluorobenzyl alcohol + NADP$^+$
S *o*-methylbenzaldehyde + NADPH <33, 52> (Reversibility: ? <33, 52> [42, 47]) [42, 47]
P *o*-methylbenzyl alcohol + NADP$^+$
S *o*-nitrobenzaldehyde + NADPH <4, 18, 38, 52> (Reversibility: r <4, 18> [2, 21]; ir <38> [31]; ? <52> [42]) [2, 21, 31, 42]
P *o*-nitrobenzylalcohol + NADP$^+$
S octyl 4-chloro-3-oxobutanoate + NADPH <38> (Reversibility: ? <38> [31]) [31]
P (R)-4-chloro-3-hydroxybutanoic acid ethyl ester + NADP$^+$
S *p*-carboxybenzaldehyde + NADPH <1, 2, 4, 5, 18> (<1> AR1 [15]) (Reversibility: r <1, 2, 4, 5, 18> [1, 2, 5, 9, 11, 14, 15, 21, 22, 36, 37, 55]) [1, 2, 5, 9, 11, 14, 15, 21, 22, 36, 37, 55]
P *p*-carboxybenzyl alcohol + NADP$^+$
S *p*-chlorobenzaldehyde + NADPH <4, 5, 38> (Reversibility: r <4, 5> [2, 5]; ir <38> [31]) [2, 5, 31]
P *p*-chlorobenzyl alcohol + NADP$^+$
S *p*-cyanobenzaldehyde + NADPH <1, 4> (Reversibility: r <1> [1]; ? <4> [2]) [1, 2]
P *p*-cyanobenzyl alcohol + NADP$^+$
S *p*-fluorobenzaldehyde + NADPH <4, 33, 52> (Reversibility: ? <4, 33, 52> [2, 42, 47]) [2, 42, 47]
P *p*-fluorobenzyl alcohol + NADP$^+$
S *p*-hydroxyphenylacetaldehyde + NADPH <1, 2, 4> (Reversibility: r <1, 2, 4> [1, 3, 11]) [1, 3, 11]
P *p*-hydroxyphenylethanol + NADP$^+$
S *p*-hydroxyphenylglycolaldehyde + NADPH <1, 2, 4> (Reversibility: r <1, 2, 4> [1, 3, 11]) [1, 3, 11]
P 4-hydroxyphenylethylalcohol + NADP$^+$
S *p*-methoxybenzaldehyde + NADPH <1, 52> (Reversibility: r <1> [1]; ? <52> [42]) [1, 42]
P *p*-methoxybenzyl alcohol + NADP$^+$
S *p*-methylbenzaldehyde + NADPH <33, 52> (Reversibility: ? <33, 52> [42, 47]) [42, 47]

P *p*-methylbenzyl alcohol + $NADP^+$
S *p*-nitroacetophenone + NADPH <4> (Reversibility: r <4> [9]) [9]
P ? + $NADP^+$
S *p*-nitrobenzaldehyde + NADPH + <1-5, 18, 24, 38, 52> (<1, 4, 5> cofactor NADPH [1, 3, 5]; <1> AR1 [15]) (Reversibility: r <1-5, 18, 24> [1-4, 5, 9, 11-16, 19, 20-22, 25, 27, 36, 41]; ir <38> [31]; ? <52> [42]) [1-4, 5, 9, 11-16, 19, 20-22, 25, 27, 31, 36, 41, 42]
P *p*-nitrobenzylalcohol + $NADP^+$
S *p*-tolualdehyde + NADPH <1> (Reversibility: <1> [1]) [1]
P toluol + $NADP^+$
S pentanal + NADPH <17, 33, 53> (Reversibility: r <17, 33, 53> [40, 43, 47]) [40, 43, 47]
P pentanol + $NADP^+$
S phenylacetaldehyde + NADPH <1, 33> (Reversibility: r <1, 33> [1, 47]) [1, 47]
P ? + $NADP^+$
S phenylethylamine + ? <1> (Reversibility: r <1> [1]) [1]
P ?
S phenylglyoxal + NADPH <4, 18> (Reversibility: r <4, 18> [9, 21, 22]) [9, 21, 22]
P ? + $NADP^+$
S propan-2-ol + $NADP^+$ <28> (<28> 2° Adh [34]) (Reversibility: r <28> [28, 34]) [28, 34]
P propanone + NADPH
S propanal + NADPH <17, 33> (Reversibility: r <17, 33> [40, 47]) [40, 47]
P propanol + $NADP^+$
S propane-1,2-diol + $NADP^+$ <1, 4, 5> (Reversibility: r <1, 4, 5> [3, 5, 15]) [3, 5, 15]
P lactaldehyde + NADPH
S propionaldehyde + NADPH <2, 5, 28, 36, 52> (Reversibility: r <2, 5, 28, 36> [5, 20, 23, 24, 28, 30]; ? <52> [42]) [5, 20, 23, 24, 28, 30, 42]
P propanol + $NADP^+$
S propiophenone+ NADPH <4> (Reversibility: ? <4> [2]) [2]
P ? + $NADP^+$
S pyridoxal + NADPH <4> (Reversibility: ? <4> [2]) [2]
P 4,5-bis(hydroxymethyl)-2-methylpyridin-3-ol + $NADP^+$
S pyruvaldehyde + NADPH <4> (Reversibility: ? <4> [2]) [2]
P ? + $NADP^+$
S pyruvate + NADPH <4> (Reversibility: ? <4> [2]) [2]
P 2-hydroxypropanoic acid + $NADP^+$
S salicylaldehyde + NADPH <52> (Reversibility: ? <52> [42]) [42]
P 2-hydroxymethylphenol + $NADP^+$
S sinapaldehyde + NADPH <52> (<52> converted less efficiently than eutypine [42]) (Reversibility: ? <52> [42]) [42]
P sinapyl alcohol + $NADP^+$
S succinic semialdehyde + NADPH <1-5> (<1> AR1 [15]) (Reversibility: r <1-5> [9, 12-16, 20, 22, 32, 36, 37, 52]) [9, 12-16, 20, 22, 32, 36, 37, 52]

P γ-hydroxybutyrate + $NADP^+$ <2> [20]
S tetradecanal + NADPH <33> (Reversibility: r <33> [47]) [47]
P tetradecanol
S trans-2-decenal + NADPH <33> (Reversibility: ? <33> [47]) [47]
P trans-2-decen-1-ol + $NADP^+$
S trans-2-hexenal + NADPH <53> (Reversibility: r <53> [43]) [43]
P trans-2-hexenol + $NADP^+$
S trans-2-hexenol + $NADP^+$ <17, 53> (Reversibility: r <17, 53> [40, 43]) [40, 43]
P trans-2-hexenal + NADPH
S trans-cinnamaldehyde + NADPH <33, 52> (<52> converted less efficiently than eutypine [42]) (Reversibility: ? <33, 52> [42, 47]) [42, 47]
P cinnamyl alcohol + $NADP^+$
S tridecanal + NADPH <33> (Reversibility: r <33> [47]) [47]
P tridecanol + $NADP^+$
S undecanal + NADPH <33> (Reversibility: ? <33> [47]) [47]
P undecanol + $NADP^+$
S vitamin A aldehyde + NADPH <33> (Reversibility: ? <33> [47]) [47]
P ? + $NADP^+$
S Additional information <1, 4, 5, 28, 26, 36, 38, 51, 52> (<1> short chain aliphathic aldehydes such as actealdehyde or propionaldehyde are no substrates, benzyl-substituted ketones such as pyruvic acid are not reduced [1]; <1> AR1 is not active with neutral sugars or with ketones, AR3 is capable of reducing neutral sugars, G2 is active with ketones, AR4 is inactive with 15-oxoprostaglandin B1 [15]; <2> enzyme from rat brain microsomes do not reduce *p*-pyridinecarboxaldehyde, which is a very active substrate for high-K_m aldehyde reductase from rat kidney cytosol [25]; <4> 5-hydroxyindol-3-ylacetaldehyde is no substrate [3]; <5> no enzymatic activity in the presence of NADPH with acetaldehyde, pyridoxal, pyridoxal phosphate, glyceraldehyde phosphate, glucose or fructose, no activity in the reverse reaction with ethanol, cyclohexanol, phenylpropyl alcohol and glycerol phosphate [5]; <5> glucose, galactose, fructose, sorbitol, galactitol, xylitol and glycerol are no substrates [18]; <2, 5> enzyme is virtually inactive for aldo-sugars [23, 29, 58]; <26> D-hexose is no substrate [33]; <28> 1° Adh specific for 1°alcohols and aldehydes, no detectable actvity towards ketones, 2°alcohols or acetyl-CoA [34]; <36> acetone is no substrate [30]; <38> D-glucuronate, D-glucose, D-xylose, D-galactose, formaldehyde, acetaldehyde, propionaldehyde, n-butyraldehyde, glycolaldehyde, acetone, acetol, lactaldehyde, dihydroacetone and acetoin are not reduced [31]; <51> methanol, isopropanol and glycerol are no substrates [39]; <52> dihydroquercitin and dihydromyricetin are no substrates [42]) [1, 3, 5, 15, 18, 23, 25, 29, 30, 31, 33, 34, 39, 42, 58]
P ?

Inhibitors

$(NH_4)_2SO_4$ <5> [23]
1,10-phenanthroline <4> [9]

2',5'-ADP <2> [11]
2,3-dimethylsuccinic acid <4, 5> [14, 17, 45]
2,4-dimethylglutaric acid <2> [11]
2-ethylhexanoate <24> [27]
2-mercaptoethanol <4> [9]
3,3'-tetramethylenglutaric acid <2, 5, 18> (<18> AR1 and AR3 [21]) [11, 18, 21, 23, 37]
3,3-dimethylglutaric acid <2> [11]
4-carboxybenzaldehyde <1> (<1> uncompetitive substrate inhibition [55]) [55]
5,5'-dithiobis-(2-nitrobenzoate) <4> [10]
AL1576 <2, 5> [36, 37, 41]
AL4114 <2> [41]
ATP-ribose <5> (<5> competitive inhibition [32]) [32]
Ag^{+} <33> (<33> 100% inhibition [47]) [47]
Blue Dextran <2> [11]
Ca^{2+} <33> (<33> 26% inhibition [47]) [47]
Cibacron Blue F3GA <2, 3> (<3> dead-end inhibitor [13]) [11, 13]
Co^{2+} <33> (<33> 91% inhibition [47]) [47]
D-glucuronate <24> (<24> high concentrations, inhibition in a non-competitive manner [27]) [27]
E79175 <2> [29]
EDTA <4> (<4> lower molecular weight form [14]) [14]
FK366 <5> [36, 37]
Hg^{2+} <33> (<33> 100% inhibition [47]) [47]
KCl <45> [35]
L-gulonic acid <1> (<1> product inhibition [16]) [16]
$LiSO_4$ <5> (<5> aldehyde reductase I inhibited by 86%, aldehyde reductase II is not effected [18]; <5> 0.4 mM completely inhibits aldehyde reductase II [23]; <5> both aldehyde reductase I and II are susceptible to inhibition [24]) [18, 23, 24]
M79175 <26> [33]
Mn^{2+} <33> (<33> 29% inhibition [47]) [47]
N-ethylmaleimide <5, 33> [23, 47]
$NADP^{+}$ <1, 5> (<1> product inhibition [16]; <5> competitive inhibition [32]) [16, 32]
NADPH <1, 4> (<1> product inhibition [16]; <4> competitive inhibitor for reductase I [9]) [9, 16]
NH_4Cl <2> [29]
NaCl <2, 5, 45> (<5> aldehyde reductase I and II [24]) [24, 29, 35]
Ni^{2+} <33> (<33> 84% inhibition [47]) [47]
ONO-2235 <2> [41]
Procion Reed HE3B <2> [11]
Zn^{2+} <33> (<33> 100% inhibition [47]) [47]
alconil <2, 26> [29, 33]
alrestatin <2, 5, 24> (<2> AY22,284A [11,14]; <24> liver ALR1 [27]) [11, 14, 18, 20, 23, 36, 37]

amaranth <5> [45]
amobarbital <4> [2]
amytal <1> [1]
arachidonic acid <5> [45]
benzodiazepine <1, 5> [14, 15]
benzoic acid <1> [16]
carbamazepine <5> [14]
chlorpromazine <1, 2, 4, 18, 56> (<18> isozyme AR3 [21]) [1, 3, 12, 21, 50]
chlorpromazine-HCl <1> [1]
chromone <5> [18]
diethyldicarbonate <5> (<5> kidney aldehyde reductase [32]) [32]
diphenic acid <4> [17]
diphenic acid <5> [45]
diphenylhydantoin <56> [50]
epalrestat <2, 26> [29, 33]
ethacrynic acid <5> [45]
ethylphenylhydantoin <5> [5]
glycerol <4> (<4> non-competitive inhibitor with D-glyceraldehyde as substrate [8]) [8]
haematin <5> [45]
hydroxylamine <4> [2]
indigo carmine <5> [45]
indocyanine green <5> [45]
indomethacin <18> [21]
iodoacetamide <33, 36> [30, 47]
iodoacetic acid <4> [2]
morin <2> [11]
p-chloromercuribenzoate <18, 36> [21, 30]
p-hydroxymercuribenzoate <5> [23]
p-mercuribenzoate <4, 5, 8-16, 33> (<4> above 1 mM [4]; <5> aldehyde reductase I is more sensitive than aldehyde reductase II [18]) [4, 6, 14, 18, 47]
phenacamide <5> [14]
phenobarbital <1, 2, 4, 5, 18> [1, 5, 9, 11, 21, 57]
phenobarbitone <24> [27]
phenolsulfophthalein <5> [45]
phenylmethylsulfonylfluoride <4> (<4> lower molecular weight form [14]) [14]
phenylpyruvic acid <3> (<3> low-molecular weight aldehyde reductase, noncompetitively inhibition [13]) [13]
ponalrestat <5, 26> [33, 36, 37]
potassium phosphate <45> [35]
pyrazole <4, 7, 8> [6, 9]
pyridine 3-aldehyde <4, 24> [22, 27]
quercetin <2, 4> [11, 17]
quercetrin <5> [23]
Rose Bengal <5> [45]
sodium amobarbital <1> [1]

sodium barbitone <1, 4> [14, 15, 49, 55]
sodium fluoride <4> [2]
sodium phenobarbital <1, 5> (<5> aldehyde reductase II activity diminished by 84% [18]) [1, 18, 23]
sodium valproate <1, 5, 24, 26> (<24> liver ALR1 [27]) [14-16, 27, 33, 45, 55, 58]
sorbinil <2, 5, 24, 26> (<24> liver ALR1 [27]) [18, 23, 24, 27, 29, 33, 37]
statil <2> [29]
sulfobromophthalein <5> [45]
tolrestat <2, 5, 26> [29, 33, 36, 37]
zopolrestat <5> [36, 37]

Cofactors/prosthetic groups
3-acetylpyridine adenine dinucleotide phosphate <5> [32]
NAD^+ <45> [35]
NADH <5> (<5> aldehyde reductase I [24]) [24]
NADPH <1, 5, 26, 38, 51, 52> (<2,5> aldehyde reductase II utilizes only NADPH as cofactor [23, 24, 29]; <38> highly specific for NADPH [31]) [1, 23, 24, 31, 33, 39, 42]

Activating compounds
$(NH_4)_2SO_4$ <45> [35]
PO_4^{3-} <4> (<4> stimulates the enzymatic reduction of 3-pyridinecarboxaldehyde [2]) [2]
SO_4^{2-} <4, 18> (<4> stimulates the enzymatic reduction of 3-pyridinecarboxaldehyde [2]) [2, 21]
adenosine <2> [11]
hexadecane <33> (<33> induces enzyme [47]) [47]
trifluperazine <2> [11]

Metals, ions
Na_2SO_4 <18> (<18> 400 mM, AR2, relative activity 150% [21]) [21]
NaCl <5> (<5> 100 mM, aldehyde reductase II slightly stimulated [18]) [18]

Turnover number (min^{-1})
1.7 <5> (glycerol) [5]
1.7 <5> (phenylethanol) [5]
2.6 <5> (propane-1,2-diol) [5]
4.2 <4> (3-pyridinemethanol) [2]
4.4 <5> (benzylalcohol) [5]
4.8 <5> (phenylethanediol) [5]
8.9 <5> (4-hydroxy-3-methoxybenzyl alcohol) [5]
36.7 <4> (glycerol) [2]
106 <4> (L-arabitol) [2]
336 <4> ((+)-camphorquinone) [2]
570 <5> (*p*-nitrobenzaldehyde) [5]
693 <4> (D-glucuronate) [2]
820 <4> (D-lactaldehyde) [2]
865 <4> (D-glyceraldehyde) [2]

998 <4> (4-cyanobenzaldehyde) [2]
1160 <4> (4-pyridinecarboxaldehyde) [2]
1300 <4> (4-carboxybenzaldehyde) [2]
1320 <4> (4-nitrobenzaldehyde) [2]
1610 <4> (3-pyridinecarboxaldehyde) [2]

Specific activity (U/mg)
0.008 <4> (<4> low-K_m reductase [3]) [3]
0.024 <55> [48]
0.05 <1> [3]
0.34 <4> (<4> high-K_m reductase [3]) [3]
0.3786 <1> [1]
0.55 <1> (<1> AR3 [15]) [15]
1.07 <4> [7]
1.09 <17> [7]
1.35 <26> [33]
1.5 <18> [7]
1.7 <4> [4]
1.7 <33> (<33> recombinant enzyme, expressed in Escherichia coli JM109 [47]) [47]
2.25 <36> [30]
2.47 <5, 19> (<5> aldehyde reductase I [18]) [7, 18]
2.5 <4> [17]
3.2 <51> [39]
4.8 <4> (<4> pyridine 3-aldehyde as substrate [17]) [17]
5.86 <18> (<18> AR2 [21]) [21]
5.9 <2> (<2> liver enzyme [20]) [20]
6 <2> (<2> brain enzyme [20]) [20]
6.53 <2> [7]
6.66 <54> (<54> native enzyme [44]) [44]
6.84 <2> [29]
8.94 <1> (<1> AR1 [15]) [15]
9.59 <2> (<2> high-K_m aldehyde reductase from brain [12]) [12]
11.02 <2> [20]
14.3 <5> (<5> pure aldehyde reductase [5]) [5]
15.7 <5> [5]
17.1 <4> [2]
18.8 <4> [26]
19.7 <18> (<18> AR1 [21]) [21]
19.9 <18> (<18> AR3 [21]) [21]
20 <2, 18> (<2> *p*-nitrobenzaldehyde as substrate [25]; <18> AR1 and AR3 [21]) [21, 25]
21 <4> [10]
24 <54> (<54> recombinant enzyme [44]) [44]
29.62 <5> (<5> aldehyde reductase II [18]) [18]
31.1 <45> [35]
35 <4> [9]

61.6 <17> [40]
69 <28> (<28> 2° Adh [34]) [34]
81.7 <38> [31]
140 <28> (<28> 1° Adh [34]) [34]
211 <2> (<2> hexadecanal as substrate [25]) [25]
560 <33> [47]
Additional information <52> (<52> 891 nkat/mg [42]) [42]

K_m-Value (mM)

0.0007 <4> (*p*-hydroxyphenylglycolaldehyde, <4> low-K_m reductase [3]) [3]
0.001 <1> (NADPH) [1]
0.001 <24> (NADPH, <24> *p*-nitrobenzaldehyde as substrate [27]) [27]
0.0012 <4> (NADPH, <4> low-K_m reductase [3]) [3]
0.0013 <4> (*p*-nitrobenzaldehyde, <4> low-K_m reductase [3]) [3]
0.0016 <24> (NADPH, <24> pyridine 3-aldehyde as substrate [27]) [27]
0.0019 <24> (NADPH, <24> D-glucuronate as substrate [27]) [27]
0.002 <1> (NADPH, <1> AR1 [15]) [15]
0.002 <18> (NADPH, <18> AR2 [21]) [21]
0.002 <53> (*m*-nitrobenzaldehyde) [43]
0.0023 <4> (*p*-hydroxyphenylacetaldehyde, <4> low-K_m reductase [3]) [3]
0.00246 <8> (1-heptanol) [6]
0.00251 <8> (1-hexanol) [6]
0.0026 <2> (*p*-nitrobenzaldehyde, <2> low-K_m reductase [52]) [52]
0.00273 <8> (1-pentanol) [6]
0.0029 <52> (*m*-nitrobenzaldehyde) [42]
0.003 <1> (NADPH, <1> G2 [15]) [15]
0.003 <18> (NADPH, <18> AR1 [21]) [21]
0.003 <18> (NADPH, <18> AR3 [21]) [21]
0.003-0.005 <2> (NADPH) [12]
0.0033 <52> (decanal) [42]
0.0033 <8> (n-butanol) [6]
0.0038 <4> (hydroxyindol-3-ylacetaldehyde, <4> low-K_m reductase [3]) [3]
0.004 <36> (NADPH, <36> n-propanol as substrate [30]) [30]
0.004 <2> (n-octadecanal) [20]
0.004 <1> (*p*-carboxybenzaldehyde) [1]
0.004 <4> (*p*-nitrobenzaldehyde, <4> low-K_m aldehyde reductase [3]) [3]
0.0045 <52> (4-pyridine carboxaldehyde) [42]
0.005 <36> (NADPH, <36> benzaldehyde as substrate [30]) [30]
0.005 <1> (*p*-cyanobenzaldehyde) [1]
0.00522 <8> (1-propanol) [6]
0.0053 <52> (4-nitrobenzaldehyde) [42]
0.0054 <52> (3-fluorobenzaldehyde) [42]
0.0056 <4> (NADPH) [9]
0.0059 <4> (indol-3-ylacetaldehyde, <4> low-K_m reductase [3]) [3]
0.006 <2> (NADPH) [29]
0.006 <1> (*p*-nitrobenzaldehyde) [1]

0.0063 <52> ((4-hydroxy-3-[3-methyl-3-butene-1-ynyl]benzaldehyde), <52> eutypine [42]) [42]
0.007 <1> (NADPH, <1> AR4 [15]) [15]
0.007 <1> (*m*-chlorobenzaldehyde) [1]
0.008 <1> (3-pyridinecarboxaldehyde, <1> AR3 [15]) [15]
0.008 <4> (NADPH) [9]
0.008 <1> (NADPH, <1> AR3 [15]) [15]
0.008 <53> (all-trans-retinal) [43]
0.008 <2> (n-hexadecanal) [20]
0.008 <5> (propionaldehyde, <5> aldehyde reductase I [18]) [18]
0.0084 <1> (*p*-nitrobenzaldehyde, <1> AR3 [15]) [15]
0.009 <53> (9-cis-retinal) [43]
0.009 <1> (menandione, <1> AR4 [15]) [15]
0.01 <54> (benzaldehyde, <54> co-substrate NADP [44]) [44]
0.01 <53> (octanal) [43]
0.012 <53> (NADPH, <53> *m*-nitrobenzaldehyde as substrate [43]) [43]
0.013 <5> (NADPH, <5> aldehyde reductase II [18]) [18]
0.013 <8> (ethanol, <8> $NADP^+$ [6]) [6]
0.014 <36> (NADPH, <36> propionaldehyde as substrate [30]) [30]
0.014 <2> (n-tetradecanal) [20]
0.015 <52> (1-hexanal) [42]
0.015 <53> (13-cis-retinal) [43]
0.016 <1> (*p*-hydroxyphenylglycolaldehyde) [1]
0.017 <1> (benzaldehyde) [1]
0.018 <1> (*p*-carboxybenzaldehyde, <1> AR1 [15]) [15]
0.019 <53> (9-cis-retinol) [43]
0.019 <5> (NADPH, <5> aldehyde reductase II [23]) [23]
0.019 <2> (n-dodecanal, <2> enzyme from brain [20]) [20]
0.0198 <8> (ethanol) [6]
0.02 <5> (3-pyridinecarboxaldehyde, <5> AR I [24]) [24]
0.02 <28> ($NADP^+$, <28> 1° Adh, ethanol as substrate [34]) [34]
0.02 <5> (NADPH, <5> AR II [24]) [24]
0.02 <53> (all-trans-retinol) [43]
0.02 <52> (*m*-methylbenzaldehyde) [42]
0.02 <2> (n-dodecanal, <2> enzyme from liver and kidney [20]) [20]
0.02 <1> (*p*-tolualdehyde) [1]
0.021 <36> (NADPH, <36> benzyl alcohol as substrate [30]) [30]
0.022 <28> ($NADP^+$, <28> secondary alcohol dehydrogenase [28]) [28]
0.022 <5> (NADPH, <5> aldehyde reductase I [18]) [18]
0.023 <5> (benzaldehyde, <5> aldehyde reductase I [18]) [18]
0.025 <28> (NADH, <28> 1° Adh, acetaldehyde as substrate [34]) [34]
0.025 <5> (*p*-carboxybenzaldehyde) [5]
0.027 <1> (*m*-methoxybenzaldehyde) [1]
0.028 <1> (succinic semialdehyde, <1> AR4 [15]) [15]
0.029 <5> (*p*-nitrobenzaldehyde, <5> aldehyde reductase I [18]) [18]
0.029 <1> (succinic semialdehyde, <1> AR1 [15]) [15]
0.03 <53> ($NADP^+$) [43]

0.03 <2> (hexadecanal) [25]
0.03 <1> (menandione, <1> G2 [15]) [15]
0.03 <55> (octanal) [48]
0.031 <2> (*p*-carboxybenzaldehyde) [11]
0.033 <45> (NADP) [35]
0.033 <28> (NADPH, <28> 2° Adh, propanone as substrate [34]) [34]
0.033 <52> (benzaldehyde) [42]
0.034 <18> (*p*-nitrobenzaldehyde, <18> AR2 [18]) [18]
0.036 <1> (succinic semialdehyde, <1> G2 [15]) [15]
0.037 <22> (succinic semialdehyde) [12]
0.038 <5> (*p*-carboxybenzaldehyde, <5> wild-type aldehyde reductase [36,37]) [36, 37]
0.04 <55> (cinnamaldehyde) [48]
0.04 <53> (hexanal) [43]
0.04 <52> (*p*-fluorobenzaldehyde) [42]
0.041 <52> (*p*-methylbenzaldehyde) [42]
0.041 <2> (*p*-nitrobenzaldehyde, <2> high-K_m reductase [52]) [52]
0.042 <18> (pyridine 3-aldehyde, <18> AR2 [18]) [18]
0.044 <24> (*p*-nitrobenzaldehyde) [27]
0.049 <52> (2-methoxybenzaldehyde) [42]
0.049 <28> ($NADP^+$, <28> 2° Adh, butan-2-ol as substrate [34]) [34]
0.05 <5> (NADPH, <5> AR I [24]) [24]
0.05 <4> (*p*-carboxybenzaldehyde) [9]
0.052 <18> (NADH, <18> AR2 [21]) [21]
0.053 <55> (benzaldehyde) [48]
0.055 <28> (NADPH, <28> 2° Adh, acetaldehyde as substrate [34]) [34]
0.056 <18> (*p*-carboxybenzaldehyde, <18> AR1 [21]) [21]
0.057 <1> (*p*-hydroxyphenylacetaldehyde) [1]
0.059 <2> (*p*-hydroxyphenylglycolaldehyde) [11]
0.06 <53> (NAD^+, <53> ethanol as substrate [43]) [43]
0.06 <52> (coniferylaldehyde) [42]
0.06 <53> (trans-2-hexenal) [43]
0.062 <18> (4-pyridinecarboxaldehyde, <18> AR2 [18]) [18]
0.062 <18> (*p*-carboxybenzaldehyde, <18> AR3 [18]) [18]
0.064 <1> (succinic semialdehyde, <1> AR3 [15]) [15]
0.066 <52> (salicylaldehyde) [42]
0.07 <28> ($NADP^+$, <28> 2° Adh, ethanol as substrate [34]) [34]
0.07 <5> (glyceraldehyde, <5> AR I [24]) [24]
0.07 <2> (n-decanal) [20]
0.07 <53> (pentanal) [43]
0.071 <1> (*p*-nitrobenzaldehyde, <1> AR1 [15]) [15]
0.078 <4> (*p*-carboxybenzaldehyde) [2]
0.08 <51> ($NADP^+$) [39]
0.08 <4> (daunorubicin, <4> reductase I [9]) [9]
0.08 <2> (*p*-nitrobenzaldehyde) [12]
0.082 <18> (*m*-nitrobenzaldehyde, <18> AR2 [18]) [18]
0.083 <5> (butyraldehyde, <5> aldehyde reductase I [18]) [18]

0.087 <52> (propionaldehyde) [42]
0.09 <53> (octanol) [43]
0.091 <38> (ethyl 4-bromo-3-oxobutanoate) [31]
0.094 <28> (NADPH, <28> 1°Adh, acetaldehyde as substrate [34]) [34]
0.096 <28> (NAD^+, <28> 1° Adh, ethanol as substrate [34]) [34]
0.099 <18> (androstane-3,17-dione, <18> AR2 [21]) [21]
0.099 <52> (cinnamaldehyde) [42]
0.1 <4> (4-carboxyphenylglyoxal) [9]
0.1 <45> (NADPH) [35]
0.1 <38> (octyl 4-chloro-3-oxobutanoate) [31]
0.1 <5> (*p*-carboxybenzaldehyde, <5> aldehyde reductase, mutant V299C [36]) [36]
0.1 <4> (succinic semialdehyde, <4> reductase I [9]) [9]
0.11 <18> (5α-dihydrotestosterone, <18> AR2 [21]) [21]
0.11 <53> (butanal) [43]
0.11 <5> (*p*-carboxybenzaldehyde, <5> aldehyde reductase, mutant R308A [37]) [37]
0.11 <18> (phenylglyoxal, <18> AR2 [18]) [18]
0.12 <2> (*p*-nitrobenzaldehyde) [41]
0.12-0.13 <51> (acetaldehyde) [39]
0.125 <52> (*o*-methylbenzaldehyde) [42]
0.126 <52> (sinapylaldehyde) [42]
0.129 <1> (*p*-nitrobenzaldehyde, <1> G2 [15]) [15]
0.13 <54> (NADPH, <54> benzaldehyde as substrate [44]) [44]
0.13 <2> (succinic semialdehyde) [12]
0.141 <1> (*p*-methoxybenzaldehyde) [1]
0.15 <28> ($NADP^+$, <28> primary alcohol dehydrogenase [28]) [28]
0.15 <53> (octanol) [43]
0.15 <52> (*p*-methoxybenzaldehyde) [42]
0.15 <5> (*p*-nitrobenzaldehyde) [5]
0.16 <5> (*p*-nitrobenzaldehyde, <5> wild-type aldehyde reductase [36, 37]) [36, 37]
0.16 <17> (pentanal, <17> coenzyme NADPH [40]) [40]
0.16 <28> (propanone, <28> 2° Adh [34]) [34]
0.169 <5> (*p*-nitrobenzaldehyde, <5> aldehyde reductase II [18]) [18]
0.17 <5> (succinic semialdehyde, <5> wild-type aldehyde reductase [36, 37]) [36, 37]
0.18 <17> (hexanal, <17> coenzyme NADPH [40]) [40]
0.2 <53> (NAD^+, <53> ethanol as substrate [43]) [43]
0.2 <4> (indol-3-ylacetaldehyde, <4> high-K_m aldehyde reductase [3]) [3]
0.2 <2> (*p*-nitrobenzaldehyde, <2> enzyme from kidney, brain or liver [20]) [20]
0.2 <5> (succinic semialdehyde, <5> aldehyde reductase, mutant R308A [37]) [37]
0.207 <52> (furaldehyde) [42]
0.21 <17> (3-methylbutanal, <17> coenzyme NADPH [40]) [40]
0.21 <2> (octanal) [25]

0.21 <5> (propionaldehyde, <5> AR I [24]) [24]
0.212 <52> (*m*-methoxybenzaldehyde) [42]
0.215 <55> (butyraldehyde) [48]
0.22 <53> (pentanol, <53> at pH 10 [43]) [43]
0.22 <36> (propionaldehyde) [30]
0.22 <53> (trans-2-hexenol) [43]
0.23 <2> (*p*-nitrobenzaldehyde) [11]
0.24 <28> (butanone, <28> 2° Adh [34]) [34]
0.25 <4> (NADPH, <4> high-K_m aldehyde reductase [3]) [3]
0.25 <28> (acetaldehyde, <28> 1° Adh [34]) [34]
0.25 <4> (*p*-nitrobenzaldehyde, <4> high-K_m aldehyde reductase [3]) [3]
0.25 <5> (*p*-nitrobenzaldehyde, <5> aldehyde reductase, mutant R308A [37]) [37]
0.258 <1> (*p*-carboxybenzaldehyde, <1> AR3 [15]) [15]
0.26 <2> (pyridine 3-aldehyde) [11]
0.262 <1> (*p*-nitrobenzaldehyde, <1> AR4 [15]) [15]
0.27 <17> (heptanal, <17> coenzyme NADPH [40]) [40]
0.27 <4> (*p*-nitrobenzaldehyde) [9]
0.29 <5> (3-pyridinecarboxaldehyde, <5> aldehyde reductase II [23]) [23]
0.29 <4> (*p*-nitrobenzaldehyde, <4> high-K_m aldehyde reductase [3]) [3]
0.31 <18> (laurylaldehyde, <18> AR2 [18]) [18]
0.31 <2> (*p*-nitrobenzaldehyde) [25]
0.32 <18> (indole-3-acetaldehyde, <18> AR2 [18]) [18]
0.32 <18> (*p*-carboxybenzaldehyde, <18> AR2 [18]) [18]
0.33 <4> ((+)-camphorquinone) [2]
0.35 <28> ($NADP^+$, <28> secondary alcohol dehydrogenase [28]) [28]
0.35 <53> (hexanol) [43]
0.35 <4> (*p*-nitrobenzaldehyde) [2]
0.36 <28> (NADPH, <28> secondary alcohol dehydrogenase [28]) [28]
0.36 <38> (ethyl 4-chloro-3-oxobutanoate) [31]
0.376 <52> (*o*-nitrobenzaldehyde) [42]
0.38 <5> (*p*-carboxybenzaldehyde, <5> aldehyde reductase, mutant I298A [36]) [36]
0.38 <5> (*p*-nitrobenzaldehyde, <5> aldehyde reductase, mutant V299C [36]; <5> mutant R311A [37]) [36, 37]
0.4 <45> (acetaldehyde) [35]
0.41 <5> (succinic semialdehyde, <5> aldehyde reductase, mutant V299C [36]) [36]
0.43 <18> (D-glyceraldehyde, <18> AR2 [21]) [21]
0.43 <5> (*p*-carboxybenzaldehyde, <5> aldehyde reductase, mutant I298C [36]) [36]
0.44 <28> (butanal, <28> 1° Adh [34]) [34]
0.46 <18> (1,2-cyclohexanedione, <18> AR3 [21]) [21]
0.5 <28> (acetylCoA, <28> 2° Adh [34]) [34]
0.5 <55> (cinnamyl alcohol) [48]
0.51 <28> (butan-2-ol, <28> 2° Adh [34]) [34]
0.51 <4> (*p*-cyanobenzaldehyde) [2]

0.52 <53> (NADH, <53> *m*-nitrobenzaldehyde as substrate [43]) [43]
0.56 <28> (propan-2-ol, <28> 2° Adh [34]) [34]
0.58 <36> (benzaldehyde) [30]
0.6 <4> (indolacetaldehyde) [9]
0.606 <5> (3-pyridinecarboxaldehyde, <5> aldehyde reductase II [18]) [18]
0.65 <4> (*p*-hydroxyphenylglycolaldehyde, <4> high-K_m aldehyde reductase [3]) [3]
0.69 <4> (4-pyridinecarboxaldehyde) [2]
0.702 <52> (*o*-fluorobenzaldehyde) [42]
0.751 <1> (3-pyridinecarboxaldehyde, <1> G2 [15]) [15]
0.77 <2> (n-octanal) [20]
0.8 <18> (1,2-cyclohexanedione, <18> AR2 [21]) [21]
0.8 <53> (NAD^+, <53> octanol as substrate [43]) [43]
0.8 <54> (NADP, <54> benzyl aclohol as substrate [44]) [44]
0.8 <36> (*p*-nitrobenzaldehyde) [30]
0.83 <28> ($NADP^+$, <28> primary alcohol dehydrogenase [28]) [28]
0.83 <17> (hexanal, <17> coenzyme NADH [40]) [40]
0.85 <4> (*p*-hydroxyphenylacetaldehyde, <4> high-K_m aldehyde reductase [3]) [3]
0.91 <2> (*p*-hydroxyphenylacetaldehyde) [11]
0.934 <1> (3-pyridinecarboxaldehyde, <1> AR1 [15]) [15]
0.94 <5> (*p*-chlorobenzaldehyde) [5]
1.1 <18> (D-galacturonate, <18> AR1 [21]) [21]
1.1 <18> (n-butyraldehyde, <18> AR2 [18]) [18]
1.1 <18> (*p*-nitrobenzaldehyde, <18> AR3 [18]) [18]
1.1 <53> (pentanol) [43]
1.18 <5> (3-pyridinecarboxaldehyde, <5> aldehyde reductase I [18]) [18]
1.18 <1> (cyclohexanone, <1> G2 [15]) [15]
1.2 <2> (4-pyridinecarboxaldehyde, <2> enzyme from kidney, brain or liver [20]) [20]
1.2 <18> (diacetyl, <18> AR2 [18]) [18]
1.2 <24> (*p*-nitrobenzaldehyde) [27]
1.2 <18> (*p*-nitrobenzaldehyde, <18> AR1 [21]) [21]
1.2 <5> (*p*-nitrobenzaldehyde, <5> aldehyde reductase, mutant I298C [36]) [36]
1.2 <4> (phenylglyoxal) [9]
1.22 <17> (hexanol, <17> coenzyme NADPH [40]) [40]
1.29 <1> (D-glucuronic acid, <1> AR1 [15]) [15]
1.3 <2> (DL-glyceraldehyde) [11]
1.3 <5> (butyraldehyde) [5]
1.3 <5> (methylglyoxal) [5]
1.3 <38> (*o*-nitrobenzaldehyde) [31]
1.3 <18> (phenylglyoxal, <18> AR3 [18]) [18]
1.4 <51> (butanol) [39]
1.4 <5> (camphorquinone) [5]
1.4 <2> (*p*-nitroacetophenone) [25]
1.43 <36> (benzyl alcohol) [30]

1.5 <4> (3-pyridinecarboxaldehyde) [9]
1.5 <24> (D-glucuronate) [27]
1.5 <4> (D-glucuronate) [9]
1.5 <2> (DL-glyceraldehyde) [12]
1.5 <5> (DL-glyceraldehyde) [5]
1.5 <51> (NAD^+) [39]
1.5 <28> (NADPH, <28> primary alcohol dehydrogenase [28]) [28]
1.5 <5> (succinic semialdehyde, <5> aldehyde reductase, mutant I298C [36]) [36]
1.6 <4> (4-carboxyacetophenone) [9]
1.6 <28> (formaldehyde, <28> 1° Adh [34]) [34]
1.6 <18> (phenylglyoxal, <18> AR1 [21]) [21]
1.69 <17> (trans-2-hexenol, <17> coenzyme NADPH [40]) [40]
1.7 <5> (DL-glyceraldehyde, <5> wild-type aldehyde reductase [36, 37]) [36, 37]
1.7 <53> (butanol, <53> at pH 10 [43]) [43]
1.8 <5> (DL-glyceraldehyde, <5> aldehyde reductase, mutant V299C [36]) [36]
1.85 <17> (2-methylbutanal, <17> coenzyme NADPH [40]) [40]
1.89 <17> (3-methylbutanal, <17> coenzyme NADH [40]) [40]
1.9 <18> (methylglyoxal, <18> AR2 [18]) [18]
2 <4> (2-chloroheptanal) [9]
2 <54> (benzyl alcohol, <54> co-substrate NADP [44]) [44]
2 <18> (n-butyraldehyde, <18> AR3 [18]) [18]
2.1 <28> (acetaldehyde, <28> 2° Adh [34]) [34]
2.2 <2> (D-glucuronate) [11]
2.2 <5> (DL-glyceraldehyde, <5> aldehyde reductase, mutant R311A [37]) [37]
2.3 <5> (DL-glyceraldehyde, <5> aldehyde reductase, mutant R308A [37]) [37]
2.4 <38> (3-pyridinecarboxaldehyde) [31]
2.4 <5> (*m*-nitrobenzaldehyde) [5]
2.48 <1> (D-glucuronic acid, <1> G2 [15]) [15]
2.5 <26> (D-glucuronate, <26> outer medulla enzyme [33]) [33]
2.5 <2> (benzalacetone) [25]
2.5 <5> (glucuronate, <5> aldehyde reductase II [23]) [23]
2.53 <5> (glucuronate, <5> AR II [24]) [24]
2.6 <4> (3-pyridinecarboxaldehyde) [2]
2.63 <5> (propionaldehyde, <5> AR II [24]) [24]
2.7 <18> (ethylacetoacetate, <18> AR2 [21]) [21]
2.76 <17> (butanal, <17> coenzyme NADPH [40]) [40]
2.8 <18> (methylglyoxal, <18> AR1 [21]) [21]
2.8 <18> (*o*-nitrobenzaldehyde, <18> AR2 [18]) [18]
2.8 <5> (*p*-carboxybenzaldehyde, <5> aldehyde reductase, mutant H112Q [36]) [36]
2.8 <5> (*p*-nitrobenzaldehyde, <5> aldehyde reductase, mutant I298A [36]) [36]

2.8 <5> (succinic semialdehyde, <5> aldehyde reductase, mutant I298A [36]) [36]
2.9 <18> (3-pyridinecarboxaldehyde, <18> AR3 [18]) [18]
2.9 <28> (butanal, <28> 2° Adh [34]) [34]
2.9 <5> (propionaldehyde) [5]
2.94 <5> (DL-glyceraldehyde, <5> aldehyde reductase II [18]) [18]
2.95 <5> (sodium D-glucuronate, <5> aldehyde reductase II [18]) [18]
2.99 <1> (D-glucuronic acid, <1> AR3 [15]) [15]
3 <18> (indole-3-acetaldehyde, <18> AR3 [18]) [18]
3 <38> (methylglyoxal) [31]
3 <18> (pyridine 3-aldehyde, <18> AR1 [21]) [21]
3.01 <17> (pentanal, <17> coenzyme NADH [40]) [40]
3.2 <18> (methylglyoxal, <18> AR3 [18]) [18]
3.22 <26> (D-glucuronate, <26> inner medulla enzyme [33]) [33]
3.3 <2> (propionaldehyde) [20]
3.34 <26> (D-glucuronate, <26> cortex enzyme [33]) [33]
3.4 <5> (benzaldehyde) [5]
3.7 <18> (indole-3-acetaldehyde, <18> AR1 [21]) [21]
3.8 <2> (acetaldehyde) [20]
3.9 <18> (*m*-nitrobenzaldehyde, <18> AR1 [21]) [21]
4 <18> (5α-dihydrotestosterone, <18> AR3 [21]) [21]
4 <18> (*m*-nitrobenzaldehyde, <18> AR3 [18]) [18]
4 <5> (*p*-carboxybenzaldehyde, <5> aldehyde reductase, mutant R311A [37]) [37]
4 <18> (pyridine 4-aldehyde, <18> AR1 [21]) [21]
4.2 <5> (D-glucuronate, <5> aldehyde reductase, mutant V299C [36]) [36]
4.2 <5> (D-glucuronate, <5> wild-type aldehyde reductase [36, 37]) [36, 37]
4.25 <17> (heptanal, <17> coenzyme NADH [40]) [40]
4.4 <38> (*m*-nitrobenzaldehyde) [31]
4.5 <2> (D-glucuronate) [12]
4.5 <18> (D-glucuronate, <18> AR1 [21]) [21]
4.5 <4> (D-glyceraldehyde) [9]
4.5 <2> (benzaldehyde) [11]
4.6 <18> (*o*-nitrobenzaldehyde, <18> AR1 [21]) [21]
4.6 <18> (*o*-nitrobenzaldehyde, <18> AR3 [18]) [18]
4.67 <26> (DL-glyceraldehyde, <26> inner medulla enzyme [33]) [33]
4.68 <26> (DL-glyceraldehyde, <26> cortex enzyme [33]) [33]
4.9 <18> (L-glyceraldehyde, <18> AR2 [21]) [21]
5 <1> (3-pyridinecarboxaldehyde, <1> AR4 [15]) [15]
5.2 <18> (pyridine 4-aldehyde, <18> AR3 [18]) [18]
5.5 <4> (3-pyridinemethanol) [2]
5.9 <5> (*p*-nitrobenzaldehyde, <5> aldehyde reductase, mutant H112Q [36]) [36]
5.92 <26> (DL-glyceraldehyde, <26> outer medulla enzyme [33]) [33]
6.1 <56> (3,4-dihydroxyphenylacetaldehyde) [50]
6.42 <2> (D-glucuronate) [29]
6.67 <5> (glyceraldehyde, <5> AR II [24]) [24]

6.7 <38> (*p*-chlorobenzaldehyde) [31]
7 <35> (2-butanol, <35> r-TBADH, wild-type, mutant D150C-TBADH [38]) [38]
7 <53> (acetaldehyde) [43]
7.17 <35> (2-butanol, <35> mutant H59Q-TBADH, [38]) [38]
7.3 <38> (diacetyl) [31]
7.5 <5> (D-glucuronate, <5> aldehyde reductase, mutant R308A [37]) [37]
7.59 <5> (glyceraldehyde, <5> aldehyde reductase II [23]) [23]
7.69 <5> (butyraldehyde, <5> aldehyde reductase II [18]) [18]
8.2 <18> (D-glyceraldehyde, <18> AR3 [21]) [21]
8.2 <53> (butanol) [43]
8.35 <2> (DL-glyceraldehyde) [29]
8.45 <35> (cyclopentanol, <35> r-TBADH, wild-type [38]) [38]
8.61 <5> (3-pyridinecarboxaldehyde, <5> AR II [24]) [24]
8.7 <35> (cyclopentanol, <35> mutant H59Q-TBADH [38]) [38]
9.09 <5> (benzaldehyde, <5> aldehyde reductase II [18]) [18]
9.2 <35> (2-butanol, <35> mutant C37A-TBADH, [38]) [38]
9.9 <18> (glycolaldehyde, <18> AR2 [18]) [18]
10 <28> (butan-1-ol, <28> 2° Adh [34]) [34]
10 <51> (ethanol) [39]
10 <4> (heptanal) [9]
10 <4> (*p*-nitroacetophenone) [9]
10.8 <18> (diacetyl, <18> AR1, AR3 [21]) [21]
10.9 <2> (succinic semialdehyde, <2> liver enzyme [12]) [12]
11.1 <18> (L-glyceraldehyde, <18> AR3 [21]) [21]
11.1 <45> (ethanol) [35]
11.5 <5> (DL-glyceraldehyde, <5> aldehyde reductase, mutant I298C [36]) [36]
12 <56> (3,4-dihydroxyphenylglycolaldehyde) [50]
12 <51> (n-propanol) [39]
12.5 <5> (DL-glyceraldehyde, <5> aldehyde reductase I [18]) [18]
13 <18> (androstane-3,17-dione, <18> AR3 [21]) [21]
14 <4> (butane 2,3-dione) [9]
14.1 <5> (succinic semialdehyde, <5> aldehyde reductase, mutant R311A [37]) [37]
14.28 <5> (propionaldehyde, <5> aldehyde reductase II [18]) [18]
14.54 <5> (propionaldehyde, <5> aldehyde reductase II [23]) [23]
16.4 <28> (ethanol, <28> primary alcohol dehydrogenase [28]) [28]
16.4 <28> (propan-2-ol, <28> secondary alcohol dehydrogenase [28]) [28]
16.71 <35> (cyclopentanol, <35> mutant D150C-TBADH [38]) [38]
17.7 <17> (2-methylbutanal, <17> coenzyme NADH [40]) [40]
18 <5> (succinic semialdehyde, <5> aldehyde reductase, mutant H112Q [36]) [36]
19 <28> (n-propanol, <28> 2° Adh [34]) [34]
19.33 <35> (cyclopentanol, <35> mutant C37S-TBADH [38]) [38]
19.5 <5> (D-glucuronate, <5> aldehyde reductase, mutant I298C [36]) [36]
19.6 <18> (D-glucuronolactone, <18> AR2 [21]) [21]

23 <28> (n-propanol, <28> 1° Adh [34]) [34]
23.1 <17> (butanal, <17> coenzyme NADH [40]) [40]
24 <2> (ethyl acetoacetate, <2> enzyme from brain or liver [20]) [20]
25 <28> (ethanol, <28> 1° Adh [34]) [34]
25 <2> (ethyl acetoacetate, <2> enzyme from kidney [20]) [20]
26 <18> (ethylacetoacetate, <18> AR3 [21]) [21]
26.3 <5> (D-glucuronate, <5> aldehyde reductase, mutant I298A [36]) [36]
27 <56> (5-hydroxyindoleacetaldehyde) [50]
27 <17> (propanal, <17> coenzyme NADH [40]) [40]
29 <2> (chloral hydrate) [11]
33 <5> (DL-glyceraldehyde, <5> aldehyde reductase, mutant I298A [36]) [36]
34 <28> (n-butanol, <28> 1° Adh [34]) [34]
36.77 <1> (cyclohexanone, <1> AR4 [15]) [15]
38.9 <17> (propanal, <17> coenzyme NADPH [40]) [40]
41 <1> (xylose, <1> AR3 [15]) [15]
42 <2> (D-glucuronate, <2> enzyme from brain [20]) [20]
43 <2> (D-glucuronate, <2> enzyme from kidney, or liver [20]) [20]
44.8 <28> (acetaldehyde, <28> secondary alcohol dehydrogenase [28]) [28]
47.1 <28> (n-butanol, <28> primary alcohol dehydrogenase [28]) [28]
48 <5> (D-glucuronate, <5> aldehyde reductase, mutant H112Q [36]) [36]
49 <28> (ethanol, <28> secondary alcohol dehydrogenase [28]) [28]
66 <28> (ethanol, <28> 2° Adh [34]) [34]
67 <18> (D-glucuronolactone, <18> AR1 [21]) [21]
70 <53> (ethanol) [43]
76 <5> (DL-glyceraldehyde, <5> aldehyde reductase, mutant H112Q [36]) [36]
100 <36> (n-propanol) [30]
111 <18> (ribose, <18> AR2 [21]) [21]
150 <4> (n-butyraldehyde) [9]
158 <17> (acetaldehyde) [40]
160 <18> (xylose, <18> AR2 [21]) [21]
170 <53> (2-butanol) [43]
210 <28> (acetaldehyde, <28> primary alcohol dehydrogenase [28]) [28]
215 <5> (D-glucose, <5> AR I [24]) [24]
270 <18> (D-xylose, <18> AR1 [21]) [21]
364 <5> (D-glucuronate, <5> aldehyde reductase, mutant R311A [37]) [37]
500 <5> (D-xylose, <5> aldehyde reductase II [18]) [18]
543 <5> (D-galactose, <5> aldehyde reductase II [18]) [18]
600 <53> (ethanol) [43]
670 <2> (D-xylose) [11]
775 <4> (glycerol) [2]
845 <26> (D-xylose, <26> outer medulla enzyme [33]) [33]
1152 <26> (D-xylose, <26> inner medulla enzyme [33]) [33]
1360 <4> (L-arabitol) [2]
1497 <26> (D-xylose, <26> cortex enzyme [33]) [33]
2080 <2> (D-xylose) [29]

pH-Optimum

5.9 <4> (<4> aldehyde reductase I [9]) [9]
6 <2, 5, 18> (<2> for reduction of *p*-pyridinecarboxaldehyde [20]; <5> aldehyde reductase II [18]; <5> aldehyde reductaseI [24]; <18> AR2, sharp acidic pH [21]) [18, 20, 21]
6-6.5 <18> (<18> AR1 and AR3 [21]) [21]
6-7 <2> [41]
6.2 <4> (<4> aldehyde reductase II [9]) [9]
6.5-9 <28> (<28> $NAD(P)^+$ reduction [34]) [34]
6.6 <5> (<5> reduction of *p*-nitrobenzaldehyde in presence of NADPH [5]) [5]
6.7-6.8 <1> (<1> aldehyde reductase with *p*-nitrobenzaldehyde and NADPH [1]; <5> aldehyde reductase II [18]) [1, 18]
6.8-7 <51> (<51> ethanol as substrate [39]) [39]
7 <1, 4, 5, 28, 33, 38> (<1> aldehyde reductase II [57]; <4> reduction of 3-pyridinecarboxaldehyde [2]; <5> aldehyde reductase II [23, 24]; <28> aldehyde-reduction [34]; <33> n-heptanal reduction [47]) [2, 23, 24, 31, 34, 47, 57]
7-8 <36> (<36> propionaldehyde as substrate [30]) [30]
7.5 <54> (<54> optimum for aldehyde reduction [44]) [44]
8.5 <5> [32]
8.8 <45> (<45> ethanol oxidation [35]) [35]
9 <28> (<28> alcohol-oxidation [34]) [34]
9-9.5 <5, 36> (<5> oxidation of benzyl alcohol in presence of $NADP^+$ [5]; <36> n-propanol as substrate [30]) [5, 29, 30]
9.5 <33> (<33> n-heptanol oxidation [47]) [47]
10 <4> (<4> oxidation of 3-pyridinemethanol [2]) [2]
10.5 <54> (<54> optimum for alcohol oxidation [44]) [44]

pH-Range

5.5-10 <5> [36]
5.8-8 <2> [12]
6-10 <38> [31]
6.2-7.5 <52> [42]

Temperature optimum (°C)

35 <5> (<5> aldehyde reductase II [23]) [23]
60 <38> [31]
80 <28, 45> (<28> secondary alcohol dehydrogenase with 2-propanol as substrate [28]; <45> ethanol oxidation [35]) [28, 35]
85 <28, 51> (<28> primary enzyme, 1-butanol as substrate [28]; <51> ethanol as substrate [39]) [28, 39]

4 Enzyme Structure

Molecular weight

25000 <4> (<4> gel filtration, ultracentrifugation, specific activity corresponding to molecular weight [2, 4, 14]) [2, 4, 14]
25500 <1> (<1> G2, gel filtration [15]) [15]
26300 <1> (<1> gel filtration [15]) [15]
28500 <4> (<4> gel filtration, reductase I [10, 14]) [10, 14]
29000 <4> (<4> gel filtration, high-K_m reductase [3]) [3]
30000 <1, 4, 5, 17, 18> (<5, 17, 18> gel filtration [5, 7]; <1> AR1, gel filtration [15]; <18> AR1 and AR3, gel filtration [21]) [5, 7, 10, 15, 21]
30000-40000 <2> [12]
30200 <4> (<4> gel filtration, sedimentation equilibrium determination [2]; <4> ultracentrifugation [14]) [2-4, 14]
31000 <1> (<1> AR3, gel filtration [15]) [15]
32000 <1, 2, 5, 18, 38> (<1> gel filtration [57]; <2> gel filtration [20]; <5> aldehyde reductase II, gel filtration [18]; <18> AR2, gel filtration [21]; <38> gel filtration [31]) [18, 20, 21, 31, 57]
32500 <5> (<5> gel filtration, Sephadex G-100 [18, 23]) [18, 23]
32700 <5> (<5> aldehyde reductase II, gel filtration [24]) [24]
33000 <2, 4> (<2,4> gel filtration [4, 7, 14]) [4, 7, 10, 14]
33600 <5> (<5> sedimentation analysis [5]) [5]
34000 <4, 19> (<4> liver enzyme, aldehyde reductase I, gel filtration [9]; <19> gel filtration [7]) [7, 9]
35000 <4, 5> (<4> liver enzyme, reductase II, gel filtration [9, 10, 14]) [9, 10, 14, 37]
36000 <4, 52> (<4> gel filtration, low-K_m reductase [3, 14]) [3, 14, 42]
36200 <5> (<5> gel filtration [5,10]) [5, 10]
36300 <5> (<5> ultracentrifugation [5]) [5]
37000 <17> (<17> gel filtration [40]) [40]
37100 <55> (<55> calculated from amino acid sequence [48]) [48]
38000 <5, 55> (<5, 55> SDS-PAGE [48, 58]) [48, 58]
38800 <2> (<2> gel filtration [10]) [10]
39000 <18> (<18> AR1and AR3, SDS-PAGE [21]) [21]
39500 <18> (<18> AR2, SDS-PAGE [21]) [21]
40000 <26> (<26> gel filtration [33]) [33]
40200 <4> (<4> SDS-PAGE [14]) [14]
41000 <2> (<2> SDS-PAGE [29]) [29]
43700 <4> (<4> sedimentation equilibrium [10]; <4> ultracentrifugation [14]) [10, 14]
44000 <5> (<5> gel filtration [5]) [5]
45000 <5> [52]
56000 <1> (<1> AR4, gel filtration [15]) [15]
70000 <56> (<56> gel filtration [50]) [50]
72000 <5> (<5> gel filtration [18]) [18]
74000 <5> (<5> aldehyde reductase I, gel filtration [24]) [24]

75000 <2> [25]
80000 <53> (<53> gel filtration [43]) [43]
80500 <54> (<54> pH-dependent enzyme structure, pH 7.5 [44]) [44]
90000 <5> [52]
96000 <32, 40> [34]
130000 <57> (<57> gel filtration [53]) [53]
158000 <51> (<51> gel filtration [39]) [39]
160000 <28, 33> (<28> 2° Adh, gel filtration [34]; <33> gel filtration [47]) [34, 47]
170000 <28> (<28> 1° Adh, primary-alcohol dehydrogenase, gel filtration [34]) [34]
172000 <28> (<28> secondary alcohol dehydrogenase, gel filtration [28]) [28]
175000 <49> [35]
184000 <28> (<28> primary alcohol dehydrogenase, gel filtration [28]) [28]
197000 <54> (<54> pH-dependent enzyme structure, pH 10.5 [44]) [44]
200000 <45> (<45> gel filtration [35]) [35]

Subunits

dimer <5, 46, 53, 54, 60> (<5> 1 * 35000 + 1 * 42000,aldehyde reductase I, SDS-PAGE [18]; <5> 1 * 32500 + 1 * 39000 [24]; <53> 2 * 44000, SDS-PAGE [43]) [24, 35, 43, 44, 48, 52]
monomer <1, 2, 4-6, 17, 18, 38, 51, 52> (<1> 1 * 33500, AR1, SDS-PAGE [15]; <1> 1 * 34700, AR3, SDS-PAGE [15]; <2> 1 * 34000, SDS-PAGE [7]; <2> 1 * 37000, SDS-PAGE [20]; <4> 1 * 35000 [9]; <4> 1 * 41700, SDS-PAGE [10,14]; <4> 1 * 38000, SDS-PAGE [2,4,14]; <4> 1 * 36900, SDS-PAGE [4,7,14]; <5> 1 * 32500, SDS-PAGE [23]; <5> 1 * 39400, SDS-PAGE [5]; <5> 1 * 45000 [52]; <17> 1 * 33000, SDS-PAGE [7]; <18> 1 * 32600, SDS-PAGE [7]; <18> 1 * 39000, AR1 and AR3, SDS-PAGE [18]; <18> 1 * 39500, AR1, SDS-PAGE [21]; <19> 1 * 38500, SDS-PAGE [7]; <4> 1 * 36700, SDS-PAGE [14]; <4> 1 * 40200, SDS-PAGE [14]; <5> 1 * 33000, aldehyde reductase II, SDS-PAGE [18]; <5> 1 * 32500, SDS-PAGE [24]; <17> 1 * 37000, SDS-PAGE [40]; <38> 1 * 32000, SDS-PAGE [31]; <51> 1 * 36000 [42]) [2, 4, 5, 7-10, 12, 14, 15, 17, 18, 20, 21, 23, 24, 26, 31, 32, 40, 42, 52, 57, 58]
tetramer <28, 33, 35, 42, 45, 49, 51, 54, 57> (<28> 4 * 40000, 2° Adh, secondary-alcohol dehydrogenase, SDS-PAGE [34]; <28> 4 * 42000, 2° Adh, secondary alcohol dehydrogenase, SDS-PAGE [28, 34]; <28> 4 * 44500, primary alcohol dehydrogenase, SDS-PAGE [28]; <33> 4 * 40000, SDS-PAGE [47]; <45> 4 * 48000, SDS-PAGE [35]; <51> 4 * 46700 [39]; <54> 4 * 45000, pH 10.5 [44]; <57> 4 * 30000, SDS-PAGE [53]) [28, 34, 35, 38, 39, 44, 47, 53]

5 Isolation/Preparation/Mutation/Application

Source/tissue

brain <1-5, 18, 21-27, 56> (<1> midbrain, hypothalamus and pons [1]; <56> pons, medulla and midbrain [50]) [1, 3, 5, 9, 11-15, 18-21, 23, 25, 27, 50-52, 54]

epimastigote <36> [30]
erythrocyte <5> [18, 23]
heart <1-5, 18, 21, 23-27> (<1> muscle [20]) [5, 19-21, 41]
intestinal mucosa <2> [1]
kidney <1-5, 21, 23-27> (<2> renal cortex [33]; <3,4> cortex [1, 2, 10, 41]; <26> renal cortex, outer medulla and inner medulla [33]) [1, 2, 4, 5, 7-11, 13-21, 25, 27, 33, 41, 45, 55]
lens <1-5, 21, 23-27> [5, 17, 19, 56, 57]
liver <1-5, 7, 21-27, 53> [2-5, 7, 9-15, 17-23, 25-27, 32, 38, 41, 43, 45, 48]
lung <2-5, 18, 21, 23-27> [19]
muscle <4-6, 18> (<5> skeletal muscle [5]; <18> leg muscle [21]) [5, 14, 21]
placenta <2-5, 21, 23-27> [18, 19, 23, 24]
seminal vesicle <2-5, 21, 23-27> [18, 19]
stomach <53> [43]
testis <2-5, 21, 23-27> [19, 29, 58]
tuber <15> [6]

Purification

<1> [1, 12, 15, 20, 57]
<2> [7, 9-12, 14, 25, 29]
<4> (aldehyde reductase I and II from brain, no recuctase II activity could be detected in pig kidney [9]; aldehyde reductase I [26]) [2-4, 9, 10, 14, 15, 17, 26, 54]
<5> (aldehyde reductase I and II [18, 24]; aldehyde reductase II [23, 56]; wild-type aldehyde reductase and mutant A H112Q [36]) [5, 10, 18, 23, 24, 36, 37, 52, 56]
<7> (reductase I [9]) [9]
<8> (partially [6]) [6]
<17> [7, 40]
<18> [7, 21]
<19> [7]
<21> [18]
<24> (aldehyde reductase I [9]; ALR1 [27]) [9, 27]
<26> [33]
<28> (1° Adh and 2°Adh [28, 34]) [28, 34]
<33> (wild-type and recombinant protein [47]) [47]
<35> [38]
<36> [30]
<37> [30]
<38> [31]
<45> [35]
<50> [35]
<51> [39]
<53> [43]
<54> [44]
<55> [48]
<56> [50]

<57> [53]
<59> (partial [40]) [40, 47]
<60> [48]
<66> [44]

Crystallization

<4> [49]
<5> [37, 38]
<7> [38]
<35> [38, 46]
<38> [31]

Cloning

<32> (1° Adh cloned and characterized [34]) [34]
<33> (alrA gene PCR amplified, cloned and highly expressed in Escherichia coli JM109 [47]) [47]
<35> (gene coding for TBADH cloned and overexpressed in Escherichia coli [38]) [38]
<40> (1° Adh cloned and characterized [34]) [34]
<46> (ADH gene cloned and overexpressed in Escherichia coli [44]) [44]
<51> (gene cloned and sequenced [39]) [39]
<53> (frog stomach ADH cDNA cloned [43]) [43]
<54> (gene encoding NADPH group III alcohol dehydrogenase cloned and overexpressed in Escherichia coli [44]) [44]
<55> (gene encoding ADHC cloned and sequenced [48]) [48]
<60> (gene encoding NADP-dependent alcohol dehydrogenase cloned and sequenced [48]) [48]
<66> (ADH gene cloned and overexpressed in Escherichia coli [44]) [44]

Application

agriculture <52> (<52> detoxification of eutypine toxin from Eutypa lata, the causal agent of Eutypa dieback in the grapevine Vitis vinifera [42]) [42]
medicine <2, 4, 5, 14, 55, 60, 62, 63> (<2> ability of aldehyde reductase from liver to reduce certain cancer chemotherapeutic antibiotics suggested a role for this enzyme in drug metabolism [11]; <4, 5> daunorubicin, a cancer chemotherapeutic agent can be reduced by aldehyde reductase [9, 32]; <5> brain aldehyde reductase activity is closely associated with pharmalogical actions of drugs affecting the central nervous system, anti-convulsant drugs have inhibitory effects, more work is needed for the mechanism of action of psychoactive drugs and interactions of barbiturates with aldehyde reductase [14]; <5> inhibitors alrestatin and sorbinil, possible method of preventing neuropathy and cataractogenesis associated with diabetes [18]; <5> drugs for testing function of liver and kidney [45]; <55> enzyme may be an interesting target for development of new anti-mycobacterial agents [48]; <14, 55, 60, 62, 63> enzyme as a target for anti-tuberculosis and other drugs [48]) [9, 11, 14, 18, 32, 45, 48]
synthesis <17, 28, 29, 33, 38> (<17, 28, 29> industrial ethanol production [34]; <28> potential use for industrial production of ethanol by fermentation,

thermophilic fermentations offer the potential to separate ethanol from continous cultures at process temperature and reduced pressure during growth [28, 34]; <33> reduction of industrially important compounds cinnamyl aldehyde and anisaldehyde, industrial bioconversion of useful alcohols and aldehydes [47]; <38> useful for asymmetric production of L-carnitine [31]) [28, 31, 34, 47]
synthesis <33> (<33> develpoment of conversion processes for petrochemicals and oil-contaminated environments, cinnamyl aldehyde and cinnamyl alcohol used in flavor and perfume industry, anisaldehyde is used for perfume and toilet soaps, decylalcohol is used in the manufacture of plasticizers, a production system for this enzyme may be useful for industrial application as a biocatalyst in the future [47]) [47]

6 Stability

pH-Stability

6-8 <2> [12]
6-10 <38> (<38> stable at 55°C for 10 min [31]) [31]
7 <38> (<38> stable below 60°C for 10 min [31]) [31]
8-9 <33> [47]

Temperature stability

40-50 <2> (<2> at 40°C for 10 min enzyme retains about 90% of its activity towards D-glucuronate, after 10 min at 50% activity decreases to less than 5% [12]) [12]
50 <5, 18> (<5> completely inactivated between 5 min [23]; <18> AR1 and AR3 loses about 30% activity after 10 min, AR2 loses more than 60% [21]) [21, 23]
55 <38> (<38> stable at pH 6.0-10.0 for 10 min [31]) [31]
70 <28, 33, 46> (<28> no loss of activity during incubation for 2 h [28]; <33> retains 90% activity after 40 min [47]; <46> half-life 5 h [35]) [28, 35, 47]
80 <28, 51> (<28> secondary alcohol dehydrogenase with 2-propanol as substrate [28]; <51> half-life of 16 min [39]) [28, 39]
85 <28> (<28> primary enzyme, 1-butanol as substrate [28]) [28]
85 <45> (<45> 2 h, 50% activity lost [35]) [35]
96 <45> (<45> 15 min, 50% activity lost [35]) [35]
98 <35> (<35> half-life 1-2 min [35]) [35]

General stability information

<4>, presence of *p*-mercuribenzoate at pH 7, no loss of activity [4]
<5>, presence of NADPH or $NADP^+$ prevents loss of activity [5]
<59>, unstable [47]

Storage stability

<1>, -20°C AR1 stable for at least 6 months [15]
<1>, -20°C AR3 loses most of its activity [15]

<1>, 0-2°C, enzyme activity retained for up to 3 months if frozen immediately after preparation, stable for several days, freezing and thawing is quite detrimental to retention of enzyme activity [1]
<1>, 4°C AR1 loses all activity after storage for 1 month [15]
<1>, 4°C AR3 stable for at least 1 month in presence of NaN_3 [15]
<1>, 4°C in 0.01% NaN_3 stable for at least 1 month, the gel-filtered enzyme is stable for approximately 2 weeks [15]
<2>, 0°C when stored frozen there is negligible loss of activity after 6 months, but repeated freezing and thawing results in loss of activity [12]
<2>, 4°C loses activity only slowly over several weeks [12]
<4>, -20°C, stable for several months if stored as an ammonium sulfate precipitate [2, 9]
<4>, 0-4°C stable for 2 or 3 weeks in 44 mM sodium phosphate buffer, pH 7 [10]
<4>, 0-4°C stable for 2-3 months as an $(NH_4)_2SO_4$ precipitate [10]
<4>, 20°C reductase I stable for more than 1 month [9]
<4>, 4°C reductase I and II are stable for several months [9]
<4>, 4°C, 6 mg/ml, stable for several months with 10 mM mercaptoethanol [17]
<5>, 0°C, freezing of the purified enzymes results in almost complete loss of activity, purified aldehyde reductase II is stable in 10 mM sodium phosphate buffer, pH 7, for 2-3 month, but appears to lose activity rapidly thereafter, purified aldehyde reductase I retains enzyme activity in the same buffer for about 3 months, after which enzyme activity declines [18]
<5>, 4°C stable for several months but loss of activity with a concomitant enzyme precipitation occurs upon freezing and thawing [5]
<8>, 0°C stable in the frozen state for 3-4 months without any appreciable loss in activity [6]
<18>, -20°C can be stored in 20% glycerol/10 mM phosphate buffer, pH 8, 0.5 mg/ml without any loss of activity for 1 month [21]
<18>, 4°C purified enzymes are stable for at least 2 days, then gradually loses activity during storage [21]
<45>, 23°C, loses all of its activity within 1 week when stored anaerobically [35]
<45>, 4°C, about 1 week, 50% activity lost [35]
<51>, 4°C half-life 6-7 days when stored aerobically [39]

References

[1] Tabakoff, B.; Erwin, G.: Purification and characterization of a reduced nicotinamide adenine dinucleotide phosphate-linked aldehyde reductase from brain. J. Biol. Chem., **245**, 3263-3268 (1970)
[2] Bosron, W.F.; Prairie, R.L.: Triphosphopyridine nucleotide-linked aldehyde reductase I. Purification and properties of the enzyme from pig kidney cortex. J. Biol. Chem., **217**, 4480-4485 (1972)

[3] Turner, A.J.; Tipton, K.F.: The characterization of two reduced nicotinamide-adenine dinucleotide phosphate-linked aldehyde reductases from pig brain. Biochem. J., **130**, 765-772 (1972)

[4] Flynn, T.G.; Shires, J.; Walton, D.J.: Properties of the nicotinamide adenine dinucleotide phosphate-dependent aldehyde reductase from pig kidney. Amino acid composition, reactivity of cysteinyl residues, and stereochemistry of D-glyceraldehyde reduction. J. Biol. Chem., **250**, 2933-2940 (1975)

[5] Wermuth, B.; Muench, J.D.B.; von Wartburg, J.-P.: Purification and properties of NADPH-dependent aldehyde reductase from human liver. J. Biol. Chem., **252**, 3821-3828 (1977)

[6] Hartz, T.K.; Houston, M.R.; Lockwood, L.B.: Substrate specificity and kinetic properties of $NADP^+$-dependent alcohol dehydrogenase of Phycomyces blakesleeanus. Mycologia, **70**, 586-593 (1978)

[7] Davidson, W.S.; Weihrauch, L.; Flynn, T.G.: Purification and compositional relatedness of aldehyde reductase from several species. Biochem. Soc. Trans., **576**, 940-943 (1978)

[8] Davidson, W.S.; Flynn, T.G.: Kinetics and mechanism of action of aldehyde reductase from pig kidney. Biochem. Soc. Trans., **576**, 943-945 (1978)

[9] Branlant, G.M; Biellmann, J.-F.: Purification and some properties of aldehyde reductases from pig liver. Eur. J. Biochem., **105**, 611-621 (1980)

[10] Morpeth, F.F.; Dickinson, F.M.: Some properties of pig kidney-cortex aldehyde reductase. Biochem. J., **191**, 619-626 (1980)

[11] Turner, A.J.; Hryszko, J.: Isolation and characterization of rat liver aldehyde reductase. Biochim. Biophys. Acta, **613**, 256-265 (1980)

[12] Rivett, A.J.; Smith, I.L.; Tipton, K.F.: Purification of the high-K_m aldehyde reductase from rat brain and liver and from ox brain. Biochem. J., **197**, 473-481 (1981)

[13] Sawada, H.; Hara, A.; Nakayama, T.; Hayashibara, M.: Kinetic mechanisms in the reduction of aldehydes and ketones catalyzed by rabbit liver aldehyde reductases and hydroxysteroid dehydrogenases. J. Biochem., **92**, 185-191 (1982)

[14] Flynn, T.G.: Aldehyde reductases: Monomeric NADPH-dependent oxidoreductases with multifunctional potential. Biochem. Pharmacol., **31**, 2705-2712 (1982)

[15] Daly, A.K.; Mantle, T.J.: Purification and characterization of the multiple forms of aldehyde reductase in ox kidney. Biochem. J., **205**, 373-380 (1982)

[16] Daly, A.K.; Mantle, T.J.: The kinetic mechanism of the major form of ox kidney aldehyde reductase with D-glucuronic acid. Biochem. J., **205**, 381-388 (1982)

[17] Branlant, G.: Properties of an aldose reductase from pig lens. Comparative studies of an aldehyde reductase from pig lens. Eur. J. Biochem., **129**, 99-104 (1982)

[18] Petrash, J.M.; Srivastava, S.K.: Purification and properties of human liver aldehyde reductases. Biochim. Biophys. Acta, **707**, 105-114 (1982)

[19] Markus, H.B.; Raducha, M.; Harris, H.: Tissue distribution of mammalian aldose reductase and related enzymes. Biochem. Med., **29**, 31-45 (1983)

[20] Takahashi, N.; Saito, T.; Tomita, K.: Purification and properties of an NADPH-linked aldehyde reductase from rat kidney. Biochim. Biophys. Acta, **748**, 444-452 (1983)

[21] Hara, A.; Deyashiki, Y.; Nakayama, T.; Sawada, H.: Isolation and characterization of multiforms of aldehyde reductase in chicken kidney. Eur. J. Biochem., **133**, 207-214 (1983)

[22] Magnien, A.; Branlant, G.: The kinetics and mechanism of pig-liver aldehyde reductase. Comparative studies with pyridine-3-aldehyde and *p*-carboxybenzaldehyde. Eur. J. Biochem., **131**, 375-381 (1983)

[23] Das, B.; Srivastava, S.K.: Purification and properties of aldose reductase and aldehyde reductase II from human erythrocyte. Arch. Biochem. Biophys., **238**, 670-679 (1985)

[24] Das, B.; Srivastava, S.K.: Purification and properties of aldehyde reductases from human placenta. Biochim. Biophys. Acta, **840**, 324-333 (1985)

[25] Takahashi, N.; Saito, T.; Goda, Y.; Tomita, K.: Characterization of microsomal NADPH-dependent aldehyde reductase from rat brain. J. Biochem., **99**, 513-519 (1986)

[26] Kovar, J.; Plocek, J.: Purification of aldehyde reductase 1 from pig liver. J. Chromatogr., **351**, 371-375 (1986)

[27] De Jongh, K.S.; Schofield, P.J.; Edwards, M.R.: Kinetic mechanism of sheep liver NADPH-dependent aldehyde reductase. Biochem. J., **242**, 143-150 (1987)

[28] Bryant, F.O.; Wiegel, J.; Ljungdahl, L.G.: Purification and properties of primary and secondary alcohol dehydrogenases from Thermoanaerobacter ethanolicus. Appl. Environ. Microbiol., **2**, 460-465 (1988)

[29] Kawasaki, N.; Tanimoto, T.; Tanaka, A.: Characterization of aldose reductase and aldehyde reductase from rat testis. Biochim. Biophys. Acta, **996**, 30-36 (1989)

[30] Arauzo, S.; Cazzulo, J.J.: The NADP-linked aldehyde reductase from Trypanosoma cruzi subcellular localization and some properties. FEMS Microbiol. Lett., **58**, 283-286 (1989)

[31] Yamada, H.; Shimizu, S.; Kataoka, M.; Sakai, H.; Miyoshi, T.: A novel NADPH-dependent aldehyde reductase, catalyzing asymmetric reduction of β-keto acid esters, from Sporobolomyces salmonicolor: purification and characterization. FEMS Microbiol. Lett., **70**, 45-48 (1990)

[32] Bhatnagar, A.; Das, B.; Liu, S.-Q.; Srivastava, S.K.: Human liver aldehyde reductase: pH dependence of steady-state kinetic parameters. Arch. Biochem. Biophys., **287**, 329-336 (1991)

[33] Ohta, M.; Tanimoto, T.; Tanaka, A.: Localization, isolation and properties of three NADPH-dependent aldehyde reducing enzymes from dog kidney. Biochim. Biophys. Acta, **1078**, 395-403 (1991)

[34] Burdette, D.; Zeikus, J.G.: Purification of acetaldehyde dehydrogenase and alcohol dehydrogenases from Thermoanaerobacter ethanolicus 39E and characterization of the secondary-alcohol dehydrogenase (2° Adh) as a bifunctional alcohol dehydrogenase-acetyl-CoA reductive thioesterase. Biochem. J., **302**, 163-170 (1994)

[35] Ma, K.; Robb, F.T.; Adams, M.W.: Purification and characterization of NADP-specific alcohol dehydrogenase and glutamate dehydrogenase from the hyperthermophilic archaeon Thermococcus litoralis. Appl. Environ. Microbiol., **60**, 562-568 (1994)
[36] Barski, O.A.; Gabbay, K.H.; Grimshaw, C.E.; Bohren, K.M.: Mechanism of human aldehyde reductase: characterization of the active site pocket. Biochemistry, **34**, 11264-11275 (1995)
[37] Barski, O.A.; Gabbay, K.H.; Bohren, K.M.: The C-terminal loop of aldehyde reductase determines the substrate and inhibitor specificity. Biochemistry, **35**, 14276-14280 (1996)
[38] Bogin, O.; Peretz, M.; Burstein, Y.: Thermoanaerobacter brockii alcohol dehydrogenase: characterization of the active site metal and its ligand amino acids. Protein Sci., **6**, 450-458 (1997)
[39] Li, D.; Stevenson, K.J.: Purification and sequence analysis of a novel NADP(H)-dependent type III alcohol dehydrogenase from Thermococcus strain AN1. J. Bacteriol., **179**, 4433-4437 (1997)
[40] van Iersel, M.F.; Eppink, M.H.; van Berkel, W.J.; Rombouts, F.M.; Abee, T.: Purification and characterization of a novel NADP-dependent branched-chain alcohol dehydrogenase from Saccharomyces cerevisiae. Appl. Environ. Microbiol., **63**, 4079-4082 (1997)
[41] Udovikova, E.A.; Wojtczak, L.: Mitochondrial aldehyde reductase: identification and characterization in rat liver and kidney cortex. Int. J. Biochem. Cell Biol., **30**, 597-608 (1998)
[42] Colrat, S.; Latche, A.; Guis, M.; Pech, J.C.; Bouzayen, M.; Fallot, J.; Roustan, J.P.: Purification and characterization of a NADPH-dependent aldehyde reductase from mung bean that detoxifies eutypine, a toxin from Eutypa lata1. Plant Physiol., **119**, 621-626 (1999)
[43] Peralba, J.M.; Cederlund, E.; Crosas, B.; Moreno, A.; Julia, P.; Martinez, S.E.; Persson, B.; Farr s, J.; Pares, X.; Jornvall, H.: Structural and enzymatic properties of a gastric NADP(H)-dependent and retinal-active alcohol dehydrogenase. J. Biol. Chem., **274**, 26021-26026 (1999)
[44] Antoine, E.; Rolland, J.L.; Raffin, J.P.; Dietrich, J.: Cloning and over-expression in Escherichia coli of the gene encoding NADPH group III alcohol dehydrogenase from Thermococcus hydrothermalis. Characterization and comparison of the native and the recombinant enzymes. Eur. J. Biochem., **264**, 880-889 (1999)
[45] Shinoda, M.; Mori, S.; Shintani, S.; Ishikura, S.; Hara, A.: Inhibition of human aldehyde reductase by drugs for testing the function of liver and kidney. Biol. Pharm. Bull., **22**, 741-744 (1999)
[46] Li, C.; Heatwole, J.; Soelaiman, S.; Shoham, M.: Crystal structure of a thermophilic alcohol dehydrogenase substrate complex suggests determinants of substrate specificity and thermostability. Proteins, **37**, 619-627 (1999)
[47] Tani, A.; Sakai, Y.; Ishige, T.; Kato, N.: Thermostable NADP(+)-dependent medium-chain alcohol dehydrogenase from Acinetobacter sp. strain M-1: purification and characterization and gene expression in Escherichia coli. Appl. Environ. Microbiol., **66**, 5231-5235 (2000)

[48] Galamba, A.; Soetaert, K.; Buyssens, P.; Monnaie, D.; Jacobs, P.; Content, J.: Molecular and biochemical characterization of Mycobacterium smegmatis alcohol dehydrogenase C. FEMS Microbiol. Lett., **196**, 51-56 (2001)
[49] Ye, Q.; Hyndman, D.; Green, N.; Li, X.; Korithoski, B.; Jia, Z.; Flynn, T.G.: Crystal structure of an aldehyde reductase Y50F mutant-NADP complex and its implications for substrate binding. Proteins, **44**, 12-19 (2001)
[50] Bronaugh, R.L.; Erwin, V.: Partial purification and characterization of NADPH-linked aldehyde reductase from monkey brain. J. Neurochem., **21**, 809-815 (1973)
[51] Reyes, E.; Erwin, V.: Distribution and properties of NADPH-linked aldehyde reductases from rat brain synaptosomes. Neurochem. Res., **2**, 87-97 (1977)
[52] Cash, C.; Maitre, M.; Mandel, P.: Purification from human brain and some properties of two NADPH-linked aldehyde reductases which reduce succinic semialdehyde to 4-hydroxybutyrate. J. Neurochem., **33**, 1169-1175 (1979)
[53] Lo, H.S.; Chang, C.J.: Purification and properties of NADP-linked, alcohol dehydrogenase from Entamoeba histolytica. J. Parasitol., **68**, 372-377 (1982)
[54] Cromlish, J.A.; Yoshimoto, C.K.; Flynn, T.G.: Purification and characterization of four NADPH-dependent aldehyde reductases from pig brain. J. Neurochem., **44**, 1477-1484 (1985)
[55] Worrall, D.M.; Daly, A.K.; Mantle, T.J.: Kinetic studies on the major form of aldehyde reductase in ox kidney: a general kinetic mechanism to explain substrate-dependent mechanisms and the inhibition by anticonsulvants. J. Enzyme Inhib., **1**, 163-168 (1986)
[56] Das, B.; Song, H.P.; Ansari, N.H.; Hair, G.A.; Srivastava, S.K.: Purification and properties of aldose reductase and aldehyde reductase II from human lens. Lens Res., **4**, 309-335 (1987)
[57] Das, B.; Hair, G.A.; Ansari, N.H.; Srivastava, S.K.: Isolation and characterization of aldose reductase and aldehyde reductase II from bovine lens. Lens Res., **5**, 233-247 (1988)
[58] Tanimoto, T.; Ohta, M.; Tanaka, A.; Ikemoto, I.; Machida, T.: Purification and characterization of human testis aldose and aldehyde reductase. Int. J. Biochem., **23**, 421-428 (1991)

Homoserine dehydrogenase 1.1.1.3

1 Nomenclature

EC number
1.1.1.3

Systematic name
L-homoserine:NAD(P)$^+$ oxidoreductase

Recommended name
homoserine dehydrogenase

Synonyms
HDH [Swissprot]
HSD
HSDH

CAS registry number
9028-13-1

2 Source Organism

<1> *Pisum sativum* (pea [1, 7, 10]) [1, 7, 10]
<2> *Hordeum vulgare* (barley [1]) [1]
<3> *Zea mays* (corn [2-5]) [2-5]
<4> *Ricinus communis* (castor bean [7]) [7]
<5> *Rhodospirillum rubrum* [8]
<6> *Triticum aestivum* (wheat [7]) [7]
<7> *Thermophilic bacterium* [11]
<8> *Thermus flavus* [12]
<9> *Corynebacterium glutamicum* [13]
<10> *Escherichia coli* (also shows aspartase kinase activity [14]) [6, 9, 17, 19]
<11> *Spinacia oleracea* (spinach [14]; also shows aspartase kinase activity [14]) [14]
<12> *Saccharomyces cerevisiae* [15, 16, 18, 20, 21]
<13> *Serratia marcescens* (also shows aspartase kinase activity [22]) [22]

3 Reaction and Specificity

Catalyzed reaction
L-homoserine + NAD(P)$^+$ = L-aspartate 4-semialdehyde + NAD(P)H

Reaction type

oxidation
redox reaction
reduction

Natural substrates and products

S L-aspartate 4-semialdehyde + NAD(P)H <1, 2, 4, 6, 10, 12> (<1, 2> third reaction in the pathway between aspartate and the amino acids threonine, isoleucine, methionine [1]; <10, 12> kinetic mechanism [6, 20]) (Reversibility: <1, 2, 4, 6, 10, 12> r [1, 6, 7, 20]) [1, 6, 7, 20]
P L-homoserine + $NAD(P)^+$

Substrates and products

S L-aspartate 4-semialdehyde + NAD(P)H <1, 2, 4, 7, 8, 10-13> (Reversibility: r <1, 2, 4, 7, 8, 10-13> [1, 6, 7, 9-12, 14, 17, 20, 22]) [1, 6, 7, 9-12, 14, 15, 17, 20-22]
P L-homoserine + $NAD(P)^+$ <1, 2, 4, 7, 8, 10-13> [1, 6, 7, 9-12, 14, 15, 17, 20-22]
S L-homoserine + $NAD(P)^+$ <1, 2, 4, 6 13> (Reversibility: r <1, 2, 4, 6-8, 10-13> [1, 6, 7, 9-14, 17, 20-22]) [1, 6, 7, 9-14, 20-22]
P L-aspartate 4-semialdehyde + NAD(P)H <7, 9, 10, 12, 13> [6, 11, 13, 17, 19, 20-22]

Inhibitors

(S)-2-amino-4-oxo-5-hydroxypentanoic acid <12> (<12> RI-331 [16,20]) [16, 20]
D-threonine <7> (<7> slight [11]) [11]
DL-allo-threonine <7> [11]
EGTA <11> [14]
H-(1,2,4-triazol-3-yl)-DL-alanine <12> [20]
L-cysteine <1, 2, 7, 9-11> (<1, 2> slight inhibition of chloroplast isozyme, strong inhibition of cytoplasmic isozyme [1]; <7> slight [11]) [1, 10-14, 17]
L-serine <10> [17]
L-threonine <1-4, 6, 7, 9-13> (<1, 4, 6> sensitive and insensitive isozymes [7, 10]; <1, 4, 6> degree of inhibition depends on age of plant [7]; <8> not inhibitory [12]; <12> weakly inhibits reverse but not forward reaction [20]) [1, 3-7, 10, 11, 13, 14, 17, 18, 20, 22]
$NADP^+$ <10> [19]
Tris <10> [9]
aspartate <11> [14]
homoserine <10> [19]
lysine <11> [14]
methionine <1, 11, 12> (<12> weakly inhibits reverse but not forward reaction [20]) [10, 14, 18, 20]
p-chloromercuribenzoate <5> [8]
trifluoperazine <11> [14]

Cofactors/prosthetic groups

NAD^+ <7> (<7> low activity [11]) [11]

NADH <1, 2, 4, 6, 12> (<1, 4, 6> threonine sensitive isozyme can use NADPH or NADH, threonine insensitive isozyme can use NADPH only [7]) [1, 7, 15]
$NADP^+$ <1, 7, 10> (<7> preferred [11]) [6, 10, 11]
NADPH <1-4, 6, 10, 12> (<1,2> preferred [1]; <1, 4, 6> threonine sensitive isozyme can use NADPH or NADH, threonine insensitive isozyme can use NADPH only [7]) [1, 4, 6, 7, 10, 15]

Activating compounds

calmodulin <11> (<11> in the presence of above 0.1 mM Ca^{2+} [14]) [14]

Metals, ions

Cs^+ <1, 4, 6> (<1, 4, 6> activation of threonine sensitive enzyme [7]) [7]
K^+ <1, 4, 6-8> (<7> required [11]; <1, 4, 6> activation of threonine sensitive isozyme [7]; <8> activation [12]) [7, 11, 12]
Li^+ <1, 4, 6, 7> (<1, 4, 6> activation of threonine sensitive isozyme [7]; <7> can partially replace K^+ [11]) [7, 11]
NH_4^+ <1, 4, 6, 7> (<1, 4, 6> activation of threonine sensitive isozyme [7]; <7> can partially replace K^+ [11]) [7, 11]
Na^+ <1, 4, 6-8> (<1, 4, 6> activation of threonine sensitive form [7]; <7> can partially replace K^+ [11]; <8> activation [12]) [7, 11, 12]
Rb^+ <7> (<7> can partially replace K^+ [11]) [11]

Specific activity (U/mg)

0.17 <6> [7]
2.8 <1> [7]
6.8 <4> [7]
23.7 <12> [21]
51 <12> [18]
343 <3> [4]
360 <3> [4]
Additional information <9> [13]

K_m-Value (mM)

0.013 <10> (L-homoserine) [6]
0.027 <1> (NADPH, <1> isozyme II [1]) [1]
0.031 <6> (NADPH, <6> threonine sensitive isozyme [7]) [7]
0.032-0.036 <3, 6> (NADPH, <3, 6> threonine resistant isozyme [4, 7]) [4, 7]
0.04 <3> (NADPH, <3> threonine sensitive isozyme [4]) [4]
0.04 <1> (aspartate 4-semialdehyde, <1> isozyme II [1]) [1]
0.043 <6> (NADPH, <6> threonine sensitive isozyme [7]) [7]
0.046 <1> (NADPH, <1> isozyme I [1]) [1]
0.066 <1> (L-aspartate 4-semialdehyde, <1> isozyme I [1]) [1]
0.073 <10> ($NADP^+$) [6]
0.08 <1> (L-aspartate 4-semialdehyde, <1> threonine insensitive isozyme [10]) [10]
0.09 <10> (NADPH) [6]
0.1 <11> (DL-aspartate 4-semialdehyde) [14]
0.1 <2> (L-aspartate 4-semialdehyde, <2> isozyme I [1]) [1]

0.13-0.15 <3, 6> (L-aspartate 4-semialdehyde, <3, 6> threonine resistant isozyme [4, 7]) [4, 7]
0.17 <10> (L-aspartate 4-semialdehyde) [6]
0.17-0.18 <7, 8> ($NADP^+$) [11, 12]
0.24-0.25 <3, 6> (L-aspartate 4-semialdehyde, <3, 6> threonine sensitive isozyme [4, 7]) [4, 7]
0.3 <11> (NADH) [14]
0.36-0.4 <2> (L-aspartate 4-semialdehyde, <2> isozyme II [1]) [1]
0.46 <1> (NADH, <1> isozyme II [1]) [1]
0.5 <11> (L-aspartate 4-semialdehyde) [14]
0.5 <1> (L-aspartate 4-semialdehyde, <1> threonine sensitive isozyme [10]) [10]

pH-Optimum
9 <8> [12]
9.8 <7> [11]

pH-Range
8-10 <7> [11]

Temperature optimum (°C)
70 <7> [11]
Additional information <8> (<8> above 50°C temperature dependent conformational change [12]) [12]

Temperature range (°C)
20-70 <7> (<7> no activity at 80°C [11]) [11]

4 Enzyme Structure

Molecular weight
34000 <1> (<1> gel filtration, isozyme II, cytoplasm [1]) [1]
69000 <2> (<2> gel filtration, isozyme II, cytoplasm [1]) [1]
70000 <3> (<3> gel filtration, threonine insensitive isozyme [4]) [4]
75000 <1, 6> (<1, 6> gel filtration, threonine insensitive isozyme [7]) [7]
80000 <4> (<4> disc gel electrophoresis, threonine insensitive isozyme [7]) [7]
81000 <12> (<12> gel filtration [18]) [18]
110000 <5> (<5> gel filtration, sedimentation equilibrium centrifugation [8]) [8]
168000-174000 <1, 2> (<1, 2> gel filtration, isozyme I, chloroplasts [1]) [1]
190000 <3> (<3> gel filtration threonine sensitive isozyme [4]) [4]
220000 <5, 11> (<5> gel filtration, aggregated form [8]; <11> gel filtration [14]) [8, 14]
280000 <4> (<4> disc gel electrophoresis, threonine sensitive isozyme [7]) [7]
290000 <6> (<6> gel filtration, threonine sensitive isozyme [7]) [7]

Subunits

? <3> (<3> x * 85000, SDS-PAGE, threonine sensitive isozyme [2]) [2]
dimer <3, 5, 12> (<3> 2 * 38000, SDS-PAGE, threonine resistant isozyme [4]; <3> x * 89000 + x * 93000, SDS-PAGE, threonine sensitive isozyme [4]; <5> 2 * 55000, SDS-PAGE [8]; <12> crystal structure [15]; <12> 2 * 40000, SDS-PAGE [18]) [4, 8, 15, 18]
tetramer <3, 11> (<3> x * 89000 + x * 93000, SDS-PAGE, threonine sensitive isozyme [4]; <11> 4 * 55000, SDS-PAGE [14]) [4, 14]

5 Isolation/Preparation/Mutation/Application

Source/tissue

cell suspension culture <3> [4]
leaf <1, 2, 11> [1, 14]
seedling <1, 3, 4, 6> [2, 3, 5-7, 10]

Localization

chloroplast <1, 2> (<1, 2> isozyme I [1]) [1]
cytoplasm <1, 2> (<1, 2> isozyme II [1]) [1]

Purification

<1> (separation of the two isozymes using Blue-Sepharose chromatography [1]) [1, 7, 10]
<2> (separation of the two isozymes using Matrex Gel Red A affinity chromatography [1]) [1]
<3> (threonine sensitive and threonine insensitive isozymes [4]) [2-4]
<4> [7]
<5> [8]
<6> [7]
<7> [11]
<8> (partial [12]) [12]
<10> [17]
<11> [14]
<12> (overexpressed in Escherichia coli [21]) [18, 21]

Renaturation

<12> (denatures in 3 M guanidine-HCl and refolds by simple dilution [18]) [18]

Crystallization

<12> [15]

Cloning

<12> (overexpression in Escherichia coli [21]) [21]

Engineering

G378E <9> (<9> feedback resistance of the enzyme [13]) [13]
H309A <12> (<12> decrease of catalytic activity and elimination of substrate inhibition [21]) [21]

6 Stability

Temperature stability

70 <8> (<8> up to [12]) [12]
80 <8> (<8> slow inactivation, protection by K^+, Na^+ [12]) [12]
90 <8> (<8> rapid inactivation [12]) [12]

Storage stability

<1>, -25°C, 50 mM potassium phosphate buffer, pH 7.2, 15% v/v glycerol, 1 mM EDTA, 1 mM threonine, 14 mM 2-mercaptoethanol [10]
<3>, -20°C, 0.05 M potassium phosphate buffer, pH 7.5, 1.0 mM EDTA, 2.0 mM dithioerythritol, 5.0 mM L-threonine, 20% v/v glycerol, at least 2 years [3]
<3>, -20°C, 30 mM potassium phosphate buffer, pH 7.5, 50% v/v glycerol, threonine resistant isozyme at least 2 months stable, threonine sensitive isozyme at least 4 months stable [4]

References

[1] Sainis, J.K.; Mayne, R.G.; Wallsgrove, R.M.; Lea, P.J.; Miflin, B.J.: Localisation and characterisation of homoserine dehydrogenase isolated from barley and pea leaves. Planta, **152**, 491-496 (1981)
[2] Krishnaswamy, S.; Bryan, J.K.: Use of monoclonal antibodies for the purification and characterization of the threonine-sensitive isozyme of maize homoserine dehydrogenase. Arch. Biochem. Biophys., **246**, 250-262 (1986)
[3] Krishnaswamy, S.; Bryan, J.K.: Ligand-induced interconversions of maize homoserine dehydrogenase among different states. Arch. Biochem. Biophys., **222**, 449-463 (1983)
[4] Walter, T.J.; Connelly, J.A.; Gengenbach, B.G.; Wold, F.: Isolation and characterization of two homoserine dehydrogenases from maize suspension cultures. J. Biol. Chem., **254**, 1349-1355 (1979)
[5] Di Camelli, C.A.; Bryan, J.K.: Comparison of sensitive and desensitized forms of maize homoserine dehydrogenase. Plant Physiol., **65**, 176-183 (1980)
[6] Wedler, F.C.; Ley, B.W.: Kinetic and regulatory mechanisms for (Escherichia coli) homoserine dehydrogenase-I. J. Biol. Chem., **268**, 4880-4888 (1993)
[7] Grego, S.; Tricoli, D.; Di Marco, G.: Comparison of homoserine dehydrogenase from different plant sources. Phytochemistry, **19**, 1619-1623 (1980)
[8] Epstein, C.C.; Datta, P.: Homoserine dehydrogenase of Rhodospirillum rubrum. Physical and chemical characterization. Eur. J. Biochem., **82**, 453-461 (1978)
[9] Ogilvie, J.W.; Whitaker, S.C.: Reaction of Tris with aldehydes. Effect of Tris on reactions catalyzed by homoserine dehydrogenase and glyceraldehyde-3-phosphate dehydrogenase. Biochim. Biophys. Acta, **445**, 525-536 (1976)

[10] Aarnes, H.; Rognes, S.E.: Threonine-sensitive aspartate kinase and homoserine dehydrogenase from Pisum sativum. Phytochemistry, **13**, 2717-2724 (1974)
[11] Cavari, B.Z.; Grossowicz: Properties of homoserine dehydrogenase in a thermophilic bacterium. Biochim. Biophys. Acta, **302**, 183-190 (1973)
[12] Saiki, T.; Shinshi, H.; Arima, K.: Studies on homoserine dehydrogenase from an extreme thermophile, Thermus flavus AT-62. Partial purification and properties. J. Biochem., **74**, 1239-1248 (1973)
[13] Morbach, S.; Kelle, R.; Winkels, S.; Sahm, H.; Eggeling, L.: Engineering the homoserine dehydrogenase and threonine dehydratase control points to analyse flux towards L-isoleucine in Corynebacterium glutamicum. Appl. Microbiol. Biotechnol., **45**, 612-620 (1996)
[14] Pavagi, S.; Kochhar, S.; Kochhar, V.K.; Sane, P.V.: Purification and characterization of homoserine dehydrogenase from spinach leaves. Biochem. Mol. Biol. Int., **36**, 649-658 (1995)
[15] DeLaBarre, B.; Thompson, P.R.; Wright, G.D.; Berghuis, A.M.: Crystal structures of homoserine dehydrogenase suggest a novel catalytic mechanism for oxidoreductases. Nature Struct. Biol., **7**, 238-244 (2000)
[16] Yamaki, H.; Yamaguchi, M.; Imamura, H.; Suzuki, H.; Nishimura, T.; Saito, H.; Yamaguchi, H.: The mechanism of antifungal action of (S)-2-amino-4-oxo-5-hydroxypentanoic acid, RI-331: The inhibition of homoserine dehydrogenase in Saccharomyces cerevisiae. Biochem. Biophys. Res. Commun., **168**, 837-843 (1990)
[17] Hama, H.; Kayahara, T.; Tsuda, M.; Tsuchiya, T.: Inhibition of homoserine dehydrogenase I by L-serine in Escherichia coli. J. Biochem., **109**, 604-608 (1991)
[18] Yumoto, N.; Kawata, Y.; Noda, S.; Tokushige, M.: Rapid purification and characterization of homoserine dehydrogenase from Saccharomyces cerevisiae. Arch. Biochem. Biophys., **285**, 270-275 (1991)
[19] Angeles, T.S.; Viola, R.E.: The kinetic mechanism of the bifunctional enzyme aspartokinase - Homoserine dehydrogenase I from Escherichia coli. Arch. Biochem. Biophys., **283**, 96-101 (1990)
[20] Jacques, S.L.; Ejim, L.J.; Wright, G.D.: Homoserine dehydrogenase from Saccharomyces cerevisiae: kinetic mechanism and stereochemistry of hydride transfer. Biochim. Biophys. Acta, **1544**, 42-54 (2001)
[21] Jacques, S.L.; Nieman, C.; Barcich, D.; Broadhead, G.; Kinach, R.; Honek, J.F.; Wright, G.D.: Characterization of yeast homoserine dehydrogenase, an antifungal target: the invariant histidine 309 is important for enzyme integrity. Biochim. Biophys. Acta, **1544**, 28-41 (2001)
[22] Omori, K.; Komatsubara, S.: Role of serine 352 in the allosteric response of Serratia marcescens aspartokinase I-homoserine dehydrogenase I analyzed by using site-directed mutagenesis. J. Bacteriol., **175**, 959-965 (1993)

(R,R)-Butanediol dehydrogenase 1.1.1.4

1 Nomenclature

EC number
1.1.1.4

Systematic name
(R,R)-butane-2,3-diol:NAD^+ oxidoreductase

Recommended name
(R,R)-butanediol dehydrogenase

Synonyms
(R)-2,3-butanediol dehydrogenase <8> [7]
(R)-diacetyl reductase
1-amino-2-propanol dehydrogenase
1-amino-2-propanol oxidoreductase
2,3-butanediol dehydrogenase
D-(-)-butanediol dehydrogenase
D-1-amino-2-propanol dehydrogenase
D-1-amino-2-propanol:NAD^+ oxidoreductase (enzyme from E. coli is identical with glycerol dehydrogenase)
D-aminopropanol dehydrogenase
EC 1.1.1.74 (formerly)
L-2,3-butanediol dehydrogenase <5> (<5> L-BDH [2]) [2]
aminopropanol oxidoreductase
butylene glycol dehydrogenase
butyleneglycol dehydrogenase
dehydrogenase, D-aminopropanol
dehydrogenase, D-butanediol
diacetyl reductase (acetoin)
meso-2,3-butanediol dehydrogenase <1> [2, 5]

CAS registry number
37250-09-2

2 Source Organism

<1> *Klebsiella pneumonia* [2, 5]
<2> *Paracoccus denitrificans* [3]
<3> *Acetobacter suboxydans* [4]
<4> *Bacillus polymyxa* [4]

<5> *Brevibacterium saccharolyticum* [2]
<6> *Aeromonas hydrophila* [4]
<7> *Bacillus subtilis* [4]
<8> *Saccharomyces cerevisiae* [6, 7, 10]
<9> *Lactococcus lactis* (subsp. lactis D10 [8]; 2,3-butanediol dehydrogenase 1 and 2 [8]) [8]
<10> *Enterobacter aerogenes* (authors suggest systematic name: -glycol:NAD^+ oxidoreductase [9]) [9]
<11> *Micrococcus sp.* [1]
<12> *Bacillus sp.* [1]
<13> *Pseudomonas sp.* [1]
<14> *Enterobacter sp.* [1]
<15> *Brevibacillus sp.* [1]

3 Reaction and Specificity

Catalyzed reaction
(R,R)-butane-2,3-diol + NAD^+ = (R)-acetoin + NADH

Reaction type
oxidation
redox reaction
reduction

Substrates and products
S (2R,3R)-2,3-butanediol + NAD^+ <8> (Reversibility: r <8> [6, 10]) [6, 10]
P (3R)-acetoin + NADH <8> [6, 10]
S (3R,3S)-acetoin + NADH <8> (Reversibility: r <8> [6, 10]) [6, 10]
P (2R,3R)-2,3-butanediol + *meso*-2,3-butanediol + NAD^+ <8> [6, 10]
S (R,S)-3-hydroxy-2-pentanone + NADH <8> (Reversibility: r <8> [7]) [7]
P 2,3-pentanediol + NAD^+ <8> [7]
S (R,S)-4-hydroxy-3-pentanone + NADH <8> (Reversibility: r <8> [7]) [7]
P 2,3-pentanediol + NAD^+ <8> [7]
S 1,2-butanediol + NAD^+ <8> (Reversibility: ? <8> [7]) [7]
P ?
S 1,2-propanediol + NAD^+ <8> (Reversibility: ? <8> [7]) [7]
P ?
S 1-hydroxy-2-butanone + NADH <8> (Reversibility: ? <8> [7]) [7]
P ?
S 1-hydroxy-2-propanone <8, 8> (Reversibility: r <8> [6]) [6]
P ?
S 2,3-butanediol + NAD^+ <2, 3, 4, 6-15> (<3, 4, 6, 7, 11-15> D(-)-2,3-butandiol [1, 4]; <2, 8> 2,3-butanediol without specification of stereochemistry [3, 7]; <8> oxidation occurs selectively at the (R)-center of 2,3-butanediol [7]) (Reversibility: r <3, 4, 6-10> [4, 7-9]) [1, 3, 4, 7-9]
P acetoin + NADH <3, 4, 6-10> [4, 7-9]

S 2,3-pentanediol + NAD^+ <8> (<8> racemate [7]) (Reversibility: r <8> [7]) [7, 10]
P 4-hydroxy-3-pentanone + 3-hydroxy-2-pentanone + NADH <8> [7]
S L(+)-2,3-butanediol + NAD^+ <5> (<5> L-BDH [2]) (Reversibility: ? <5> [2]) [2]
P L(+)-acetoin + NADH <5> [2]
S L-acetoin + NADH <1> (Reversibility: ? <1> [5]) [5]
P L-2,3-butanediol + NAD^+ <1> [5]
S acetoin + NADH <3, 4, 6-10> (Reversibility: r <3, 4, 6-10> [4, 7-9]) [4, 7-9]
P 2,3-butanediol + NAD^+ <3, 4, 6, 7, 9> (<9> enzyme 1, only *meso*-2,3-butanediol is formed, enzyme 2 gives mixture of *meso*- and optical isomers [8]) [4, 8]
S diacetyl + NADH <8-10> (Reversibility: r <9> [8]) [7-9]
P 2,3-butanediol + NAD^+ <8, 9> (<9> enzyme 1, only *meso*-2,3-butanediol is formed, enzyme 2 gives mixture of *meso*- and optical isomers [8]) [7, 8]
S ethyl pyruvate + NADH <10> (Reversibility: ? <10> [9]) [9]
P ?
S hydroxyacetone + NADH <8> (Reversibility: ? <8> [7]) [7]
P 1,2-propanediol <8> [7]
S *meso*-2,3-butanediol + NAD^+ <1, 8, 11-15> (<1> *meso*-2,3-butanediol dehydrogenase [2]) (Reversibility: r <8> [10]) [1, 2, 10]
P D-(-)-acetoin + NADH <1> (<8> (3S)-acetoin [10]) [2, 10]
S methyl glyoxal + NADH <10> (Reversibility: ? <10> [9]) [9]
P ?
S methyl pyruvate + NADH <10> (Reversibility: ? <10> [9]) [9]
P ?
S Additional information <8> (<8> (2S,3S)-2,3-butanediol, (3R/3S)-acetoin, glycerol, sorbitol or xylitol are no substrates [10]) [10]
P ?

Inhibitors

2-mercaptoethanol <1, 5> [2]
4-methylpyrazole <8> [6]
EDTA <8> [7]
borate buffer <4> [4]
dipicolinate <8> [7]
o-phenanthroline <8> [7]

Cofactors/prosthetic groups

NAD^+ <1-7, 8, 10-15> (<8, 10> no activity with $NADP^+$ as cofactor [6, 9, 10]) [1-4, 6, 9, 10]
NADH <3, 4, 6, 7> [4]

Metals, ions

Mg^{2+} <8> (<8> activation [7]) [7]
Mn^{2+} <8> (<8> activation [7]) [7]
Additional information <3, 4, 6, 7> (<3, 4, 6, 7> no requirement [4]) [4]

Specific activity (U/mg)

0.21 <4> [4]
0.56 <2> [3]
0.6 <3> [4]
5.1 <9> (acetoin, enzyme 1 [8]) [8]
7.55 <9> (acetoin, enzyme 2 [8]) [8]
35.55 <9> (diacetyl, enzyme 2 [8]) [8]
35.6 <9> (diacetyl, enzyme 1 [8]) [8]
270 <10> [9]
968 <8> [10]

K_m-Value (mM)

0.055 <8> (NADH) [6]
0.1 <4> (NADH) [4]
0.16 <10> (NAD^+) [9]
0.2 <4> (NAD^+) [4]
0.4 <10> (acetoin) [9]
0.53 <4> (acetoin) [4]
0.55 <8> (NAD^+) [6]
1.6 <10> (diacetyl) [9]
2.6 <10> (2,3-butanediol) [9]
3.3 <4> (D-(-)2,3-butanediol) [4]
4.5 <8> ((3R,3S)-acetoin) [6]
6 <10> (2,3-pentanediol) [9]
6.25 <4> (*meso*-2,3-butanediol) [4]
14 <8> ((2R,3R)-2,3-butanediol) [6]
14.5 <8> (1-hydroxy-2-propanone) [6]
18 <10> (methyl pyruvate) [9]
20 <10> (ethyl pyruvate) [9]
38 <4> (1,2-propanediol) [4]
75 <10> (methyl glyoxal) [9]
87 <4> (diacetyl) [4]

pH-Optimum

5.8 <9> (<9> substrate reduction, enzyme 1 [8]) [8]
6.1 <9> (<9> substrate reduction, enzyme 2 [8]) [8]
6.7 <8> (<8> substrate reduction [7]) [7]
7 <3, 8> (<3,8> acetoin reduction [4,10]) [4, 10]
7-8 <8> [6]
7.2 <8> (<8> substrate oxidation [7]) [7]
8 <8> (<8> substrate oxidation [10]) [10]
8.5 <1, 9> (<9> substrate oxidation, enzyme 2 [8]) [2, 8]
9 <3> (<3> D(-)-2,3-butanediol oxidation [4]) [4]
10 <9, 10> (<9> substrate oxidation, enzyme 1 [8]) [8, 9]

pH-Range

5-9 <8> [6]

Temperature optimum (°C)

32 <1> [2]
60-62 <3, 4> [4]

4 Enzyme Structure

Molecular weight

61000 <10> (<10> gel filtration [9]) [9]
81800 <8> (<8> gel filtration [10]) [10]
100000-107000 <4> (<4> gel filtration [4]) [4]
140000 <8> (<8> gel filtration [7]) [7]
170000 <9> (<9> enzyme 1 and 2, gel filtration [8]) [8]

Subunits

dimer <10> (<10> 2 * 28000, SDS-PAGE [9]) [9]
dimer <8> (<8> 2 * 41000, SDS-PAGE [10]) [10]
tetramer <8> (<8> 4 * 35000, SDS-PAGE [7]) [7]

5 Isolation/Preparation/Mutation/Application

Purification

<4> [4]
<8> [6, 10]
<9> (partially; 2,3-butanediol dehydrogenase 1 and 2 [8]) [8]
<10> [9]

Crystallization

<1> [2]
<5> [2]
<8> [7]

Cloning

<1> (expression in Escherichia coli [5]) [5]
<8> (overexpression in Saccharomyces cerevisiae [6, 10]) [6, 10]

6 Stability

pH-Stability

7-7.6 <4> [4]

Temperature stability

60-65 <4> [4]

Storage stability

<2>, very sensitive to storage, even at -20°C [3]
<4>, stable at -18°C [4]

References

[1] Sadaharu, U.; Masuda, H.; Miroyuhi, M.: Stereospecific and electrophoretic natures of bacterial 2,3-butanediol dehydrogenases. J. Ferment. Technol., **61**, 467-471 (1983)

[2] Otagiri, M.; Kurisu, G.; Ui, S.; Takusagawa, Y.; Ohkuma, M.; Kudo, T.; Kusunoki, M.: Crystal structure of meso-2,3-butanediol dehydrogenase in a complex with NAD^+ and inhibitor mercaptoethanol at 1.7 A resolution for understanding of chiral substrate recognition mechanisms. J. Biochem., **129**, 205-208 (2001)

[3] Nokhal, T.-H.; Schlegel, H.-G.: Fermentation enzymes in strains of Paracoccus denitrificans. Arch. Microbiol., **145**, 197-201 (1986)

[4] Höhn-Bentz, H.; Radler, F.: Bacterial 2,3-butanediol dehydrogenases. Arch. Microbiol., **116**, 197-203 (1978)

[5] Ui, S.; Mimura, A.; Ohkuma, M.; Kudo, T.: Formation of a chiral acetoinic compound from diacetyl by Escherichia coli expressing meso-2,3-butanediol dehydrogenase. Lett. Appl. Microbiol., **28**, 457-460 (1999)

[6] Gonzalez, E.; Fernandez, M.R.; Larroy, C.; Pares, X.; Biosca, J.A.: Characterization and functional role of Saccharomyces cerevisiae 2,3-butanediol dehydrogenase. Chem.-Biol. Interact., **130**, 425-434 (2001)

[7] Heidlas, J.; Tressl, R.: Purification and characterization of a (R)-2,3-butanediol dehydrogenase from Saccharomyces cerevisiae. Arch. Microbiol., **154**, 267-273 (1990)

[8] Crow, V.L.: Properties of 2,3-butanediol dehydrogenase from Lactococcus lactis subsp. lactis in relation to citrate fermentation. Appl. Environ. Microbiol., **56**, 1656-1665 (1990)

[9] Carballo, J.; Martin, R.; Bernardo, A.; Gonzalez, J.: Purification, characterization and some properties of diacetyl(acetoin) reductase from Enterobacter aerogenes. Eur. J. Biochem., **198**, 327-332 (1991)

[10] Gonzalez, E.; Fernandez, M.R.; Larroy, C.; Sola, L.; Pericas, M.A.; Pares, X.; Biosca, J.A.: Characterization of a (2R,3R)-2,3-butaendiol dehydrogenase as the Saccharomyces cerevisiae YAL060W gene product. J. Biol. Chem., **275**, 35876-35885 (2000)

Acetoin dehydrogenase 1.1.1.5

1 Nomenclature

EC number
1.1.1.5

Systematic name
acetoin:NAD^+ oxidoreductase

Recommended name
acetoin dehydrogenase

Synonyms
Acetoin dehydrogenase [Swissprot]
diacetyl reductase

CAS registry number
9028-49-3

2 Source Organism

<1> *Mesocricetus auratus* [1, 16]
<2> *Bos taurus* [3-5, 12]
<3> *pigeon* [2, 4, 5, 7, 9]
<4> *Escherichia coli* [5, 8]
<5> *Klebsiella pneumoniae* [6]
<6> *Aerobacter aerogenes* [10, 11, 13]
<7> *Kluyveromyces marxianus* (strain NRRL Y-1196, lactose-fermenting yeast [14]) [14]
<8> *Saccharomyces cerevisiae* (top-fermenting yeasts: BB1, BB2, BB12; bottom-fermenting yeasts: BB6, BB7, BB9, BB11 [15]) [15]
<9> *Gallus gallus* (chicken [17]) [17]
<10> *Streptococcus lactis* (strain 18-16 [17]) [17]
<11> *Lactobacillus kefir* [18]
<12> *Candida boidinii* [18]
<13> *Lactobacillus plantarum* [18]
<14> *Lactobacillus brewis* [18]
<15> *Lactobacillus buchneri* [18]
<16> *Leuconostoc cremoris* [18]
<17> *Hansenula polymorpha* [18]
<18> *Kluyveromyces lactis* [18]
<19> *Torulopsis candia* [18]
<20> *Staphylococcus aureus* [5]

3 Reaction and Specificity

Catalyzed reaction

diacetyl + NADH = acetoin + NAD^+

Reaction type

oxidation
redox reaction
reduction

Natural substrates and products

S 2,3-butanedione + NAD(P)H <1-4, 6-11> (<7> highest activity with acetoin as substrate [14]; <9> liver enzyme does not accept acetoin as substrate [17]) (Reversibility: ir <2-4, 9-11> [3, 4, 8, 11, 12]; r <7> [14]; ? <8> [15]) [1, 3-5, 8, 11, 14-18]
P 3-hydroxy-2-butanone + $NAD(P)^+$ <1, 4, 6-11> [1, 8, 11, 14-18]

Substrates and products

S 1,4-dibromo-2,3-butanedione + NAD(P)H <1> (Reversibility: ? <1> [16]) [16]
P ?
S 1-phenyl-1,2-propanedione + NAD(P)H <1> (Reversibility: ? <1> [16]) [16]
P ?
S 2,3-butanediol + NAD^+ <6> (Reversibility: r <6> [10]) [10]
P acetoin + NADH <6> [10]
S 2,3-butanedione + NAD(P)H <1-4, 6-11> (<7> highest activity with acetoin as substrate [14]; <9> liver enzyme does not accept acetoin as substrate [17]) (Reversibility: ir <2-4, 9-11> [3, 4, 8, 11, 12]; r <7> [14]; ? <8> [15]) [1, 3-5, 8, 11, 14-18]
P 3-hydroxy-2-butanone + $NAD(P)^+$ <1, 4, 6-11> [1, 8, 11, 14-18]
S 2,3-heptanedione + NAD(P)H <1> (Reversibility: ? <1> [16]) [16]
P ?
S 2,3-hexanedione + NAD(P)H <1> (Reversibility: ? <1> [16]) [16]
P ?
S 2,3-pentanedione + NAD(P)H <1, 2, 8-10> (Reversibility: ? <1, 8-10> [15-17]) [5, 15-17]
P ?
S 2,5-hexanedione + NADH <11> (Reversibility: ir <11> [18]) [18]
P ?
S 2R,3R-butanediol + $NAD(P)^+$ <1> (Reversibility: ? <1> [16]) [16]
P ?
S 2S,3S-butanediol + $NAD(P)^+$ <1> (Reversibility: ? <1> [16]) [16]
P ?
S 3,4-hexanedione + NAD(P)H <1> (Reversibility: ? <1> [16]) [16]
P ?
S 3-hydroxy-2-butanone + NADH <1, 6, 7, 10> (Reversibility: ? <1, 6, 7, 10> [1, 10, 14, 16, 17]) [1, 10, 14, 16, 17]

P 2,3-butanedione + NAD(P)H
S 4-benzoylpyridine + NAD(P)H <1> (Reversibility: ? <1> [16]) [16]
P 3,4-dihydroxyhexane + $NAD(P)^+$
S 4-hydroxy-3-hexanone + NAD(P)H <1> (Reversibility: ? <1> [16]) [16]
P ?
S 4-nitro-benzaldehyde + NADPH <1> (Reversibility: ? <1> [1]) [1]
P 4-nitrobenzylalcohol + $NADP^+$
S D-arabinose + NAD(P)H <1> (Reversibility: ? <1> [16]) [16]
P ?
S D-arabitol + $NAD(P)^+$ <1> (Reversibility: ? <1> [16]) [16]
P ?
S D-erythrose + NAD(P)H <1> (Reversibility: ? <1> [16]) [16]
P ?
S D-ribulose + NAD(P)H <1> (Reversibility: ? <1> [16]) [16]
P ?
S D-threitol + $NAD(P)^+$ <1> (Reversibility: ? <1> [16]) [16]
P ?
S D-threose + NAD(P)H <1> (Reversibility: ? <1> [16]) [16]
P ?
S D-xylulose + NAD(P)H <1> (Reversibility: ? <1> [16]) [16]
P ?
S DL-glyceraldehyde + NAD(P)H <1, 9, 10> (Reversibility: ? <1, 9, 10> [16, 17]) [16, 17]
P glycerol + $NAD(P)^+$
S L-arabitol + $NAD(P)^+$ <1> (Reversibility: ? <1> [16]) [16]
P ?
S L-erythrose + NAD(P)H <1> (Reversibility: ? <1> [16]) [16]
P ?
S L-erythrulose + NAD(P)H <1> (Reversibility: ? <1> [16]) [16]
P ?
S L-threitol + $NAD(P)^+$ <1> (Reversibility: ? <1> [16]) [16]
P ?
S L-threose + NAD(P)H <1> (Reversibility: ? <1> [16]) [16]
P ?
S L-xylulose + NAD(P)H <1> (Reversibility: ? <1> [16]) [16]
P ?
S R-1,3-butanediol + $NAD(P)^+$ <1> (Reversibility: ? <1> [16]) [16]
P ?
S S-1,3-butanediol + $NAD(P)^+$ <1> (Reversibility: ? <1> [16]) [16]
P ?
S acetaldehyde + NAD(P)H <8> (Reversibility: ? <8> [15]) [15]
P ethanol + $NADP^+$
S acetoacetyl N-acetylcysteamine + NADPH <4> (Reversibility: ? <4> [8]) [8]
P ?
S acetylacetone + NADH <11> (Reversibility: ir <11> [18]) [18]
P ?

S β-phenylpyruvate + NADH <11> (Reversibility: ir <11> [18]) [18]
P 2-hydroxy-3-phenylpropanoate + NAD^+
S dihydroxyacetone + NAD(P)H <1> (Reversibility: ? <1> [16]) [16]
P glycerol + $NAD(P)^+$
S ethylacetoacetate + NADPH <4, 9, 10> (Reversibility: ? <4, 9, 10> [8, 17]) [8, 17]
P ethyl-2-hydroxybutanoate + $NADP^+$
S ethylpyruvate + NAD(P)H <4, 9, 10> (Reversibility: ? <4, 9, 10> [8, 17]) [8, 17]
P ethyl-2-hydrozypropanoate + $NAD(P)^+$
S ethylpyruvate + NADH <11> (Reversibility: ir <11> [18]) [18]
P ?
S ethylthioacetoacetate + NADPH <4> (Reversibility: ? <4> [8]) [8]
P 5-ethyl-3-hydroxybutanethioate + $NADP^+$
S meso-erythritol + $NAD(P)^+$ <1> (Reversibility: ? <1> [16]) [16]
P ?
S methyl pyruvate + NAD(P)H <9, 10> (Reversibility: ? <9, 10> [17]) [17]
P methyl-2-hydroxypropanoate + $NADP^+$
S methylacetoacetate + NADPH <4> (Reversibility: ? <4> [8]) [8]
P methyl-3-hydroxybutanoate + $NADP^+$
S methylglyoxal + NAD(P)H <7-10> (Reversibility: ? <7-10> [14, 15]) [14, 15, 17]
P ?
S pyridine-4-aldehyde + NADPH <1> (Reversibility: ? <1> [1]) [1]
P pyridine-4-methanol + $NADP^+$
S pyruvate + NADH <11> (Reversibility: ir <11> [18]) [18]
P 2-hydroxypropanoate + NAD^+
S ribitol + $NAD(P)^+$ <1> (Reversibility: ? <1> [16]) [16]
P ?
S xylitol + $NAD(P)^+$ <1> (Reversibility: ? <1> [16]) [16]
P ?
S Additional information <3, 4> (<4> substrates with uncharged α-dicarbonyl structure [8]; <2> β-keto acid esters but no free β-keto acids [5]; <3> not: acetone [3, 4]; <3> pentane-3-one [3, 4]; <3> pentane-2,4-dione [3, 4]; <3> hexane-2,5-dione [3, 4]; <2> pyruvate [3]; <2> oxaloacetate [3]; <2> acetylmethylcarbinol [3]; <3> acetate [4]; <3> 2-oxoglutarate [4]) [3, 4, 8]
P ?

Inhibitors

1,10-phenanthroline <11> (<11> 1 mM, complete inhibition [18]) [18]
1,4-dithioerythritol <11> (<11> 1 mM, 72% remaining activity [18]) [18]
2,2-dinitro-5,5-dithiobenzoate <11> (<11> 1 mM, complete inhibition [18]) [18]
2,2-dipyridine <11> (<11> 1 mM, 19% remaining activity [18]) [18]
2,3-butanediol <4, 6> (<4, 6> product inhibition [5, 10, 11]; <6> noncompetitive versus 2,3-butandione [11]; <4> above 40 mM [8]) [8, 10, 11]

2-oxoglutarate <3> (<3> noncompetitive [9]) [9]
Ba^{2+} <5> (<5> strong inhibition [6]) [6]
Ca^{2+} <5> (<5> slight inhibition [6]) [6]
$CaCl_2$ <11> (<11> 1 mM, 96% remaining activity [18]) [18]
Cu^{2+} <5> (<5> strong inhibition [6]) [6]
$CuCl_2$ <11> (<11> 1 mM, complete inhibition [18]) [18]
$FeCl_2$ <11> (<11> 1 mM, complete inhibition [18]) [18]
$FeCl_3$ <11> (<11> 1 mM, 82% remaining activity [18]) [18]
Hg^{2+} <5> (<5> strong inhibition [6]) [6]
$HgCl_2$ <11> (<11> 1 mM, complete inhibition [18]) [18]
K^+ <5> (<5> slight inhibition [6]) [6]
Mg^{2+} <5> [6]
Mn^{2+} <5> [6]
$MnCl_2$ <11> (<11> 1 mM, 91% remaining activity [18]) [18]
NAD^+ <1, 2, 6> (<2,6> product inhibition, competitive versus NADH [11,12]; <2> noncompetitive with diacetyl [12]) [1, 10, 11, 12]
NADH <4> (<4> above 0.3 mM [8]) [8]
$NADP^+$ <1> (<1> product inhibition, in 100 fold excess over enzyme needed for considerable inactivation [1]) [1]
NADPH <4> (<4> above 0.15 mM [8]) [8]
Na^+ <5> (<5> slight inhibition [6]) [6]
Tris-HCl buffer <4, 5> [6, 8]
Zn^{2+} <5> (<5> strong inhibition [6]) [6]
$ZnCl_2$ <11> (<11> 1 mM, 85% remaining activity [18]) [18]
acetate <2, 6> (<6> only oxidation of 2,3-butanediol is inhibited by acetate [11]) [5, 10, 11]
acetoin <1, 2, 6> (<1, 2, 6> product inhibition [1, 10, 12]; <1> uncompetitive with respect to NADPH [1]; <2> noncompetitive versus NADH [12]) [1, 10, 12]
acetone <3> (<3> competitive for diacetyl [9]) [4, 9]
ethanol <2> (<2> inhibition reversible [5]) [5]
ethylenediaminetetraacetate <11> (<11> 1 mM, 94% remaining activity [18]) [18]
hexane-2,5-dione <3> (<3> noncompetitive [9]) [9]
p-chloromercuriphenylsulfonate <1> [1]
p-hydroxymercuribenzoate <11> (<11> 1 mM, 79% remaining activity [18]) [18]
pentane-3-one <3> (<3> competitive for diacetyl [9]) [9]
phenylglyoxal <1> [1]
phenylhydrazine <11> (<11> 1 mM, complete inhibition [18]) [18]
salicylic acid <2> (<2> 1 mM reduces activity to 60% [3]) [3]

Cofactors/prosthetic groups

NAD^+ <6> [10]
NADH <2-4, 10, 11> (<10> bacterial enzyme only accepts NADH [17]; <11> NADH is preferred, cannot be replaced by NADPH [18]) [3-5, 17, 18]

NADPH <1-4, 8, 9> (<1> transfer of pro-4S hydrogen to diacetyl, i.e. B-specific [1]; <8, 9> NADPH and NADH are both accepted [15, 17]; <2-4> NADH and NADPH equally effective [3-5]; <4> highly specific for NADPH [8]) [1, 3-5, 8, 15, 17]

Activating compounds

N-ethylmaleimide <11> (<11> 1 mM, 112% relative activity [18]) [18]
Triton X-100 <11> (<11> 1 mM, 113% relative activity [18]) [18]

Metals, ions

Additional information <1, 2, 5> (<5> no activation by monovalent and divalent cations at 1 mM [6]; <1> Zn^{2+}, Cu^{2+}, Fe^{2+},Fe^{3+} cannot be detected in enzyme preparation [1]; <2> does not require metallic activators or other dialyzable cofactors [3]) [1, 3, 6]

Specific activity (U/mg)

0.75 <2> [3]
2.45 <4> [5, 8]
12 <3> [4, 5, 9]
16 <2> [5]
42.2 <9> [17]
59 <5> [6]
90 <1> (<1> 80-98 units/mg protein [1]) [1]
91 <7> [14]
121.5 <10> [17]
183 <1> (<1> recombinant enzyme [16]) [16]
198 <1> (<1> purified from liver [16]) [16]
366 <11> (<11> isozyme E2 [18]) [18]
1061 <11> (<11> isozyme E1 [18]) [18]

K_m-Value (mM)

0.008 <4> (NADPH, <4> with ethylacetoacetate or acetoacetyl N-acetylcysteamine [8]) [8]
0.009 <6> (NADH, <6> in phosphate buffer, pH 7.0 [10]) [10]
0.02 <10> (NADH) [17]
0.02 <4> (NADPH, <4> with diacetyl [5, 8]) [5, 8]
0.03 <11> (NADH, <11> isozyme E1 [18]) [18]
0.036 <11> (NADH, <11> isozyme E2 [18]) [18]
0.04 <2> (2,3-butanedione) [5, 12]
0.045 <7> (NADH) [14]
0.06 <10> (3-hydroxy-2-butanone) [17]
0.067 <2> (2,3-butanedione) [4]
0.1 <2, 3> (NADH) [5, 9, 12]
0.116 <3> (NADH, <3> measured at pH 6.3 [7]) [7]
0.118 <1> (NADH, <1> NADPH, 0.003 mM [16]) [16]
0.18 <6> (NAD^+, <6> with 2,3-butanediol, in phosphate buffer, pH 7.0 [10]) [10]
0.2 <1> (2,3-butanedione) [16]
0.45 <7> (3-hydroxy-2-butanone) [14]

0.46 <4> (NADH) [5, 8]
0.501 <9> (NADH, <9> NADPH, 0.017 mM [17]) [17]
0.53 <6> (3-hydroxy-2-butanone, <6> in phosphate buffer, pH 7.0 [10]) [10]
1.8 <9> (2,3-butanedione) [17]
2.5 <7> (2,3-butanedione) [14]
3 <3> (2,3-butanedione, <3> measured at pH 6.3 [7]) [7]
3.1 <3> (2,3-butanedione) [5, 9]
3.3 <1> (3-hydroxy-2-butanone) [16]
3.5 <3> (2,3-butanedione) [4]
3.61 <10> (2,3-butanedione) [17]
4.44 <4> (2,3-butanedione) [8]
7.15 <4> (acetoacetyl N-acetylcysteamine) [8]
7.4 <1> (3-hydroxy-2-butanone) [1]
11.3 <6> (2,3-butanediol, <6> in phosphate buffer, pH 7.0, K_m in acetate buffer: 122 mM, K_m increases with an increase of acetate in the assay [10]) [10]
15.4 <4> (ethylacetoacetate) [8]
67 <11> (2,3-butanedione, <11> isozyme E2 [18]) [18]
310 <11> (2,3-butanedione, <11> isozyme E1 [18]) [18]

pH-Optimum

4.2 <11> (<11> isozyme E1, in acetate buffer [18]) [18]
4.8 <6> (<6> 3-hydroxy-2-butanone, in morpholinoethane sulfonate buffer [10]) [10]
5 <11> (<11> isozyme E2, standard deviation 0.6, in phosphate buffer [18]) [18]
5.9 <3> (<3> between pH 5.6-6.2 [4]) [4, 4]
6.1 <2, 3> [3, 5, 9]
6.1 <9> (<9> maximal activity [17]) [17]
6.4 <10> (<10> maximal activity [17]) [17]
6.5 <4> (<4> between pH 6-7 [8]) [8, 8]
7 <7> (<7> maximal activity [14]) [14]
7.5 <5> [6]
9.5 <6> (<6> 2,3-butanediol in glycine buffer [10]) [10]

pH-Range

5-8.3 <9, 10> (<9, 10> broad pH activity range [17]) [17]
5.4-7.6 <3> (<3> in phosphate buffer, acetone precipitate, over 60 units activity in this range [7]) [7]
5.5-7.5 <2> (<2> less than 50% of maximal activity above and below this range [3]) [3]
5.5-8.5 <5> (<5> over 50% of maximal relative activity, 60% of maximal relative activity at pH 5.5 and 8.5 [6]) [6]

Temperature optimum (°C)

37 <5> [6]
40 <2, 7> (<7> for 2,3-butanedione [14]) [3, 14]
60 <11> (<11> crude cell extract, maximum activity [18]) [18]

Temperature range (°C)

3.5-30 <2> (<2> activity rises in this range [3]) [3]
18-46 <5> (<5> less than 50% of maximal activity above and below this range [6]) [6]
35-40 <9, 10> (<9, 10> increased activity [17]) [17]

4 Enzyme Structure

Molecular weight

10000 <4> (<4> gel filtration [8]) [5, 8]
50000 <7> (<7> gel filtration [14]) [14]
66000 <11> (<11> isozyme E1, estimated 66000-71000 Da [18]) [18]
74000 <11> (<11> isozyme E2, gel filtration, standard deviation 5000 Da [18]) [18]
76000 <2> (<2> gel filtration [3]) [3, 5]
81000 <9> (<9> gel filtration [17]) [17]
92000 <1> (<1> sucrose density gradient centrifugation [1]) [1]
100000 <1> (<1> gel filtration [16]) [16]
105000 <10> (<10> estimated to be 105000-110000 [17]) [17]
110000 <3> (<3> gel filtration [4,5]) [4, 5, 9]
Additional information <11> (<11> SDS-PAGE yields 77000 Da for both isozymes [18]) [18]

Subunits

? <10> (<10> ? * 30000 [17]) [17]
tetramer <1> (<1> 4 * 23000, SDS-PAGE; 4 * 24000, gel filtration in presence of 6 M guanidine HCl [1]) [1]
tetramer <1> (<1> 4 * 31000, SDS-PAGE, 25655 Da calculated from cDNA-sequence [16]) [16]
trimer <9> (<9> 3 * 28000, SDS-PAGE [17]) [17]

Posttranslational modification

glycoprotein <1> [1]

5 Isolation/Preparation/Mutation/Application

Source/tissue

liver <1-3, 9> [1-5, 7, 9, 12, 16, 17]

Localization

cytoplasm <10> [17]
mitochondrion <2> [5]
soluble <2, 3, 5> [5, 6]

Purification

<1> [1]
<1> (recombinant protein from Escherichia coli, hamster liver [16]) [16]

<2> [3, 5, 12]
<3> [4, 5, 9]
<4> [5, 8]
<5> [6]
<6> [13, 10, 11]
<7> [14]
<9> (liver [17]) [17]
<10> [17]
<11, 12> [18]

Renaturation
<1> [1]

Cloning
<1> (recombinant in Escherichia coli [16]) [16]

6 Stability

pH-Stability
5-9 <11> (<11> both enzymes remain fairly stable, E1: pH 4.0 to 10.5 [18]) [18]
5.4-7.6 <3> (<3> 25°C, 0.05 M phosphate buffer [7]) [7]
7-8 <5> [6]

Temperature stability
37 <5> (<5> no loss of activity between 20 and 37°C, pH 7.5, 15 min [6]) [6]
Additional information <11> (<11> stability of highly purified enzyme is poorer, incubation for 5 min at 55°C or 60°C results in about 50% deactivation about 50% [18]) [18]

General stability information
<2>, inactivation by dialysis against H_2O, no reactivation by addition of Zn^{2+}, Cu^{2+}, Mg^{2+}, Fe^{2+}, or Fe^{3+} [3]
<3>, storage in water results in loss of activity of about 75% in 24 h at 0°C [4]
<6>, diluted solution without any compounds loses activity rapidly [10]
<11>, Mg^{2+} stabilizes enzyme, 5 mM optimal concentration [18]
<2, 3>, 0°C, in 0.5 M sodium-potassium phosphate buffer or in 0.1 M phosphate buffer, pH 6.1, at least 24 h stable [3, 4]

Storage stability
<2>, -15°C, N_2-atmosphere [12]
<2>, -15°C, lyophilized after chromatography on DEAE-22, several months [3]
<2>, -18°C, after acetone precipitation and drying of cake after filtration, stable for at least 3 months [5]
<3>, -15°C, lyophilized, at least 6 months, only partial retention of activity [4]

<3>, -18°C, after acetone precipitation, centrifugation and redissolving of precipitate in minimum of destilled water, freeze dried enzyme stable for several months with high retention of activity [5]
<4>, -20°C, after precipitation with solid ammonium sulfate and collection by centrifugation [5]
<5>, 10°C, McIlvaine buffer, pH 9.0, 24 h, more than 90% of activity remains [6]
<6>, 0.1 mM NAD^+, 25 mM 2-mercaptoethanol, 0.25% albumin, 20% glycerol, pH 7.0 [10]
<11>, 4°C, 1 M sterilized buffer, one week, E1: pH 9.0, E2: pH 7.0 [18]

References

[1] Sawada, H.; Akira, H.; Nakayama, T.: Kinetic and structural properties of diacetyl reductase from hamster liver. J. Biochem., **98**, 1349-1357 (1985)
[2] Martin, R.; Burgos, J.: Diacetyl reductase in animal tissues and its intracellular distribution. Biochim. Biophys. Acta, **212**, 356-358 (1970)
[3] Burgos, J.; Martin, R.: Purification and some properties of diacetyl reductase from beef liver. Biochim. Biophys. Acta, **268**, 261-270 (1972)
[4] Diez, V.; Burgos, J.; Martin, R.: Pigeon liver diacetyl reductase: purification and some properties. Biochim. Biophys. Acta, **350**, 253-262 (1974)
[5] Sarmiento, R.M.; Burgos, J.: Diacetyl reductase. Methods Enzymol., **89**, 516-523 (1982)
[6] Shimizu, H.; Hanaichi, Y.; Okada, A.; Tomoyeda, M.: Formation and some properties of diacetyl reductase from Klebsiella pneumoniae. Agric. Biol. Chem., **41**, 527-532 (1977)
[7] Martin, R.; Diez, V.; Burgos, J.: Pigeon liver diacetyl reductase. Effects of pH on the kinetic parameters of the reaction. Biochim. Biophys. Acta, **429**, 293-300 (1976)
[8] Silber, P.; Chung, H.; Gargiulo, P.; Schulz, H.: Purification and properties of a diacetyl reductase from Escherichia coli. J. Bacteriol., **118**, 919-927 (1974)
[9] Burgos, J.; Martin, R.; Diez, V.: Pigeon liver diacetyl reductase. Kinetic and thermodynamic studies with NADH as coenzyme. Biochim. Biophys. Acta, **364**, 9-16 (1974)
[10] Larsen, S.H.; Stormer, F.C.: Diacetyl (acetoin) reductase from Aerobacter aerogenes. Kinetic mechanism and regulation by acetate of the reversible reduction of acetoin to 2,3-butanediol. Eur. J. Biochem., **34**, 100-106 (1973)
[11] Johansen, L.; Larsen, S.H.; Stormer, F.C.: Diacetyl (acetoin) reductase from Aerobacter aerogenes. Kinetic studies of the reduction of diacetyl to acetoin. Eur. J. Biochem., **34**, 97-99 (1973)
[12] Martin, R.; Burgos, J.: Kinetic studies of beef liver diacetyl reductase. Biochim. Biophys. Acta, **289**, 13-18 (1972)
[13] Bryn, K.; Hetland, O. Stormer, F.C.: The reduction of diacetyl and acetoin in Aerobacter aerogenes. Evidence for one enzyme catalyzing both reactions. Eur. J. Biochem., **18**, 116-119 (1971)

[14] Schwarz, J.G.; Hang, Y.D.: Purification and characterization of diacetyl reductase from Kluyveromyces marxianus. Lett. Appl. Microbiol., **18**, 272-276 (1994)
[15] Murphy, C.A.; Large, P.J.; Wadforth, C.; Dack, S.J.; Boulton, C.A.: Strain-dependent variation in the NADH-dependent diacetyl reductase activities of lager- and ale-brewing yeasts. Biotechnol. Appl. Biochem., **23**, 19-22 (1996)
[16] Ishikura, S.; Isaji, T.; Usami, N.; Kitahara, K.; Nakagawa, J.; Hara, A.: Molecular cloning, expression and tissue distribution of hamster diacetyl reductase. Identity with L-xylulose reductase. Chem.-Biol. Interact., **130-132**, 879-889 (2001)
[17] Gibson, T.D.; Parker, S.M.; Woodward, J.R.: Purification and characterization of diacetyl reductase from chicken liver and Streptococcus lactis and enzymic determination of diacetyl and diketones. Enzyme Microb. Technol., **13**, 171-178 (1991)
[18] Hummel, W.; Boermann, F.; Kula, M.-R.: Purification and characterization of an acetoin dehydrogenase from Lactobacillus kefir suitable for the production of (+)-acetoin. Biocatalysis, **2**, 293-308 (1989)

Glycerol dehydrogenase 1.1.1.6

1 Nomenclature

EC number
1.1.1.6

Systematic name
glycerol:NAD$^+$ 2-oxidoreductase

Recommended name
glycerol dehydrogenase

Synonyms
GLDH [Swissprot]
NAD-linked glycerol dehydrogenase
glycerin dehydrogenase

CAS registry number
249285-11-8
9028-14-2 (this number also refers to EC 1.1.99.22)

2 Source Organism

<1> *Schizosaccharomyces pombe* (strain 972h- [2]) [1, 2]
<2> *Candida valida* (strain H 112 [3]) [3]
<3> *Cellulomonas sp.* (strain NT3060 [4, 7, 14]) [4, 7, 8, 14, 16]
<4> *Bacillus megaterium* [5]
<5> *Geotrichum candidium* [6]
<6> *Klebsiella aerogenes* (strain 2103 [9]) [9]
<7> *Aerobacter aerogenes* (strain 1033 [10]; strain P14 [11]) [10-12]
<8> *Escherichia coli* (strain ECFS [13]) [13, 17]
<9> *Clostridium butyricum* (strain E5, E5 mutant D [15]) [15]
<10> *Bacillus stearothermophilus* [19]
<11> *Clostridium beijerinckii* [18]
<12> *Clostridium difficile* [18]

3 Reaction and Specificity

Catalyzed reaction
glycerol + NAD$^+$ = dihydroxyacetone + NADH

Reaction type

oxidation
redox reaction
reduction

Natural substrates and products

- **S** glycerol + NAD^+ <1-4, 7> (Reversibility: r <1, 3> [16]) [1-5, 7, 11, 12, 14-16, 19]
- **P** dihydroxyacetone + NADH <1, 3> [1, 4, 7, 15, 16, 19]

Substrates and products

- **S** 1,2,3-butanetriol + NAD^+ <1> (Reversibility: ? <1> [1]) [1]
- **P** 1,3-dihydroxybutane-2-one + NADH <1> [1]
- **S** 1,2-butanediol + NAD^+ <1> (Reversibility: r <1> [1]) [1]
- **P** 1-hydroxybutane-2-one + NADH <1> [1]
- **S** 1,2-propanediol + NAD^+ <1-4, 7> (Reversibility: r <1, 3> [1, 4]) [1, 3-5, 7, 11, 12, 14, 16]
- **P** hydroxyacetone + NADH <1, 3> [1, 4]
- **S** 1,3-butanediol + NAD^+ <3> (<3> relative activity 6,5% [7]) (Reversibility: ? <3> [7]) [7]
- **P** ?
- **S** 1,3-dichloro-2-propanol + NAD^+ <7> (Reversibility: ? <7> [11]) [11]
- **P** 1,3-dichloro-2-propanone + NADH
- **S** 1,3-propanediol + NAD^+ <3, 7, 9> (<3> relative activity 18% [7]) (Reversibility: ? <3, 7, 9> [7, 15, 12]) [7, 12, 15]
- **P** ?
- **S** 1,3-propanediol + NAD^+ <3, 7> (<7> relative rate of activity 37% [12]) (Reversibility: ? <3, 7> [7, 12]) [7, 12]
- **P** ?
- **S** 1,4-butanediol + NAD^+ <3, 7> (<7> relative activity 17% [12]) (Reversibility: ? <3, 7> [7, 12]) [7, 12]
- **P** ?
- **S** 1-butanol + NAD^+ <9> (Reversibility: ? <9> [15]) [15]
- **P** 1-butanal + NADH
- **S** 1-chloro-2,3-propanediol + NAD^+ <7> (Reversibility: ? <7> [11]) [11]
- **P** ?
- **S** 2,3-butanediol + NAD^+ <1, 3, 7, 9> (Reversibility: r <1, 3> [1, 4]) [1, 4, 7, 12, 14, 15]
- **P** 3-hydroxybutane-2-one + NADH <1, 3> [1, 4]
- **S** 3-amino-1,2-propanediol + NAD^+ <3> (Reversibility: ? <3> [16]) [16]
- **P** ?
- **S** 3-bromo-1,2-propanediol + NAD^+ <3> (Reversibility: ? <3> [16]) [16]
- **P** ?
- **S** 3-chloro-1,2-propanediol + NAD^+ <3> (Reversibility: ? <3> [16]) [16]
- **P** ?
- **S** 3-hydroxypropionaldehyde + NADH <9> (Reversibility: ? <9> [15]) [15]
- **P** propan-1,3-diol + NAD^+
- **S** 3-mercapto-1,2-propanediol + NAD^+ <3> (Reversibility: ? <3> [16]) [16]

P ?
S DL-α-glycerolphosphate + NAD$^+$ <3, 7> (<3> relative activity 16% [7], <7> relative activity 11% [12]) (Reversibility: ? <3, 7> [7, 12]) [7, 12]
P ?
S DL-glyceraldehyde + NAD$^+$ <1-3> (<7> relative rate of reduction 14% [12]) (Reversibility: ? <1, 2, 3, 7> [1, 3, 7, 12]) [1, 3, 7, 12]
P 3-hydroxypyruvaldehyde + NADH <1, 3> [1, 7]
S DL-glyceraldehyde + NADH <9> (Reversibility: ? <9> [15]) [15]
P glycerol + NAD$^+$
S N-butyraldehyde + NADH <9> (Reversibility: ? <9> [15]) [15]
P 1-butanol + NAD$^+$
S R-1-amino-2-propanol + NAD$^+$ <3> (Reversibility: ? <3> [16]) [16]
P ?
S S-1-amino-2-propanol + NAD$^+$ <3> (Reversibility: ? <3> [16]) [16]
P ?
S acetaldehyde + NADH <9> (Reversibility: ? <9> [15]) [15]
P ethanol + NAD$^+$
S β-glycerol-phoshate + NAD$^+$ <7> (<7> relative activity 2% [12]) (Reversibility: ? <7> [12]) [12]
P ?
S diglycerol + NAD$^+$ <7> (<7> relative activity 21% [12]) (Reversibility: ? <7> [12]) [12]
P ?
S erythrit + NAD$^+$ <4> (Reversibility: ? <4> [5]) [5]
P ?
S ethanediol + NAD$^+$ <1, 3> (Reversibility: r <3> [7]) [1, 7]
P glycolaldehyde + NADH <1, 3> (<3> reduction at concentration of 50 mM [7]) [1, 7]
S ethanol + NAD$^+$ <7> (<7> relative activity 1% [12]) (Reversibility: ? <7> [12]) [12]
P acetaldehyde + NADH
S ethylene glycol + NAD$^+$ <3, 7> (<7> relative activity 20% [12]) (Reversibility: ? <3, 7> [7, 12]) [7, 12]
P ?
S glycerol + NAD$^+$ <1-4, 7> (Reversibility: r <1, 3> [16]) [1-5, 7, 11, 12, 14-16, 19]
P dihydroxyacetone + NADH <1, 3> [1, 4, 7, 15, 16, 19]
S glycerol-α-monochlorohydrin + NAD$^+$ <3, 7> (<3> relative activity 71% [7]; <7> relative activity 26% [12]) (Reversibility: ? <3, 7> [7, 12]) [7, 12]
P ?
S glycerol-α-monomethyl ether + NAD$^+$ <7> (Reversibility: ? <7> [11]) [11]
P ?
S hydroxy-2-propanone + NADH <7> (<7> relative activity 27% [12]) (Reversibility: ? <7> [12]) [12]
P propylene glycol + NAD$^+$

S i-inositol + NAD^+ <7> (<7> relative activity 18% [12]) (Reversibility: ? <7> [12]) [12]
P ?
S isopropanol + NAD^+ <7> (<7> relative activity 17% [12]) (Reversibility: ? <7> [12]) [12]
P ?
S methylglyoxal + NADH <1, 7> (<7> relative rate of reduction 56% [12]) (Reversibility: ? <1, 7> [1, 12]) [1, 12]
P lactaldehyde + NAD^+ <1> [1]
S propionaldehyde + NADH <9> (Reversibility: ? <9> [15]) [15]
P 1-propanol + NAD^+
S sorbitol + NAD^+ <7> (<7> relative activity 3% [12]) (Reversibility: ? <7> [12]) [12]
P ?

Inhibitors

1,10-phenanthroline <3, 4> (<3> 1 mM [7]) [5, 7]
1,3-propanediol <3> (<3> competitive inhibitor versus glycerol in direction of glycerol oxidation, noncompetitive inhibitor versus NAD^+, competitive inhibitor versus dihydroxyacetone in direction of glycerol reduction, noncompetitive inhibitor versus NADH [16]) [16]
2,2'-bipyridyl <4, 7> (<7> powerful inhibition [11]) [5, 11]
2-amino-2-(hydroxymethyl)-1,3-propanediol <2> (<2> strong inhibition, IC50 2 mM [3]) [3]
2-mercaptoethanol <1, 2> (<2> strong inhibition, IC50 0.3 mM [3]; <7> strongly inhibitory at 100 mM, but slight activation in lower concentration, 1 mM [10]) [1, 3, 10]
3,4-dimercaptotoluol <4> [5]
8-hydroxyquinoline <4> [5]
8-quinolinol <7> (<7> 1 mM, 10 mM [10]) [10, 11]
ADP <5> (<5> 10% inhibition [6]) [6]
AMP <5> (<5> 13% inhibition [6]) [6]
ATP <5> (<5> 6% inhibition [6]) [6]
Anhitol 24B <3> [14]
Cd^{2+} <3> (<3> strong inhibition [7]) [7]
Cu^{2+} <3, 8> (<3> strong inhibition [7]) [7, 8, 13]
D-xylose <3> [14]
EDTA <1, 4, 7> (<1> 10 mM [1]; <7> 3 mM, 30 mM [10]; <7> slight inhibition, 10 mM [11]) [1, 5, 10, 11]
Fe^{3+} <3, 8> [7, 13]
K^+ <3> (<3> inhibition of reduction at lower concentration [7]) [7]
L-cysteine <1, 2> (<2> slight inhibition [3]) [1, 3]
Li^+ <1> (<1> competitive, forward reaction [1]) [1]
N-ethylmaleimide <2, 7> (<2> rapid inactivation, IC50 0.2 mM [3]) [3, 10]
NADH <3> (<3> competitive inhibitor versus NAD^+ in direction of glycerol oxidation, noncompetitive inhibitor versus glycerol [16]) [4, 16]
$NADP^+$ <1, 2, 5> (<1> inhibits oxidation [1]) [1, 3, 6]

Na^+ <1> (<1> competitive, forward reaction [1]) [1]
Zn^{2+} <3, 8> (<3> strong inhibitory [7]; <8> over 90% inhibition at 0.01 mM [13]) [7, 13]
cetylpyridinium chloride <3> (<3> strong inhibition at low concentration [14]) [14]
cetyltrimethylammonium bromide <3> (<3> strong inhibition at low concentration [14]) [14]
diethyldithiocarbamate <7> (<7> slight inhibition, 2 mM, 10 mM [11]) [11]
dihydroxyacetone <3> (<3> noncompatible inhibitor, at saturated level of glycerol [4]) [4]
dihydroxyacetone phosphate <2> [3]
diphosphate <4> [5]
dithioerythritol <7> (<7> 10 mM [10]) [10]
dithiothreitol <1, 2> (<2> slight inhibition [3]) [1, 3]
ethylene glycol <2> (<2> strong inhibition, IC50 4 mM [3]) [3]
glycerol 3-phosphate <2> [3]
imidazol <4> [5]
iodoacetamide <7> (<7> strong inhibition, 10 mM [10]) [10]
meso-erythritol <3> [14]
p-chloromercuribenzoate <3, 8> (<8> 95% inhibition at 0.01 mM [13]) [7, 13]
pyridin-2,6-dicarboxylic acid <4> [5]
sodium dodecyl sulfate <3> [14]
sodium tetraborate <3> (<3> 53% inhibition [7]) [7]
sucrose <3> [14]

Cofactors/prosthetic groups

NAD^+ <1-4, 7> [1-5, 7, 11, 12, 14-16, 19]
NADH <1-4, 7> [1-5, 7, 11, 12, 14-16, 19]

Activating compounds

2-mercaptoethanol <7> (<7> 1 mM, slight activation [10]) [10]
8-quinolinol <7> (<7> 1 mM, slight activation [10]) [10]
N-ethylmaleimide <1> [1]
NH_4^+ <1, 3, 7> (<3> oxidation and reduction greatly increased [7]) [1, 4, 7, 10-12]
dithioerythritol <7> (<7> 1 mM, slight activation [10]) [10]
iodoacetamide <1> [1]
multivalent anions <2> (<2> citrate, phosphate, sulfate [3]) [3]
o-phenanthroline <7> (<7> 1.85 mM [10]) [10]
p-chloromercuribenzoic acid <1> [1]

Metals, ions

Ca^{2+} <9> (<9> 1 mM, relative activity 64.5% [15]) [15]
K^+ <1, 3, 7> (<3> oxidation greatly increased [4,7]; <3> required for enzymatic activity and glycerol binding [16]) [1, 4, 7, 10, 11, 16]
Li^+ <1, 3> (<3> oxidation slightly increased [7]) [1, 7]

Mn^{2+} <1, 9> (<1> increases activity significantly [1]; <9> 1 mM, relative activity 100% [15]) [1, 15]
Na^+ <3> (<3> oxidation and reduction slightly increased [7]) [7]
Rb^+ <3, 7> (<3> oxidation greatly increased [7]) [7, 10, 11]
Tl^+ <7> [10]
Zn^{2+} <1, 4, 10> (<1> increases activity [1]; <4> 1.6 mol Zn^{2+} bound per subunit [5]) [1, 5, 19]

Specific activity (U/mg)

9.3 <1> [1]
16 <7> [11]
35 <7> [10]
41.2 <5> [6]
49 <6> [9]
58.1 <4> [5]
82 <7> [12]
128 <3> [4]
138 <3> [7]
448 <2> [3]

K_m-Value (mM)

0.014 <7> (NADH) [12]
0.048 <2> (NADH) [3]
0.06 <1> (dihydroxyacetone) [1]
0.08 <9> (NADH, <9> mutant D 0.12 mM [15]) [15]
0.089 <3> (NAD^+) [7]
0.12 <1> (NADH) [1]
0.13 <1> (NAD^+) [1]
0.14 <2> (NAD^+) [3]
0.15 <7> (NAD^+) [12]
0.3 <4> (NAD^+) [5]
0.385 <3> (dihydroxyacetone) [4, 7]
0.5 <1> (glycerol) [1]
0.77 <2> (dihydroxyacetone) [3]
1.18 <9> (dihydroxyacetone, <9> mutant D 2.0 mM [15]) [15]
1.3 <7> (dihydroxyacetone) [12]
1.4 <4> (glycerol) [5]
4.07 <9> (NAD^+, <9> mutant D 4.7 mM [15]) [15]
5.1 <3> (glycerol) [16]
10.9 <3> (glycerol) [4]
10.9 <3> (glycerol) [7]
39 <7> (glycerol) [12]
58 <2> (glycerol) [3]
74.3 <7> (glycerol, <7> in the presence of NH_4^+ [10]) [10]
91.7 <9> (glycerol, <9> mutant D 76.3 mM [15]) [15]

pH-Optimum

6 <1, 3> (<1, 3> reduction of dihydroxyacetone [1, 7]) [1, 4, 7]

7.17 <9> (<9> glycerol, mutant D pH 7.49 [15]) [15]
8 <2> (<2> oxidation of glycerol [3]) [3]
8.6 <9> (<9> dihydroxyacetone, mutant D pH 8.31 [15]) [15]
9 <4> [5]
9 <3> (<3> butanediol [4]) [4]
9 <3, 7> (<3, 7> oxidation of glycerol [7, 11]) [4, 7, 11]
10 <8> [13]

pH-Range
5-6 <1> (<1> dihydroxyacetone reduction [1]) [1]
5.5-10 <3> [7]
6.8-11 <7> (<7> oxidation [12]) [12]
7.5-8.5 <2> (<2> reduction [3]) [3]
10-12 <1, 2> (<1> glycerol oxidation [1]; <2> pH 9.5-10.5, oxidation in diphosphate buffer [3]) [1, 3]

Temperature optimum (°C)
36 <2> [3]
50 <3> (<3> oxidation of glycerol [4, 7]; <3> oxidation of butanediol [4]) [4, 7]
65 <3> (<3> oxidation of propanediol [4]) [4]

Temperature range (°C)
25-65 <3> [4, 7]

4 Enzyme Structure

Molecular weight
39500 <10> [19]
76000-80000 <2> (<2> gel filtration, PAGE under nondenaturing conditions [3]) [3]
156000-160000 <4> (<4> density gradient centrifugation, gel filtration [5]) [5]
180000 <6> (<6> gel filtration [9]) [9]
181000 <9> (<9> gel filtration [15]) [15]
324000 <3> (<3> 39500 Da per subunit, gel filtration [16]) [16]
336000-390000 <3> (<3> gel filtration [4, 8]; <3> sedimentation equilibrium method [4]) [4, 8, 4, 8]
400000 <1> (<1> gel filtration [1]) [1]

Subunits
dimer or tetramer <6> (2 or 4 * 40000, SDS-PAGE [9]) [9]
homodimer <2> (2 * 38000, SDS-PAGE [3]) [3]
homooctamer <1> (<1> 8 * 47000, SDS-PAGE [1]) [1]
homooctamer <3> (<3> 8 * 39500, sedimentation equilibrium, SDS-PAGE [16]) [16]
homooctamer <3> (<3> 8 * 42000-43000, SDS-PAGE [4, 8]) [4, 8]
tetramer <4> (<4> 4 * 38000, SDS-PAGE [5]) [5]

5 Isolation/Preparation/Mutation/Application

Localization

cytoplasm <1-12> (<1-12> soluble, isolation from crude cell extract [1-19]) [1-19]

Purification

<1> [1, 2]
<2> [3]
<3> [4, 7, 8, 14, 16]
<4> [5]
<5> [6]
<6> [9]
<7> [10-12]
<8> [13, 17]
<9> [15]
<10> [19]
<11> [18]
<12> [18]

Crystallization

<3> [4, 8]
<10> (<10> S305C mutant [19]) [19]

Engineering

S305C <10> (<10> S305C mutant used for crystallisation [19]) [19]

6 Stability

pH-Stability

5.5-10 <3> [7]
7 <7> (<7> stable [12]) [12]
7.5 <2> (<2> most stable [3]) [3]

Temperature stability

50 <1> (<1> 10 min stable [2]) [2]
70 <4> (<4> 30 min stable [5]) [5]
75 <8> (<8> 10 min stable, E. coli [13]) [13]

Organic solvent stability

acetone <3> (<3> inactivation, enzyme regains activity after removal of ketone [14]) [14]
catechol <3> (<3> i.e. 1,2-butanediol, inactivation, enzyme regains activity after removal of alcohol [14]) [14]
diacetyl <3> (<3> inactivation, enzyme regains activity after removal of ketone [14]) [14]
dimethylformamide <3> (<3> inactivation [14]) [14]
dimethylsulfoxide <3> (<3> inactivation [14]) [14]

dioxane <3> (<3> inactivation, enzyme regains activity after removal of ketone [14]) [14]
ethanol <3> (<3> 10% v/v: 25-75% loss of activity, depending on substrate, enzyme regains activity after removal of alcohol [14]) [14]
isobutanol <3> (5% v/v: 0-50% loss of activity, depending on substrate, enzyme regains activity after removal of alcohol [14]) [14]
methanol <3> (<3> 10% v/v: 25-75% loss of activity, 5% v/v, 50-95% loss of activity, depending on substrate, enzyme regains activity after removal of alcohol [14]) [14]
n-amyl alcohol <3> (2% v/v: 0-50% loss of activity, depending on substrate, enzyme regains activity after removal of alcohol [14]) [14]
n-butanol <3> (<3> 5% v/v: 0-50% loss of activity, depending on substrate, enzyme regains activity after removal of alcohol [14]) [14]
n-propanol <3> (<3> 5% v/v: 25-60% loss of activity, depending on substrate, enzyme regains activity after removal of alcohol [14]) [14]

General stability information

<4>, $(NH_4)_2SO_4$ stabilizes [5]
<4>, KCl stabilizes [5]
<4>, $ZnCl_2$ stabilizes [5]
<4>, glycerol stabilizes [5]

Storage stability

<1>, -70°C, long term storage [1]
<1>, 4°C, in 2 M $(NH_4)_2SO_4$, 1 M glycerol, several months [1]
<2>, -15°C, with 5% culture filtrate or 5% boiled crude cell extract, 10% Ficoll, pH 7.5, stable for 1 month [3]
<7>, frozen, retains about 42% of initial activity for 2 months [12]
<8>, 0-5°C, 50 to 55% loss of original activity within 48 h [13]

References

[1] Marshall, J.H.; May, J.W.; Sloan, J.: Purification and properties of glycerol: NAD+ 2-oxidoreductase (glycerol dehydrogenase) from Schizosaccharomyces pombe. J. Gen. Microbiol., **131**, 1581-1588 (1985)
[2] Kong, Y.-C.; May, J.W.; Marshall, J.H.: Glycerol oxidation and triose reduction by pyridine nucleotide-linked enzyme in the fission yeast Schizosaccharomyces pombe. J. Gen. Microbiol., **131**, 1571-1579 (1985)
[3] Gärtner, G.; Kopperschläger, G.: Purification and properties of glycerol dehydrogenase from Candida valida. J. Gen. Microbiol., **130**, 3225-3233 (1984)
[4] Nishise, H.; Nagao, A.; Tani, Y.; Yamada, H.: Further characterization of glycerol dehydrogenase from Cellulomonas sp. NT3060. Agric. Biol. Chem., **48**, 1603-1609 (1984)
[5] Scharschmidt, M.; Pfleiderer, G.; Metz, H.; Brümmer, W.: Isolierung und Charakterisierung von Glycerol-Dehydrogenase aus Bacillus megaterium. Hoppe-Seyler's Z. Physiol. Chem., **364**, 911-921 (1983)

[6] Sasaki, I.; Itoh, N.; Goto, H.; Yamamoto, R.; Tanaka, H.; Yamshita, K.; Yamashita J.; Horio, T.: Specific affinity of glycerol dehydrogenase from Geotrichum candidium for 10-carboxydecyl-Sepharose: its application to chromatography for purification of the enzyme. J. Biochem., **91**, 211-217 (1982)
[7] Yamada H.; Nagao, A.; Nishise, H.; Tani, Y.: Glycerol dehydrogenase from Cellulomonas sp. NT 3060: purification and characterization. Agric. Biol. Chem., **46**, 2333-2339 (1982)
[8] Yamada, H.; Nagao, A.; Tani, Y.: Crystallisation of glycerol dehydrogenase. Agric. Biol. Chem., **44**, 471-472 (1980)
[9] Ruch, F.E.; Lin, E.C.C.; Kowit, J.D.; Tang, C.-T.; Goldberg, A.L.: In vivo inactivation of glycerol dehydrogenase in Klebsiella aerogenes: properties of active and inactivated proteins. J. Bacteriol., **141**, 1077-1085 (1980)
[10] McGregor, W.G.; Philips, J.; Suelter, C.H.: Purification and kinetic characterization of a monovalent cation-activated glycerol dehydrogenase from Aerobacter aerogenes. J. Biol. Chem., **249**, 3132-3139 (1974)
[11] Lin, E.C.C.; Magasanik, B.: The activation of glycerol dehydrogenase from Aerobacter aerogenes by monovalent cations. J. Biol. Chem., **235**, 1820-1823 (1960)
[12] Burton, R.M.: Glycerol dehydrogenase from Aerobacter aerogenes. Methods Enzymol., **1**, 397-400 (1955)
[13] Asnis, R.A.; Brodie, A.F.: A glycerol dehydrogenase from Escherichia coli. J. Biol. Chem., **203**, 153-159 (1953)
[14] Nishise, H.; Maehashi, S.; Yamada, H.; Tani, Y.: Effects of organic solvents and chemical modification on the activity of glycerol dehydrogenase. Agric. Biol. Chem., **51**, 3347-3353 (1987)
[15] Malaoui, H.; Marczak, R.: Separation and characterization of the 1,3-propanediol and glycerol dehydrogenase activities from Clostridium butyricum E5 wild-type and mutant D. J. Appl. Microbiol., **90**, 1006-1014 (2001)
[16] Leichus, B.N.; Blanchard, J.S.: Isotopic analysis of the reaction catalyzed by glycerol dehydrogenase. Biochemistry, **33**, 14642-14649 (1994)
[17] Altaras, N.E.; Cameron, D.C.: Metabolic engineering of 1,2-propanediol pathway in Escherichia coli. Appl. Environ. Microbiol., **65**, 1180-1185 (1999)
[18] Liyanage, H.; Kashket, S.; Young, M.; Kashket, E.R.: Clostridium beijerinckii and Clostridium difficile detoxify methylglyoxal by a novel mechanism involving glycerol dehydrogenase. Appl. Environ. Microbiol., **67**, 2004-2010 (2001)
[19] Burke, J.; Ruzheinikov, S.N.; Sedelnikova, S.; Baker, P.J.; Holmes, D.; Muir, N.M.; Gore, M.G.; Rice, D.W.: Purification, crystallisation and quaternary structure analysis of a glycerol dehydrogenase S305C mutant from Bacillus stearothermophilus. Acta Crystallogr. Sect. D, **57**, 165-167 (2001)

Propanediol-phosphate dehydrogenase 1.1.1.7

1 Nomenclature

EC number
1.1.1.7

Systematic name
propane-1,2-diol-1-phosphate:NAD^+ oxidoreductase

Recommended name
propanediol-phosphate dehydrogenase

Synonyms
1,2-propanediol-1-phosphate:NAD^+ oxidoreductase
PDP dehydrogenase
dehydrogenase, propanediol phosphate
propanediol phosphate dehydrogenase

CAS registry number
9028-15-3

2 Source Organism

<1> *Oryctolagus cuniculus* (rabbit [1, 2]) [1, 2]

3 Reaction and Specificity

Catalyzed reaction
propane-1,2-diol 1-phosphate + NAD^+ = hydroxyacetone phosphate + NADH

Reaction type
oxidation
redox reaction
reduction

Natural substrates and products
S propane-1,2-diol-1-phosphate + NAD^+ <1> (Reversibility: r <1> [1]) [1, 2]
P hydroxyacetone phosphate + NADH <1> [1]

Substrates and products
S hydroxyacetone phosphate + NADH <1> (Reversibility: r <1> [1]) [1]
P propane-1,2-diol-1-phosphate + NAD^+

Inhibitors

Br^- <1> [1, 2]
PO_4^{3-} <1> [2]
SO_3^{2-} <1> [1, 2]
SO_4^{2-} <1> [1, 2]
formate <1> [1, 2]
hydroxyacetone phosphate <1> (<1> product inhibition, competitive [2]) [2]

Specific activity (U/mg)

0.3 <1> (<1> 0.1-0.43 [1]) [1]

K_m-Value (mM)

0.05 <1> (NADH) [2]
0.1 <1> (hydroxyacetone phosphate) [2]
4.4 <1> (DL-propane-1,2-diol 1-phosphate) [2]
5.5 <1> (hydroxyacetone phosphate) [1]

pH-Optimum

7.7 <1> (<1> hydroxyacetone phosphate reduction [1]) [1]
Additional information <1> (<1> dehydrogenation of propane-1,2-diol 1-phosphate has alkaline pH-optimum [1]) [1]

pH-Range

5.4-9.6 <1> (<1> hydroxyacetone phosphate reduction, values strongly dependent on buffer system [1]) [1]

5 Isolation/Preparation/Mutation/Application

Source/tissue

muscle <1> [1]

Purification

<1> (partial [1]) [1]

6 Stability

Storage stability

<1>, 4°C, dialyzed against Tris-buffer, pH 7.7, 80% loss of activity in 1 month [1]
<1>, 4°C, dialyzed against distilled water for 12-24 h [2]

References

[1] Sellinger, O.Z.; Miller, O.N.: The metabolism of acetol phosphate, II. 1,2-Propanediol-1-phosphate dehydrogenase. J. Biol. Chem., **234**, 1641-1646 (1959)
[2] Miller, O.N.: 1,2-Propanediol phosphate (PDP) dehydrogenase. Methods Enzymol., **9**, 336-338 (1966)

Glycerol-3-phosphate dehydrogenase (NAD^+) 1.1.1.8

1 Nomenclature

EC number
1.1.1.8

Systematic name
sn-glycerol-3-phosphate:NAD^+ 2-oxidoreductase

Recommended name
glycerol-3-phosphate dehydrogenase (NAD^+)

Synonyms
L-α-glycerol phosphate dehydrogenase
L-α-glycerophosphate dehydrogenase
L-glycerol phosphate dehydrogenase
L-glycerophosphate dehydrogenase
NAD-L-glycerol-3-phosphate dehydrogenase
NAD-α-glycerophosphate dehydrogenase
NAD-dependent glycerol phosphate dehydrogenase
NAD-dependent glycerol-3-phosphate dehydrogenase
NAD-linked glycerol 3-phosphate dehydrogenase
NADH-dihydroxyacetone phosphate reductase
α-glycerol phosphate dehydrogenase (NAD)
α-glycerophosphate dehydrogenase (NAD)
dehydrogenase, glycerol phosphate
glycerol 1-phosphate dehydrogenase
glycerol phosphate dehydrogenase (NAD)
glycerophosphate dehydrogenase (NAD)
hydroglycerophosphate dehydrogenase

CAS registry number
9075-65-4

2 Source Organism

<1> *Saccharomyces cerevisiae* (strain H44-3D [1]; strain YSH 11-6B [30]) [1, 20, 30]
<2> *Debaryomyces hansenii* (van Rij strain 26, salt-tolerant [24]) [24]
<3> *Homo sapiens* [2, 21, 31]
<4> *Rattus norvegicus* (Sprague-Dawley [7]) [2, 7]
<5> *Oryctolagus cuniculus* (New Zealand white [6]) [2, 3, 5, 6, 8, 12, 14, 16, 28]

<6> *Bombus sp.* (insects [2]; bumblebee species [9]) [9]
<7> *Apis mellifera* (honey bee [3, 9]) [3, 9]
<8> *Coturnix coturnix japonica* (japanese quail [4]) [4]
<9> *Gallus gallus* [10]
<10> *Drosophila melanogaster* (fruit fly, Samarkand wild-type [19]) [11, 17, 19]
<11> *Drosophila virilis* [18]
<12> *Ceratitis capitata* (strain Wiedemann, mediterranean fruit fly [13]) [13]
<13> *Spinacia oleracea* (spinach, cv Long Standing Bloomsdale [22]) [15, 22]
<14> *Chlamydomonas reinhardtii* (strain 11/32-90, green alga [23]) [23]
<15> *Aspergillus nidulans* [25]
<16> *Leishmania mexicana* (strain Leishmania mexicana mexicana [29]) [26, 27, 29]
<17> *Trypanosoma brucei* (strain Trypanosoma brucei brucei [29]) [27, 29, 32]
<18> *Vespula germanica* (yellow jacket [9]) [9]
<19> *Bos taurus* [2]
<20> *Mus musculus* [2]
<21> *Locusta migratoria* [2]

3 Reaction and Specificity

Catalyzed reaction

sn-glycerol 3-phosphate + NAD^+ = dihydroxyacetone phosphate + NADH (<3-5> one active site per enzyme molecule [2]; <1, 14> equilibrium random-bi-bi reaction mechanism [23, 30]; <2> ordered ternary-complex mechanism, NADH binds first [24])

Reaction type

oxidation
redox reaction
reduction

Natural substrates and products

S sn-glycerol 3-phosphate + NAD^+ <1-17> (<1> reverse reaction favoured direction [20]; <1> oxidation at 3% the reaction rate of dihydroxyacetone phosphate reduction at pH 7.0 [1]; <13, 14> oxidation at 25% the rate of dihydroxyacetone phosphate reduction at optimal pH [15, 23]; <5> favoured reaction of heart isozyme II6.1 [12]; <3-5> equilibrium far to the side of α-glycerophosphate at neutral pH [2]) (Reversibility: r <1-14> [1, 3-7, 9-24]) [1-5, 7-32]

P dihydroxyacetone phosphate + NADH <1-17> [1, 2, 4-7, 9-21, 23-32]

Substrates and products

S sn-glycerol 3-phosphate + NAD^+ <1-17> (Reversibility: r <1-14> [1, 3-7, 9-24]) [1-5, 7-32]

P dihydroxyacetone phosphate + NADH <1-17> [1, 2, 4-7, 9-21, 23-32]

S Additional information <1, 2, 5, 13, 14> (<1, 2, 13, 14> no substrates: dihydroxyacetone [1, 15, 23, 24]; <1, 13> glycerol [1, 15]; <2, 5, 13, 14>

glyceraldehyde-3-phosphate [5, 15, 23, 24]; <13> DL-glyceraldehyde, phosphohydroxypyruvate [15]; <13, 14> fructose-6-phosphate, fructose-1,6-bisphosphate [15, 23]; <14> glucose-6-phosphate, acetaldehyde, oxaloacetate [23]) [1, 2, 5, 15, 23, 24]

P ?

Inhibitors

$(NH_4)_2SO_4$ <3, 13> (<13> inhibition of chloroplastic and cytosolic form [22]) [21, 22]
2,3-dimercaptopropanol <5> (<5> competitive inhibitor to glycerophosphate [2]) [2]
2-amino-2-hydroxymethylpropane-1,3-diol <13> (<13> i.e. Tris [22]) [22]
2-hydroxy-1,2,3-nonadecanetricarboxylic acid <16> (<16> trypanocidal drug, IC50 0.00055 mM [27]; <17> IC50 0.0011 mM [27]) [27]
ADP <1> (<1> at physiological concentration, 10 mM 90% inhibition [1]) [1]
ADP-ribose <5> (<5> allosteric inhibitor [28]) [28]
ATP <1, 10> (<1> at physiological concentration, 10 mM 95% inhibition [1]; <10> weak inhibitor of both isozymes, i.e. GPDH-1 and -3 [17]) [1, 17, 19]
Cl^- <1, 14> (<1> cations, i.e. H^+, K^+, Na^+ associated with Cl^- do not effect reduction of dihydroxyacetone phosphate [20]; <14> competitive inhibitor with respect to dihydroxyacetone phosphate [23]) [20, 23]
KCl <2> (<2> inactivation [24]) [24]
L-glycerol 3-phosphate <5-7, 13, 14> (<5, 7> inhibits dihydroxyacetone phosphate reduction [3]; <6> inhibitor of bumble bee enzyme, non-competitive [9]; <13> over 0.025 mM, strong inhibitor of cytosolic and chloroplastic enzyme [22]; <14> competitive inhibitor to dihydroxyacetone phosphate, non-competitive to NADH [23]) [3, 9, 22, 23]
N-ethylmaleimide <4, 5, 7, 14> (<4> reversible by dithiothreitol, 5 mM inhibits more than 90% of the enzyme activity [7]; <5> 0.1 mM complete inhibition [2]; <7> 0.1 mM inhibits enzyme over 60% [9]; <14> 1 mM results in 50% inhibition, reversible by NADH, inhibition increases in the presence of dihydroxyacetone phosphate and/or glycerol 3-phosphate [23]) [2, 7, 9, 23]
NAD^+ <1, 2, 14> (<1, 2, 14> competitive inhibitor to NADH at physiological concentration [1, 23, 24, 30]; <14> non-competitive to dihydroxyacetone phosphate [23]) [1, 23, 24, 30]
NADH <13, 14> (<13, 14> at high concentration [15, 23]) [15, 23]
NADH-X <5> (<5> allosteric inhibitor [28]) [28]
NADPH <3, 16, 17> (<16, 17> inhibits in presence of saturing concentration of NADH and dihydroxyacetone phosphate [27]) [27, 31]
NaCl <1, 2, 16, 17> (<1> inactivates irreversibly [20]; <2> inactivation [24]; <16, 17> strongly inhibits, 50% activity at 250 mM NaCl [27]) [20, 24, 27]
SO_4^{2-} <2> (<2> at high concentration, $MgSO_4$ most inhibitory [24]) [24]
TEA-buffer <1> [20]
Triton X-100 <13> (<13> 0.006 mM, 50% inhibition of chloroplastic form and 52% inhibition of cytosolic form [22]) [22]
adenosine diphosphate ribose <5, 7> (<7> 0.094 mM slightly inhibits NADH binding by honeybee GPDH, but not by rabbit GPDH [3]) [3]

dihydroxyacetone phosphate <7, 11, 14, 16, 17> (<16, 17> substrate inhibition at 0.2 mM [27]; <14> substrate inhibition above 1 mM, NADH above 0.2 mM [23]; <11> 0.5 mM, GPDHs more susceptible to inhibition [18]; <7> high levels of substrate, i.e. dihydroxyacetone phosphate result in inhibition [9] <14> substrate inhibition only at high concentration, over 1 mM [23]) [9, 18, 23, 27]
fructose-1,6-bisphosphate <1> (<1> at physiological concentration, non-competitive [1]) [1]
glycerol <2> (<2> slight inhibition [24]) [24]
glycerol phosphate <13> (<13> 0.025 mM, strong inhibition of chloroplastic and cytosolic form, endproduct inhibition [22]) [22]
iodoacetamide <3> (<3> effective inhibitor [31]) [31]
iodoacetate <3, 4, 5> (<3> 50 mM, incubated for 60 min [31]; <5> 0.1 mM 23% inhibition [2]; <4> reversible by dithiothreitol, 10 mM inhibits to 50% [7]) [2, 7, 31]
large peptide factor <13> (<13> chloroplast enzyme [22]) [22]
linoleic acid <13> (<13> 0.003 mM, 40% inhibition of chloroplastic form and 43% inhibition of cytosolic form [22]) [22]
malate <2> (<2> at high concentration [24]) [24]
melarsen oxide <16, 17> (<16, 17> active principle of the trypanocidal drug melarsoprol and cymelarsen [27]; <17> cymelarsen, IC50 0.0015-0.005 mM [27]) [27]
nucleic acids <16> (<16> strongly inhibit [27]) [27]
octyl glucose <13> (<13> 0.003 mM, 41% inhibition of chloroplastic form and 61% inhibition of cytosolic form [22]) [22]
p-chloromercuribenzoate <3, 4, 5, 7, 10, 17> (<4, 5> 10 nM, strong inhibition [7, 12]; <7> 0.0001 mM inhibits enzyme over 60% [9]; <4> reversible by dithiothreitol, 0.001 mM inhibits more than 90% of the enzyme activity [7]; <3> effective inhibitor [31]; <10> 0.05 mM, 100% inhibition [11]; <17> 0.1 mM, reversed by thiols, e.g. dithiothreitol [27]) [7, 9, 11, 12, 27, 31]
p-mercuribenzoate <5> (<5> 0.1 mM complete inhibition [2]) [2]
palmitic acid <13> (<13> 0.003 mM, 40% inhibition of chloroplastic form and 43% inhibition of cytosolic form [22]) [22]
phenylmethyl sulfonyl fluoride <13> (<13> 0.5 mM 50% inhibition of cytosolic form, little effect on chloroplastic form [22]) [22]
phosphate <1, 6, 13> (<1> competitive inhibitor against dihydroxyacetone phosphate, non-competitive inhibitor against NADH [30]; <6> competitive inhibition [9]; <13> slight inhibition of cytosolic form, at higher concentrations above 20-30 mM both forms are inhibited [22]) [9, 20, 22, 30]
phosphatidyl choline <13> (<13> 0.003 mM, 42% inhibition of chloroplastic form and 43% inhibition of cytosolic form [22]) [22]
phosphogluconate <13> (<13> cytosolic isozyme [22]) [22]
reduced thioredoxin <13> (<13> stimulation of chloroplastic form [22]) [22]
sedoheptulose 1,7-bisphosphate <13> (<13> 0.5 mM, chloroplastic and cytosolic form inhibited by 50% [22]) [22]
small peptide factor <13> (<13> cytosolic enzyme [22]) [22]

suramin <16> (<16> trypanocidal drug, potent inhibitor, IC50 0.0002 mM [27]; <17> IC50 0.00044 mM [27]) [27]
thylakoid fraction <13> (<13> 0.02 mM 97% inhibition [22]) [22]
Additional information <1, 13> (<1> no inhibition by NADPH, acetaldehyde, glycerol, ethanol [1]; <13> cyclic-AMP [22]) [1, 22]
Additional information <1, 12> (<12> high ionic strength above 0.03 M [13]) [13, 20]
Additional information <5> (<5> coenzyme analogs: acetylpyridine-NAD^+, deaminoacetylpyridine-NAD^+, pyridinealdehyde-NAD^+, deaminopyridinealdehyde-NAD^+, potent inhibitors [2]) [2]

Cofactors/prosthetic groups

NAD^+ <1-14, 17-20> [1, 2, 4-21, 23-27, 29-32]
NADH <1-5, 7, 9-14, 17-20> (<3-5> 1 mol per mol enzyme [2]; <5, 7> 2 mol per mol enzyme [3, 5, 9]; <5> spectrophoto-/fluorimetric monitoring [5]) [1-3, 5-32]
Additional information <1-3, 5, 7, 9, 13, 14> (<5> deamino-NAD^+ 60.3% as effective as NAD^+ [2]; <7> deamino-NAD^+ 68% as effective as NAD^+ [9]; <7> 3-acetylpyridine-NAD^+ 1.5% as effective as NAD^+ [9]; <14> activity with dihydroxyacetone phosphate and NADPH 5% of that with NADH [23]; <1-3, 5, 9, 13> no activity with NADPH [1, 10, 12, 15, 20, 24, 31]; <5, 13> not $NADP^+$ [12, 15]) [1, 2, 9, 10, 12, 15, 20, 23, 24, 31]

Activating compounds

ATP <13> (<13> slight activation [22]) [22]
EDTA <5> (<5> without EDTA the rates are about 60% of the maximal rate [2]) [2]
bovine serum albumin <13> (<13> activity increased [22]) [22]
dihydrolipoic acid <13> (<13> stimulation of chloroplastic form, 147% of enzyme activity [22]) [22]
dipotassium malate <2> (<2> slight activation at low ionic strength [24]) [24]
dithiothreitol <13> (<13> stimulation of chloroplastic form, 150% of enzyme activity [22]) [22]
fructose 2,6-bisphosphate <13> (<13> 0.025 mM maximum stimulation of 2.3-fold of cytosolic form, not chloroplastic form, no stimulation by fructose-1,6-bisphosphate [22]) [22]
glycerol <3> (<3> competitive activator with respect to dihydroxyacetone phosphate above 30 mM [31]) [31]
glycine buffer <5> (<5> 0.9 M, slight activation [2]) [2]
phosphate <13> (<13> around 5 mM, activation, chloroplastic form [22]) [22]
phosphogluconate <13> (<13> 0.005 mM slightly stimulates chloroplastic form [22]) [22]
potassium glutamate <2> (<2> up to 0.1 M, activation, high activity also at high ionic strength [24]) [24]

Metals, ions

KCl <2> (<2> 0.1-0.3 M, chloride salts become strongly inhibitory at higher concentrations, replacing K^+ with Na^+ or NH_4^+ yields similar results [24]) [24]
$MgCl_2$ <2> (<2> 0.1-0.3 M [24]) [24]
$MgSO_4$ <2> (<2> 0.1-0.3 M, slight activation at low ionic strength, most inhibitory at higher concentration [24]) [24]

Turnover number (min^{-1})

1920 <12> (L-glycerol-3-phosphate, <12> per subunit of 33500 Da [13]) [13]
11280 <12> (dihydroxyacetone phosphate, <12> per subunit of 33500 Da [13]) [13]
20670 <5> (dihydroxyacetone phosphate, <5> reduction of synthetic dihydroxyacetone phosphate, 20°C, pH 7, rabbit muscle [2]) [2]
23400 <5> (dihydroxyacetone phosphate, <5> reduction of synthetic dihydroxyacetone phosphate, 20°C, pH 7, rabbit muscle, under conditions where competitive inhibition by fructose 1,6-diphosphate are not excluded) [2]

Specific activity (U/mg)

0.028 <15> [25]
7.1 <8> (<8> standard deviation 0.2, crude preparation, testes [4]) [4]
33 <8> (<8> isozyme P-V, liver [4]) [4]
34 <14> [23]
37.5 <13> [15]
41.1 <1> [1]
50.1 <1> [30]
70.7 <8> (<8> standard deviation 5.1, crude preparation, liver [4]) [4]
71.8 <11> [18]
77 <5> (<5> isozyme I6.5, after affinity chromatography, after preparative isoelectric-focusing: 175 units/mg [16]) [16]
80 <17> (<17> estimated [29]) [29]
88.2 <3> [21]
96 <8> (<8> isozyme P I, liver [4]) [4]
103.5 <10> [11]
115 <5> (<5> heart [12]) [12]
119 <5> (<5> 62 units/mg calculated by using protein concentration determined by dry-weight measurements [5]) [5]
123 <8> (<8> isozyme P-IV, liver [4]) [4]
152.4 <4> [7]
154 <16> [27]
156 <10> (<10> isozyme GPDH-3 [17,19]) [17, 19]
158 <1> [20]
162 <5> (<5> isozyme I5.9 [16]) [16]
170 <17> (<17> standard deviation 45 units/mg [32]) [32]
179.8 <10> (<10> isozyme GPDH-1 [17, 19]) [17, 19]
180 <5> (<5> mammary gland [12]) [12]
198.9 <12> [13]
200 <2> [24]

234 <5> (<5> adipose tissue, 147 units/mg refer to protein determination by Kjeldahl nitrogen analysis [6]) [6]
250 <3> [31]
253 <8> (<8> isozyme P-II, liver [4]) [4]
295 <5> [14]
310 <9> [10]
340 <7> [9]
806 <8> (<8> isozyme P-III, liver [4]) [4]

K_m-Value (mM)

0.0043 <5> (NADH, <5> adipose tissue [6]) [6]
0.0044 <5> (NAD^+, <5> adipose tissue [6]) [6]
0.005 <3> (NADH, <3> measured at pH 7.5 [21]) [21]
0.00526 <11> (NADH, <11> α-GPDHs [18]) [18]
0.006 <9> (NADH) [10]
0.0066 <2> (NADH, <2> standard deviation 0.003 mM [24]) [24]
0.01 <5> (NAD^+, <5> I6.5 [16]) [16]
0.01 <5> (NAD^+, <5> muscle [6]) [6]
0.01 <13> (NADH) [15]
0.01 <17> (NADH) [27]
0.01 <4> (NADH, <4> less than 0.01 mM [7]) [7]
0.01 <5> (NADH, <5> I6.5 [16]) [16]
0.01 <5> (NADH, <5> muscle [6]) [6]
0.01 <7> (NADH, <7> less than 0,01 mM [9]) [9]
0.012 <14> (dihydroxyacetone phosphate) [23]
0.016 <14> (NADH) [23]
0.018 <12> (NADH) [13]
0.022 <5> (NADH) [14]
0.023 <1> (NADH) [1]
0.024 <16> (NADH) [27]
0.026 <3> (dihydroxyacetone phosphate, <3> measured at pH 7.5 [21]) [21]
0.03 <4> (NAD^+) [7]
0.036 <3> (NAD^+, <3> measured at pH 9.0 [21]) [21]
0.075 <5> (dihydroxyacetone phosphate) [14]
0.078 <10> (NADH, <10> GPDH-1 [19]) [19]
0.083 <5> (NAD^+) [14]
0.1 <17> (dihydroxyacetone phosphate) [27]
0.1003 <11> (NAD^+, <11> α-GPDHs, standard deviation 0.0041 mM [18]) [18]
0.11 <5> (α glycerophosphate) [2]
0.12 <16> (dihydroxyacetonephosphate) [27]
0.13 <2> (dihydroxyacetone phosphate, <2> standard deviation 0.03 mM [24]) [24]
0.143 <3> (sn-glycerol-3-phosphate, <3> measured at pH 9.0 [21]) [21]
0.15 <5> (dihydroxyacetone phosphate, <5> muscle [6]) [6]
0.17 <4> (dihydroxyacetone phosphate) [7]
0.19 <5> (DL-glycerol-3-phosphate, <5> muscle [6]) [6]

0.19 <13> (NAD^+) [15]
0.2 <5> (dihydroxyacetone phosphate, <5> I6.5 [16]) [16]
0.2144 <11> (dihydroxyacetone phosphate, <11> α-GPDHs, standard deviation 0.0215 mM [18]) [18]
0.23 <9> (dihydroxyacetone phosphate) [10]
0.23 <7> (dihydroxyacetone phosphate, <7> in MOPS buffer, 0.33 mM in Tris-histidine [9]) [9]
0.25 <5> (dihydroxyacetone phosphate, <5> adipose tissue [6]) [6]
0.26 <12> (NAD^+) [13]
0.3 <5> (DL-glycerol-3-phosphate, <5> I6.5 [16]) [16]
0.3 <2> (NAD^+, <2> standard deviation 0.06 mM [24]) [24]
0.3 <4> (α-glycerol phosphate) [7]
0.3 <13> (dihydroxyacetone phosphate, <13> 0.30-0.35 mM [15]) [15]
0.32 <17> (NAD^+) [27]
0.33 <8> (α-glycerophosphate, <8> testes [4]) [4]
0.33 <12> (dihydroxyacetonephosphate) [13]
0.38 <5> (NAD^+) [2]
0.38 <10> (NAD^+, <10> GPDH-1 [19]) [19]
0.4 <5> (DL-glycerol-3-phosphate, <5> adipose tissue [6]) [6]
0.42 <16> (NAD^+) [27]
0.46 <5> (dihydroxyacetone phosphate) [2]
0.5 <9> (NAD^+) [10]
0.52 <10> (dihydroxyacetone phosphate, <10> GPDH-1 [19]) [19]
0.54 <1> (dihydroxyacetone phosphate) [1]
0.59 <10> (glycerol 3-phosphate, <10> GPDH-1 [19]) [19]
0.74 <12> (L-glycerol 3-phosphate) [13]
0.909 <5> (glycerol 3-phosphate) [14]
1.2 <2> (DL-glycerol-3-phosphate, <2> standard deviation 0.5 mM [24]) [24]
1.6 <13> (glycerol phosphate) [15]
1.7 <16> (glycerol-3-phosphate) [27]
1.89 <11> (glycerol 3-phosphate, <11> α-GPDHs, standard deviation 0.25 mM [18]) [18]
2 <9> (glycerol 3-phosphate) [10]
2.25 <17> (glycerol-3-phosphate) [27]
3.39 <8> (α-glycerophosphate, <8> liver [4]) [4]
Additional information <8> (α-glycerophosphate, <8> K_m values from five different isoelectric focused components [4]) [4]
Additional information <10> (<10> K_m values of isozyme GPDH-3 [19]) [19]
Additional information <11> (<11> K_m values of two further allelic isozymes, α-GPDHf and α-GPDHm [18]) [18]
Additional information <5> (<5> K_m values of isozyme I5.9 [16]) [16]
Additional information <5> (<5> various K_m-values of 4 isozymes of heart, muscle, liver and mammary gland [12]) [12]

pH-Optimum

6.5 <16, 17> (<16, 17> glycerol-3-phosphate formation [27]) [27]

6.6 <7, 12> (<7> broad pH optimum around pH 6.6 [9]; <12> optimum pH for NADH oxidation around pH 6.6 [13]) [9, 13]
6.8 <1> (<1> most active [30]) [30]
6.8 <10, 13, 14> (<14> dihydroxyacetone phosphate reduction [23]; <13> reduction of dihydroxyacetone-P [15]; <10> reduction of dihydroxyacetone-P, isozyme GPDH-1 [19]) [15, 19, 23]
7.2 <4> (<4> reduction of dihydroxyacetone phosphate, purified enzyme, decline in activity is more rapid at higher pH values, 7.3-7.5 crude homogenate of enzyme [7]) [7]
7.4 <10> (<10> reduction of dihydroxyacetone-phosphate, isozyme GPDH-3 [19]) [19]
7.5 <3> (<3> dihydroxyacetone phosphate reduction [31]) [31]
7.5 <5> (<5> reduction of dihydroxyacetone phosphate, in 0.2 M phosphate buffer [2]) [2]
7.6 <1> (<1> maximum rate of dihydroxyacetone phosphate reduction [1]) [1]
9 <3> (<3> glycerol-3-phosphate oxidation [31]) [31]
9.3 <8, 14> (<8> liver, component P-III, pH optimum of testis enzyme also around pH 9.3 [4]; <14> glycerol-phosphate oxidation [23]) [4, 23]
9.5 <13> (<13> oxidation of glycerol-phosphate [15]) [15]
9.6 <16, 17> (<16,17> dihydroxyacetone phosphate formation [27]) [27]
10 <12> (<12> optimum pH for NAD^+ reduction about pH 10.0 [13]) [13]
10.2 <5> (<5> oxidation of glycerol 3-phosphate, in 0.03 glycine buffer [2]) [2]
Additional information <8> (<8> optimum pH values for components P-I, P-II, PIV, P-V range between pH 8.2 and 8.6, liver [4]) [4]

pH-Range

5-7.5 <16, 17> (<16, 17> highest activity, glycerol-3-phosphate formation [27]) [27]
5-8.1 <13> (<13> about half-maximal activity at pH 5.0 and 8.1, reduction of dihydroxyacetone phosphate [15]) [15]
5.2-7.8 <12> (<12> about half-maximal activity between pH 5.2 and 7.8, reduction of dihydroxyacetone phosphate [13]) [13]
5.8-7.2 <13> (<13> reduction of dihydroxyacetone phosphate occured over a broad range between 5.8 to 7.2 [15]) [15]
6-6.5 <11> (<11> maximal activities of the three allelic forms are obtained within a pH range of 6.0-6.5 for dihydroxyacetone phosphate reduction [18]) [18]
6.5-8.2 <4> (<4> about half-maximal activity at pH 6.5 and 8.2, reduction of dihydroxyacetone phosphate [7]) [7]
6.7-8.7 <3> (<3> about 80% of maximal activity at pH 6.7 and 8.7, reduction of dihydroxyacetone phosphate, decrease of activity below and above these pH values [21]) [21]
6.8-7.2 <1> (<1> activity falls off rapidly as pH is increased [20]) [20]
7-8 <1> (<1> dihydroxyacetone reduction, activity decreases rapidly below pH 7.0 and above pH 8.0, more rapidly at acidic values [1]) [1]

7.5-8 <9> (<9> pH-rate profile for the reduction of dihydroxyacetone phosphate shows rather sharp optimum between pH 7.5 and 8.0 [10]) [10]
7.9-8.2 <2> (<2> maximum activity in pH interval 7.9-8.2 [24]) [24]
8-10 <8> (<8> about half-maximal activity at pH 8.0 and 10.0, oxidation of glycerol 3-phosphate, isozymes P-III, liver) [4]
8-11 <12> (<12> about half-maximal activity at pH 8.0 and 11.0, oxidation of L-glycerol-3-phosphate [13]) [13]
8.1-9.4 <3> (<3> about 80% of maximal activity at pH 8.1 and 9.4, oxidation of glycerol-3-phosphate, decrease of activity below and rapidly above these pH values [21]) [21]
8.6-10 <13> (<13> about half-maximal activity at pH 8.6 and 10.0, oxidation of glycerol 3-phosphate [15]) [15]
10-10.5 <11> (<11> maximal activities of the three allelic forms are obtained within a pH range of 10.0-10.5 for glycerol 3-phosphate oxidation [18]) [18]

Temperature optimum (°C)
25 <8> (<8> isozymes P-I, P-V, liver [4]) [4]
30 <11> (<11> isozyme GPDHf, activity falls off at temperature above 30°C [18]) [18]
31 <12> [13]
35 <8> (<8> isozymes P-II, P-III, P-IV, liver [4]) [4]
40 <11> (<11> isozyme GPDHm and GPDHs [18]) [18]
51 <1> [30]

Temperature range (°C)
4.8-9.9 <7> (<7> stable for 15 min at 21°C [9]) [9]

4 Enzyme Structure

Molecular weight
62000 <3> (<3> gel filtration [21]) [6, 21]
63500 <13> (<13> gel filtration [15]) [15]
65000 <11> (<11> gel filtration [18]) [18]
66000 <5, 10> (<5> 61000-71000 Da, gel filtration [6]; <10> gel filtration [17]) [6, 17]
67000 <10> (<10> gel filtration [11]) [11]
68000 <1> (<1> gel filtration [1]) [1]
69000 <17> (<17> 66000-72000 Da, gel filtration [32]) [32]
72000 <4> (<4> gel filtration [7]) [7]
72500 <12> (<12> 69000-76000 Da, gel filtration [13]) [13]
73000 <7> (<7> calculated by amino acid analysis [9]) [9]
74000 <3> (<3> electrophoresis in a non-denaturing system [31]) [31]
75000 <4, 5> (<4> sucrose density gradient ultracentrifugation [7]; <5> calculated from subunit molecular weight and physical plus chemical properties [6]) [6, 7]
76000 <5, 7> (<5> meniscus depletion method [5]; <7> gel filtration [9]) [5, 9]

76500 <7> (<7> 74000-79000 Da, SDS-PAGE under non-denaturing conditions [9]) [9]
78000 <4, 5> (<4> gel filtration [7]; <5> calculated from partial specific volume, sedimentation and diffusion data [2]) [2, 7]
78600 <5> (<5> fluorometric titrations [5]) [5]
79500 <7> (<7> meniscus depletion method [9]) [9]
Additional information <5, 7> (<5, 7> amino acid analysis [3]) [3]

Subunits

? <2, 14> (<14> x * 33000, calculated from Stokes radius [23]; <2> x * 42000, SDS-PAGE [24]; <5> x * 37500, isozyme from heart (II6.1), calculated by amino acid composition [12]; <1> x * 31000, gel filtration, estimated from elution volume on sepharose 6B [20]; <1> x * 43000-47000, SDS-PAGE [30]; <16> x * 37000, SDS-PAGE [27]; <16> x * 39140, calculated by cDNA-sequence [29]; <17> x * 37651, calculated by cDNA-sequence [29]) [12, 20, 23, 24, 27, 29, 30]
dimer <1, 3, 5, 7, 10-12, 17> (<10> 2 * 31700, SDS-PAGE [11]; <10> 2 * 32000, SDS-PAGE, confirmed by amino acid composition [17]; <12> 2 * 33500, SDS-PAGE [13]; <11> 2 * 35000-37000, SDS-PAGE [18]; <5> 2 * 37000, SDS-PAGE [16]; <5> 2 * 37500, gel filtration in 6 M guanidinium chloride, confirmed by amino acid composition [6]; <3> 2 * 38000, SDS-PAGE [21]; <7> 2 * 39000, gel filtration in guanidinium chloride [9]; <1> 2 * 42000, SDS-PAGE [1]; <5> determined after carboxypeptidase A digestion [3]; <3> 2 * 35000, SDS-PAGE [31]; <17> 2 * 37200-37900 [32]) [1, 3, 6, 9, 11, 13, 16-18, 21, 31, 32]
Additional information <5> (<5> 37800-39300, rabbit muscle, minimal molecular weight calculated by titration with NADH at pH 6.0 [5]) [5]

Posttranslational modification

side-chain modification <16> (<16> modification of Cys345 in hydrophobic pocket [26]) [26]

5 Isolation/Preparation/Mutation/Application

Source/tissue

Brown Pearce carcinoma cell <5> [12]
Ehrlich ascites carcinoma cell <3> [2]
adipose tissue <3, 5> (<5> renal-adipose tissue [6]) [6, 31]
brain <4, 5> [7, 16]
breast cancer cell <3> [2]
kidney <3, 4> [2]
larva <10> (<10> late third instar [17]) [17]
leaf <13> [15, 22]
leukocyte <3, 4> (<3, 4> also other blood components [2]) [2]
liver <3-5, 8> [2, 4, 12, 14]
mammary gland <5> [12]

muscle <4, 5, 6, 7, 9, 10, 19, 20> (<5> skeletal-muscle [6]; <9> breast muscle [10]; <19> bovine smooth muscles [2]; <6, 7, 10> thoracic flight muscle [2, 3, 9, 11, 19]; <5> heart muscle [12, 16]) [2, 5, 6, 8, 9, 10-12, 16, 19]
mycelium <15> [25]
placenta <3> [21]
sarcoma cell <3, 5, 20> (<20> Crocker sarcoma [2]) [2]
testis <8> [4]

Localization
chloroplast <13, 14> (<13, 14> stroma, soluble [22, 23]) [15, 22, 23]
cytoplasm <4, 5, 21> [2, 12]
glycosome <16, 17> [27, 29]
mitochondrion <16> (<16> determined by isopycnic gradient centrifugation [29]) [29]

Purification
<1> (affinity chromatography [1, 20]) [1, 20, 30]
<2> (2 isozymes [24]) [24]
<3> (affinity chromatography [21, 31]) [21, 31]
<4> (ion-exchange chromatography combined with affinity elution, 2 isozymes that differ in charge by analytical PAGE [7]) [7]
<5> (preparative isoelectric focusing [6]; several isozymes, distinct in isoelectric points [12, 14]; 2 isozymes with distinct isoelectric points in brain [16]; sequential affinity chromatography [14]) [5, 6, 12, 14, 16, 28]
<6> (6 to 8 isozymes of bumblebee flight muscle enzyme [9]) [9]
<7> [9]
<8> (4 isozymes in testes and 5 isozymes in liver, distinct in isoelectric focusing [4]) [4]
<9> [10]
<10> (affinity chromatography [11, 17, 19]; 2 allelic forms of the enzyme [11]; 3 isozymes in adults and 1 in larvae, distinct in isoelectric focusing [17]; 3 isozymes, product of the same gene mapped to left arm of chromosome II, isozyme 1: flight muscle, isozyme 3: larval and adult fat body, isozyme 2: heterodimer of 1 and 3 [19]) [11, 17, 19]
<11> (3 allelic forms distinguishable by electrophoresis [18]) [18]
<12> [13]
<13> (affinity chromatography [15]; 2 isozymes [22]) [15, 22]
<14> (dye-affinity chromatography [23]) [23]
<15> [25]
<16> [26, 27, 29]
<17> [27, 29, 32]

Crystallization
<5, 7, 9, 16> [2, 5, 9, 10, 26]

Cloning
<15> [25]
<16> (recombinant in Escherichia coli [27, 29]) [27, 29]
<17> (recombinant in Escherichia coli [29]) [29]

6 Stability

pH-Stability

4.8-9.9 <7> (<7> stable for 15 min at 21°C [9]) [9]
5 <12> (<12> inactivation after 1 h at 20°C [13]) [13]
5-9 <16, 17> (<16, 17> under diluted conditions [27]) [27]
5.7-8.5 <5> (<5> around this range, most stable [2]) [2]
6-9 <12> (<12> stable for 16 h at 20°C [13]) [13]
7 <5, 11, 14> (<5> less stable [2]; <11> rate of inactivation increases below pH 7 [18]; <14> stable, increasing from 7.0 to 8.0 causes 80% loss of activity [23]) [2, 18, 23]
10 <12> (<12> inactivation after 1 h at 20°C [13]) [13]
Additional information <4, 14> (<4> more stable at alkaline pH, more stable in phosphate buffer than in Tris-glycine buffer [7]; <14> increasing pH from 7.0 to 8.0 causes 80% loss of activity [23]) [7, 23]

Temperature stability

20 <5, 12> (<5> stable for at least 30 min [2]; <12> between pH 6.0 and 9.0, stable for 1 h [13]) [2, 13]
21 <7> (<7> 15 min stable [9]) [9]
30 <12> (<12> 30 min stable, pH 6.6 [13]) [13]
35 <11> (<11> most stable at pH 8, activity reduced, GPDHf to a greater extent [18]) [18]
45 <4, 12> (<12> pH 6.6, 5 min stable [13]; <4> at least 5 min stable [7]) [7, 13]
48 <5> (<5> $t_{1/2}$: 2 min, heart isozyme II6.1 [12]; <5> $t_{1/2}$: 2 min, brain isozyme I5.9 [12, 16]; <5> $t_{1/2}$: 7.5 min, isozyme I6.5 [16]; <5> $t_{1/2}$: 15 min, muscle isozyme I6.5 [12]) [12, 16]
50 <4, 10> (<5> $t_{1/2}$: 2 min in heart and 10 min in muscle, mammary gland and liver [12]; <4> rapid denaturation, prevented by 0.1 mg/ml bovine serum albumin, by over 20% ammonium sulfate or by 0.1 M phosphate buffer, pH 7.5 [7]; <10> crude extract at least 20 min stable [11]) [7, 11]
55 <7, 12> (<5> $t_{1/2}$: 1 min [2]; <7> 5 min stable in Tris-buffer plus EDTA and bovine serum albumin [9]; <12> inactivation after 30 min [13]) [2, 9, 13]
60 <3, 4, 5, 12, 19, 20> (<5> complete inactivation after 1 min [2]; <4, 12> inactivation after 5 min [7, 13]) [2, 7, 13]
61 <7> (<7> complete inactivation after 5 min [9]) [9]

General stability information

<2>, substrates stabilize dilute preparation [24]
<3>, ammonium sulfate inactivates [21]
<4>, bovine serum albumin stabilizes [7]
<4>, lyophilization inactivates [7]
<4>, phosphate buffer stabilizes [7]
<4>, repeated freezing and thawing results in rapid loss of activity, to some extent restorable at room temperature [7]
<5>, charcoal inactivates rabbit enzyme, restorable by thiamic acid [2]

<5>, even cold dilute solutions in 0.2 M ammonium sulfate at pH 5.8 are stable for weeks [2]
<5>, redistilled water prevents denaturation in absence of salts [2]
<9>, NADH stabilizes preparations of low ionic strength [10]
<10>, NAD^+ stabilizes [11]
<10>, NADH stabilizes [19]
<10>, phenylmethane-sulfonyl fluoride stabilizes during purification [17]
<13>, freezing inactivates completely, ammonium sulfate prevents inactivation [15]
<14>, 2-mercaptoethanol does not stabilize [23]
<14>, dithiothreitol does not stabilize [23]
<2, 14>, 20% polyethylene glycol stabilizes during purification [23, 24]
<4, 9>, ammonium sulfate stabilizes [7, 9, 10]
<4, 9, 12>, 2-mercaptoethanol stabilizes during purification [7, 10, 13]
<5, 7, 10, 14>, EDTA stabilizes [2, 9, 11, 23]
<7, 10>, dithiothreitol stabilizes [9, 11, 17]

Storage stability

<1>, -20-0°C, stable in 75% ammonium sulfate [20]
<1>, -20°C, in 0.1 M Tris-HCl, 2 mM 2-mercaptoethanol, pH 7.0, 50% polyethylene glycol 2000, 5 mM NADH or dihydroxyacetone phosphate, unstable [1]
<1>, -80°C, in 0.1 M Tris-HCl, 2 mM 2-mercaptoethanol, pH 7.0, 50% polyethylene glycol 2000, 5 mM NADH or dihydroxyacetone phosphate, unstable [1]
<1>, 4°C, 1 M Tris-HCl, 2 mM 2-mercaptoethanol, pH 7.0, eluted from Sephadex G-100 column, 90% activity remained after several weeks [1]
<1>, 4°C, stable for several weeks in 1 M Tris/HCl buffer with 2-mercaptoethanol, pH 7.0 [1]
<2>, 0°C, 30 mM Tris-H_2SO_4, 2.5 mM Na_2EDTA, 5 mM dithiothreitol, 0,05% NaN_3 (pH 8, 4°C), 0.1 mM phenylmethylsulfonylfluoride, 15 17.5% w/v poly (ethylene glycol) 4000, after six days loss of activity insignificant [24]
<2>, 0°C, crude extract stable for 6 days in presence of polyethylene glycol [24]
<3>, -20°C, 50% glycerol, retains activity for several months [31]
<5>, 4°C, dilute solution, pH 5.8, stable for weeks [2]
<7>, 2°C, crystals stable at least a month in 0.1 M Tris buffer, pH 7.6 with added EDTA, dithiothreitol and ammonium sulfate [9]
<7>, crystals, 2°C in 0.1M Tris, 10 mM EDTA, 2 mM dithiothreitol, pH 7.6, $(NH_4)_2SO_4$ in final concentration of 1.7 M [9]
<9>, crystallized, in ammonium sulfate stable for months [10]
<10>, 5°C, isozyme 1: stable for over a month in 10 mM sodium phosphate buffer, pH 6.5, with added EDTA, dithiothreitol and NADH, isozyme 3: 50% loss of activity under the same conditions [19]
<13>, -20°C, stable as $(NH_4)_2SO_4$-precipitate [15]
<13>, 4°C, purified and dilute preparation stable for several days in presence of NADH [15]

<14>, -20°C, stable for at least 2 months [23]
<14>, 0°C, $t_{1/2}$: 50 min [23]
<14>, 4°C, $t_{1/2}$: 6 h in absence of EDTA and polyethyleneglycol [23]
<2, 5>, 4°C, stable [14, 24]
<5, 14>, 4°C, 20% v/v, glycerol, purified stable [12, 23]

References

[1] Albertyn, J.; Van Tonder, A.; Prior, B.A.: Purification and characterization of glycerol-3-phosphate dehydrogenase of Saccharomyces cerevisiae. FEBS Lett., **308**, 130-132 (1992)

[2] Baranowski, T.: α-Glycerophosphate dehydrogenase. The Enzymes, 2nd Ed. (Boyer, P.D. ed.), 7, 85-96 (1963)

[3] Brosemer, R.W.; Kuhn, R.W.: Comparative structural properties of honey-bee and rabbit α-glycerophosphate dehydrogenases. Biochemistry, **8**, 2095-2105 (1969)

[4] Yamada, M.: Studies on L-glycerol-3-phosphate dehydrogenases in Japanese quail, Coturnix coturnix japonica. I. Multiple components and some enzymatic properties of nicotinamide adenine dinucleotide-linked L-glycerol-3-phosphate dehydrogenase from liver and testes. J. Biochem., **72**, 1081-1086 (1972)

[5] Bentley, P.; Dickinson, F.M.; Jones, I.G.: Purification and properties of rabbit muscle L-glycerol 3-phosphate dehydrogenase. Biochem. J., **135**, 853-859 (1973)

[6] Warkentin, D.L.; Fondy, T.P.: Isolation and characterization of cytoplasmic L-glycerol-3-phosphate dehydrogenase from rabbit renal adipose tissue and its comparison with the skeletal muscle enzyme. Eur. J. Biochem., **36**, 97-109 (1973)

[7] McGinnis, J.F.; DeVellis, J.: Purification and characterization of rat brain glycerol phosphate dehydrogenase. Biochim. Biophys. Acta, **364**, 17-27 (1974)

[8] Bentley, P.; Dickinson, F.M.: A study of the coenzyme-binding characteristics of rabbit muscle L-glycerol 3-phosphate dehydrogenase. Biochem. J., **143**, 11-17 (1974)

[9] Fink, S.C.; Brosemer, R.W.: Glycerol-3-phosphate dehydrogenase from the honey bee. Methods Enzymol., **41B**, 240-245 (1975)

[10] White, H.B.: Glycerol phosphate dehydrogenase of chicken breast muscle. Methods Enzymol., **41B**, 245-249 (1975)

[11] Collier, G.E.; Sullivan, D.T.; MacIntyre, R.J.: Purification of α-glycerophosphate dehydrogenase from Drosophila melanogaster. Biochim. Biophys. Acta, **429**, 316-323 (1976)

[12] Ostro, M.J.; Fondy, T.P.: Isolation and characterization of multiple molecular forms of cytosolic NAD-linked glycerol-3-phosphate dehydrogenase from normal and neoplastic rabbit tissues. J. Biol. Chem., **252**, 5575-5583 (1977)

[13] Fernández-Sousa, J.M.; Gavilanes, J.G.; Municio, A.M.; Pérez-Aranda, A.: L-glycerol-3-phosphate dehydrogenase from the insect Ceratitis capitata. Purification, physicochemical and enzymic properties. Biochim. Biophys. Acta, **481**, 6-24 (1977)

[14] McLoughlin, D.J.; MacQuarrie, R.: Purification and properties of the major isozymic form of cytoplasmic glycerol-3-phosphate dehydrogenase from rabbit liver. Biochim. Biophys. Acta, **527**, 204-211 (1978)

[15] Santora, G.T.; Gee, R.; Tolbert, N.E.: Isolation of a sn-glycerol 3-phosphate: NAD oxidoreductase from spinach leaves. Arch. Biochem. Biophys., **196**, 403-411 (1979)

[16] Kornbluth, R.; Tracy, P.S.; Fondy, T.P.: Isoenzymic forms of NAD-linked glycerol-3-phosphate dehydrogenase from rabbit brain. Biochim. Biophys. Acta, **568**, 273-286 (1979)

[17] Niesel, D.W.; Bewley, G.C.; Miller, S.G.; Armstrong, F.B.; Lee, C.-Y.: Purification and structural analysis of the soluble sn-glycerol-3-phosphate dehydrogenase isozymes in Drosophila melanogaster. J. Biol. Chem., **255**, 4073-4080 (1980)

[18] Narise, S.: Purification and biochemical properties of allelic forms of cytoplasmic glycerol-3-phosphate dehydrogenase from Drosophila virilis. Biochim. Biophys. Acta, **615**, 289-298 (1980)

[19] Niesel, D.W.; Bewley, G.C.; Lee, C.-Y.; Armstrong, F.B.: sn-Glycerol-3-phospate dehydrogenase (soluble) from Drosophila melanogaster. Methods Enzymol., **89**, 296-301 (1982)

[20] Merkel, J.R.; Straume, M.; Sajer, S.A.; Hopfer, R.L.: Purification and some properties of sn-glycerol-3-phosphate dehydrogenase from Saccharomyces cerevisiae. Anal. Biochem., **122**, 180-185 (1982)

[21] Zolnierowicz, S.; Swierczynski, J.; Zelewski, L.: Isolation and properties of glycerol-3-phosphate oxidoreductase from human placenta. Eur. J. Biochem., **154**, 161-166 (1986)

[22] Gee, R.W.; Byerrum, R.U.; Gerber, D.W.; Tolbert, N.E.: Differential inhibition and activation of two leaf dihydroxyacetone phosphate reductases. Plant Physiol., **87**, 379-383 (1988)

[23] Klöck, G.; Kreuzberg, K.: Kinetic properties of a sn-glycerol-3-phosphate dehydrogenase purified from the unicellular alga Chlamydomonas reinhardtii. Biochim. Biophys. Acta, **991**, 347-352 (1989)

[24] Nilsson, A.; Adler, L.: Purification and characterization of glycerol-3-phosphate dehydrogenase (NAD+) in the salt-tolerant yeast Debaryomyces hansenii. Biochim. Biophys. Acta, **1034**, 180-185 (1990)

[25] Fillinger, S.; Ruijter, G.; Tamás, M.J.; Visser, J.; Thevelein, J.M.; d'Enfert, C.: Molecular and physiological characterization of the NAD-dependent glycerol 3-phosphate dehydrogenase in the filamentous fungus Aspergillus nidulans. Mol. Microbiol., **39(1)**, 145-157 (2001)

[26] Suresh, S.; Turley, S.; Opperdoes, F.R.; Michels, P.A.M.; Hol, W.G.J.: A potential target enzyme for trypanocidal drugs revealed by the crystal structure of NAD-dependent glycerol-3-phosphate dehydrogenase from Leishmania mexicana. Structure, **8**, 541-552 (2000)

[27] Marché, S.; Michels, P.A.M.; Opperdoes, F.R.: Comparative study of Leishmania mexicana and trypanosoma brucei NAD-dependent glycerol-3-phosphate dehydrogenase. Mol. Biochem. Parasitol., **106**, 83-91 (2000)
[28] Praphakar, P.; Laboy, J.I.; Wang, J.; Budker, T.; Din, Z.Z.; Chobanian, M.; Fahien, L.A.: Effect of NADH-X on cytosolic glycerol-3-phosphate dehydrogenase. Arch. Biochem. Biophys., **360**, 195-205 (1998)
[29] Kohl, L.; Drmota, T.; Do Thi, C.-D.; Callens, M.; Van Beeumen, J.; Opperdoes, F.R.; Michels, P.A.M.: Clonig and characterization of the NAD-linked glycerol-3-phosphate dehydrogenase of Trypanosoma brucei brucei and Leishmania mexicana mexicana and expression of the trypanosome enzyme in Escherichia coli. Mol. Biochem. Parasitol., **76**, 159-173 (1996)
[30] Cai, J.; Pietzsch, M.; Theobald, U.; Rizzi, M.: Fast purification and kinetic studies of the glycerol-3-phosphate dehydrogenase from the yeast Saccharomyces cerevisiae. Biotechnology, **49**, 19-27 (1996)
[31] Koekemoer, T.C.; Litthauer, D.; Oelofsen, W.: Isolation and characterization of adipose tissue glycerol-3-phasphate dehydrogenase. Int. J. Biochem. Cell Biol., **27**, 625-632 (1995)
[32] Misset, O.; Bos, O.J.M.; Opperdoes, F.R.: Glycolytic enzymes of Trypanosoma brucei. Simultaneous purification, intraglycosomal concentration and physical properties. Eur. J. Biochem., **157**, 441-453 (1986)

D-Xylulose reductase 1.1.1.9

1 Nomenclature

EC number
1.1.1.9

Systematic name
xylitol:NAD^+ 2-oxidoreductase (D-xylulose-forming)

Recommended name
D-xylulose reductase

Synonyms
2,3-cis-polyol(DPN) dehydrogenase (C3-5)
NAD-dependent xylitol dehydrogenase
XDH [Swissprot]
erythritol dehydrogenase
pentitol-DPN dehydrogenase
reductase, D-xylulose
xylitol dehydrogenase
xylitol-2-dehydrogenase

CAS registry number
9028-16-4

2 Source Organism

<1> *Cavia porcellus* [1, 6]
<2> *Aerobacter aerogenes* [2]
<3> *Penicillium chrysogenum* [3]
<4> *Candida shehatae* [4, 14]
<5> *Pachysolen tannophilus* (strain Boidin et Adzet [5]) [5, 8]
<6> *Morganella morganii* [7]
<7> *Providencia stuartii* [7]
<8> *Serratia marcescens* [7]
<9> *Pichia stipitis* [9]
<10> *Candida tropicalis* [10, 13]
<11> *Candida tenuis* (CBS 4435 [11]) [11]
<12> *Fusarium oxysporum* [12]
<13> *Saccharomyces cerevisiae* [15]

3 Reaction and Specificity

Catalyzed reaction

xylitol + NAD^+ = D-xylulose + NADH

Reaction type

redox reaction

Natural substrates and products

S xylitol + NAD^+ <4-8> (<4> inducible pathway of xylose catabolism [4]) (Reversibility: r, <4> [4]) [4, 5, 7]

P ?

Substrates and products

S D-arabitol + NAD^+ <2> (Reversibility: ? <2> [2]) [2]

P ?

S D-ribitol + NAD^+ <1, 4-8, 10> (<8> oxidation at 11% of the rate of xylitol oxidation [7]; <5> 85% of the rate of xylitol oxidation [8]) (Reversibility: r <4> [4]) [1, 4, 6-8, 13]

P D-ribulose + NADH <1> [6]

S D-ribulose + NADH <4> (Reversibility: r <4> [4]) [4]

P D-ribitol + NAD^+ <4> [4]

S D-sorbitol + NAD^+ <1, 4-10> (<4> oxidation at about 45% [4]; <7> rate of xylitol oxidation at 56% [7]; <6> rate of xylitol oxidation at 67% [7]; <5> rate of xylitol oxidation at 95% [8]; <8> rate of xylitol oxidation at 118% [7]) (Reversibility: ? <1, 4-10> [1, 4, 6-9, 13]) [1, 4, 6-9, 13]

P D-fructose + NADH <1> [6]

S D-xylulose + NADH <1-4, 6-8, 10, 13> (Reversibility: r <1-4, 6-8, 10, 13> [1-4, 7, 13, 15]) [1-4, 7, 13, 15]

P xylitol + NAD^+ <1-4, 6-8, 10, 13> [1-4, 7, 13, 15]

S L-arabitol + NAD^+ <1, 2> (Reversibility: ? <1, 2> [1, 2]) [1, 2]

P ?

S L-erythrulose + NADH <1, 2> (<1> reduction at the same rate as D-xylulose [1]) (Reversibility: r <2> [2]) [1, 2]

P erythritol + NAD^+ <1, 2> [2, 6]

S L-iditol + NAD^+ <1> (Reversibility: ? <1> [1]) [1]

P ?

S L-xylulose + NADH <2> (Reversibility: ? <2> [2]) [2]

P L-xylitol + NAD^+

S dihydroxyacetone + NADH <2> [2]

P glycerol + NAD^+

S erythritol + NAD^+ <2> (Reversibility: r <2> [2]) [2]

P L-erythrulose + NADH <2> [2]

S glycerol + NAD^+ <2, 5> (<5> poor substrate [8]) (Reversibility: ? <2, 5> [2, 8]) [2, 8]

P ?

S xylitol + NAD^+ <1-13> (Reversibility: r <1-4, 6-8, 10, 13> [1-4, 7, 13, 15]) [1-4, 6-15]

P D-xylulose + NADH <1, 3-8, 11, 13> [1, 3, 4, 6-8, 10-13, 15]

S Additional information <1, 2, 4-8> (<2> specific for polyols of 5 or less carbon bearing cis-hydroxy groups in C_2 and C_3 [2]; <1, 2> no oxidation of threitol, xylitol, sorbitol, D-iditol [1, 2]; <4, 6-8> no oxidation of ribitol [4, 7]; <4, 5> no oxidation of mannitol [4, 8]; <5> no oxidation of inositol, meso-erythritol and D-(+)-arabitol [8]; <6-8> no reduction of L-xylulose [7]; <1> no reduction of D-ribulose [1]; <1> no reduction of D-ribose and D-galacturonic acid [1]; <1, 6-8> no reduction of D/L-xylose [1, 7]; <2> no reduction of D-fructose, L-sorbose and D-tagatose [2]) (Reversibility: ? <1, 2, 4-8> [1, 2, 4, 7, 8]) [1, 2, 4, 7, 8]

P ?

Inhibitors

$AgNO_3$ <5> [8]
$CuSO_4$ <5> [8]
D-xylulose <1> (<1> high concentration [1]) [1]
EDTA <2> [2]
$FeSO_4$ <5> [8]
$HgCl_2$ <5> (<5> complete inhibition [8]) [8]
$MnSO_4$ <5> [8]
NH_4Cl <5> (<5> at 310 mM [8]) [8]
NaCl <5> (<5> at 160 mM [8]) [8]
PCMB <1> (<1> complete inhibition [1]) [1]
$Pb(CH_3COO)_2$ <5> [8]
$ZnSO_4$ <5> (<5> complete inhibition [8]) [8]
cysteine <5> (<5> at 294 mM [8]) [8]
iodoacetate <5> (<5> complete inhibition [8]) [8]
Additional information <1> (<1> no inhibition by EDTA or iodoacetate [1]) [1]

Cofactors/prosthetic groups

NAD^+ <1-10> (<1> stereochemistry of hydrogen transfer from xylitol, ribitol or sorbitol to NAD^+ is of the (R)- or A-type [6]; <10> xylitol oxidation favors NAD^+ over $NADP^+$, but xylulose reduction favors NADPH over NADH [13]) [1-10, 12, 13]
NADH <1-8> [1-7]
$NADP^+$ <1, 2, 4, 5, 10> (<1, 2, 4> at high concentrations of xylitol 1.5% as effective as NAD^+ [1, 2, 4]; <1, 2, 4> no activity with $NADP^+$ [1, 2, 4]; <5> no oxidation of ribitol or sorbitol with $NADP^+$ [8]; <10> xylitol oxidation favors NAD^+ over $NADP^+$, but xylulose reduction favores NADPH over NADH [13]) [1, 2, 4, 8, 10, 13]
Additional information <1, 2, 4> (<4> very low activity with NADPH [4]; <1, 2> no activity with NADPH [1, 2]) [1, 2, 4]

Activating compounds

D-glucose <12> (<12> induces activity [12]) [12]
D-lyxose <11> (<11> induces activity [11]) [11]
D-xylose <11-13> (<11-13> induces activity [11, 12, 15]) [11, 12, 15]

GSH <5> (<5> activation [8]) [8]
L-arabinose <11> (<11> induces activity [11]) [11]
cysteine <1, 5> (<5> slight activation [8]; <1> of partially purified preparation [1]) [1, 8]
glycerol <12> (<12> induces activity [12]) [12]
glycine <5> (<5> slight activation [8]) [8]

Metals, ions

KCl <5> (<5> slight activation [8]) [8]
Mg^{2+} <2, 5> (<2> activation at high concentration [2]) [2, 8]
Mn^{2+} <2> (<2> requirement [2]) [2]
NH_4Cl <5> (<5> activation [8]) [8]
NaCl <5> (<5> activation [8]) [8]
Additional information <1, 2> (<1> no evidence for metal requirement [1]; <2> no activation by Zn^{2+}, Cu^{2+}, Ca^{2+}, Ni^{2+}, Co^{2+}, Fe^{2+} [2]) [1, 2]

Specific activity (U/mg)

0.11 <6> [7]
0.11 <8> [7]
0.15-0.18 <4> (<4> depending on nitrogen source in culture medium [14]) [14]
0.22 <7> [7]
0.68 <12> [12]
0.9 <5> [5]
2 <11> [11]
3.1 <2> [2]
15.95 <1> [1]
29.9 <5> [8]
130 <10> [13]
Additional information <1, 13> [1, 15]

K_m-Value (mM)

0.037 <4> (NADH) [4]
0.038 <10> (NAD^+) [13]
0.1 <13> (NAD^+) [15]
0.16 <4> (sorbitol) [4]
0.24 <13> (NADH) [15]
0.66 <1> (D-xylulose) [1]
1.1 <13> (D-xylulose) [15]
7.1 <8> (xylitol) [7]
11 <5> (xylitol) [8]
12.6 <7> (xylitol) [7]
13.8 <4> (xylulose) [4]
16.4 <6> (xylitol) [7]
18.5 <4> (xylitol) [4]
25 <13> (xylitol) [15]
30 <5> (D-sorbitol) [8]
49.8 <10> (xylitol) [13]

50 <5> (ribitol) [8]
496 <8> (ribitol) [7]

pH-Optimum
6.7 <2> (<2> reduction of erythrulose [2]) [2]
7 <1> (<1> substrate reduction [1]) [1]
7.2 <4> (<4> xylitol formation [4]) [4]
8 <10> [13]
8.6 <2, 4> (<2> oxidation of erythritol [2]; <4> D-xylulose formation [4]) [2, 4]
9.1-10 <5> (<5> substrate oxidation, plateau [8]) [8]

pH-Range
4-8 <1> (<1> about 75% of maximal reduction activity at pH 4.0 and 8.0 [1]) [1]
8.7-10.5 <5> (<5> about 80% of maximal oxidation activity at pH 8.7 and 10.5 [8]) [8]

Temperature optimum (°C)
50 <10> [13]
55 <5> [8]

Temperature range (°C)
35-68 <5> (about half-maximal activity at 35°C and 68°C) [8]

4 Enzyme Structure

Molecular weight
82000 <4> (<4> gel filtration [4]) [4]
120000 <5> (<5> gel filtration [8]) [8]
130000 <8> (<8> PAGE [7]) [7]
142000 <6> (<6> PAGE [7]) [7]
155000 <7> (<7> PAGE [7]) [7]
172000 <5> (<5> gel filtration [5]) [5]

Subunits
dimer <4> (<4> 2 * 40000, SDS-PAGE [4]) [4]
tetramer <5> (<5> 4 * 40000, SDS-PAGE [8]; <5> 2 * 40400 + 2 * 41800, SDS-PAGE [5]) [5, 8]

5 Isolation/Preparation/Mutation/Application

Source/tissue
liver <1> [1, 6]

Localization
cytoplasm <1, 2, 4-8> [1, 2, 4-8]

Purification

<1> (partial [1]) [1]
<2> (partial [2]) [2]
<4> [4]
<5> [5]
<6> (partial [7]) [7]
<7> (partial [7]) [7]
<8> (partial [7]) [7]
<10> (partial [13]) [13]

Cloning

<9> [9]
<13> (overexpression in Saccharomyces cerevisiae [15]) [15]

6 Stability

pH-Stability

6.8-7.4 <2> (<2> 90% of maximal activity retained at pH 6.8 and 7.4 at 2°C [2]) [2]
7 <2> (<2> 24 h stable at 2°C [2]) [2]
7-9 <5> (<5> stable [8]) [8]

Temperature stability

25 <5> (<5> and below stable [8]) [8]
60 <5> (<5> complete inactivation within 5 min [8]) [8]

Organic solvent stability

acetone <1> (<1> stable to precipitation with 50% v/v [1]) [1]

General stability information

<1>, crude water extract from liver acetone-powder unstable [1]
<2>, 2-mercaptoethanol stabilizes during purification [2]
<2>, EDTA stabilizes during purification [2]
<1, 5>, freezing/thawing inactivates, 8 cycles lead to 89% loss of activity and 11 cycles to complete inactivation [1, 8]

Storage stability

<1>, -10°C, acetone precipitate stable for weeks [1]
<1>, 0°C, partially purified stable for weeks [1]
<2>, 2°C, 90% of maximal activity retained at pH 6.8 and 7.4 after 24 h [2]
<4>, 20°C, stable with 50% v/v glycerol [4]
<5>, -15°C, 60% loss of activity after 1 month [8]
<5>, -2°C, purified stable for a month [8]
<5>, 0°C, 34% loss of activity within a month [8]

References

[1] Hickman, J.; Ashwell, G.: A sensitive and stereospecific enzymatic assay for xylulose. J. Biol. Chem., **234**, 758-761 (1959)

[2] Jakoby, W.B.; Fredericks, J.: Erythritol dehydrogenase from Aerobacter aerogenes. Biochim. Biophys. Acta, **48**, 26-32 (1961)

[3] Chiang, C.; Knight, S.G.: A new pathway of pentose metabolism. Biochem. Biophys. Res. Commun., **3**, 554-559 (1960)

[4] Yang, V.W.; Jeffries, T.W.: Purification and properties of xylitol dehydrogenase from the xylose-fermenting yeast Candida shehatae. Appl. Biochem. Biotechnol., **26**, 197-206 (1990)

[5] Bolen, P.L.; Roth, K.A.; Freer, S.N.: Affinity purifications of aldose reductase and xylitol dehydrogenase from the xylose-fermenting yeast Pachysolen tannophilus. Appl. Environ. Microbiol., **52**, 660-664 (1986)

[6] Alizade, M.A.; Brendel, K.; Gaede, K.: Chirality of xylitol-oxidizing enzymes from mammalian liver. FEBS Lett., **67**, 41-44 (1976)

[7] Doten, R.C.; Mortlock, R.P.: Inducible xylitol dehydrogenases in enteric bacteria. J. Bacteriol., **162**, 845-848 (1985)

[8] Morimoto, S.; Matsuo, M.; Azuma, K.; Sinskey, A.J.: Purification and properties of D-xylulose reductase from Pachysolen tannophilus. J. Ferment. Technol., **64**, 219-225 (1986)

[9] Persson, B.; Hallborn, J.; Walfridsson, M.; Hahn-Haegerdal, B.; Keraenen, S.; Penttilae, M.; Joernvall, H.: Dual relationships of xylitol and alcohol dehydrogenases in families of two protein types. FEBS Lett., **324**, 9-14 (1993)

[10] Ikeuchi, T.; Kiritani, R.; Azuma, M.; Ooshima, H.: Effect of D-glucose on induction of xylose reductase and xylitol dehydrogenase in Candida tropicalis in the presence of NaCl. J. Basic Microbiol., **40**, 167-175 (2000)

[11] Kern, M.; Haltrich, D.; Nidetzky, B.; Kulbe, K.D.: Induction of aldose reductase and xylitol dehydrogenase activities in Candida tenuis CBS 4435. FEMS Microbiol. Lett., **149**, 31-37 (1997)

[12] Singh, A.; Schuegerl, K.: Induction and regulation of D-xylose catabolizing enzymes in Fusarium oxysporum. Biochem. Int., **28**, 481-488 (1992)

[13] Takamizawa, K.; Uchida, S.; Hatsu, M.; Suzuki, T.; Kawai, K.: Development of a xylitol biosensor composed of xylitol dehydrogenase and diaphorase. Can. J. Microbiol., **46**, 350-357 (2000)

[14] Palnitkar, S.; Lachke, A.: Effect of nitrogen sources on oxidoreductive enzymes and ethanol production during D-xylose fermentation by Candida shehatae. Nat. Chem. Lab. Commun., **5153**, 258-260 (1991)

[15] Richard, P.; Toivari, M.H.; Penttilae, M.: Evidence that the gene YLR070c of Saccharomyces cerevisiae encodes a xylitol dehydrogenase. FEBS Lett., **457**, 135-138 (1999)

L-Xylulose reductase 1.1.1.10

1 Nomenclature

EC number
1.1.1.10

Systematic name
xylitol:NADP$^+$ 4-oxidoreductase (L-xylulose-forming)

Recommended name
L-xylulose reductase

Synonyms
reductase, L-xylulose
xylitol dehydrogenase <6> [9]
Additional information <5> (<5> xylitol:NAD$^+$ 2-oxidoreductase, EC 1.1.1.9, is termed as xylitol dehydrogenase [8]) [8]

CAS registry number
9028-17-5

2 Source Organism

<1> *Erwinia uredovora* (xylitol utilizing mutant strains, e.g. DM101, of strain 19321 [1]) [1]
<2> *Erwinia sp.* (strain 4D2P [2]) [2]
<3> *Saccharomyces cerevisiae* [3]
<4> *Cavia porcellus* (guinea pig [4-7]) [4-7]
<5> *Aspergillus niger* [8]
<6> *Pichia stipitis* [9]
<7> *hamster* [10]

3 Reaction and Specificity

Catalyzed reaction
xylitol + NADP$^+$ = L-xylulose + NADPH (<7> probably identical with EC 1.1.1.5 diacetyl reductase [10])

Reaction type
oxidation
redox reaction
reduction

Natural substrates and products

S L-xylulose + NADPH <5> (Reversibility: r <5> [8]) [8]
P xylitol + $NADP^+$ <5> [8]

Substrates and products

S D-sorbitol + $NADP^+$ <5> (Reversibility: ? <5> [8]) [8]
P ?
S DL-threitol + NAD^+ <1> (Reversibility: r <1> [1]) [1]
P D-threose + NADH <1> [1]
S DL-threitol + $NADP^+$ <7> (<7> probably identical with erythrulose reductase, EC 1.1.1.162 and diacetyl reductase, EC 1.1.1.5 [10]) (Reversibility: r <7> [10]) [10]
P D-erythrulose + NADPH <7> [10]
S L-xylulose + NADPH <4, 5, 7> (<7> also reduces 2,3-butadione [10]) (Reversibility: r <4, 5, 7> [6-8, 10]) [6-8, 10]
P xylitol + $NADP^+$ <4, 5, 7> [6-8, 10]
S xylitol + $NADP^+$ <1, 2, 4-7> (Reversibility: r <4, 5, 7> [4-8, 10]; ? <1, 2, 6> [1, 2, 7, 9]) [1, 2, 4-10]
P L-xylulose + NADPH <1, 2, 4-7> (<1> strain DM101 and DM122, product identification is not exact, could also be D-xylulose [1]) [1, 2, 4-10]
S Additional information <1, 4, 5> (<1> minimal oxidative activity with D-arabitol, mannitol, 1,2-propanediol and 2,3-butanediol [1]; <5> minimal reductive activity also with D-xylulose, D-erythrose and dihydroxyacetone [8]; <4> no activity in the mitochondrial fraction with D-xylulose, D-sorbose, L-sorbose, L-erythrulose, D-fructose, D-sorbitol, L-arabitol, D-gulitol, D-talitol [7]) [1, 7, 8]
P ?

Inhibitors

aminopyrine <4> (<4> 70 mM [7]) [7]
fluoride <4> (<4> 50 mM [7]) [7]
iodoacetate <4> (<4> not inhibitory [4]; <4> complete inhibition of L-xylulose reductive activity at 5 mM [6]) [6]
p-chloromercuribenzoate <4> [4]
phosphate buffer <4> [7]
Additional information <4> (<4> not: EDTA [4]) [4]

Cofactors/prosthetic groups

NAD^+ <1, 3-6> (<4> much less effective than $NADP^+$ [6]; <5> no activity with [8]; <4, 6> preference for, coenzyme-binding domain is a β-α-β-fold typical for NAD^+ dehydrogenases [5, 9]) [1, 3, 6, 8, 9]
$NADP^+$ <3-5> (<4> B-specific hydrogen transfer from xylitol to $NADP^+$ [5]; <4> much more effective than NAD^+ [6]) [3-6, 8]
NADPH <5> (<5> preference for [8]) [8]
Additional information <5> (<5> determination of Ki of $NADP^+$ and NADPH in competition with each other [8]) [8]

Metals, ions

Mg^{2+} <4> [6]

Specific activity (U/mg)

0.23-0.32 <1> (<1> xylitol dehydrogenase activity [1]) [1]
33 <5> (<5> xylitol oxidation, $NADP^+$ [8]) [8]
200 <5> (<5> L-xylulose reduction, NADPH [8]) [8]

K_m-Value (mM)

0.03 <5> (NADPH) [8]
0.13 <5> ($NADP^+$) [8]
0.29 <4> (L-xylulose) [4]
17 <5> (L-xylulose) [8]
48 <1> (xylitol) [1]
72 <2> (xylitol) [2]
925 <5> (xylitol) [8]

pH-Optimum

6.5 <5> (<5> reduction of L-xylulose [8]) [8]
7 <4> [4]
7.5 <4> (<4> approximately [6]) [6]
10-11 <5> (<5> oxidation of xylitol [8]) [8]

pH-Range

5-8.5 <5> (<5> reduction of L-xylulose [8]) [8]
8-11.5 <5> (<5> oxidation of xylitol [8]) [8]

4 Enzyme Structure

Molecular weight

102000 <2> (<2> 4D2P, nondenaturing PAGE [2]) [2]
136000 <1> (<1> non-denaturing PAGE [1]) [1]
250000 <5> (<5> gel filtration [8]) [8]

Subunits

octamer <5> (<5> 8 * 32000, SDS-PAGE [8]) [8]
tetramer <7> (<7> 4 * 40000, about, sequence calculation [10]) [10]

5 Isolation/Preparation/Mutation/Application

Source/tissue

kidney <4, 7> [7, 10]
liver <4, 7> [4-7, 10]

Purification

<1> (partial from strain DM101 [1]) [1]
<2> (strain 4D2P, partial [2]) [2]
<4> (partial [4]) [4]
<5> [8]
<7> [10]

Cloning

<1> (Tn5 transposon transfer [1]) [1]

Engineering

Additional information <6> (<6> several mutants per site-directed mutagenesis to change the coenzyme binding specificity, mutants showed reduced NAD^+ specificity [9]) [9]

6 Stability

pH-Stability

4 <4> (<4> 2°C, 2 h stable [6]) [6]
9 <4> (<4> room temperature, several h stable [6]) [6]

Temperature stability

50 <4> (<4> 20 min stable [6]) [6]

General stability information

<4>, inactivation during dialysis [6]

Storage stability

<4>, 0°C, mitochondria, 0.15 M KCl, 0.01 M $NaHCO_3$, minimum 2 days [7]
<4>, 2°C, aqueous solution, many weeks [6]
<4>, frozen, ammonium sulfate precipitate, several days [4]
<5>, 4°C, Tris-HCl 20 mM, pH 7.0, at leat several weeks [8]

References

[1] Doten, R.C.; Mortlock, R.P.: Characterization of xylitol-utilizing mutants of Erwinia uredovora. J. Bacteriol., **161**, 529-533 (1985)
[2] Doten, R.C.; Mortlock, R.P.: Inducible xylitol dehydrogenases in enteric bacteria. J. Bacteriol., **162**, 845-848 (1985)
[3] Lowe, C.R.; Mosbach, K.; Dean, P.D.G.: Some applications of insolubilised cofactors to the purification of pyridine nucleotide-dependent dehydrogenases. Biochem. Biophys. Res. Commun., **48**, 1004-1010 (1972)
[4] Hickman, J.; Ashwell, G.: A sensitive and stereospecific enzymatic assay for xylulose. J. Biol. Chem., **234**, 758-761 (1959)
[5] Alizade, M.A.; Brendl, K.; Gaede, K.: Chirality of xylitol-oxidizing enzymes from mammalian liver. FEBS Lett., **67**, 41-44 (1976)
[6] Hollmann, S.; Touster, O.: The L-xylulose-xylitol enzyme and other polyol dehydrogenases of guinea pig liver mitochondria. J. Biol. Chem., **225**, 87-102 (1957)
[7] Touster, O.; Reynolds, V.H.; Hutcheson, R.M.: The reduction of L-xylulose to xylitol by guinea pig liver mitochondria. J. Biol. Chem., **221**, 697-709 (1954)

[8] Witteveen, C.F.B.; Weber, F.; Busink, R.; Visser, J.: Isolation and characterization of two xylitol dehydrogenases from Aspergillus niger. Microbiology, **140**, 1679-1685 (1994)
[9] Metzger, M.H.; Hollenberg, C.P.: Amino acid substitutions in the yeast Pichia stipitis xylitol dehydrogenase coenzyme-binding domain affect the coenzyme specificity. Eur. J. Biochem., **228**, 50-54 (1995)
[10] Ishikura, S; Isaji, T.; Usami, N.; Kitahara, K.; Nakagawa, J.; Hara, A.: Molecular cloning, expression and tissue distribution of hamster diacetyl reductase. Identity with L-xylulose reductase. Chem.-Biol. Interact., **130-132**, 879-889 (2001)

D-Arabinitol 4-dehydrogenase 1.1.1.11

1 Nomenclature

EC number
1.1.1.11

Systematic name
D-arabinitol:NAD 4-oxidoreductase

Recommended name
D-arabinitol 4-dehydrogenase

Synonyms
D-arabitol dehydrogenase
arabitol dehydrogenase
dehydrogenase, D-arabinitol

CAS registry number
9028-18-6

2 Source Organism

<1> *Klebsiella aerogenes* (i.e. Klebsiella pneumoniae, gene expression in Escherichia coli K12 [1]) [1]
<2> *Aerobacter aerogenes* (strain PRL-R3 [2-4]; strain P14 (guanine auxotroph), derived from strain 1033 [5]) [2-5]
<3> *Candida tropicalis* [6, 7]
<4> *Pichia stipitis* (strain CBS 6054 [8]) [8]
<5> *Candida albicans* (strain B331, WO-1, 1006, 1001 [9]) [9]

3 Reaction and Specificity

Catalyzed reaction
D-arabinitol + NAD^+ = D-xylulose + NADH

Reaction type
oxidation
redox reaction
reduction

Natural substrates and products

S D-arabinitol + NAD$^+$ <1-2, 4> (<4> D-arabitol i.e. D-arabinitol [8]) (Reversibility: r <2> [3-5]) [1-5, 8]
P D-xylulose + NADH <2, 4> [2, 4, 5, 8]
S D-arabinitol + NAD$^+$ <3-5> (<4> recombinant protein from Saccharomyces cerevisiae [8]) (Reversibility: ? <3-5> [6-9]) [6-9]
P D-ribulose + NADH <3-5> [6-9]
S Additional information <4> (<4> D-arabinitol dehydrogenases (ArDH) from fungi and yeast oxidize D-arabinitol to D-ribulose, whereas bacterial ArDH oxidizes D-arabinitol to D-xylulose [8]) [8]
P ?

Substrates and products

S D-arabinitol + NAD$^+$ <1-2, 4> (<4> D-arabitol i.e. D-arabinitol [8]) (Reversibility: r <2> [3-5]) [1-5, 8]
P D-xylulose + NADH <2, 4> [2, 4, 5, 8]
S D-arabinitol + NAD$^+$ <3-5> (<4> recombinant protein from Saccharomyces cerevisiae [8]) (Reversibility: ? <3-5> [6-9]) [6-9]
P D-ribulose + NADH <3-5> [6-9]
S D-glucitol + NAD$^+$ <4> (<4> recombinant protein from Saccharomyces cerevisiae and enzyme from Pichia stipitis [8]) (Reversibility: ? <4> [8]) [8]
P ?
S D-mannitol + NAD$^+$ <1, 2> (Reversibility: ? <1, 2> [1, 5]) [1, 5]
P D-fructose + NADH <2> [5]
S L-xylulose + NADH <3> (<3> 3% activity [6]) (Reversibility: ? <3> [6]) [6]
P ?
S galactitol + NAD$^+$ <3, 4> (<3> 1.6% activity [6]; <4> recombinant protein from Saccharomyces cerevisiae [8]) (Reversibility: ? <3, 4> [6, 7, 8]) [6, 7, 8]
P ?
S xylitol + NAD$^+$ <3, 4> (<3> 5% activity [6]; <4> recombinant protein from Saccharomyces cerevisiae and enzyme from Pichia stipitis [8]) (Reversibility: ? <3, 4> [6-8]) [6-8]
P ?
S Additional information <3> (<3> no activity with D-mannitol [6]) [6]
P ?
S Additional information <4> (<4> D-arabinitol dehydrogenases (ArDH) from fungi and yeast oxidize D-arabinitol to D-ribulose, whereas bacterial ArDH oxidizes D-arabinitol to D-xylulose [8]) [8]
P ?

Inhibitors

1,10-phenanthroline <2> (<2> 2 mM, 37% inhibition [5]) [5]
2,2'-dipyridyl <2> (<2> 2 mM, 19% inhibition [5]) [5]
diethyldithiocarbamate <2> (<2> 10 mM, 28% inhibition [5]) [5]

Cofactors/prosthetic groups

NAD^+ <1-3> [1-7]
NADH <2, 3> [3-7]
Additional information <2, 4> (<2, 4> no activity with NADPH [3, 5, 8]) [3, 5, 8]

Metals, ions

Mg^{2+} <3> (<3> 1-5 mM increases activity about 50%, also enhances stability [6]) [6]
Additional information <4> (<4> no divalent cations are required for activity [8]) [8]

Specific activity (U/mg)

0.17 <5> (<5> lysates of Escherichia coli, transformed with enzyme gene [9]) [9]
0.2 <4> (<4> D-xylulose, of recombinant protein [8]) [8]
0.37 <5> (<5> lysates of Saccharomyces cerevisiae, transformed with enzyme gene [9]) [9]
0.78 <4> [8]
7.5 <2> [5]
37.1 <1> [1]
56 <2> [4]
68.25 <4> (<4> D-arabinitol, of recombinant protein [8]) [8]
96 <3> [6]
187 <3> (<3> of recombinant protein [7]) [7]
Additional information <2> (<2> grown on medium with D-arabinitol, specific activity of crude cell extract: 1.13-3.47 [2]) [2]

K_m-Value (mM)

0.036 <2> (NADH) [3, 4]
0.04 <1> (NAD^+) [1]
0.12 <3> (NAD^+) [6]
0.2 <4> (NAD^+) [8]
1.14 <2> (D-xylulose) [4]
2 <2> (D-xylulose) [3]
3.5 <2> (D-arabinitol) [5]
6.1 <2> (D-mannitol) [5]
20 <1> (D-arabinitol) [1]
20 <4> (D-arabinitol) [8]
39.8 <3> (D-arabinitol) [6]
70 <1> (D-mannitol) [1]

pH-Optimum

6.5 <3> (<3> optimal activity, reduction [6]) [6]
7 <2> [3, 4]
9 <2> (<2> D-arabinitol, D-mannitol [5]) [5]

pH-Range

6-8.5 <2> (<2> pH 6.0: about 60% of maximum activity, pH 8.5: about 60% of activity maximum [4]) [4]
7.5-10 <2> (<2> at pH 7.5 and 10.0 about 50% relative activity, D-arabinitol [5]) [5]
10-11 <3> (<3> enzyme activity increases with increasing pH, measurements above pH 11.0 not feasible [6]) [6]

Temperature optimum (°C)

25 <2> (<2> enzyme assay at [2]) [2]
28 <1> (<1> enzyme assay at [1]) [1]

4 Enzyme Structure

Molecular weight

43000 <1> (<1> gel filtration, ultracentrifugation [1]) [1]
44000 <1> (<1> sedimentation equilibrium [1]) [1]
46000 <1> (<1> calculated from amino acid composition [1]) [1]
110000 <4> (<4> native gradient PAGE [8]) [8]

Subunits

? <3> (<3> x * 30748, gene sequence [7]) [7]
? <3> (<3> x * 31000, SDS-PAGE [6,7]) [6, 7]
? <5> (<5> x * 30643, calculated from gene sequence [9]) [9]
monomer <1> (<1> 1 * 46500, SDS-PAGE [1]) [1]
tetramer <4> (<4> 4 * 30000, calculated from gene sequence and molecular weight of native enzyme [8]) [8]

5 Isolation/Preparation/Mutation/Application

Purification

<1> (recombinant protein from Escherichia coli [1]) [1]
<2> (strain PRL-R3 [3-4]; strain P14 [5]) [3-5]

Cloning

<1> (gene over-expressed in Escherichia coli K12 [1]) [1]
<3> (gene over-expressed in Escherichia coli BW31M [7]) [7]
<4> (gene over-expressed in Saccharomyces cerevisiae S700 [8]) [8]
<5> (gene over expressed in S. cerevisiae BWG 1-7A and in Escherichia coli JM109, DH5-α [9]) [9]

6 Stability

pH-Stability

6-9.5 <3> (<3> more labile at increasing pH from 6.0-9.5 [6]) [6]

Temperature stability

43 <2> (<2> half-life: 50 min [5]) [5]
45 <1, 2> (<2> half-life: 7 min [5]; <1> 20 min [1]) [1, 5]

General stability information

<1>, glycerol stabilizes [1]
<3>, Mg^{2+} stabilizes [6]

Storage stability

<1>, -20°C, protein concentration 50 mg/ml, 50% v/v glycerol, stable 3 months, loss of activity less than 50% [1]
<2>, -15°C, after ammonium sulfate fractionation, pH 7.0, stable for at least one year without loss in activity, higher activity with potassium phosphate buffer than with glycine-HCl [3]
<2>, frozen after second calcium phosphate gel preparation, stable for 3 years [4]

References

[1] Neuberger, M.S.; Patterson, R.A.; Hartley, B.S.: Purification and properties of Klebsiella aerogenes D-arabitol dehydrogenase. Biochem. J., **183**, 31-42 (1979)

[2] Wilson, B.L.; Mortlock, R.P.: Regulation of D-xylose and D-arabitol catabolism by Aerobacter aerogenes. J. Bacteriol., **113**, 1404-1411 (1973)

[3] Fossitt, D.D.; Wood, W.A.: D-Arabitol dehydrogenase. Methods Enzymol., **9**, 184-187 (1966)

[4] Wood, W.A.; McDonough, M.J.; Jacobs, L.B.: Ribitol and D-arabitol utilization by Aerobacter aerogenes. J. Biol. Chem., **236**, 2190-2195 (1961)

[5] Lin, E.C.C.: An inducible D-arabitol dehydrogenase from Aerobacter aerogenes. J. Biol. Chem., **236**, 31-36 (1961)

[6] Quong, M.W.; Miyada, C.G.; Switchenko, A.C.; Goodman, T.C.: Identification, purification, and characterization of a D-arabinitol-specific dehydrogenase from Candida tropicalis. Biochem. Biophys. Res. Commun., **196**, 1323-1329 (1993)

[7] Murray, J.S.; Wong, M.L.; Miyada, C.G.; Switchenko, A.C.; Goodman, T.C.; Wong, B.: Isolation, characterization and expression of the gene that encodes D-arabinitol dehydrogenase in Candida tropicalis. Gene, **155**, 123-128 (1995)

[8] Hallborn, J.; Walfridsson, M.; Penttilae, M.; Keraenen, S.; Hahn-Haegerdal, B.: A short-chain dehydrogenase gene from Pichia stipitis having D-arabinitol dehydrogenase activity. Yeast, **11**, 839-847 (1995)

[9] Wong, B.; Murray, J.S.; Castellanos, M.; Croen, K.D.: D-Arabitol metabolism in Candida albicans: studies of the biosynthetic pathway and the gene that encodes NAD-dependent D-arabitol dehydrogenase. J. Bacteriol., **174**, 6314-6320 (1993)

L-Arabinitol 4-dehydrogenase 1.1.1.12

1 Nomenclature

EC number
1.1.1.12

Systematic name
L-arabinitol:NAD 4-oxidoreductase (L-xylulose-forming)

Recommended name
L-arabinitol 4-dehydrogenase

Synonyms
L-arabitol dehydrogenase
dehydrogenase, L-arabinitol
pentitol-DPN dehydrogenase

CAS registry number
9028-19-7

2 Source Organism

<1> *Penicillium chrysogenum* [1, 4]
<2> *Aspergillus niger* [2]
<3> *Aspergillus nidulans* [3]

3 Reaction and Specificity

Catalyzed reaction
L-arabinitol + NAD^+ = L-xylulose + NADH

Reaction type
oxidation
redox reaction
reduction

Natural substrates and products
S L-arabinitol + NAD^+ <1-3> (Reversibility: r <1, 3> [1, 3]) [1-4]
P L-xylulose + NADH <1, 3> [1, 3, 4]

Substrates and products
S D-adonitol + NAD^+ <1> (Reversibility: ? <1> [1]) [1]
P ?

S L-arabinitol + NAD^+ <1-3> (Reversibility: r <1, 3> [1, 3]) [1-4]
P L-xylulose + NADH <1, 3> [1, 3, 4]
S xylitol + NAD^+ <1> (Reversibility: ? <1> [1]) [1]
P ?

Cofactors/prosthetic groups
NAD^+ <1> [1]
NADH <1> [1]

References

[1] Chiang, C.; Knight, S.G.: A new pathway of pentose metabolism. Biochem. Biophys. Res. Commun., **3**, 554-559 (1960)
[2] Witteveen, C.F.B.; Busink, R.; Van de Vondervoort, P.; Dijkema, C.; Swart, K.; Visser, J.: L-Arabinose and D-xylose catabolism in Aspergillus niger. J. Gen. Microbiol., **135**, 2163-2171 (1989)
[3] De Vries, R.P.; Flipphi, M.J.A.; Witteveen, C.F.B.; Visser, J.: Characterization of an Aspergillus nidulans L-arabitol dehydrogenase mutant. FEMS Microbiol. Lett., **123**, 83-90 (1994)
[4] Chiang, C.; Knight, S.G.: L-Arabinose metabolism by cell-free extracts of Penicillium chrysogenum. Biochim. Biophys. Acta, **46**, 271-278 (1961)

L-Arabinitol 2-dehydrogenase 1.1.1.13

1 Nomenclature

EC number
1.1.1.13

Systematic name
L-arabinitol:NAD^+ 2-oxidoreductase (L-ribulose-forming)

Recommended name
L-arabinitol 2-dehydrogenase

Synonyms
L-arabinitol (ribulose-forming) dehydrogenase
L-arabinitol dehydrogenase (ribulose-forming)

CAS registry number
9028-20-0

2 Source Organism

<1> *Penicillium chrysogenum* [1]

3 Reaction and Specificity

Catalyzed reaction
L-arabinitol + NAD^+ = L-ribulose + NADH

Reaction type
oxidation
redox reaction
reduction

Natural substrates and products
S L-arabinitol + NAD^+ <1> (Reversibility: r <1> [1]) [1]
P L ribulose + NADH <1> [1]
S Additional information <1> (<1> enzyme of pentose metabolism [1]) [1]
P ?

Substrates and products
S L-arabinitol + NAD^+ <1> (Reversibility: r <1> [1]) [1]
P L-ribulose + NADH <1> [1]
S Additional information <1> (<1> enzyme of pentose metabolism [1]) [1]
P ?

Cofactors/prosthetic groups

NAD^+ <1> [1]

Temperature optimum (°C)

8.5 <1> (<1> reaction proceeds to completion in vitro [1]) [1]

References

[1] Chiang, C.; Knight, S.G.: L-Arabinose metabolism by cell-free extracts of Penicillium chrysogenum. Biochim. Biophys. Acta, **46**, 271-278 (1961)

L-Iditol 2-dehydrogenase 1.1.1.14

1 Nomenclature

EC number
1.1.1.14

Systematic name
L-iditol:NAD^+ 2-oxidoreductase

Recommended name
L-iditol 2-dehydrogenase

Synonyms
L-iditol 2-dehydrogenase [Swissprot]
L-iditol dehydrogenase (sorbitol)
L-iditol:NAD oxidoreductase
L-iditol:NAD^+ 5-oxidoreductase
NAD^+-dependent sorbitol dehydrogenase
NAD-dependent sorbitol dehydrogenase
NAD-sorbitol dehydrogenase
Protein tms1 [Swissprot]
dehydrogenase, L-iditol
glucitol dehydrogenase
polyol dehydrogenase
sorbitol dehydrogenase

CAS registry number
9028-21-1

2 Source Organism

<1> *Malus domestica* (apple; cv Granny Smith [3]; cv. Ourin [17]) [3, 17]
<2> *Rattus norvegicus* [1, 4, 5, 6, 8, 10]
<3> *Ovis aries* (ram [11]) [1, 2, 7, 9, 11, 12, 18, 20, 22, 23, 29]
<4> *Homo sapiens* [4, 6, 21]
<5> *Canis familiaris* [4]
<6> *Mus musculus* [4]
<7> *Cavia porcellus* (guinea pig) [11]
<8> *Equus caballus* [12]
<9> *Pichia stipitis* [13, 14]
<10> *Pseudomonas sp.* [15]
<11> *Bacillus subtilis* [16]

<12> *Rhodobacter sphaeroides* (strain Si4 (DSM8371) [19]) [19]
<13> *Bos taurus* [24, 27]
<14> *Bacillus fructosus* [25]
<15> *Gluconobacter suboxydans* (strain IFO3257 [26]) [26]
<16> *Pyrus serotina* (japanese pear) [28]
<17> *Candida boidinii* ((Kloeckera sp.) No. 2201 [30]) [30]

3 Reaction and Specificity

Catalyzed reaction

L-iditol + NAD^+ = L-sorbose + NADH

Reaction type

oxidation
redox reaction
reduction

Natural substrates and products

S sorbitol + NAD^+ <2, 4> (Reversibility: r <4> [6]) [6, 8]
P D-fructose + NADH <4> [6]
S xylitol + NAD^+ <9> (Reversibility: r <9> [13]) [13]
P D-xylulose + NADH <9> [13]

Substrates and products

S D-arabinose + NADH <10, 17> (<10,17> low activity [15, 30]; <12, 14> not [19,25]) (Reversibility: ? <10, 17> [15, 30]) [15, 30]
P ?
S D-arabitol + NAD^+ <2, 3, 10, 12, 17> (<1, 2, 8, 9, 14, 15> not [3, 10, 12, 13, 25, 26]; <2, 3, 10, 12, 17> D-arabinitol is identical, low activity [1, 11, 15, 19, 30]) (Reversibility: ? <2, 3, 10, 12, 17> [1, 11, 15, 19, 30]) [1, 11, 15, 19, 30]
P ?
S D-fructose + NADH <1-4, 8-10, 12, 13, 15, 17> (Reversibility: ? <1-4, 8-10, 12, 13, 15, 17> [1, 3, 5, 8, 12, 13, 15, 17, 19-21, 24, 26, 27, 29, 30]) [1, 3, 5, 8, 12, 13, 15, 17, 19 21, 24, 26, 27, 29, 30]
P D-sorbitol + NAD^+
S D-galactitol + NAD^+ <2, 8, 10-12> (<8> low activity [12]; <1, 2, 9, 13, 15, 16> not [1, 3, 5, 13, 24, 26, 28]) (Reversibility: r <10, 12> [15, 19]; ? <2, 8, 11> [8, 12, 16]) [8, 12, 15, 16, 19]
P D-tagatose + NADH <10, 12> [15, 19]
S D-glucose + NADH <17> (<17> low activity [30]; <1, 10, 12, 14> not [3, 15, 19, 25]) (Reversibility: ? <17> [30]) [30]
P ?
S D-glycero-D-gluco-heptitol + NAD^+ <3> (Reversibility: ? <3> [9]) [9]
P ?
S D-mannitol + NAD^+ <1-3, 7-13, 15, 17> (<2, 7, 10-13, 15> low activity [10, 11, 15, 16, 19, 24, 26]; <9, 14, 16> not [13, 25, 28]) (Reversibility: ?

<1-3, 7-13, 15, 17> [1, 3, 5, 8, 10-13, 15-17, 19, 24, 26, 30]) [1, 3, 5, 8, 10-13, 15-17, 19, 24, 26, 30]

P ?

S D-mannose + NADH <10> (<10> low activity [15]; <1, 2, 12, 14, 15> not [1, 3, 5, 19, 25, 26]) (Reversibility: ? <10> [15]) [15]

P ?

S D-psicose + NADH <4> (<15> not [26]) (Reversibility: ? <4> [21]) [21]

P ?

S D-raffinose + NADH <3> (<3> low activity [29]) (Reversibility: ? <3> [29]) [29]

P ?

S D-ribitol + NAD^+ <1-3, 7-9, 11, 13, 16, 17> (<10, 12, 15> not [15, 19, 26]) (Reversibility: ? <1-3, 7-9, 11, 13, 16, 17> [1, 3, 5, 8-13, 16, 17, 24, 28, 30]) [1, 3, 5, 8-13, 16, 17, 24, 28, 30]

P ?

S D-ribose + NADH <17> (<17> low activity [30]; <2, 10, 12, 15> not [1, 5, 15, 19, 26]) (Reversibility: ? <17> [30]) [30]

P ?

S D-ribulose + NADH <2, 3, 4, 13, 17> (<10, 15> not [15, 26]) (Reversibility: ? <2, 3, 4, 13, 17> [8, 21, 24, 29, 30]) [8, 21, 24, 29, 30]

P ?

S D-sorbitol + NAD^+ <1-4, 7-17> (Reversibility: r <3, 10, 12, 14, 15> [11, 15, 19, 25, 26]; ? <1, 2, 4, 7-9, 11, 13, 16, 17> [1, 3, 5, 8-10, 12, 13, 16-18, 20, 23, 24, 28, 30]) [1, 3, 5, 8-13, 15-21, 23-26, 28, 30]

P D-fructose + NADH <3, 10, 12, 14, 15> [11, 15, 19, 25, 26]

S D sorbose + NADH <4> (<4> low activity [21]) (Reversibility: ? <4> [21]) [21]

P ?

S D-tagatose + NADH <4, 10, 12> (<1, 8, 15> not [3, 12, 26]) (Reversibility: r <10, 12> [15, 19]; ? <4> [21]) [15, 19, 21]

P D-galactitol + NAD^+

S D-xylitol + NAD^+ <1-3, 7-13, 15-17> (<10, 12> low activity [15, 19]) (Reversibility: ? <1-3, 7-13, 15-17> [1, 3, 5, 8-13, 15-17, 19, 24, 26, 28, 30]) [1, 3, 5, 8-13, 15-17, 19, 24, 26, 28, 30]

P ?

S D-xylose + NADH <3, 17> (<3, 17> low activity [29, 30]; <1, 2, 10, 12, 15> not [1, 3, 5, 15, 19, 26]) (Reversibility: ? <3, 17> [29, 30]) [29, 30]

P ?

S D-xylulose + NADH <3, 4, 9, 10, 12, 15, 17> (<10, 12> low activity [15, 19]) (Reversibility: ? <3, 4, 9, 10, 12, 15, 17> [13, 15, 19, 21, 26, 29, 30]) [13, 15, 19, 21, 26, 29, 30]

P ?

S L-arabitol + NAD^+ <1-3, 7, 8, 12, 16> (<1, 8, 16> low activity [12, 17, 28]; <9> not [13]) (Reversibility: ? <1-3, 7, 8, 12, 16> [3, 8, 9, 11, 12, 17, 19, 28]) [3, 8, 9, 11, 12, 17, 19, 28]

P ?

S L-erythrulose + NADH <4> (Reversibility: ? <4> [21]) [21]

P ?

S L-iditol + NAD^+ <2-4, 7-13, 16, 17> (<17> low activity [30]) (Reversibility: ? <2-4, 7-13, 16, 17> [5, 6, 9-13, 15, 16, 19, 24, 28, 30]) [5, 6, 9-13, 15, 16, 19, 24, 28, 30]

P ?

S L-rhamnose + NADH <17> (<17> low activity [30]) (Reversibility: ? <17> [30]) [30]

P ?

S L-sorbose + NADH <1, 2, 4, 8-10, 12, 13, 17> (<10,12,17> low activity [15,19,26,30]) (Reversibility: ? <1, 2, 4, 8-10, 12, 13, 17> [1, 3, 5, 8, 12, 13, 15, 19, 21, 24, 30]) [1, 3, 5, 8, 12, 13, 15, 19, 21, 24, 30]

P ?

S L-threitol + NAD^+ <1, 13, 16> (<13> low activity [24]) (Reversibility: ? <1, 13, 16> [3, 17, 24, 28]) [3, 17, 24, 28]

P ?

S erythritol + NAD^+ <9, 16> (<16> low activity [28]; <1, 8, 10> not [3, 12, 15]) (Reversibility: ? <9, 16> [13, 28]) [13, 28]

P ?

S ethanol + NAD^+ <1, 16> (<16> low activity [28]; <3, 10, 11> not [2, 15, 16]) (Reversibility: ? <1, 16> [17, 28]) [17, 28]

P acetaldehyde + NADH

S glycerol + NAD^+ <3> (<3> low activity [29]; <1, 9, 10, 12, 16> not [3, 13, 15, 17, 19, 28]) (Reversibility: ? <3> [29]) [29]

P ?

S sorbitol + NAD^+ <2, 4> (Reversibility: r <4> [6]) [6, 8]

P D-fructose + NADH <4> [6]

S xylitol + NAD^+ <9> (Reversibility: r <9> [13]) [13]

P D-xylulose + NADH <9> [13]

Inhibitors

(2S)-1-(3-thio-2-methylpropionyl)-L-proline <3> [18]
1,10-phenanthroline <3, 13> [2, 9, 18, 24]
1,2,3,4,5,6-cyclohexanehexolphosphoric acid <3> (<3> at pH 9.9 [18]) [18]
1,2-dithioethane <3> [18]
1,7-phenanthroline <3> [18]
1-thiobutane <3> [18]
1-thioglycol <3> [18]
1-thiosorbitol <3> [18]
2,2'-dipyridyl <3> [18]
2,3-dimercaptopropanol <2> [8, 10]
2,3-dithiopropanol <3> [18]
2-hydroxybenzoic acid <3> (<3> at pH 9.9 [18]) [18]
2-mercaptoethanol <3> [18]
2-methypropanoic acid amide <3> (<3> at pH 9.9 [18]) [18]
2-propylpentanoic acid <3> (<3> at pH 9.9 [18]) [18]
2-thiobutane <3> [18]
3,4-dihydroxyphenyl-1-hydroxyacetic acid <3> (<3> at pH 7.4 [18]) [18]

3,4-dihydroxyphenyl-ethanediol <3> (<3> at pH 7.4 [18]) [18]
3-hydroxy-4-methoxy-phenyl-1-hydroxy-acetic acid <3> (<3> at pH 7.4 [18]) [18]
4,4'-dipyridyl <3> [18]
4-hydroxy-3-methoxy-phenyl-1-hydroxy-acetic acid <3> (<3> at pH 7.4 [18]) [18]
4-hydroxy-3-methoxy-phenylacetic acid <3> (<3> at pH 7.4 [18]) [18]
4-hydroxy-3-methoxy-phenylethanediol <3> (<3> at pH 7.4 [18]) [18]
4-hydroxy-3-methoxy-phenylethyl alcohol <3> (<3> at pH 7.4 [18]) [18]
8-hydroxyquinoline <3> [18]
ADP <9> (<9> 2 mM, 72% activity [13]) [13]
AMP <9> (<9> 2 mM, 83% activity [13]) [13]
ATP <9> (<9> 2 mM, 54% activity [13]) [13]
Ag^+ <14> [25]
$AgNO_3$ <1, 2> [3, 10]
$Ba(CH_3COO)_2$ <17> (<17> 1 mM reduces activity to about 20% [30]) [30]
$CaCl_2$ <1> [17]
$CdCl_2$ <10> (<10> 1 mM [15]) [15]
Cibacron Blue 3G-A <3> [18, 21]
$CoCl_2$ <10, 17> (<10> 1 mM [15]) [15, 30]
Cu^{2+} <14> [25]
$CuSO_4$ <3, 10> (<10> 1 mM [15]) [9, 15]
D-fructose <2> (<2> above 250 mM, substrate inhibition [1]) [1]
D-sorbitol <2> (<2> above 49 mM, substrate inhibition [1]) [1]
DL-2-thiovaline <3> [18]
EDTA <2, 3, 13, 17> (<2> at pH 7.4, not at pH 9.0, fructose reduction [8]) [2, 8, 24, 30]
Eriochrome Black T <3> [18, 21]
$FeSO_4$ <10> (<10> 1 mM [15]) [15]
Hg^{2+} <14> [25]
$HgCl_2$ <1, 2, 16> [3, 10, 17, 28]
ICI 171071 <3> [18]
ICI 222155 <3> [18]
$MgCl_2$ <1, 16> [17, 28]
$MgSO_4$ <10> (<10> 1 mM [15]) [15]
$MnSO_4$ <16> [28]
N-bromosuccinic acid <10> (<10> 1 mM [15]) [15]
N-bromosuccinimide <12> [19]
N-ethylmaleimide <10> (<10> 1 mM [15]) [15]
NAD^+ <2, 8, 9> (<2, 8> above 7 mM, substrate inhibition [1, 12]; <9> above 1 mM, substrate inhibition [14]) [1, 12, 14]
NADH <2> (<2> above 0.4 mM, substrate inhibition [1]) [1]
Na_2SO_4 <17> [30]
$NiCl_2$ <10, 17> (<10> 1 mM [15]; <17> 1 mM reduces activity to about 20% [30]) [15, 30]
$Pb(CH_3COO)_2$ <17> [30]
$ZnCl_2$ <1> [17]

$ZnSO_4$ <10, 17> (<10> 1 mM [15]) [15, 30]
acetic acid amide <3> (<3> at pH 9.9 [18]) [18]
alrestatin <3> [18]
berberine <3> [18]
bis-diethylthiocarbamoyldisulfide <3> [18]
butanedioic acid <3> (<3> at pH 9.9 [18]) [18]
cyanamide <3> (<3> at pH 9.9 [18]) [18]
cyanide <3> [9]
cysteamine <3> (<3> significant only at pH 9.9 [18]) [18]
cysteine <1-3, 16> (<1> strong effect on sorbitol oxidation, slight effect on fructose reduction, $ZnSO_4$ reverses inhibition [3]; <2> 4 mM does not inhibit sorbitol oxidation [10]; <3> at alkaline pH [9]; <16> inhibits fructose reduction at pH 9.0, stimulates at pH 7.4 [28]) [3, 8, 9, 18, 28]
decanoic acid <3> (<3> at pH 9.9 [18]) [18]
diethyldicarbonate <3> [2]
diethylthiocarbamate <3> (<3> at pH 9.9 [18]) [18]
dithioerythritol <3> (<3> protection at low concentration, inhibition at high concentration, 100 mM [2]) [2]
dithiothreitol <3, 17> (<3> protection at low concentration, inhibition at high concentration, 100 mM [2]) [2, 18, 30]
ethanedioic acid <3> (<3> at pH 9.9 [18]) [18]
glutathione <3> [9]
imidazole <3> [18]
iodoacetamide <3> (<3> NADH 0.4 mM protects [2]) [2]
iodoacetate <1, 2, 12> (<1> strong effect on sorbitol oxidation, weaker effect on fructose reduction [3]) [3, 8, 19]
isonicotinic acid hydrazide <3> [18]
o-phenanthroline <17> [30]
p-chloromercuribenzoate <1, 3, 10, 14> (<10> 1 mM [15]) [3, 9, 15, 25]
p-hydroxymercuribenzoate <2> [8]
ponalrestat <3> [18]
propanedioic acid <3> (<3> at pH 9.9 [18]) [18]
pyrazole <3> [18]
quercetin <3, 13> (<3> at pH 7.4 [18]) [18, 24]
semicarbazide <17> (<17> 10 mM [30]) [30]
thioacetate <3> (<3> at pH 9.9 [18]) [18]
thiocyanate <3> (<3> at pH 9.9 [18]) [18]
thiourea <3> (<3> at pH 9.9 [18]) [18]
tolrestat <3> [18]
urea <3> (<3> at pH 9.9 [18]) [18]

Cofactors/prosthetic groups

NAD^+ <1-17> (<2, 8, 9, 12, 13, 15> specific for [1, 5, 7, 12, 13, 19, 24, 26]; <17> preferred [30]) [1-30]
NADH <1-17> (<2, 8, 9, 12, 13> specific for [1, 5, 12, 13, 19, 24]; <17> preferred [30]) [1-30]

$NADP^+$ <3, 17> (<1, 2, 3> cannot replace NAD^+ [3, 5, 8, 11]; <3> less effective than NAD^+ [2]) [2, 30]
NADPH <2, 3, 17> (<1> less effective than NADH [3]) [2, 8, 30]

Activating compounds

(R)-(-)-2-octanol <3> (<3> at pH 9.9 [18]) [18]
(S)-(+)-2-octanol <3> (<3> at pH 9.9 [18]) [18]
1-hexanol <3> [18]
1-octanol <3> [18]
2,2,2-tribromoethanol <3> [18]
2,2,2-trichloroethanol <3> [18]
2,2,2-trifluoroethanol <3> [18]
2-monobromoethanol <3> [18]
2-monochloroethanol <3> [18]
2-monofluoroethanol <3> [18]
NaBr <3> [18]
NaCl <3> [18]
NaI <3> [18]
SDS <3> [18]
$ZnSO_4$ <1, 2, 16> (<1, 16> reverses cysteine inhibition [3, 28]; <2> 3.5 mM reverses EDTA inhibition [8]) [3, 8, 28]
cysteine <2> (<2> fructose reduction at pH 7.4, inhibits at pH 9.0 [8]) [8]
dithiothreitol <10> (<10> 1 mM 24% stimulation [15]) [15]
semicarbazide <17> (<17> 1 mM [30]) [30]
sodium azide <17> (<17> 10 mM stimulates 2fold [30]) [30]
sodium taurodeoxycholate <3> [18]

Metals, ions

Fe^{3+} <17> (<17> 10 mM $FeCl_3$ stimulates activity 2 fold [30]) [30]
Zn <11, 13> (<11> 1 mol zinc per subunit [16]; <13> 0.83 mol per subunit [24]) [16, 24]

Specific activity (U/mg)

1.105 <1> [17]
2.7 <14> [25]
7.47 <2> [1, 5]
13.6 <11> [16]
14.4 <3> [9]
30.94 <17> (<17> partially purified [30]) [30]
42.3 <12> [19]
52 <13> [24]
55 <16> [28]
58.1 <9> [13]
89.5 <10> [15]
136.5 <15> [26]

K_m-Value (mM)

0.007 <8> (NADH) [12]

0.041 <8> (NAD^+) [12]
0.052 <13> (NADH) [24]
0.06 <12> (NADH) [19]
0.067 <2> (NADH) [1]
0.082 <2> (NAD^+) [1]
0.089 <13> (NAD^+) [24]
0.09 <10> (NADH) [15]
0.13 <12> (NADH) [19]
0.16 <9> (NAD^+) [13]
0.18 <3> (D-xylitol) [9]
0.21 <15> (NAD^+) [26]
0.22 <14> (NAD^+) [25]
0.24 <8> (D-sorbitol) [12]
0.24 <10> (NAD^+) [15]
0.28 <2> (D-sorbitol) [8]
0.28 <15> (NADH) [26]
0.35 <2> (D-sorbitol) [5]
0.38 <2> (D-sorbitol) [1]
0.4 <2> (D-xylitol) [8]
0.6 <3> (NAD^+) [9]
1 <4> (D-xylulose) [21]
1.1 <3> (D-sorbitol) [9]
1.3 <2> (D-ribitol) [8]
1.5 <12> (D-galactitol) [19]
1.8 <3> (D-ribitol) [9]
3.1 <10> (D-galactitol) [15]
3.2 <13> (D-xylitol) [24]
5 <15> (D-sorbitol) [26]
6.2 <12> (D-glucitol) [19]
9 <3> (L-iditol) [11]
9.1 <10> (D-glucitol) [15]
9.8 <3> (D-sorbitol) [11]
10 <10> (D-tagatose) [15]
11 <11> (D-sorbitol) [16]
11 <14> (D-sorbitol) [25]
13 <12> (D-tagatose) [19]
14 <11> (D-xylitol) [16]
16 <2> (D-mannitol) [8]
16.7 <2> (D-galactitol) [8]
18 <13> (D-sorbitol) [24]
18 <11> (L-iditol) [16]
20 <15> (D-fructose) [26]
20 <2> (L-arabitol) [8]
20 <13> (L-iditol) [24]
25 <4> (L-erythrulose) [21]
26 <9> (D-xylitol) [13]
34 <4> (D-ribulose) [21]

37 <1> (D-xylitol) [3]
40.3 <1> (D-sorbitol) [17]
54 <2> (D-ribulose) [8]
67 <13> (D-ribitol) [24]
77 <2> (D-fructose) [8]
86 <1> (D-sorbitol) [3]
110 <2> (D-fructose) [5]
110 <2> (L-sorbose) [8]
125 <11> (D-ribitol) [16]
136 <2> (D-fructose) [1]
140 <4> (D-fructose) [21]
160 <12> (D-fructose) [19]
176 <10> (D-fructose) [15]
180 <13> (D-fructose) [24]
215 <1> (D-fructose) [17]
260 <13> (D-ribulose) [24]
268 <13> (L-threitol) [24]
330 <8> (D-fructose) [12]
381 <13> (D-mannitol) [24]
500 <11> (D-galactitol) [16]
555 <11> (D-mannitol) [16]
670 <13> (L-sorbose) [24]
800 <4> (L-sorbose) [21]
1500 <1> (D-fructose) [3]

pH-Optimum

5.6 <17> (<17> D-fructose reduction [30]) [30]
5.7-7 <1> (<1> D-fructose reduction [3]) [3]
5.9 <2> (<2> D-fructose reduction [1, 5]) [1, 5]
6 <1, 15> (<1,15> D-fructose reduction [17, 26]) [17, 26]
6-7.2 <12> (<12> D-fructose reduction [19]) [19]
6.1 <2> (<2> D-fructose reduction [6]) [6]
6.7 <10> (<10> D-fructose reduction [15]) [15]
7 <16> (<16> D-fructose reduction [28]) [28]
7 <3> (<3> D-fructose reduction [9]) [9]
7.4 <3, 13> (<3,13> D-fructose reduction [20, 24]) [20, 24]
7.4-8 <2> (<2> D-fructose reduction [8]) [8]
8.1-8.5 <2> (<2> sorbitol oxidation [1, 5]) [1, 5]
8.8 <8> (<8> D-fructose reduction [12]) [12]
9 <9, 13> (<9,13> sorbitol oxidation [13, 24]) [13, 24]
9-10 <1> (<1> sorbitol oxidation [3]) [3]
9-10 <2> (<2> sorbitol oxidation [8]) [8]
9-10.5 <15> (<15> sorbitol oxidation [26]) [26]
9.5 <17> (<17> sorbitol oxidation [30]) [30]
9.6 <1> (<1> sorbitol oxidation [17]) [17]
9.6 <8> (<8> sorbitol oxidation [12]) [12]
9.9 <3> (<3> sorbitol oxidation [20]) [20]

10 <3> (<3> sorbitol oxidation [9]) [9]
11 <10, 12> (<10,12> sorbitol oxidation [15, 19]) [15, 19]

pH-Range

4.7-8 <1> (<1> at pH 4.7 and 8 about 50% of activity maximum, D-fructose reduction [3]) [3]
5-7 <1, 2> (<1> 30% of maximal activity at pH 5.0 and 60% of maximal activity at pH 7.0, fructose reduction [17]; <2> at pH 5.0 and pH 7.0 about 60% of maximal activity, fructose reduction [5]) [5, 17]
5-7.5 <16> (<16> less than 20% of maximal activity below pH 5, less than 80% above pH 7.5, fructose reduction [28]) [28]
6-7.2 <12> (<12> less than 50% of maximal activity above and below, sugar reduction [19]) [19]
6-9 <2> (<2> at pH 6 about 70% of maximal activity, at pH 9 about 40%, fructose reduction [8]) [8]
7-10 <9> (<9> 40% of maximal activity at 7 and 70% of maximal activity at 10, sorbitol oxidation [13]) [13]
7.5-9 <16> (<16> less than 20% of maximal activity below pH 7.5, 100% at pH 9.0, sorbitol oxidation [28]) [28]
7.5-9.5 <2> (<2> at pH 7.5 and pH 9.5 about 70% of maximal activity, sorbitol oxidation [5]) [5]
7.5-10 <2> (<2> at pH 7.5 about 70% of maximal activity, at pH 10 about 100%, sorbitol oxidation [8]) [8]
8-11 <1> (<1> at pH 8 and 11 about 50% of activity maximum, sorbitol oxidation [3]) [3]
9-10.5 <1> (<1> less than 70% of maximal activity above and below, sorbitol oxidation [17]) [17]
9.2-11.6 <12> (<12> less than 50% of maximal activity above and below, sorbitol oxidation [19]) [19]

Temperature optimum (°C)

35 <9> [13]
37 <17> [30]
60 <1> [3]

Temperature range (°C)

15-50 <9> (<9> 50% of maximal activity at 15°C, 40% of maximal activity at 50°C [13]) [13]
35-65 <1> (<1> 35°C: about 50% of activity maximum, 65°C: about 95% of activity maximum [3]) [3]

4 Enzyme Structure

Molecular weight

61000 <12> (<12> Stokes' radius, sedimentation coefficient [19]) [19]
63000 <9> (<9> gel filtration [13]) [13]
65800 <10> (<10> Stokes' radius, sedimentation coefficient [15]) [15]

95000 <2> (<2> gel filtration [1,5]) [1, 5]
97000 <2> (<2> disc gel electrophoresis [1,5]) [1, 5]
106000 <3> (<3> gel filtration [12]) [12]
110000 <8> (<8> gel filtration [12]) [12]
115000 <3> (<3> ultracentrifugal analysis [9]) [9]
120000 <1> (<1> gel filtration [17]) [17]
140000 <3> (<3> gel filtration [2]) [2]
153000 <11> (<11> gel filtration [16]) [16]
156000 <13> (<13> gel filtration [24]) [24]

Subunits

? <3, 8> (<3,8> x * 26000, SDS-PAGE [12]) [12]
dimer <10> (<10> 2 * 32300, SDS,PAGE [15]) [15]
dimer <12> (<12> 2 * 29000, SDS-PAGE [19]) [19]
dimer <9> (<9> 2 * 32000, SDS-PAGE [13]) [13]
tetramer <11> (<11> 4 * 39000, SDS-PAGE [16]) [16]
tetramer <13> (<13> 4 * 39000, SDS-PAGE [24]) [24]
tetramer <15> (<15> 4 * 26000, SDS-PAGE [26]) [26]
tetramer <2> (<2> 4 * 35000-40000, SDS-PAGE) [1]

5 Isolation/Preparation/Mutation/Application

Source/tissue

brain <2> [8]
coagulating gland <2> [10]
fruit <1, 16> [17, 28]
lens <13> [24, 27]
liver <2, 3, 8> (<3> commercial preparation [2,7]) [1, 2, 5, 7, 9, 12, 18, 20, 22, 23, 29]
mesocarp tissue culture <1> [3]
prostate gland <2> (<2> ventral and dorsal [10]) [10]
seminal vesicle <2> [10]
serum <2, 4-6> [4, 6]
spermatozoon <3, 7> [11]

Purification

<2> (partial [8]) [1, 5, 8]
<3> (further purification from commercial preparation [2]) [2, 9]
<8> [12]
<9> [13]
<10> [15]
<11> [16]
<12> [19]
<13> [24]
<14> [25]
<15> [26]
<16> [28]

Crystallization

<3> [9]
<14> [25]

Cloning

<11> [16]

6 Stability

pH-Stability

4.5 <1> (<1> at lower pH denaturation occurs [3]) [3]
4.5-11 <15> (<15> several days at 4°C [26]) [26]
5-9 <14> (<14> 30 min at 30°C [25]) [25]
6 <4> (<4> loses zinc at lower pH [21]) [21]
7-8 <17> (<17> 12 h at 4°C [30]) [30]

Temperature stability

4 <10> (<10> 50% loss of activity after 4 days [15]) [15]
18 <13> (<13> 50% loss of activity after 40 min, α-crystallin protects fully against activity loss [27]) [27]
20 <8, 10> (<8> 30% loss of activity after 3 days, 70% loss of activity after 6 days, 100% loss of activity after 10 days [12]; <10> 50% loss of activity after 32 hours [15]) [12, 15]
21 <4> (<4> 50% loss of activity after 2 days, non-purified enzyme in serum [6]) [6]
22 <2> (<2> 5 h, stable [1, 5]) [1, 5]
30 <10, 12> (<10> 50% loss of activity after 6 hours [15]; <12> 50% loss of activity after 8 days in presence of 30% sucrose [19]) [15, 19]
40 <10, 12, 14, 15> (<10> 50% loss of activity after 5 min [15]; <12> 50% loss of activity after 18 hours in presence of 30% sucrose [19]; <14, 15> 30 min, stable [25, 26]) [15, 19, 25, 26]
45 <13> (<13> 50% loss of activity after 20 min, α-crystallin protects fully against activity loss [27]) [27]
50 <10, 12, 14, 15> (<10> 50% loss of activity after 1 min [15]; <12> 50% loss of activity after 2 min in presence of 30% sucrose [19]; <14> 5% loss of activity after 30 min [25]; <15> denaturation after 5 min [26]) [15, 19, 25, 26]
55 <13> (<13> 50% loss of activity after 10 min, α-crystallin protects fully against activity loss [27]) [27]
60 <2, 13> (<2> 10 min, 65% loss of activity [1, 5]; <13> 50% loss of activity after 5 min, α-crystallin reduces activity loss by 50% [27]) [1, 5, 27]

General stability information

<2>, 0.1 mg/ml bovine serum albumin stabilizes at -20°C [5]
<2>, 0.1 mg/ml bovine serum albumin stabilizes at 4°C [1]
<2>, 0.1 mg/ml bovine serum albumin stabilizes at room temperature [5]
<2>, 278 mM maltose stabilizes at 4°C [1]
<2>, 6.8 M glycerol stabilizes at 4°C [1]

<2>, bovine serum albumin stabilizes at 4°C [8]
<3>, dithioerythritol protccts at 1 mM, inhibits at 100 mM [2]
<3>, dithiothreitol protects at 1 mM at 4°C [2]
<3>, freezing to -20°C, loss of activity [2]
<12>, 1 mM dithiothreitol reduces stability [19]
<13>, α-crystallin stabilizes [27]
<13>, bovine serum albumin stabilizes [27]
<13>, freezing, loss of activity [24]
<10, 12>, 30% sucrose stabilizes [15, 19]

Storage stability

<2>, -20°C, 70% loss of activity after 1 week, 0.1 mg/ml bovine serum albumin stabilizes [1, 5]
<2>, -85°C, good preservation [8]
<3>, immobilized enzyme loses 20% of activity after 10 days at room temperature [2]
<4>, -18°C, 2 days, non-purified enzyme in serum [6]
<8>, -20°C, stable for 4 months [12]
<9>, -60°C, 5 mM 2-mercaptoethanol, several months [13]
<12>, 4°C, 20 mM potassium phosphate buffer, pH 7.0, 3 months [19]
<13>, 4°C, 2 mM dithiothreitol, 0.1 mM NADH, 15 days [24]
<15>, 4°C, several days [26]

References

[1] Leissing, N.; McGuinness, E.T.: Sorbitol dehydrogenase from rat liver. Methods Enzymol., **89**, 135-140 (1982)

[2] Jeffery, J.; Cummins, L.; Carlquist, M.; Jörnvall, H.: Properties of sorbitol dehydrogenase and characterization of a reactive cysteine residue reveal unexpected similarities to alcohol dehydrogenases. Eur. J. Biochem., **120**, 229-234 (1981)

[3] Negm, F.B.; Loescher, W.H.: Detection and characterization of sorbitol dehydrogenase from apple callus tissue. Plant Physiol., **64**, 69-73 (1979)

[4] Dooley, J.F.; Turnquist, L.J.; Racich, L.: Kinetic determination of serum sorbitol dehydrogenase activity with a centrifugal analyzer. Clin. Chem., **25**, 2026-2029 (1979)

[5] Leissing, N.; Mc Guinness, E.T.: Rapid affinity purification and properties of rat liver sorbitol dehydrogenase. Biochim. Biophys. Acta, **524**, 254-261 (1978)

[6] Gerlach, U.; Hiby, W.: Sorbit-Dehydrogenase. Methods Enzym. Anal., 3rd Ed. (Bergmeyer, H.U., ed.), **1**, 601-606 (1974)

[7] Myers, J.S.; Jakoby, W.B.: Effect of polyhydric alcohols on kinetic parameters of enzymes. Biochem. Biophys. Res. Commun., **51**, 631-636 (1973)

[8] Rehg, J.E.; Torack, R.M.: Partial purification and characterization of sorbitol dehydrogenase from rat brain. J. Neurochem., **28**, 655-660 (1977)

[9] Smith, M.G.: Polyol dehydrogenase. Biochem. J., **83**, 135-144 (1962)

[10] Williams-Ashman, H.G.; Banks, J.; Wolfson, S.K.: Oxidation of polyhydric alcohols by the prostate gland and seminal vesicle. Arch. Biochem. Biophys., **72**, 485-494 (1957)
[11] King, T.E.; Mann, T.: Sorbitol metabolism in spermatozoa. Proc. R. Soc. Lond. B Biol. Sci., **151**, 226-243 (1959)
[12] Bailey, J.P.; Renz, C.; McGuinness, E.T.: Sorbitol dehydrogenase from horse liver: Purification, characterization and comparative properties. Comp. Biochem. Physiol. B, **69**, 909-914 (1981)
[13] Rizzi, M.; Harwart, K.; Erlemann, P.; Bui-Thanh, N.-A.; Dellweg, H.: Purification and properties of the NAD^+-xylitol-dehydrogenase from the yeast Pichia stipitis. J. Ferment. Bioeng., **67**, 20-24 (1989)
[14] Rizzi, M.; Harwart, K.; Bui-Thanh, N.-A.; Dellweg, H.: A kinetic study of the NAD^+-xylitol-dehydrogenase from the yeast Pichia stipitis. J. Ferment. Bioeng., **67**, 25-30 (1989)
[15] Schneider, K.-H.; Giffhorn, F.: Sorbitol dehydrogenase from Pseudomonas sp.: Purification, characterization and application to quantitative determination of sorbitol. Enzyme Microb. Technol., **13**, 332-337 (1991)
[16] Ng, K.; Ye, R.; Wu, X.-C.; Wong, S.-L.: Sorbitol dehydrogenase from Bacillus subtilis, purification, characterization and gene cloning. J. Biol. Chem., **267**, 24989-24994 (1992)
[17] Yamaguchi, H.; Kanayama, Y.; Yamaki, S.: Purification and properties of NAD-dependent sorbitol dehydrogenase from apple fruit. Plant Cell Physiol., **35**, 887-892 (1994)
[18] Lindstad, R.I.; Hermansen, L.F.; McKinley-McKee, J.S.: Inhibition and activation studies on sheep liver sorbitol dehydrogenase. Eur. J. Biochem., **221**, 847-854 (1994)
[19] Schauder, S.; Schneider, K.-H.; Giffhorn, F.: Polyol metabolism of Rhodobacter sphaeroides: biochemical characterization of a short-chain sorbitol dehydrogenase. Microbiology, **141**, 1857-1863 (1995)
[20] Lindstad, R.I.; McKinley-McKee, J.S.: Effect of pH on sheep liver sorbitol dehydrogenase steady-state kinetics. Eur. J. Biochem., **233**, 891-898 (1995)
[21] Maret, W.: Human sorbitol dehydrogenase - a secondary alcohol dehydrogenase with distinct pathophysiological roles, pH dependent kinetic studies. Enzymology and Molecular Biology of Carbonyl Metabolism 6 (Weiner et al. ed.), 383-393 (1996)
[22] Kvernmo, T.; Winberg, J.-O.; McKinley-McKee, J.S.: Reversible and irreversible inhibition of sheep liver sorbitol dehydrogenase with Cibaron Blue 3GA and Eriochrome Black T. Int. J. Biochem.Cell Biol., **28**, 303-309 (1996)
[23] Lindstad, R.I.; McKinley-McKee, J.S.: Reversible inhibition of sheep liver sorbitol dehydrogenase by thiol compounds. Eur. J. Biochem., **241**, 142-148 (1996)
[24] Marini, I.; Bucchioni, L.; Borella, P.; Del Corso, A.; Mura, U.: Sorbitol dehydrogenase from bovine lens: purification and properties. Arch. Biochem. Biophys., **340**, 383-391 (1997)
[25] Takayuki, U.: Purification and properties of NAD^+-dependent sorbitol dehydrogenase from Bacillus fructosus. Biosci. Biotechnol. Biochem., **63**, 573-574 (1999)

[26] Adachi, O.; Toyama, H.; Theeragool, G.; Lotong, N.; Matsushita, K.: Crystallization and properties of NAD-dependent D-sorbitol dehydrogenase from Gluconobacter suboxydans IFO 3257. Biosci. Biotechnol. Biochem., **63**, 1589-1595 (1999)
[27] Marini, I.; Moschini, R.; Del Corso, A.; Mura, U.: Complete protection by α-crystallin of lens sorbitol dehydrogenase undergoing thermal stress. J. Biol. Chem., **275**, 32559-32565 (2000)
[28] Oura, Y.; Yamada, K.; Shiratake, K.; Yamaki, S.: Purification and characterization of NAD^+-dependent sorbitol dehydrogenase from Japanese pear fruit. Phytochemistry, **54**, 567-572 (2000)
[29] Boguslawski, G.; Bertch, S.W.: A method for the assay of glucose isomerase activity in complex fermentation mixtures. J. Appl. Biochem., **2**, 367-372 (1980)
[30] Vongsuvanlert, V.; Tani, Y.: Characterization of D-sorbitol dehydrogenase involved in D-sorbitol production of a methanol yeast, Candida boidinii (Kloeckera sp.) No. 2201. Agric. Biol. Chem., **52**, 419-426 (1988)

D-Iditol 2-dehydrogenase 1.1.1.15

1 Nomenclature

EC number
1.1.1.15

Systematic name
D-iditol:NAD$^+$ 2-oxidoreductase

Recommended name
D-iditol 2-dehydrogenase

Synonyms
D-sorbitol dehydrogenase

CAS registry number
9028-22-2

2 Source Organism

<1> *Pseudomonas sp.* [1]

3 Reaction and Specificity

Catalyzed reaction
D-iditol + NAD$^+$ = D-sorbose + NADH (Also converts xylitol into L xylulose and L-glucitol into L-fructose)

Reaction type
oxidation
redox reaction
reduction

Natural substrates and products

S D-iditol + NAD$^+$ <1> (<1> less stable enzyme, dulcitol-inducible, accounts for oxidation of certain fully hydroxylated polyols containing a D-threo-configuration adjacent to a primary alcohol [1]) (Reversibility: ? <1> [1]) [1]
P D-sorbose + NADH <1> [1]

Substrates and products

S D-iditol + NAD$^+$ <1> (<1> less stable enzyme, dulcitol-inducible, accounts for oxidation of certain fully hydroxylated polyols containing a D-threo-configuration adjacent to a primary alcohol [1]) (Reversibility: ? <1> [1]) [1]

P D-sorbose + NADH <1> [1]

Cofactors/prosthetic groups

NAD^+ <1> [1]

5 Isolation/Preparation/Mutation/Application

Source/tissue

cell culture <1> [1]

References

[1] Shaw, D.R.D.: Polyol dehydrogenases. 3. Galactitol dehydrogenase and D-iditol dehydrogenase. Biochem. J., **64**, 394-405 (1956)

Galactitol 2-dehydrogenase 1.1.1.16

1 Nomenclature

EC number
1.1.1.16

Systematic name
galactitol:NAD^+ 2-oxidoreductase

Recommended name
galactitol 2-dehydrogenase

Synonyms
dulcitol dehydrogenase

CAS registry number
9028-23-3

2 Source Organism

<1> *Rhodobacter sphaeroides* [1]
<2> *Pseudomonas cepacia* [2]
<3> *Rhizobium trifolii* [3]
<4> *Pseudomonas sp.* [4]

3 Reaction and Specificity

Catalyzed reaction
galactitol + NAD^+ = D-tagatose + NADH (Also converts other alditols containing an L-threo-configuration adjacent to a primary alcohol group into the corresponding sugars)

Reaction type
oxidation
redox reaction
reduction

Natural substrates and products
S D-arabitol + NAD^+ <1> (Reversibility: ? <1> [1]) [1]
P ? + NADH <1> [1]
S D-ribitol + NAD^+ <3> (Reversibility: ? <3> [3]) [3]
P ? + NADH <3> [3]
S D-ribulose + NADH <1> (Reversibility: ? <1> [1]) [1]

P ? + NAD^+ <1> [1]
S D-threitol + NAD^+ <1> (<1> best substrate tested [1]) (Reversibility: ? <1> [1]) [1]
P ? + NADH <1> [1]
S D-xylulose + NADH <1> (Reversibility: ? <1> [1]) [1]
P ? + NAD^+ <1> [1]
S L-arabitol + NAD^+ <4> (Reversibility: ? <4> [4]) [4]
P L-ribulose + NADH <4> [4]
S L-erythrulose + NADH <1> (Reversibility: ? <1> [1]) [1]
P ? + NAD^+ <1> [1]
S L-glucitol + NAD^+ <1> (Reversibility: ? <1> [1]) [1]
P ? + NADH <1> [1]
S L-iditol + NAD^+ <4> (Reversibility: r <4> [4]) [4]
P L-sorbose + NADH <4> [4]
S galactitol + NAD^+ <1> (Reversibility: r <1> [1]) [1]
P L-tagatose + NADH <1> [1]
S galactitol + NAD^+ <2, 3> (Reversibility: ? <2, 3> [2, 3]) [2, 3]
P tagatose + NADH <2, 3> [2, 3]
S galactitol + NAD^+ <4> (Reversibility: r <4> [4]) [4]
P D-tagatose + NADH <4> [4]
S sorbitol + NAD^+ <3, 4> (Reversibility: ? <3> [3]; r <4> [4]) [3, 4]
P D-fructose + NADH <3, 4> [3, 4]
S xylitol + NAD^+ <3> (Reversibility: ? <3> [3]) [3]
P ? + NADH <3> [3]
S xylitol + NAD^+ <1> (<1> high activity [1]) (Reversibility: r <1> [1]) [1]
P L-xylulose + NADH <1> [1]

Substrates and products

S (2R)-butanol + NAD^+ <1> (Reversibility: ? <1> [1]) [1]
P 2-butanone + NADH <1> [1]
S (2S)-butanol + NAD^+ <1> (Reversibility: ? <1> [1]) [1]
P 2-butanone + NADH <1> [1]
S (2S,3S)-butanediol + NAD^+ <1> (Reversibility: ? <1> [1]) [1]
P ? + NADH <1> [1]
S 1,2-butanediol + NAD^+ <1> (Reversibility: ? <1> [1]) [1]
P ? + NADH <1> [1]
S 1,2-hexanediol + NAD^+ <1> (Reversibility: ? <1> [1]) [1]
P ? + NADH: <1> [1]
S 1,2-pentanediol + NAD^+ <1> (Reversibility: ? <1> [1]) [1]
P ? + NADH <1> [1]
S 1,2-propanediol + NAD^+ <1> (Reversibility: ? <1> [1]) [1]
P ? + NADH <1> [1]
S 2-hexanol + NAD^+ <1> (Reversibility: ? <1> [1]) [1]
P 2-hexanone + NADH <1> [1]
S 2-propanol + NAD^+ <1> (Reversibility: ? <1> [1]) [1]
P acetone + NADH <1> [1]
S D-arabitol + NAD^+ <1> (Reversibility: ? <1> [1]) [1]

P ? + NADH <1> [1]
S D-ribitol + NAD^+ <3> (Reversibility: ? <3> [3]) [3]
P ? + NADH <3> [3]
S D-ribulose + NADH <1> (Reversibility: ? <1> [1]) [1]
P ? + NAD^+ <1> [1]
S D-threitol + NAD^+ <1> (<1> best substrate tested [1]) (Reversibility: ? <1> [1]) [1]
P ? + NADH <1> [1]
S D-xylulose + NADH <1> (Reversibility: ? <1> [1]) [1]
P ? + NAD^+ <1> [1]
S L-arabitol + NAD^+ <4> (Reversibility: ? <4> [4]) [4]
P L-ribulose + NADH <4> [4]
S L-erythrulose + NADH <1> (Reversibility: ? <1> [1]) [1]
P ? + NAD^+ <1> [1]
S L-glucitol + NAD^+ <1> (Reversibility: ? <1> [1]) [1]
P ? + NADH <1> [1]
S L-iditol + NAD^+ <4> (Reversibility: r <4> [4]) [4]
P L-sorbose + NADH <4> [4]
S acetoin + NADH <1> (Reversibility: ? <1> [1]) [1]
P ? + NAD^+ <1> [1]
S dihydroxyacetone + NADH <1> (Reversibility: ? <1> [1]) [1]
P ? + NAD^+ <1> [1]
S galactitol + NAD^+ <1> (Reversibility: r <1> [1]) [1]
P L-tagatose + NADH <1> [1]
S galactitol + NAD^+ <2, 3> (Reversibility: ? <2, 3> [2, 3]) [2, 3]
P tagatose + NADH <2, 3> [2, 3]
S galactitol + NAD^+ <4> (Reversibility: r <4> [4]) [4]
P D-tagatose + NADH <4> [4]
S glycerol + NAD^+ <1> (Reversibility: r <1> [1]) [1]
P DL-glyceraldehyde + NADH <1> [1]
S sorbitol + NAD^+ <3, 4> (Reversibility: ? <3> [3]; r <4> [4]) [3, 4]
P D-fructose + NADH <3, 4> [3, 4]
S xylitol + NAD^+ <3> (Reversibility: ? <3> [3]) [3]
P ? + NADH <3> [3]
S xylitol + NAD^+ <1> (<1> high activity [1]) (Reversibility: r <1> [1]) [1]
P L-xylulose + NADH <1> [1]

Inhibitors

2,2'-dipyridyl <1> (<1> 13% inhibition at 1 mM, restored by addition of 2 mM $MgCl_2$ [1]) [1]
EDTA <1> (<1> 68% inhibition at 1 mM, restored by addition of 2 mM $MgCl_2$ [1]) [1]
$FeSO_4$ <1> [1]
$ZnSO_4$ <1> [1]

Cofactors/prosthetic groups

NAD^+ <1-4> (<1, 4> strictly specific [1, 4]) [1-4]
NADH <1> (<1> strictly specific [1]) [1]

Metals, ions

Co^{2+} <1> (<1> 21% activation at 1 mM [1]) [1]
Mg^{2+} <1> (<1> 1 mM required for optimal activity [1]) [1]
Mn^{2+} <1> (<1> 1 mM required for optimal activity [1]) [1]
Ni^{2+} <1> (<1> 25% activation at 1 mM [1]) [1]

K_m-Value (mM)

0.004 <1> (NADH) [1]
0.012 <1> (NAD^+) [1]
0.2 <1> (1,2-hexanediol) [1]
48 <1> (dihydroxyacetone) [1]
62 <1> (acetoin) [1]
85 <1> (D-threitol) [1]
144 <1> (L-erythrulose) [1]
240 <1> (galactitol) [1]

pH-Optimum

4 <1> (<1> reduction of acetoin [1]) [1]
10.5 <1> (<1> 1,2-hexanediol oxidation [1]) [1]

4 Enzyme Structure

Molecular weight

110000 <1> (<1> sucrose density gradient centrifugation [1]) [1]

Subunits

tetramer <1> (<1> 4 * 28000, SDS-PAGE [1]) [1]

5 Isolation/Preparation/Mutation/Application

Source/tissue

cell culture <1-3> [1-3]

Purification

<1> (to homogeneity, chromatography steps [1]) [1]
<3> (partial [3]) [3]
<4> (partial [4]) [4]

6 Stability

pH-Stability

6.5 <1> (<1> 63% of the reduction activity [1]) [1]
9 <1> (<1> 81% of the oxidation activity [1]) [1]

Temperature stability

37 <1> (<1> 50% loss of activity after 8 days [1]) [1]
50 <1> (<1> 50% loss of activity after 12 min [1]) [1]

General stability information

<1>, sucrose stabilizes enzyme preparation [1]

Storage stability

<1>, 4°C, 20 mM Bis-Tris, pH 6.5, 1 mM $MgCl_2$, 30% sucrose, stable for 3 months [1]

<4>, -20°C, 5% loss of activity within 5 months [4]

References

[1] Schneider, K.-H.; Jäkel, G.; Hoffmann, R.; Giffhorn, F.: Enzyme evolution in Rhodobacter sphaeroides: Selection of a mutant expressing a new galactitol dehydrogenase and biochemical characterization of the enzyme. Microbiology, **141**, 1865-1873 (1995)

[2] Allenza, P.; Lee, Y.N.; Lessie, T.G.: Enzymes related to fructose utilization in Pseudomonas cepacia. J. Bacteriol., **150**, 1348-1356 (1982)

[3] Primrose, S.B.; Ronson, C.W.: Polyol metabolism by Rhizobium trifolii. J. Bacteriol., **141**, 1109-1114 (1980)

[4] Shaw, D.R.D.: Galactitol dehydrogenase. Methods Enzymol., **5**, 323-325 (1962)

Mannitol-1-phosphate 5-dehydrogenase 1.1.1.17

1 Nomenclature

EC number
1.1.1.17

Systematic name
D-mannitol-1-phosphate:NAD$^+$ 5-oxidoreductase

Recommended name
mannitol-1-phosphate 5-dehydrogenase

Synonyms
D-mannitol-1-phosphate dehydrogenase
dehydrogenase, mannitol 1-phosphate
fructose 6-phosphate reductase
hexose reductase
mannitol 1-phosphate dehydrogenase

CAS registry number
9028-24-4

2 Source Organism

<-3> no activity in *Nocardia erythropolis* [11]
<-2> no activity in *Mycobacterium smegmatis* [11]
<-1> no activity in *Actinoplanes missouriensis* [11]
<1> *Klebsiella pneumoniae* [1]
<2> *Cryptococcus neoformans* [2]
<3> *Bacillus stearothermophilus* [3]
<4> *Brochothrix thermosphacta* (formerly *Microbacterium thermosphactum*) [4]
<5> *Streptococcus mutans* [5, 12]
<6> *Enterococcus faecalis* [6]
<7> *Staphylococcus carnosus* [6]
<8> *Escherichia coli* [7, 9, 15]
<9> *Aspergillus parasiticus* [8]
<10> *Aspergillus niger* [8, 10]
<11> *Micromonospora sp.* (strains BC-1751, BC-2079) [11]
<12> *Streptomyces lactamdurans* [11]
<13> *Aspergillus nidulans* [13]
<14> *Aerobacter aerogenes* [14]

3 Reaction and Specificity

Catalyzed reaction

D-mannitol 1-phosphate + NAD^+ = D-fructose 6-phosphate + NADH

Reaction type

intramolecular phosphate transfer
oxidation
redox reaction
reduction

Natural substrates and products

S D-mannitol 1-phosphate + NAD^+ <1-14> (<5, 6> part of the phosphoenolpyruvate phosphotransferase system [5, 6]; <5, 8, 10> highly specific for both substrates [9, 10, 12, 15]; <10> random bi-bi kinetic with two dead-end complexes [10]; <8> favours reverse reaction [15]) (Reversibility: ? <1, 3, 5-7, 11, 12> [1, 3, 5, 6, 9, 11]; r <2, 4, 8-10, 13, 14> [2, 4, 7, 8, 10, 12-15]) [1-15]

P D-fructose 6-phosphate + NADH <1-14> [1-15]

S D-sorbitol 6-phosphate + NAD^+ <2> (<2> 38.8% of the activity compared to D-mannitol 1-phosphate [2]) (Reversibility: r <2> [2]) [2]

P D-glucose 6-phosphate + NADH <2> [2]

S ethanol + NAD^+ <2> (<2> 56% of the activity compared to D-mannitol 1-phosphate [2]) (Reversibility: r <2> [2]) [2]

P acetaldehyde + NADH <2> [2]

S hexitol phosphate + NAD^+ <8> (Reversibility: ? <8> [9]) [9]

P ? + NADH <8> [9]

Substrates and products

S D-mannitol 1-phosphate + 3-acetylpyridine-NAD^+ <14> (<14> 7% of the activity compared to NAD^+ [14]) (Reversibility: ? <14> [14]) [14]

P D-fructose 6 phosphate + ? <14> [14]

S D-mannitol 1-phosphate + NAD^+ <1-14> (<5, 6> part of the phosphoenolpyruvate phosphotransferase system [5, 6]; <5, 8, 10> highly specific for both substrates [9, 10, 12, 15]; <10> random bi-bi kinetic with two dead-end complexes [10]; <8> favours reverse reaction [15]) (Reversibility: ? <1, 3, 5-7, 11, 12> [1, 3, 5, 6, 9, 11]; r <2, 4, 8-10, 13, 14> [2, 4, 7, 8, 10, 12-15]) [1-15]

P D-fructose 6-phosphate + NADH <1-14> [1-15]

S D-mannitol 1-phosphate + acetylpyridine adenine dinucleotide <10> (<10> 50% of the activity compared to NAD^+ [10]) (Reversibility: ? <10> [10]) [10]

P D-fructose 6-phosphate + ? <10> [10]

S D-mannitol 1-phosphate + dichlorophenolindophenol <11> (<11> higher activity than for NAD^+ [11]) (Reversibility: ? <11> [11]) [11]

P D-fructose 6-phosphate + reduced dichlorophenolindophenol <11> [11]

S D-mannitol 1-phosphate + nicotinamide 6-(2-hydoxyethylamino) purine dinucleotide <14> (<14> 75% of the activity compared to NAD^+ [14]) (Reversibility: ? <14> [14]) [14]

P D-fructose 6-phosphate + ? <14> [14]

S D-mannitol 1-phosphate + nicotinamide hypoxanthine dinucleotide <14> (<14> 84% of the activity compared to NAD^+ [14]) (Reversibility: ? <14> [14]) [14]

P D-fructose 6-phosphate + ? <14> [14]

S D-sorbitol 6-phosphate + NAD^+ <2> (<2> 38.8% of the activity compared to D-mannitol 1-phosphate [2]) (Reversibility: r <2> [2]) [2]

P D-glucose 6-phosphate + NADH <2> [2]

S ethanol + NAD^+ <2> (<2> 56% of the activity compared to D-mannitol 1-phosphate [2]) (Reversibility: r <2> [2]) [2]

P acetaldehyde + NADH <2> [2]

S hexitol phosphate + NAD^+ <8> (Reversibility: ? <8> [9]) [9]

P ? + NADH <8> [9]

Inhibitors

1,10-phenanthroline <8> (<8> weak inhibition [7]) [7]

2,2'-bipyridyl <8> (<8> weak inhibition [7]) [7]

4,7-phenanthroline <8> (<8> weak inhibition [7]) [7]

5,5'-dithiobis-(2-nitrobenzoate) <8> (<8> complete inactivation at 0.24 mM [7]) [7]

ADP <4, 5> (<4> competitive inhibition [4]; <5> 20% inhibition of mannitol 1-phosphate oxidation [12]) [4, 12]

ATP <4, 5> (<4> competitive inhibition [4]; <5> 53% inhibition of mannitol 1-phosphate oxidation [12]) [4, 12]

Cd^{2+} <9, 10> (<9> cooperative binding [8]; <10> competitive inhibition [8]) [8]

Co^{2+} <9, 10> (<9> cooperative binding [8]; <10> competitive inhibition [8]) [8]

Cu^{2+} <9, 10> (<9,10> irreversible inactivation [8]) [8]

D-fructose 1,6-diphosphate <5> (<5> 12% inhibition of mannitol 1-phosphate oxidation [12]) [12]

D-fructose 6-phosphate <5, 10> (<10> competitive inhibition with respect to D-mannitol 1-phosphate [10]; <5> 56% inhibition of mannitol 1-phosphate oxidation [12]) [10, 12]

D-glucose 6-phosphate <10> (<10> competitive inhibition with respect to D-fructose 6-phosphate, noncompetitive inhibition with respect to NADH [10]) [10]

D-mannitol 1-phosphate <10> (<10> competitive inhibition with respect to D-fructose 6-phosphate, noncompetitive inhibition with respect to NADH [10]) [10]

GTP <5> (<5> 11% inhibition of mannitol 1-phosphate oxidation [12]) [12]

Hg^{2+} <9, 10> (<9,10> irreversible inactivation [8]) [8]

NAD^+ <10> (<10> competitive inhibition with respect to NADH, noncompetitive inhibition with respect to D-fructose 6-phosphate [10]) [10]

NADH <10> (<10> competitive inhibition with respect to NAD^+ [10]) [10]
Ni^{2+} <9, 10> (<9> cooperative binding [8]; <10> competitive inhibition [8]) [8]
Zn^{2+} <9, 10> (<9> cooperative binding of Zn^{2+}, competitive type if KCl concentration is greater 100 mM, sigmoidal type at lower salt concentrations [8]; <10> competitive with respect to D-fructose 6-phosphate, noncompetitive inhibition with respect to NADH [8]) [8]
adenosine <5> (<5> 48% inhibition of fructose 6-phosphate reduction [12]) [12]
adenosine diphosphoribose <10> (<10> competitive inhibition with respect to NADH, noncompetitive inhibition with respect to D-fructose 6-phosphate [10]) [10]
cytidine <5> (<5> 17% inhibition of fructose 6-phosphate reduction [12]) [12]
diethyldicarbonate <8> (<8> 60% inhibition at 1 mM [9]) [9]
diethyldithiocarbamate <8> (<8> 95% inactivation, prevented by presence of 2-mercaptoethanol [7]) [7]
guanosine <5> (<5> 10% inhibition of fructose 6-phosphate reduction [12]) [12]
phosphoenolpyruvate <5> (<5> 14% inhibition of mannitol 1-phosphate oxidation [12]) [12]
thymidine <5> (<5> 8% inhibition of fructose 6-phosphate reduction [12]) [12]
uridine <5> (<5> 16% inhibition of fructose 6-phosphate reduction [12]) [12]

Cofactors/prosthetic groups

NAD^+ <1-14> (<2, 5, 8, 10, 13, 14> strictly specific [2, 9, 10, 12-15]; <4> higher affinity for NADH [4]) [1-15]
NADH <2-5, 8-10, 13, 14> (<4> higher affinity than for NAD^+ [4]; <8, 13, 14> strictly specific [13-15]) [2-4, 7, 8, 10, 12-15]

Metals, ions

Mn^{2+} <2> (<2> 2fold activation at 5 mM, no activation of ethanol oxidation [2]) [2]
Zn^{2+} <8> (<8> 1.14 Zn atoms/monomer [7]) [7]

Specific activity (U/mg)

0.03 <12> (<12> crude extract, D-mannitol 1-phosphate oxidation [11]) [11]
0.03-0.05 <11> (<11> crude extract, D-mannitol 1-phosphate oxidation [11]) [11]
0.09 <4> (<4> crude extract, D-mannitol 1-phosphate oxidation [4]) [4]
0.1 <2> (<2> crude extract, D-mannitol 1-phosphate oxidation [2]) [2]
0.13 <13> (<13> crude extract, dehydrogenase activity [13]) [13]
0.28 <13> (<13> crude extract, reductase activity [13]) [13]
0.43 <5> [12]
18 <4> (<4> purified enzyme, D-fructose 6-phosphate reduction [4]) [4]
116 <2> (<2> purified enzyme D-mannitol 1-phosphate oxidation [2]) [2]

150 <10> [8, 10]
190 <9> [8]
502 <8> [7]
715 <8> [9]
Additional information <8, 14> (<8, 14> 1 unit defined as change of 0.01 in optical density at 340 nm during 30-90 second interval [14, 15]; <14> 240 U/mg protein [14]; <8> 860 U/mg protein [15]) [14, 15]

K_m-Value (mM)

0.005 <10> (NADH, <10> cosubstrate: D-fructose 6-phosphate [10]) [10]
0.016 <5> (NADH, <5> cosubstrate: D-fructose 6-phosphate [12]) [12]
0.03 <4> (NADH, <4> higher affinity for NADH [4]) [4]
0.038 <10> (D-mannitol 1-phosphate, <10> cosubstrate: NAD^+ [10]) [10]
0.05 <4> (D-mannitol 1-phosphate, <4> cosubstrate: NAD^+ [4]) [4]
0.05 <4> (NAD^+) [4]
0.055 <2> (D-mannitol 1-phosphate, <2> cosubstrate: NAD^+ [2]) [2]
0.06 <5> (NAD^+, <5> cosubstrate: D-mannitol 1-phosphate [12]) [12]
0.083 <10> (NAD^+, <10> cosubstrate: D-mannnitol 1-phosphate [10]) [10]
0.099 <2> (NADH, <2> cosubstrate: D-fructose 6-phosphate [2]) [2]
0.11 <2> (NAD^+, <2> cosubstrate: D-mannitol 1-phosphate [2]) [2]
0.14 <14> (NADH, <14> cosubstrate: D-fructose 6-phosphate [14]) [14]
0.15 <5> (D-mannitol 1-phosphate, <5> cosubstrate: NAD^+ [12]) [12]
0.2 <8> (NAD^+, <8> cosubstrate: hexitol phosphate [9]) [9]
0.25 <4> (D-fructose 6-phosphate, <4> cosubstrate: NADH [4]) [4]
0.3 <2> (D-fructose 6-phosphate, <2> cosubstrate: NADH [2]) [2]
0.38 <14> (NAD^+, <14> cosubstrate: D-mannitol 1-phosphate [14]) [14]
0.54 <10> (D-fructose 6-phosphate, <10> cosubstrate: NADH [10]) [10]
0.8 <8> (hexitol phosphate, <8> cosubstrate: NAD^+ [9]) [9]
0.94 <2> (ethanol, <2> cosubstrate: NAD^+ [2]) [2]
1.66 <5> (D-fructose 6-phosphate, <5> cosubstrate: NADH [12]) [12]
2 <14> (D-mannitol 1-phosphate, <14> cosubstrate: NAD^+ [14]) [14]
2.1 <14> (D-fructose 6-phosphate, <14> cosubstrate: NADH [14]) [14]

K_i-Value (mM)

0.7 <4> (ADP, <4> more potent inhibitor than ATP [4]) [4]
1.6 <4> (ATP) [4]
2 <5> (ADP, <5> competitive inhibition with respect to NADH, noncompetitive inhibition with respect to fructose 6-phosphate [12]) [12]
2.2 <5> (ATP, <5> strong inhibition of mannitol 1-phosphate oxidation [12]) [12]
3.33 <5> (D-fructose 6-phosphate, <5> able to inhibit the forward reaction to 100%, competitive inhibition with respect to mannitol 1-phosphate, noncompetitive inhibition with respect to NAD^+ [12]) [12]
8 <8> (1,10-phenanthroline) [7]
11 <8> (2,2'-bipyridyl) [7]
12 <8> (4,7-phenanthroline) [7]

pH-Optimum

6-7 <4> (<4> oxidation of D-mannitol 1-phosphate [4]) [4]
6.5 <5> (<5> reduction of D-fructose 6-phosphate [12]) [12]
7-7.5 <2> (<2> NADH-dependent reduction of D-fructose 6-phosphate [2]) [2]
8.5-9 <5> (<5> oxidation of D-mannitol 1-phosphate [12]) [12]
9 <8, 11> (<8> D-mannitol 1-phosphate oxidation [9]) [9, 11]
9-9.5 <2> (<2> NAD^+-dependent oxidation of D-mannitol 1-phosphate [2]) [2]
9.5 <11, 14> (<11> dichlorophenolindophenol-linked assay [11]; <14> mannitol 1-phosphate oxidation, pH depends on buffer system used [14]) [11, 14]
9.5-10 <4> (<4> reduction of D-fructose 6-phosphate [4]) [4]

pH-Range

5.5-10 <10> (<10> more than 95% of its original activity [10]) [10]
6-9 <8> (<8> no sharp optimum [15]) [15]

4 Enzyme Structure

Molecular weight

40000 <8-10, 14> (<8-10> gel filtration [8-10]; <14> sedimentation equilibrium analysis [14]) [8-10, 14]
45000 <5> (<5> gel filtration [12]) [12]
45500 <8> (<8> gel filtration [7]) [7]
50000 <4> (<4> gel filtration [4]) [4]
148000 <2> (<2> gel filtration [2]) [2]

Subunits

? <1, 2, 6> (<1> x * 42000, SDS-PAGE [1]; <2> x * 36000, SDS-PAGE [2]; <6> x * 40000, SDS-PAGE [6]) [1, 2, 6]
dimer <8, 10> (<8> 2 * 22500, SDS-PAGE [7]; <10> 2 * 22000, SDS-PAGE [10]) [7, 10]
monomer <4, 8> (<4> 1 * 50000, SDS-PAGE [4]; <8> 1 * 40000, SDS-PAGE [9]) [4, 9]

5 Isolation/Preparation/Mutation/Application

Source/tissue

cell culture <1-14> (<4,14> inducible by D-mannitol [4, 14]; <13> reductase activity affected by the presence of nitrate during growth [13]) [1-15]

Purification

<2> (to homogeneity, chromatography techniques [2]) [2]
<4> (to homogeneity, chromatography techniques [4]) [4]
<5> (to homogeneity, chromatography steps, affinity chromatography [12]) [12]

<8> (to homogeneity, chromatography techniques [7]; recombinant enzyme [9]; precipitation techniques [15]) [7, 9, 15]
<9> (to homogeneity, chromatography techniques [8]) [8]
<10> (to homogeneity, chromatography techniques [8, 10]) [8, 10]
<14> (to homogeneity, chromatography techniques [14]) [14]

Cloning

<1> (expression in Escherichia coli [1]) [1]
<6> (expression in Escherichia coli [6]) [6]
<8> (overexpression in Escherichia coli [9]) [9]

Engineering

Additional information <1> (<1> C-terminal truncation, lacking 162 amino acids of the C-terminus, 10% in vivo activity compared to the wild-type form [1]) [1]

6 Stability

Temperature stability

4 <2> (<2> stable for one week [2]) [2]
30 <9, 10> (<9> half-life: 2 h, decreasing stability in the presence of zinc ions [8]; <10> half-life: above 280 min [8]) [8]
40 <8> (<8> stable during 20 min [9]) [9]
55 <8> (<8> 45% loss of activity in 10 min [15]) [15]
60 <8> (<8> 90% loss of activity after 20 min [9]; <8> complete inactivation in 10 min [15]) [9, 15]

General stability information

<14>, loss of activity partially protected by addition of substrates [14]

Storage stability

<5>, -80°C, 0.05 M Tris-HCl, pH 7.5, presence of 2-mercaptoethanol, long term storage without loss of activity [12]
<8>, 4°C, 0.03 M imidazole-HCl, pH 7.3, several months [7]
<14>, 4°C, 0.05 M Tris-HCl, pH 7.6, 0.05 M NaCl, 5 mM 2-mercaptoethanol, stable for weeks [14]

References

[1] Otte, S.; Lengeler, J.W.: The mtl genes and the mannitol-1-phosphate dehydrogenase from Klebsiella pneumoniae KAY2026. FEMS Microbiol. Lett., **194**, 221-227 (2001)
[2] Survarna, K.; Bartiss, A.; Wong, B.: Mannitol-1-phosphate dehydrogenase from Cryptococcus neoformans is a zinc-containing long-chain alcohol/polyol dehydrogenase. Microbiology, **146**, 2705-2713 (2000)
[3] Henstra, S.A.; Tolner, B.; Ten Hoeve Duurkens, R.H.; Konings, W.N.; Robillard, G.T.: Cloning, expression, and isolation of the mannitol transport pro-

tein from the thermophilic bacterium Bacillus stearothermophilus. J. Bacteriol., **178**, 5586-5591 (1996)
[4] Singh, S.P.; Rogers, P.J.: Isolation and characterization of mannitol-1-phosphate dehydrogenase from Brochothrix thermosphacta. J. Gen. Appl. Microbiol., **39**, 327-337 (1993)
[5] Honeyman, A.L.; Curtiss III, R.: Isolation, characterization, and nucleotide sequence of the Streptococcus mutans mannitol-phosphate dehydrogenase gene and the mannitol-specific factor III gene of the phosphoenolpyruvate phosphotransferase system. Infect. Immun., **60**, 3369-3375 (1992)
[6] Fischer, R.; von Strandmann, R.P.; Hengstenberg, W.: Mannitol-specific phosphoenolpyruvate-dependent phosphotransferase system of Enterococcus faecalis: Molecular cloning and nucleotide sequences of the enzyme III Mtl gene and the mannitol-1-phosphate dehydrogenase gene, expression in Escherichia coli, and comparison of the gene products with similar enzymes. J. Bacteriol., **173**, 3709-3715 (1991)
[7] Chase, T.: Mannitol-1-phosphate dehydrogenase of Escherichia coli. Biochem. J., **239**, 435-443 (1986)
[8] Foreman, J.E.; Niehaus, W.G.: Zn^{2+}-induced cooperativity of mannitol-1-phosphate dehydrogenase from Aspergillus parasiticus. J. Biol. Chem., **260**, 10019-10022 (1985)
[9] Novotny, M.J.; Reizer, J.; Esch, F.; Saier, M.H.: Purification and properties of D-mannitol-1-phosphate dehydrogenase and D-glucitol-6-phosphate dehydrogenase from Escherichia coli. J. Bacteriol., **159**, 986-990 (1984)
[10] Kiser, R.C.; Niehaus, W.G.: Purification and kinetic characterization of mannitol-1-phosphate dehydrogenase from Aspergillus niger. Arch. Biochem. Biophys., **211**, 613-621 (1981)
[11] Mehta, R.J.; Fare, L.R.; Shearer, M.E.; Nash, C.H.: Mannitol oxidation in two Micromonospora isolates and in representative species of other Actinomycetes. Appl. Environ. Microbiol., **33**, 1013-1015 (1977)
[12] Brown, A.T.; Bowles, R.D.: Polyol metabolism by a caries-conductive Streptococcus: Purification and properties of a nicotinamide adenine dinucleotide-dependent mannitol-1-phosphate dehydrogenase. Infect. Immun., **16**, 163-173 (1977)
[13] Hankinson, O.; Cove, D.J.: Regulation of mannitol-1-phosphate dehydrogenase in Aspergillus nidulans. Can. J. Microbiol., **21**, 99-101 (1975)
[14] Liss, M.; Horwitz, S.B.; Kaplan, N.O.: D-Mannitol 1-phosphate dehydrogenase and D-sorbitol 6-phosphate dehydrogenase in Aerobacter aerogenes. J. Biol. Chem., **237**, 1342-1350 (1962)
[15] Wolff, J.B.; Kaplan, N.O.: D-Mannitol-1-phosphate dehydrogenase from E. coli. Methods Enzymol., **1**, 346-348 (1955)

Inositol 2-dehydrogenase 1.1.1.18

1 Nomenclature

EC number
1.1.1.18

Systematic name
myo-inositol:NAD^+ 2-oxidoreductase

Recommended name
inositol 2-dehydrogenase

Synonyms
inositol dehydrogenase
myo-inositol 2-dehydrogenase
myo-inositol dehydrogenase
myo-inositol:NAD2 oxidoreductase

CAS registry number
9028-25-5

2 Source Organism

<1> *Sinorhizobium fredii* (strain USDA191) [1]
<2> *Sinorhizobium meliloti* [2]
<3> *Galdieria sulphuraria* [3]
<4> *Bacillus subtilis* [4]
<5> *Bacillus subtilis* [5]
<6> *Serratia marcescens* (inducible [6]) [6]
<7> *Streptomyces hygroscopius* (formerly *Streptomyces glebosus*) [7]
<8> *Acetomonas oxydans* [8]
<9> *Klebsiella pseudomoniae* [9]
<10> *Cryptococcus melibiosum* [10]
<11> *Aerobacter aerogenes* [11]

3 Reaction and Specificity

Catalyzed reaction
myo-inositol + NAD^+ = 2,4,6/3,5-pentahydroxycyclohexanone + NADH

Reaction type
oxidation

redox reaction
reduction

Natural substrates and products

S D-2,3-diketo-4-deoxy-epi-inositol + NADH <11> (Reversibility: ? <11> [11]) [11]

P ketodeoxyinositol + NAD^+ <11> [11]

S myo-inositol + NAD^+ <1, 2, 4-7, 9, 11> (<1> involved in nitrogen fixation and nodulation of soybean [1]; <2> involved in rhizopine utilization [2]; <9> should be classified as A-type enzyme [9]) (Reversibility: ? <1, 2, 4-7, 9, 11> [1, 2, 4-7, 9, 11]) [1, 2, 4-7, 9, 11]

P 2,4,6/3,5-pentahydroxycyclohexanone + NADH <1, 2, 4-7, 9, 11> [1, 2, 4-7, 9, 11]

S myo-inositol + NAD^+ <3, 7, 10> (<3> 75 times lower activity compared to the reverse reaction [3]) (Reversibility: r <3, 10> [3, 10]; ? <7> [7]) [3, 7, 10]

P scyllo-inosose + NADH <3, 7, 10> [3, 7, 10]

Substrates and products

S D-2,3-diketo-4-deoxy-epi-inositol + NADH <11> (Reversibility: ? <11> [11]) [11]

P ketodeoxyinositol + NAD^+ <11> [11]

S D-glucose + NAD^+ <4, 5> (<4> 9% of the activity compared to myo-inositol as substrate [4]) (Reversibility: ? <4, 5> [4, 5]) [4, 5]

P D-gluconate + NADH <4, 5> [4, 5]

S D-xylose + NAD^+ <4, 5> (<4> 7% of the activity compared to myo-inositol as substrate [4]; <5> very low activity [5]) (Reversibility: ? <4, 5> [4, 5]) [4, 5]

P ? + NADH <4, 5> [4, 5]

S α-D-glucopyranose + NAD^+ <5> (<5> 4fold lower activity compared to myo-inositol as substrate, does not act with the β-anomer [5]) (Reversibility: ? <5> [5]) [5]

P D-gluconate + NADH <5> [5]

S epi-inositol + NAD^+ <3> (<3> 5% of the activity compared to myo-inositol as substrate, conversion of epi-inosose 5% of the activity compared to scyllo-inosose as substrate [3]) (Reversibility: r <3> [3]) [3]

P epi-inosose + NADH <3> [3]

S myo-inositol + 2,6-dichlorophenolindophenol <8> (Reversibility: ? <8> [8]) [8]

P 2,4,6/3,5-pentahydroxycyclohexanone + reduced 2,6-dichlorophenolindophenol <8> [8]

S myo-inositol + NAD^+ <1, 2, 4-7, 9, 11> (<1> involved in nitrogen fixation and nodulation of soybean [1]; <2> involved in rhizopine utilization [2]; <9> should be classified as A-type enzyme [9]) (Reversibility: ? <1, 2, 4-7, 9, 11> [1, 2, 4-7, 9, 11]) [1, 2, 4-7, 9, 11]

P 2,4,6/3,5-pentahydroxycyclohexanone + NADH <1, 2, 4-7, 9, 11> [1, 2, 4-7, 9, 11]

S myo-inositol + NAD^+ <3, 7, 10> (<3> 75 times lower activity compared to the reverse reaction [3]) (Reversibility: r <3, 10> [3, 10]; ? <7> [7]) [3, 7, 10]
P scyllo-inosose + NADH <3, 7, 10> [3, 7, 10]
S scyllo-inositol + NAD^+ <3> (<3> 5% of the activity compared to myo-inositol as substrate [3]) (Reversibility: ? <3> [3]) [3]
P scyllo-inosose + NADH <3> [3]

Inhibitors

2-keto-myo-inositol <11> [11]
NADH <10> (<10> competitive inhibition with NAD^+ as variable substrate, bi-bi mechanism [10]) [10]
deoxycholate <8> (<8> loss of activity within 40 h, if detergent is not removed by chromatography [8]) [8]
scyllo-inosose <10> (<10> non-competitive inhibition with myo-inositol as varied substrate, uncompetitive with NAD^+ as variable substrate, bi-bi ordered mechanism [10]) [10]

Cofactors/prosthetic groups

NAD^+ <1-7, 9, 10, 11> (<4, 10> can not be replaced by $NADP^+$ [4, 10]; <11> preferred hydrogen acceptor [11]) [1-7, 9, 10, 11]
NADH <3, 10, 11> [3, 10, 11]
$NADP^+$ <7> (<7> less than 10% of the activity compared to NAD^+ [7]) [7]

Specific activity (U/mg)

0.07 <1> (<1> after induction of gene expression by myo-inositol [1]) [1]
0.241 <2> (<2> after induction of gene expression by myo-inositol [2]) [2]
1.68 <10> [10]
12 <4> (<4> overepression in Escherichia coli [4]) [4]
34.3 <5> [5]
50 <3> [3]
280 <11> [11]

K_m-Value (mM)

0.23 <5> (NAD^+) [5]
1.3 <3> (scyllo-inosose) [3]
18 <5> (myo-inositol) [5]
56 <5> (α-D-glucose) [5]
167 <5> (D-glucose) [5]
190 <5> (D-xylose) [5]
430 <3> (myo-inositol) [3]

pH-Optimum

6.2 <8> [8]
8 <3, 6> (<3> clearly favours reduction of scyllo-inosose [3]) [3, 6]
9.5 <5> [5]

4 Enzyme Structure

Molecular weight

160000 <3-5> (<3> density gradient centrifugation [3]; <4, 5> gel filtration [4, 5]) [3-5]
180000 <3> (<3> gel filtration [3]) [3]

Subunits

? <1, 2, 4> (<1> x * 34648, calculation from ORF of cDNA sequence [1]; <2> x * 35130, calculation from ORF of cDNA sequence [2]; <4> x * 30000, SDS-PAGE [4]) [1, 2, 4]
tetramer <3, 5> (<3> 4 * 42000, SDS-PAGE [3]; <5> 4 * 39000, SDS-PAGE [5]) [3, 5]

5 Isolation/Preparation/Mutation/Application

Source/tissue

cell culture <1-11> [1-11]

Purification

<3> (to homogeneity, chromatography steps [3]) [3]
<4> (90% homogeneity, recombinant enzyme [4]) [4]
<5> (to homogeneity, chromatography steps [5]) [5]
<10> (partial [10]) [10]
<11> (partial [11]) [11]

Cloning

<4> (expression in Escherichia coli [4]) [4]

6 Stability

pH-Stability

5.5 <3> (<3> half-maximal activity [3]) [3]
6.5 <5> (<5> optimal pH-value for stability [5]) [5]
6.8 <8> (<8> most stable at [8]) [8]
8 <6> (<6> sharp decrease of activity below or above [6]) [6]
9.5 <3> (<3> half-maximal activity [3]) [3]

Temperature stability

50 <3> (<3> stable for 2 h [3]) [3]

Storage stability

<11>, -10°C, several months [11]

References

[1] Jiang, G.; Krishnan, A.H.; Kim, Y.-W.; Wacek, T.J.; Krishnan, H.B.: A functional myo-inositol dehydrogenase gene is required for efficient nitrogen fixation and competitiveness of Sinorhizobium fredii USDA191 to nodulate soybean (Glycine max [L.] Merr.). J. Bacteriol., **183**, 2595-2604 (2001)

[2] Galbraith, M.P.; Feng, S.F.; Borneman, J.; Triplett, E.W.; de Bruijn, F.J.; Rossbach, S.: A functional myo-inositol catabolism pathway is essential for rhizopine utilization by Sinorhizobium meliloti. Microbiology, **144**, 2915-2924 (1998)

[3] Stein, R.; Schnarrenberger, C.; Gross, W.: myo-Inositol dehydrogenase from the acido- and thermophilic red alga Galdieria sulphuraria. Phytochemistry, **46**, 17-20 (1997)

[4] Fujita, Y.; Shindo, K.; Miwa, Y.; Yoshida, K.-i.: Bacillus subtiltis inositol dehydrogenase-encoding gene (idh): sequence and expression in Escherichia coli. Gene, **108**, 121-125 (1991)

[5] Ramaley, R.; Fujita, Y.; Freese, E.: Purification and properties of Bacillus subtilis inositol dehydrogenase. J. Biol. Chem., **254**, 7684-7690 (1979)

[6] Fawole, M.O.: Inositol dehydrogenase from Serratia marcescens. Z. Allg. Mikrobiol., **16**, 327-328 (1976)

[7] Walker, J.B.: myo-Inositol:NAD$^+$ 2-oxidoreductase. Methods Enzymol., **43**, 433-439 (1975)

[8] Criddle, W.J.; Fry, J.C.; Keaney, M.M.: myo-Inositol dehydrogenase(s) from Acetomonas oxydans. Optimization of conditions for solubilization of membrane-bound enzyme. Biochem. J., **137**, 449-452 (1974)

[9] Alizade, M.A.; Gaede, K.: Chirality of the hydrogen transfer to NAD catalyzed by myo-inositol dehydrogenase from Klebsiella pneumoniae. Z. Naturforsch.C, **31**, 624-625 (1976)

[10] Vidal-Leiria, M.; Van Uden, N.: Inositol dehydrogenase from the yeast Cryptococcus melibiosum. Biochim. Biophys. Acta, **293**, 295-303 (1973)

[11] Berman, T.; Magasanik, B.: The pathway of myo-inositol degradation in Aerobacter aerogenes. Dehydrogenation and dehydration. J. Biol. Chem., **241**, 800-806 (1966)

L-Glucuronate reductase 1.1.1.19

1 Nomenclature

EC number
1.1.1.19

Systematic name
L-gluconate:$NADP^+$ 6-oxidoreductase

Recommended name
L-glucuronate reductase

Synonyms
D-glucuronate dehydrogenase
D-glucuronate reductase
D-glucuronic reductase
L-gulonate NAD-3-oxidoreductase
L-hexonate:NADP dehydrogenase
NADP-L-gulonate dehydrogenase
TPN-L-gulonate dehydrogenase
aldehyde reductase
aldehyde reductase II
hexonate dehydrogenase
Additional information (may be identical with EC 1.1.1.2)

CAS registry number
9028-29-9

2 Source Organism

<1> *Sus scrofa* [1-3]
<2> *Mesocricetus auratus* [4, 9]
<3> *Mus musculus* [5]
<4> *Schwanniomyces occidentalis* [6]
<5> *Bos taurus* [7, 8]
<6> *Homo sapiens* [8-12]
<7> *Rattus norvegicus* [8-10]
<8> *Canis familiaris* [10]

3 Reaction and Specificity

Catalyzed reaction

L-gulonate + NADP$^+$ = D-glucuronate + NADPH (also reduces D-galacturonate; may be identical with EC 1.1.1.2 (alcohol dehydrogenase (NADP$^+$)) e.g. in [11, 12])

Reaction type

oxidation (<6> pH dependent reaction mechanism [11]; <1> ordered bi bi mechanism [1]) [1, 11]
reduction (<6> pH dependent reaction mechanism [11]; <1> ordered bi bi mechanism [1]) [1, 11]

Natural substrates and products

S D-glucuronate + NADPH <1, 3, 6, 7, 5> (<1> formation of ascorbic acid [2]; <3> catabolism of inositol [5]) (Reversibility: r <6, 7, 5> [8-12]) [2, 5, 8-12]
P L-gulonate + NADP$^+$
S L-gulonate + NADP$^+$ <6, 7, 5> (Reversibility: r <6, 7, 5> [8-12]) [8-12]
P D-glucuronate + NADPH
S γ-hydroxybutyrate + NADP$^+$ <2> (<2> preverably in a coupled reaction in vivo with reduction of D-glucuronate [9]) (Reversibility: ? <2> [9]) [9]
P succinic semialdehyde + NADPH

Substrates and products

S 4-nitrobenzaldehyde + NADPH <1, 3> (<3> 2-, and 3-isomer also reduced [5]) (Reversibility: r <1, 3> [1, 5]) [1, 5]
P 4-nitrobenzyl alcohol + NADP$^+$
S D-aldose + NADPH <6, 7> (<6> i.e. D-xylose, D-glucose, D-galactose are reduced with 100fold less activity than D-glucuronate [10]) (Reversibility: r <6, 7> [10]) [10, 10]
P L-alditol + NADP$^+$
S D-galacturonate + NADPH <1> (Reversibility: r <1> [2, 3]) [2, 3]
P ?
S D-glucuronate + NADPH <6, 7, 5> (Reversibility: r <6, 7, 5> [8-12]) [8-12]
P L-gulonate + NADP$^+$
S D-glyceraldehyde + NADPH <1, 3> (Reversibility: r <1, 3> [1, 5]) [1, 5]
P glycerol + NADP$^+$
S L-gulonate + NADP$^+$ <1, 5-7> (Reversibility: r <1, 5-7> [3, 6, 8, 9, 11]) [1-3, 6, 8, 9, 11]
P D-glucuronate + NADPH
S γ-hydroxybutyrate + NADP$^+$ <2> (Reversibility: ? <2> [9]) [9]
P succinic semialdehyde + NADPH
S pyridine-3-aldehyde + NADPH <1, 3> (Reversibility: r <1, 3> [1, 5]) [1, 5]
P 3-hydroxypyridine + NADP$^+$
S Additional information <1-3> (<1> broad specificity [1]; <3> reduction of aldehyde greatly favored compared to oxidation of alcohol [5]; <1> re-

duction of D-glyceraldehyde: pro-4R hydrogen is transferred from NADPH, i.e. A-specific, attacks re face of carbonyl group of the D-glyceraldehyde [1]; <1> no reaction with γ-lactones of the substrates [3]; <2> coupled γ-hydroxybutyrate oxidation and D-glucuronate reduction [4]) [3-5]

P ?

Inhibitors

2,3-butanedione <1> [1]
2-mercaptoethanol <2, 3, 7> (<7> preincubation of γ-hydroxybutyrate reductase revealed no inhibitory effect [9]) [5, 9]
ATP-ribose <6> (<6> competitive versus NADPH [11]) [11]
Al 1576 <6, 7, 8> [10]
DL-β-hydroxybutyrate <7, 2> [9]
N-ethylmaleimide <7, 5> (<7, 5> no exact differentiation between organisms in the reference [8]) [8]
NaF <1, 3> [1, 5]
Zn^{2+} <3> [5]
acetoacetate <7, 2> [9]
α-ketoglutarate <7, 2> [9]
amaranth <6> [12]
barbiturates <5> [8]
branched α-ketoacids <7, 2> [9]
diethyldicarbonate <1> [1]
diphenylhydantoin <5> [8]
dithiothreitol <7, 2> (<7> preincubation of γ-hydroxybutyrate reductase revealed no inhibitory effect [9]) [9]
haematin <6> [12]
hydroxylamine <1, 3> [1, 5]
iodoacetic acid <1, 3> [1, 5]
menadione <6> (<6> competitive versus D-glucuronate [11]) [11]
p-chloromercuribenzoate <1, 7, 5> (<7, 5> no exact differentiation between organisms in the reference [8]) [1-3, 8]
phenobarbital <1, 3> [1, 5]
phenolsulfonphthalein <6> [12]
phenylacetate <7, 2> [9]
phenylglyoxal <1> [1]
ponalrestat <6, 7, 8> [10]
pyridoxal 5'-phosphate <1, 6> [1, 11]
salicylates <7, 2> [9]
sorbinil <5> [7]
statil <5> [7]
sulfobromophthalein <6> [12]
tolrestat <6, 7, 8> [10]
valproate <5> (<5> poor [7]) [7, 9]

Additional information <6> (<6> potent inhibitors, i.e. carboxylic and sulfonic acids, bind weakly to the free enzyme but strongly to the enzyme-$NADP^+$-binary complex [12]) [12]

Cofactors/prosthetic groups

$NADP^+$ <1, 2, 4-8> [3, 6, 8-12]
NADPH <1, 2, 4-8> (<1> reduction of D-glyceraldehyde: pro-4R hydrogen is transferred from NADPH, i.e. A-specific, attacks re face of carbonyl group of the D-glyceraldehyde [1]) [1-3, 6, 8-12]

Turnover number (min^{-1})

Additional information <6> (<6> 0.119 per mM * s, L-gulonate, steady state kinetic [11]; <6> 10.82 per mM * s, D-glucuronate, steady state kinetic [11]; <6> 1124 per mM * s, NADPH, steady state kinetic [11]) [11]

Specific activity (U/mg)

0.0025 <7, 5> (<7, 5> substrate D-glucuronate, defined as D-glucuronate reductase [8]; no exact differentiation between organisms in the reference [8]) [8]
1.339 <6> (<6> highly purified enzyme, substrate NADPH [10]) [10]
5.45 <1> [1]
Additional information <1, 3> [2, 5]

K_m-Value (mM)

0.00045 <7> (γ-hydroxybutyrate, <7> coupled reaction of γ-hydroxybutyrate oxidation and D-glucuronate reduction, brain and kidney [9]) [9]
0.0014 <2> ($NADP^+$, <2> coupled reaction of γ-hydroxybutyrate oxidation and D-glucuronate reduction [4]) [4]
0.0022 <7> (γ-hydroxybutyrate, <7> kidney [9]) [9]
0.0025 <7> (γ-hydroxybutyrate, <7> brain [9]) [9]
0.009 <1> (NADPH) [1]
0.02 <1, 2> ($NADP^+$, <1, 2> uncoupled reaction [1, 3, 4]) [1, 3, 4]
0.036 <3> (NADPH) [5]
0.0543 <7, 5> (D-glucuronate, <7,5> defined as D-glucuronate reductase [8]; no exact differentiation between organisms in the reference [8]) [8]
0.133 <3> (4-nitrobenzaldehyde) [5]
0.35 <1> (4-nitrobenzaldehyde) [1]
0.526 <3> (3-nitrobenzaldehyde) [5]
0.769 <3> (2-nitrobenzaldehyde) [5]
1.05 <3> (benzaldehyde) [5]
1.6 <8> (D-glucuronate) [10]
1.671 <3> (DL-glyceraldehyde) [5]
2.1 <8> (L-gulonate) [10]
2.6 <1> (pyridine-3-aldehyde) [1]
3.1 <6> (D-glucuronate) [10]
4.8 <1> (D-glyceraldehyde) [1]
4.8 <6> (L-gulonate) [10]
6.2 <7> (D-glucuronate) [10]
7.3 <1> (L-gulonate) [1, 3]

7.6 <7> (L-gulonate) [10]
8.4 <1> (D-glucuronate) [1]
9.1 <3> (D-glucuronate) [5]
139.9 <3> (D-arabinose) [5]
333 <3> (D-galactose) [5]
333 <3> (D-glucose) [5]
333 <3> (D-ribose) [5]
333 <3> (D-xylose) [5]
500 <3> (2-deoxyglucose) [5]

pH-Optimum

6-7 <3> (<3> DL-glyceraldehyde [5]) [5]
6.2-6.6 <3> (<3> D-glucuronate [5]) [5]
7 <2> (<2> shifted pH-optimum for the coupled reaction of oxidation of γ-hydroxybutyrate and reduction of D-glucuronate in vitro [9]) [9]
7.5 <5, 2, 7> (<7, 5> no exact differentiation between organisms in the reference [8]; <2> in vivo for coupled reaction of γ-hydroxybutyrate oxidation and D-glucuronate reduction [9]) [8, 9]
8.5 <6> (<6> substrate L-gulonate or D-glucuronate [11]) [11]
8.6-9.3 <1> [2, 3]
9 <2> (<2> substrate γ-hydroxybutyrate, at pH 7.0 activity is reduced by 50% [9]) [9]

pH-Range

7-10 <6, 7, 2> [9, 11]

Temperature optimum (°C)

38 <7, 5> (<7, 5> no exact differentiation between organisms in the reference [8]) [8]

4 Enzyme Structure

Molecular weight

33000 <1> (<1> SDS-PAGE, gel filtration [1]) [1]
35000 <3> (<3> gel filtration [5]) [5]
36000 <6> (<6> gel filtration [10]) [10]
39000 <6> (<6> gel filtration [10]) [10]
60000-70000 <7, 5> (<7, 5> no exact differentiation between organisms in the reference, gel filtration [8]) [8]

Subunits

dimer <7, 5> (<7, 5> 2 * 36000, SDS-PAGE, no exact differentiation between organisms in the reference [8]) [8]
monomer <1, 3, 6> (<1> 1 * 36700, SDS-PAGE [1]; <3> 1 * 33000, SDS-PAGE [5]; <6> 1 * 39000, SDS-PAGE [10]) [1, 5, 10]

5 Isolation/Preparation/Mutation/Application

Source/tissue

brain <7> [9]
kidney <1, 6, 7, 8> (<6, 7> cortex [10]) [1-3, 9, 10, 12]
lens <7, 5, 6> [8]
liver <2, 3, 6, 7> [4, 5, 9, 11, 12]
muscle <7> [9]

Purification

<1> [1-3]
<2> [4, 9]
<3> [5]
<5> (no exact differentiation between organisms in the reference [8]) [8]
<6> [10]
<7> (no exact differentiation between organisms in the reference [8]) [8]

6 Stability

pH-Stability

5-10.5 <6> (<6> irreversible loss of activity below pH 5.0 and above pH 11 [11]) [11]
7 <1> (<1> most stable [3]) [3]
8 <7, 5> (<7, 5> rapid decrease of actvity above pH 8.0 [8]; no exact differentiation between organisms in the reference [8]) [8]

Temperature stability

40 <1> (<1> 15 min, 50% loss of activity, ammonium sulfate fraction [3]) [3]
50 <1> (<1> 15 min, complete inactivation, ammonium sulfate fraction [3]) [3]

Organic solvent stability

acetone <1> (<1> inactivates [3]) [3]
ethanol <1> (<1> inactivates [3]) [3]

General stability information

<1>, EDTA 1 mM protects against heat denaturation at 40°C [3]
<1>, lyophilization destroys activity [3]

Storage stability

<1>, -5°C, 2 weeks [2, 3]

References

[1] Flynn, T.G.; Cromlish, J.A.; Davidson, W.S.: Aldehyde reductase (L-hexonate: NADP dehydrogenase) from pig kidney. Methods Enzymol., **89**, 501-506 (1982)

[2] Bublitz, C.; Lehninger, A.L.: Ascorbic acid synthesis in animal tissue I. TPN-L-gulonate dehydrogenase. Methods Enzymol., **6**, 334-337 (1963)
[3] York, J.L.; Grollman, A.P.; Bublitz, C.: TPN-L-gulonate dehydrogenase. Biochim. Biophys. Acta, **47**, 298-306 (1961)
[4] Kaufman, E.E.; Nelson, T.: Kinetics of coupled γ-hydroxybutyrate oxidation and D-glucuronate reduction by an $NADP^+$-dependent oxidoreductase. J. Biol. Chem., **256**, 6890-6894 (1981)
[5] Tulsiani, D.R.P.; Touster, O.: Resolution and partial characterization of two aldehyde reductases of mammalian liver. J. Biol. Chem., **252**, 2545-2550 (1977)
[6] Sivak, A.; Hoffmann-Ostenhof, O.: Enzymes of *meso*-inositol catabolism in the yeast Schwanniomyces occidentalis. Biochim. Biophys. Acta, **53**, 426-428 (1961)
[7] Poulsom, R.: Inhibition of hexonate dehydrogenase and aldose reductase from bovine retina by sorbinil, statil, M79175 and valproate. Biochem. Pharmacol., **35**, 2955-2959 (1986)
[8] Goode, D.; Lewis, M.E.; Crabbe, M.J.C.: Accumulation of xylitol in the mammalian lens is realted to glucoronate metabolism. FEBS Lett., **395**, 174-178 (1996)
[9] Kaufman, E.E.; Nelson, T.: An overview of γ-hydroxybutyrate catabolism: the role of the cytosolic $NADP^+$-dependent oxidoreductase EC 1.1.1.19 and of a mitochondrial hydroxyacid-oxoacid transhydrogenase in the initial, rate-limiting step in this pathway. Neurochem. Res., **16**, 965-974 (1991)
[10] Sato, S.; Kador, P.F.: Human kidney aldose and aldehyde reductases. J. Diab. Comp., **7**, 179-187 (1993)
[11] Bhatnagar, A.; Das, B.; Liu, S.-Q.; Srivastava, S.K.: Human liver aldehyde reductase: pH dependence of steady-state kinetic parameters. Arch. Biochem. Biophys., **287**, 329-336 (1991)
[12] Shinoda, M.; Mori, S.; Shintani, S.; Ishikura, S.; Hara, A.: Inhibition of human aldehyde reductase by drugs for testing the function of liver and kidney. Biol. Pharm. Bull., **22**, 741-744 (1999)

Glucuronolactone reductase 1.1.1.20

1 Nomenclature

EC number
1.1.1.20

Systematic name
L-gulono-1,4-lactone:$NADP^+$ 1-oxidoreductase

Recommended name
glucuronolactone reductase

Synonyms
GRase
gulonolactone dehydrogenase
reductase, glucuronolactone

CAS registry number
9028-30-2

2 Source Organism

<1> *Rattus norvegicus* [1, 2]

3 Reaction and Specificity

Catalyzed reaction
L-gulono-1,4-lactone + $NADP^+$ = D-glucorono-3,6-lactone + NADPH

Reaction type
oxidation
reduction

Natural substrates and products
S L-gulono-1,4-lactone + $NADP^+$ <1> (Reversibility: ? <1> [1]) [1]
P D-glucorono-3,6-lactone + NADPH

Substrates and products
S D-galacturonic acid + NADPH <1> (Reversibility: ? <1> [1]) [1]
P ?
S D-glucuronic acid + NADPH <1> (Reversibility: ? <1> [1]) [1]
P L-gulonic acid + $NADP^+$

S D-glucurono-3,6-lactone + NADPH <1> (<1> oxidation activity is very low [1]) (Reversibility: r <1> [1]) [1]
P D-gulono-1,4-lactone + $NADP^+$
S D-iduronic acid + NADPH <1> (Reversibility: ? <1> [1]) [1]
P ?
S L-gulono-1,4-lactone + $NADP^+$ <1> (Reversibility: ? <1> [1]) [1]
P D-glucorono-3,6-lactone + NADPH
S erythrose + NADPH <1> (Reversibility: ? <1> [1]) [1]
P glycerol + $NADP^+$
S glyceraldehyde + NADPH <1> (Reversibility: ? <1> [1]) [1]
P ?

Inhibitors
Hg^{2+} <1> [1]
NaF <1> [1]
Zn^{2+} <1> [1]
barbital <1> [1]
barbiturates <1> [1]
dithiothreitol <1> [1]
iodoacetic acid <1> [1]
phenobarbital <1> [1]

Cofactors/prosthetic groups
NADH <1> (<1> NADPH is 145-times as effective as NADH [1]) [1]
$NADP^+$ <1> [1]
NADPH <1> (<1> NADPH is 145-times as effective as NADH [1]) [1]

Specific activity (U/mg)
Additional information <1> (<1> 13450 U/min * mg, 1 unit is defined as change of absorbance of 0.01/min at 340 nm [1]) [1]

K_m-Value (mM)
0.045 <1> (NADPH, <1> with D-glucuronic acid [1]) [1]
4 <1> (D-galacturonic acid) [1]
6 <1> (D-glucuronic acid, <1> L-iduronic acid [1]) [1]
9 <1> (D-glucuronolactone) [1]

pH-Optimum
5.5-7.5 <1> (<1> DL-glyceraldehyde [1]) [1]
6-7 <1> (<1> D-glucuronic acid [1]) [1]
9.5 <1> (<1> broad, oxidation of glycerol and L-gulonolactone [1]) [1]

4 Enzyme Structure

Molecular weight
41000 <1> (<1> gel filtration [1]) [1]

Subunits
monomer <1> (<1> 1 * 40000, SDS-PAGE [1]) [1]

5 Isolation/Preparation/Mutation/Application

Source/tissue

kidney <1> [1]
liver <1> [2]

Purification

<1> [1]

6 Stability

Temperature stability

40 <1> (<1> 30 min, 20% loss of activity [1]; <1> 60 min, 30% loss of activity [1]; <1> 180 min, 50% loss of activity [1]) [1]

General stability information

<1>, freezing and thawing: 3 times causes 25% decrease of activity, 5 times causes 40% decrease of activity [1]

Storage stability

<1>, -80°C, 10 days, 30% loss of activity [1]
<1>, 0°C, 2 months, less than 10% loss of activity [1]

References

[1] Hayashi, S.; Watanabe, M.; Kimura, A.: Enzymatic determination of free glucoronic acid with glucuronolactone reductase. I. Isolation and purification of glucoronolactone reductase from rat kidney. J. Biochem., **95**, 223-232 (1984)

[2] Nandi, A.; Chatterjee, I.B.: Interrelation of xanthine oxidase and dehydrogenase and L-gulonolactone oxidase in animal tissue. Indian J. Exp. Biol., **29**, 574-578 (1991)

Aldehyde reductase 1.1.1.21

1 Nomenclature

EC number
1.1.1.21

Systematic name
alditol:NAD(P) 1-oxidoreductase

Recommended name
aldehyde reductase

Synonyms
20-α-HSD [Swissprot]
20-α-hydroxysteroid dehydrogenase [Swissprot]
ALR2
Aldehyde reductase [Swissprot]
Aldo-keto reductase family 1 member B7 [Swissprot]
FR-1 protein [Swissprot]
Fibroblast growth factor regulated protein [Swissprot]
MVDP [Swissprot]
NADPH-aldopentose reductase
NADPH-aldose reductase
TPN-polyol dehydrogenase
VAS deferens androgen-dependent protein [Swissprot]
alditol:NADP oxidoreductase
alditol:NADP 1-oxidoreductase
aldose reductase
aldose xylose reductase
polyol dehydrogenase (NADP)

CAS registry number
9028-31-3

2 Source Organism

<1> *Homo sapiens* [1, 3-6, 8, 17, 18, 21, 25, 26, 28]
<2> *Euonymus japonica* [2]
<3> *Celastrus orbiculatus* [2]
<4> *Oryctolagus cuniculus* [9]
<5> *Rattus norvegicus* [9, 25, 27]
<6> *Sus scrofa* [9]

<7> *Bos taurus* (calf [10-12]; 2 enzymes AR I (ALR 1) and AR II (ALR 2) [10, 20]; ox, low-K_M enzyme subtype [15, 16]) [7, 9-12, 14-16, 20]
<8> *Euonymus sp.* (distribution in various species) [2]
<9> *Candida utilis* [13]
<10> *Galdieria sulphuraria* [19]
<11> *Candida tenuis* [22, 24]
<12> *Candida intermedia* (2 isoenzymes ALR 1 und ALR 2 [23]) [23]

3 Reaction and Specificity

Catalyzed reaction

alditol + $NAD(P)^+$ = aldose + NAD(P)H (<1, 11> reaction mechanism [18, 22, 24, 28])

Reaction type

oxidation
reduction

Natural substrates and products

S D-galactose + NADPH <2, 3, 8> (Reversibility: r <2, 3, 8> [2]) [2]
P ?
S aldose + NAD(P)H <1-9> (<2, 3, 8> wide specificity [2, 19, 21]) (Reversibility: r <1-9> [1-17, 19, 21]) [1-17, 19, 21]
P alditol + $NAD(P)^+$
S pentose + NADP <10> (<10> with hexoses and sugar phosphates as substrate the activity is only 5-10% of that retained with pentoses [19]; <10> D-isoforms of pentoses show less than 10% activity compared to the L-isoforms [19]) (Reversibility: r <10> [19]) [19]
P pentiol + NADPH
S Additional information <1> (<1> enzyme of sorbitol (polyol) pathway [4]; <1> enzyme involved in metabolism of corticosteroids [5]) [4, 5]

Substrates and products

S 11-deoxyisocorticosterone + NAD(P)H <1> (Reversibility: ? <1> [4]) [4]
P ?
S 2,3-bornanedione + NAD(P)H <1> (Reversibility: ? <1> [4]) [4]
P ?
S 2-D-deoxy-2-fluoro-galactose + NADPH <11> (Reversibility: r <11> [22]) [22]
P ?
S 2-D-deoxygalactose + NADPH <1, 2, 11> (Reversibility: r <2, 11> [2, 22]; ? <1> [1]) [1, 2, 22]
P ?
S 2-deoxyribose + NADPH <2> (Reversibility: r <2> [2]) [2]
P ?
S 20-isocortisol + NAD(P)H <1> (Reversibility: ? <1> [4, 5]) [4, 5]
P ?

S 20α-isocorticosterone + NAD(P)H <1> (Reversibility: ? <1> [4]) [4]
P ?
S 4-hydroxyphenylacetaldehyde + NAD(P)H <1> (Reversibility: ? <1> [4, 5]) [4, 5]
P 4-hydroxyphenylethyl alcohol + NAD(P)$^+$
S 4-hydroxyphenylglycolaldehyde + NAD(P)H <1> (Reversibility: ? <1> [4, 5]) [4, 5]
P ?
S 4-nitrobenzaldehyde + NAD(P)H <1, 2, 4, 7> (Reversibility: ? <1, 2, 4, 7> [2, 4, 5, 9, 12, 17, 18]) [2, 4, 5, 9, 12, 17, 18]
P 4-nitrobenzyl alcohol + NAD(P)$^+$
S D-arabinose + NADPH <10> (Reversibility: r <10> [19]) [19]
P arabitol + NADP$^+$
S D-arabinose + NADPH <10> (Reversibility: r <10> [19]) [19]
P arabitol + NADP$^+$
S D-erythrose + NAD(P)H <1, 9> (Reversibility: r <1, 9> [8, 13]) [8, 13]
P erythritol + NAD(P)$^+$
S D-fructose + NAD(P)H <7, 10> (Reversibility: r <7, 10> [14, 19]) [14, 19]
P D-fructitol + NAD(P)$^+$
S D-fucose + NADPH <10> (Reversibility: r <10> [19]) [19]
P fucitol + NADP$^+$
S D-galactose + NAD(P)H <1, 2, 5, 9, 10> (Reversibility: r <1, 2, 5, 9, 10> [1, 2, 4, 5, 8, 13, 19, 21, 25, 26]) [1, 2, 4, 5, 8, 13, 19, 21, 25, 26]
P dulcitol + NAD(P)$^+$
S D-glucose + NAD(P)H <1, 4, 5, 7, 9, 10> (<2> not [2]; <7> low activity [12]) (Reversibility: r <1, 4, 5, 7, 9, 10> [1, 4-6, 8, 9, 12, 13, 17, 19, 21, 25, 26, 28]) [1, 4-6, 8, 9, 12, 13, 17, 19, 21, 25, 26, 28]
P sorbitol + NAD(P)$^+$ <1> [6, 17]
S D-glucuronate + NAD(P)H <1, 4, 5, 7> (Reversibility: r <1, 4, 5, 7> [1, 4, 5, 9, 11, 15, 17, 21, 25, 26]) [1, 4, 5, 9, 11, 15, 17, 21, 25, 26]
P ?
S D-glucuronolactone + NADPH <1, 7> (Reversibility: ? <1, 7> [8, 11]) [8, 11]
P ?
S D-glyceraldehyde + NAD(P)H <9> (Reversibility: r <9> [13]) [13]
P glycerol + NAD(P)$^+$
S D-lyxose + NADPH <10> (Reversibility: r <10> [19]) [19]
P arabitol + NADP$^+$
S D-mannose NAD(P)H <9, 10> (Reversibility: r <9, 10> [13, 19]) [13, 19]
P mannitol + NAD(P)$^+$
S D-ribose + NAD(P)H <1, 7, 9, 10> (<2> not [2]) (Reversibility: r <1, 7, 9, 10> [1, 8, 11, 13, 19]) [1, 8, 11, 13, 19]
P ribitol + NAD(P)$^+$
S D-ribose + NADPH <10> (Reversibility: r <10> [19]) [19]
P adonitol + NADP$^+$
S D-ribose 5-phosphate + NADPH <10> (Reversibility: r <10> [19]) [19]
P D-ribitol 5-phosphate + NADP$^+$

S D-tagatose + NADPH <10> (Reversibility: r <10> [19]) [19]
P ?
S D-xylose + NAD(P)H <1, 4, 5, 7, 9-12> (<7> low activity [12]) (Reversibility: r <1, 4, 5, 7, 9-12> [4, 5, 8, 9, 11-15, 19, 21-26]) [4, 5, 8, 9, 11-15, 19, 21-26]
P xylitol + NAD(P)$^+$
S D-xylose 5-phosphate + NADPH <10> (Reversibility: r <10> [19]) [19]
P D-xylitol 5-phosphate + NADP$^+$
S DL-glyceraldehyde + NAD(P)H <1, 2, 5, 7, 10, 11> (Reversibility: r <1, 2, 5, 7, 10, 11> [2, 5, 11, 14, 17-22, 25-27]) [2, 5, 11, 14, 17-22, 25-27]
P glycerol + NAD(P)$^+$
S L-arabinose + NAD(P)H <1, 2, 9, 10> (Reversibility: r <1, 2, 9, 10> [2, 8, 13, 19]) [2, 8, 13, 19]
P L-arabitol + NAD(P)$^+$
S L-arabinose + NADPH <10> (Reversibility: r <10> [19]) [19]
P arabitol + NADP$^+$
S L-fucose + NADPH <10> (Reversibility: r <10> [19]) [19]
P fucitol + NADP$^+$
S L-lyxose + NADPH <10> (Reversibility: r <10> [19]) [19]
P arabitol + NADP$^+$
S L-rhamnose + NADPH <10> (Reversibility: r <10> [19]) [19]
P L-rhamnitol + NADP$^+$
S L-ribose + NADPH <10> (Reversibility: r <10> [19]) [19]
P adonitol + NADP$^+$
S L-sorbose + NADPH <10> (Reversibility: r <10> [19]) [19]
P L-sorbitol + NADP$^+$
S L-xylose + NAD(P)H <1, 10> (Reversibility: r <1, 10> [5, 19]) [5, 19]
P L-xylitol + NAD(P)$^+$
S acetaldehyde + NAD(P)H <2, 7> (Reversibility: ? <2, 7> [2, 12]) [2, 12]
P ethanol + NAD(P)$^+$
S acrolein + NADPH <11> (Reversibility: ? <11> [22]) [22]
P 2-propen-1-ol + NADP$^+$
S alditol + NAD(P)$^+$ <1-9, 11, 12> (<11> 10% oxidation activity compared to reduction reaction [24]) (Reversibility: r <1-9, 11> [1-17, 19, 21-28]) [1-17, 19, 21-28]
P aldose + NAD(P)H
S aldose + NAD(P)H <1-9, 11, 12> (<1, 2, 3, 7, 8, 11, 12> wide specificity [2, 11, 13, 17, 19, 21-24, 28]) (Reversibility: r <1-9, 11> [1-17, 19, 21-27]) [1-17, 19, 21-28]
P alditol + NAD(P)$^+$
S benzaldehyde + NAD(P)H <2, 7> (Reversibility: ? <2, 7> [2, 12]) [2, 12]
P benzyl alcohol + NAD(P)$^+$
S butyraldehyde + NAD(P)H <1, 2, 7> (Reversibility: ? <1, 2, 7> [2, 4, 12]) [2, 4, 12]
P butanol + NAD(P)$^+$
S decanal + NAD(P)H <1> (Reversibility: ? <1> [4]) [4]
P decanol + NAD(P)$^+$

S dihydroxyacetone + NAD(P)H <7> (Reversibility: ? <7> [12]) [12]
P glycerol + NAD(P)$^+$
S glycoaldehyde + NADPH <2> (Reversibility: ? <2> [2]) [2]
P ethylene glycol + NADP$^+$
S glyoxal + NADPH <2> (Reversibility: ? <2> [2]) [2]
P ?
S hexanal + NADPH <2, 10> (Reversibility: r <2, 10> [2, 19]) [2, 19]
P hexanol + NADP$^+$
S indole-3-acetaldehyde + NAD(P)H <1> (Reversibility: ? <1> [4, 5]) [4, 5]
P indole-3-ethanol + NAD(P)$^+$
S isobutyraldehyde + NADPH <7> (Reversibility: ? <7> [11]) [11]
P isobutanol + NADP$^+$
S methylglyoxal + NADPH <2, 5, 11> (<2> i.e. pyruvaldehyde [2]) (Reversibility: ? <2, 5, 11> [2, 22, 27]) [2, 22, 27]
P ?
S n-pentanal + NAD(P)H <2, 7> (Reversibility: ? <2, 7> [2, 12]) [2, 12]
P n-pentanol + NAD(P)$^+$
S pentose + NADPH <10-12> (<10> with hexoses and sugar phosphates as substrate the activity was only 5-10% of that with pentoses [19]) (Reversibility: r <10-12> [19, 21-24]) [19, 21-24]
P pentitol + NADP$^+$
S phenylglyoxal + NAD(P)H <1, 2> (Reversibility: ? <1, 2> [2, 5]) [2, 5]
P ?
S propanal + NADPH <11> (Reversibility: r <11> [22]) [22]
P n-propanol + NADP$^+$
S propionaldehyde + NAD(P)H <1, 2, 7> (Reversibility: ? <1, 2, 7> [1, 2, 12]) [1, 2, 12]
P propanol + NAD(P)$^+$
S pyridine-3-aldehyde + NAD(P)H <5, 7> (<7> low-K_M enzyme subtype [15, 16]) (Reversibility: r <5, 7> [15, 16, 27]) [15, 16, 27]
P 3-hydroxypyridine + NAD(P)$^+$
S pyridine-4-aldehyde + NAD(P)H <5> (Reversibility: r <5> [27]) [27]
P 4-hydroxypyridine + NAD(P)$^+$
S pyridinecarboxaldehyde + NAD(P)H <1, 2> (<1, 2> 2-, 3- and 4-pyridinecarboxaldehyde [1, 2]) (Reversibility: ? <1, 2> [1, 2]) [1, 2]
P pyridylmethanol + NAD(P)$^+$ (2-, 3-, and 4-isomer)
S succinic semialdehyde + NADH <1, 7> (Reversibility: ? <1, 7> [15, 17]) [15, 17]
P 4-hydroxybutanoic acid + NAD$^+$
S xylitol + NAD(P)$^+$ <1, 4, 7, 9-12> (Reversibility: r <1, 4, 7, 9-12> [4, 5, 8, 9, 11-15, 19, 21-24]) [4, 5, 8, 9, 11-15, 19, 21-24]
P D-xylose + NAD(P)H
S Additional information <1-3, 8> (<1> oxidation of alcohols by NADP$^+$ is negligible at 25°C, measurable at 37°C [5]; <1> oxidation of polyols to polyolaldehydes by NADP$^+$ proceeds very slowly [4]; <2,3,8> requirement for a free aldehyde group [2]) [2, 4, 5]
P ?

Inhibitors

$(NH_4)_2SO_4$ <1, 4> (<1,4> 0.05 mM: slight inhibition, 400 mM: activation [9,17]) [9, 17]
1,10-phenanthroline <2> [2]
2'-AMP <7> [7, 12]
2,2'-dipyridyl <2> [2]
2-[5-fluoro-2-methyl-1-*p*-(methylsulphinyl)benzylidine]indene-3-acetic acid <7> (<7> non-competitive [14]) [14]
2-mercaptoethanol <7> [12]
2-oxodecanoate <7> [11]
2-oxoglutarate <7> [11]
2-oxoheptanoate <7> [11]
2-oxohexanoate <7> [11]
2-oxoisohexanoate <7> [11]
2-oxoisopentanoate <7> [11]
2-oxononanoate <7> [11]
2-oxooctanoate <7> [11]
2-oxopentanoate <7> [11]
3'-AMP <7> [12]
3,3-tetramethylene glutaric acid <1> [1, 4, 8]
3-hydroxybutyrate <1, 7> [8, 11]
3-hydroxybutyric acid <1> [8]
5'-AMP <7> [12]
5,5-diphenylhydantoin <1, 2> [2, 6]
7-hydroxy-4-oxo-4H-chromen-2-carboxylic acid <1> [4]
AMP <11> [24]
ATP <11> [24]
Ca^{2+} <7> (<7> 10 mM inhibits complex formation of ALR 2 with α-crystallin [20]) [20]
Cl^- <1, 5, 7> (<7> at high ionic strength [15]; <1, 5> 0.2 M, wild type and recombinant from E. coli [25]) [4, 15, 25]
Cu^{2+} <4> [9]
EDTA <11> [24]
$HgCl_2$ <4, 7> [9, 12]
Li_2SO_4 <1, 2> (<1> 0.4 M: essential for full activity [1,17]; <2> 200 mM: inhibition [2]) [1, 2, 17]
Mg^{2+} <7> (<7> 10 mM inhibits complex formation of ALR 2 with α-crystallin, induces ALR 2 aggregation and precipitation [20]) [20]
Mn^{2+} <11> [24]
N-ethylmaleimide <1, 2> [1, 2]
$NADP^+$ <1, 2, 7, 11> (<2,7> competitve to NADPH [2, 14, 16]) [2, 6, 7, 14, 16, 24]
NaCl <1, 7> [8, 11, 15]
$NaNO_3$ <7> [15]
Zn^{2+} <4, 11> [9, 24]
acetoacetate <7> [11]

alrestatin <1, 4> (<1, 4> 1,3-dioxo-1H-benz[de]isoquinoline-2-(3H)acetic acid [1, 4, 9]) [1, 4, 9, 17]
ammonium acetate <7> [11]
barbital <1, 5> (<5> 26% inhibition [27]) [6, 27]
c-3',5'-AMP <7> [12]
citrate <1> (<1> inhibition is pH-dependent [18]) [18]
disodium 4,6-dioxo-10-propyl-4H,6H-pyranol[3,2-g-]-quinoline-2,8-dicarboxylate <7> (<7> non-competitive [14]) [14]
dithiothreitol <1, 7, 11> (<1> increased inhibitory effect of sorbinil [17]) [12, 17, 24]
epalrestat ((E,E)-5[2-methyl-3-phenyl-2-propenylidene-4-oxo-2-thioxo-3-thiazolidine]acetic acid) <1> (<1> recombinant from baculovirus system in Spodoptera frugiperda cells [26]) [21, 26]
ethosuccinimide <1> [6]
glyceraldehyde <7> (<7> substrate inhibition [7, 14]) [7, 14]
glycerol <2> (<2> uncompetitive to DL-glyceraldehyde [2]) [2]
hydroxylamine <2> [2]
iodoacetamide <2> (<2> slight [2]) [2]
iodoacetate <1, 7> [5, 12]
ionic detergents <11> [24]
malondialdehyde <7> (<7> at concentration above 0.8 mM substrate inhibition [14]) [14]
octanoate <7> [11]
oxaloacetate <7> [11]
p-carboxybenzaldehyde <7> (<7> substrate inhibition at higher concentration [15]) [15]
p-chloromercuribenzoate <4, 5> (<5> complete inhibition [27]) [9, 27]
p-hydroxymercuribenzoate <1, 2, 7> [1, 2, 5, 12]
p-nitrobenzaldehyde <7> (<7> at concentration above 2 mM substrate inhibition [14]) [14]
phenobarbital <1, 2, 5> [1, 2, 17, 27]
phenobarbitone <1> [5, 6]
ponalrestat <1, 5> (<1, 5> wild type and recombinant from E. coli [25]) [21, 25]
pyrazol <2> (<2> slight [2]) [2]
pyrid-3-ylcarbinol <7> [16]
pyridoxal 5'-phosphate <11> (<11> inhibits cofactor binding [22]) [22, 24]
quercetin <1, 4> [4, 5, 9, 17]
quercitrin <1, 5> [4, 27]
rutin <1> [4, 5]
sorbinil ((S)-6-fluorospiro[chroman-4,5'-imidazolidine]-2',4'-dione) <1, 4, 5> (<1, 5> wild type and recombinant from E. coli [25]; <1> recombinant from baculovirus system in Spodoptera frugiperda cells [26]) [1, 9, 17, 21, 25, 26]
sorbitol <7> (<7> competitive inhibition with $NADP^+$ and glucose [14]) [14]

statil (3-[(4-bromo-2-fluorophenyl)methyl]-3,4-dihydro-4-oxo-1-phthalazineacetic acid) <1> (<1> recombinant from baculovirus system in Spodoptera frugiperda cells [26]) [26]
sulfhydryl inhibitors <7> [12]
tolrestat (N-[[6-methoxy-5(trifluoromethyl)-1-naphthalenyl]thioxomethyl]-N-methylglycine) <1, 5> (<1, 5> wild type and recombinant from E. coli [25]; <1> recombinant from baculovirus system in Spodoptera frugiperda cells [26]) [21, 25, 26]
zopolrestat <1, 11> [22, 28]
Additional information <7> (<7> high ionic strength inhibits the enzyme to a higher extent when NADPH instead of NADH is used as a cofactor [16]) [16]

Cofactors/prosthetic groups

NADH <1, 7, 11, 12> (<1> 10% of the NADPH-dependent activity [4]; <1> 5% [5]; <7> 50% [12]; <11> used in vivo [22]; <12> ALR 2 [23]) [1, 4, 5, 12, 14-17, 21-24]
$NADP^+$ <1, 7, 9, 10, 11> (<10> NADP(H) dependent, no activity with NAD(H) in either directions [19]; <11> 2.2fold higher activity than with NAD^+ [24]) [4, 13-17, 19, 24]
NADPH <1-3, 8, 9, 11, 12> (<1, 11> NADPH-specific, pro-4R hydrogen is transferred from the nicotinamide ring of the coenzyme to the substrate [4, 18, 22]; <12> ALR 1 strictly specific for NADPH [23]; <11> 5.3fold higher activity than with NAD(H) [24]) [2, 4-6, 8, 13-18, 21-24, 28]
Additional information <9> (<9> no activity with dinucleotide cofactors [13]) [13]

Activating compounds

$(NH_4)_2SO_4$ <1, 4, 5> (<1> at 0.3 M more than 3fold activity [21]; <1,5> 0.1 M; wild type and recombinant from E. coli [25]; <1> activation [4, 8]; <4> 400 mM: activation [9]) [4, 8, 9, 21, 25]
D-xylose <9> (<9> induces enzyme activity [13]) [13]
Na_2SO_4 <1, 5, 11> (<1> at 0.3 M 3fold activity [21]; <1, 5> 0.1 M, wild type and recombinant from E. coli [25]) [21, 24, 25]
NaCl <1> (<1> increased concentration [17]) [17]
SO_4^{2-} <1, 7> (<1, 7> stimulation [5, 7, 10]; <7> ammonium and lithium salt is somewhat more effective than sodium and potassium salt [10]) [5, 7, 10]
Tween <11> (<11> Tween-20, Tween-80, Tween-100, each 0.1% (w/v) [24]) [24]
bovine serum albumin <11> (<11> 10-15% increased activity [24]) [24]
pyridoxal 5'-phosphate <1> (<1> 2.5fold activity at 0.5 mM [21]) [21]
sodium phosphate <7> (<7> at pH 5.0-5.8 and pH 7.2 [15]) [15]

Metals, ions

Li_2SO_4 <1, 2> (<1> 0.1 M activation wild type and recombinant from E. coli [25]; <1> stimulation [8]; <1> 0.4 M: essential for full expression of enzyme activity [1]; <2> 200 mM: inhibition [2]) [1, 2, 8, 25]

Turnover number (min^{-1})

11 <11> (2-deoxy-2-fluoro-galactose) [22]
11 <1> (D-glucose) [21]
14 <1> (D-galactose) [21]
17 <11> (D-xylose, <11>) [22]
25.08 <1> (D-glucuronate, <1> recombinant from baculovirus system in Spodoptera frugiperda cells [26]) [26]
25.85 <1> (DL-glyceraldehyde) [21]
28.8 <1> (D-glucose, <1> recombinant from baculovirus system in Spodoptera frugiperda cells [26]) [26]
31.5 <1> (D-galactose, <1> recombinant from baculovirus system in Spodoptera frugiperda cells [26]) [26]
33 <1> (D-glyceraldehyde) [5]
38.1 <1> (D-xylose, <1> recombinant from baculovirus system in Spodoptera frugiperda cells [26]) [26]
38.4 <1> (20α-isocorticosterone, <1> 20α-isocortisol [5]) [5]
40.38 <1> (DL-glyceraldehyde, <1> recombinant from baculovirus system in Spodoptera frugiperda cells [26]) [26]
45.6 <1> (phenylglyoxal) [5]
66 <11> (D-galactose) [22]
102 <1> (D-xylose) [21]
138 <1> (D-glucuronate) [21]
144 <11> (acrolein) [22]
780 <11> (propanal) [22]
9300 <11> (DL-glyceraldehyde) [22]
15960 <11> (methylglyoxal) [22]
Additional information <1> (DL-glyceraldehyde, <1> turnover number in mol per seconds for several mutants [18]) [18]

Specific activity (U/mg)

0.003 <9> [13]
0.024 <9> [13]
0.2 <7> (<7> purified enzyme [14]) [14]
0.475 <1> [4]
0.488 <7> [10]
1.2 <7> (<7> NADPH, purified enzyme [20]) [20]
1.32 <1> (<1> purified enzyme, NADPH [21]) [21]
1.5 <11> [24]
1.68 <1> [8]
5.24 <2> [2]
7.333 <1> (<1> purified enzyme [17]) [17]
10.2 <1> [6]
10.4 <12> (<12> isozyme ALR 2 [23]) [23]
11.2 <7> [12]
12.1 <12> (<12> recombinant expressed in Escherichia coli [23]) [23]
17.1 <10> (<10> purified enzyme [19]) [19]
26.8 <12> (<12> isozyme ALR 1 [23]) [23]

28.64 <11> (<11> purified enzyme [24]) [24]
37 <9> (<9> purified enzyme [13]) [13]
333.3 <7> (<7> purified enzyme, with NADH in sodium phosphate buffer 100 mM, pH 7.0 [15]) [15]
Additional information <7> [7, 11]

K_m-Value (mM)

0.001 <1> (20α-isocorticosterone, and 20α-isocortisol, values below) [5]
0.002 <1> (4-nitrobenzaldehyde) [5]
0.003 <1, 2> (4-nitrobenzaldehyde, <2> [2];<1> 4-hydroxyphenylglycolaldehyde [5]) [2, 5]
0.0032 <1> (NADPH) [21]
0.004 <2> (4-pyridinecarboxaldehyde) [2]
0.004 <1> (NADPH, <1> recombinant from baculovirus system in Spodoptera frugiperda cells [26]) [26]
0.005 <1> (*p*-carboxybenzaldehyde) [17]
0.007 <1> (*p*-nitrobenzaldehyde) [17]
0.01 <1> (4-hydroxyphenylacetaldehyde) [5]
0.01-0.015 <1, 2, 7> (NADPH) [1, 2, 12]
0.011 <1> (propionaldehyde) [1]
0.016 <9> (NAD(P)H) [13]
0.016 <1> (indole-3-acetaldehyde) [5]
0.018 <1> (L-glyceraldehyde, <1> recombinant from baculovirus system with Spodoptera frugiperda cells [26]) [5, 26]
0.02 <1> (DL-glyceraldehyde) [18]
0.02 <7> (isobutyraldehyde) [11]
0.02 <1> (*p*-bromobenzaldehyde) [17]
0.025 <1> (DL-glyceraldehyde) [5, 21]
0.025 <1> (benzaldehyde) [17]
0.03 <7> (DL-glyceraldehyde) [11]
0.034 <2> (hexanal) [2]
0.035 <1> (D-glyceraldehyde) [5]
0.039 <1> (pyridine-3-aldehyde) [1]
0.046 <2> (3-pyridinecarboxaldehyde) [2]
0.046 <1> (DL-glyceraldehyde) [25]
0.047 <1> (DL-glyceraldehyde, <1> recombinant from E. coli [25]) [25]
0.049 <1> (NADPH) [17]
0.08 <5> (DL-glyceraldehyde) [25]
0.087 <2> (n-pentanal) [2]
0.099 <2> (2-pyridinecarboxaldehyde) [2]
0.12 <5> (DL-glyceraldehyde, <5> recombinant from E. coli [25]) [25]
0.14 <9> ($NAD(P)^+$) [13]
0.152 <1> (D-glucuronate) [17]
0.22 <7> (NADH) [12]
0.25 <1> (NADH) [21]
0.285 <7> (succinic semialdehyde) [15]
0.295 <2> (benzaldehyde) [2]

0.4 <7> (D-erythrose) [11]
0.444 <2> (butyraldehyde) [2]
0.51 <1> (glyceraldehyde) [1]
0.62 <1> (NADH) [1]
0.68 <2> (phenylglyoxal) [2]
0.7 <7> (D-glucuronolactone) [11]
1.18 <1> (Dglucuronate) [1]
1.58 <7> (D-glucuronate) [15]
1.67 <7> (xylose) [15]
1.8 <2> (glycoaldehyde) [2]
2 <7> (D-xylose) [5, 11]
2 <11> (DL-glyceraldehyde) [22]
3.4 <2> (DL-glyceraldehyde) [2]
3.53 <1> (D-glucuronate, <1> recombinant from baculovirus system in Spodoptera frugiperda cells [26]) [26]
4 <7> (D-glucuronate) [11]
4.8 <1> (D-glucuronate) [25]
5 <1, 7> (D-glucuronate) [5, 11]
5.04 <1> (D-xylose, <1> recombinant fron baculovirus system in Spodoptera frugiperda cells [26]) [26]
5.05 <1> (D-glucuronate) [21]
5.1 <1> (D-xylose) [5]
6.2 <1> (D-xylose) [21]
6.3 <11> (methylglyoxal) [22]
6.4 <1> (D-glucuronate, <1> recombinant from E. coli [25]) [25]
7 <7> (D-ribose) [11]
7.4 <1> (D-xylose) [18]
8 <9> (D-xylose) [13]
8.2 <1> (D-xylose) [25]
8.2 <7> (NADPH) [14]
8.7 <1> (D-xylose, <1> recombinant from E. coli [25]) [25]
9.6 <10> (D-fructose) [19]
10.6 <2> (2-deoxygalactose) [2]
12 <1> (DL-glyceraldehyde, <1> mutant H110Q [18]) [18]
13 <11> (acrolein) [22]
13.5 <5> (D-xylose) [25]
13.9 <5> (D-glucuronate, <5> recombinant from E. coli [25]) [25]
16.8 <5> (D-xylose, <5> recombinant from E. coli [25]) [25]
17.5 <2> (2-deoxyribose) [2]
18.9 <2> (glyoxal) [2]
19.2 <5> (D-glucuronate) [25]
21.8 <7> (glucose) [14]
22 <10> (L-sorbose) [19]
24.8 <10> (D-tagatose) [19]
24.9 <10> (D-xylose-5-phosphate) [19]
27 <10> (D-xylose) [19]
28.1 <2> (D-erythrose) [2]

29.1 <10> (D-ribose-5-phosphate) [19]
29.77 <1> (xylose) [17]
40 <1> (D-galactose) [5]
42.9 <10> (D-galactose) [19]
42.9 <1> (D-galactose, <1> recombinant from E. coli [25]) [25]
43.3 <1> (D-galactose, <1> recombinant from baculovirus system in Spodoptera frugiperda cells [26]) [26]
46.7 <1> (D-galactose) [25]
47.3 <10> (D-glucose) [19]
49.3 <1> (D-galactose) [21]
52.3 <10> (L-rhamnose) [19]
53.4 <10> (D-mannose) [19]
56 <12> (D-xylose, <12> isozyme ALR 2, NADPH [23]) [23]
57.3 <1> (D-glucose) [21]
59.3 <10> (D-fucose) [19]
61 <12> (D-xylose, <12> isozyme ALR 1, NADPH [23]) [23]
64 <1> (DL-glyceraldehyde, <1> mutant Y48S [18]) [18]
71 <1> (DL-glyceraldehyde, <1> mutant Y48H [18]) [18]
76 <11> (D-xylose) [22]
78 <5> (D-galactose, <5> recombinant from E. coli [25]) [25]
80.7 <10> (L-fucose) [19]
83 <5> (D-galactose) [25]
84 <1> (DL-glyceraldehyde, <1> mutant H110A [18]) [18]
86.8 <1> (D-glucose) [25]
95 <1> (D-glucose) [5]
98.1 <1> (D-glucose, <1> recombinant from E. coli [25]; <1> recombinant from baculovirus system in Spodoptera frugiperda cells [26]) [25, 26]
100 <1> (L-xylose) [5]
102 <1> (ribose) [1]
111 <1> (L-xylose) [18]
113 <7> (glucose-6-phosphate) [15]
120.5 <10> (hexanal) [19]
130.2 <1> (galactose) [17]
142 <1> (galactose) [1]
155 <1> (glucose) [1]
204 <5> (D-glucose) [25]
227.1 <2> (D-galactose) [2]
228 <11> (D-galactose) [22]
245 <10> (L-ribose) [19]
281 <5> (D-glucose, <5> recombinant from E. coli [25]) [25]
287.2 <2> (L-arabinose) [2]
312.3 <10> (L-lyxose) [19]
360 <1> (L-xylose, <1> mutant H110A [18]) [18]
365 <11> (propanal) [22]
370.9 <10> (D-ribose) [19]
416.8 <10> (L-arabinose) [19]
446.7 <10> (D-arabinose) [19]

459.2 <10> (L-xylose) [19]
479.2 <10> (D-lyxose) [19]
506 <1> (D-lyxose, <1> mutant H110Q [18]) [18]
530 <9> (xylitol) [13]
730.2 <10> (DL-glyceraldehyde) [19]
861 <1> (D-xylose, <1> mutant H110Q [18]) [18]
1190 <1> (L-xylose, <1> mutant H110Q [18]) [18]
1340 <1> (D-xylose, <1> mutant H110A [18]) [18]
2630 <1> (D-lyxose) [18]
2740 <1> (D-lyxose, <1> mutant H110A [18]) [18]
Additional information <11> (<11> brief investigation of steady state kinetic [24]) [24]

pH-Optimum
5.8-6.7 <1> [8]
6 <11> (<11> reduction of D-xylose, NADH [24]) [24]
6-7 <1> (<1> aldehyde reduction [4]) [4]
6.2 <1, 7> [1, 15]
6.5 <5> [27]
6.5-7 <4> [9]
6.6 <1, 7> (<1, 7> DL-glyceraldehyde + NADPH [6, 10]) [6, 10]
6.7 <7> (<7> glyceraldehyde + NAD(P)H [12]) [12]
6.8 <10> [19]
8 <2> (<2> phosphate buffer, Tris-HCl buffer, substrate: DL-glyceraldehyde 3-pyridine carboxaldehyde [2]) [2]
8.6-9 <2> (<2> glycine-NaOH buffer, substrate: DL-glyceraldehyde 3-pyridine carboxaldehyde [2]) [2]
8.9 <11> (<11> oxidation of xylitol, NAD^+ [24]) [24]
9.7-10.4 <7> (<7> glycerol + NAD(P)H [12]) [12]
10 <7> (<7> glycerol + $NADP^+$ [10]) [10]

pH-Range
5-8 <11> (<11> below pH 5.0 and above pH 8.0 inactivation at 25°C within 3-6 days [24]) [24]
5.5-7.5 <1> (<1> at pH 5.5 and pH 7.5 about 50% of activity maximum [6]) [6]
5.5-7.6 <9> (<9> below pH 4.4 and above pH 9.7 less than 10% activity [13]) [13]
5.8-7.6 <10> (<10> 59% activity at both threshold values [19]) [19]
Additional information <7> (<7> inactive below pH 5.9 [14]) [14]

Temperature optimum (°C)
50 <11> [24]

Temperature range (°C)
25-50 <11> [24]

4 Enzyme Structure

Molecular weight

29000 <7> (<7> isozyme AR I, gel filtration [10]) [10]
30000 <7> (<7> isozyme AR II, gel filtration [10]) [10]
31000 <7> (<7> gel filtration [15]) [15]
32000 <1, 7> (<1, 7> gel filtration [14, 17]) [14, 17]
34000 <4> (<4> gel filtration [9]) [9]
34000 <7> (<7> isozyme ALR 2, gel filtration [20]) [20]
34300 <1> (<1> gel filtration [1]) [1]
35800 <1> [28]
36500 <5> [27]
37000 <7> (<7> gel filtration [7]) [7]
38000 <1> (<1> gel filtration [5]) [5]
39000 <1> (<1> sucrose density gradient centrifugation, thin layer gel filtration [8]) [8]
43000 <11> (<11> gel filtration [24]) [24]
58000 <12> (<12> ALR1 und ALR 2, theoretical molar mass 76000 Da [23]) [23]
61000 <7> (<7> gel filtration [12]) [12]
67000 <2> (<2> gel filtration [2]) [2]
69700 <9> (<9> equilibrium ultracentrifugation [13]) [13]
80000 <10, 11> (<10, 11> gel filtration [19, 22]) [19, 22]

Subunits

dimer <2, 7, 10, 11> (<7> 2 * 30500, SDS-PAGE [12]; <2, 11, 12> 2 * 36000, SDS-PAGE [2, 22, 23]; <10> 2 * 40000, SDS-PAGE [19]) [2, 12, 19, 22, 23]
monomer <1, 4, 5, 7, 11> (<1> 1 * 32500, SDS-PAGE in presence of urea and 2-mercaptoethanol [1]; <1> 1 * 38000, SDS-PAGE [5]; <7> 1 * 37000, SDS-PAGE in presence of 2-mercaptoethanol [7]; <1> 1 * 36000, SDS-PAGE [21]; <4> 1 * 40200, aldose reductase 1, SDS-PAGE [9]; <7> 1 * 41500, aldose reductase 2, SDS-PAGE [9]; <7> 1 * 30000, isozyme AR I, SDS-PAGE, monomer is in equilibrium with a less effective dimer [10]; <7> 1 * 40000, about, SDS-PAGE [14]; <7> 1 * 35100, SDS-PAGE in presence of 2-mercaptoethanol [15]; <1> 1 * 32000, SDS-PAGE in presence of 2-mercaptoethanol [17]; <11> 1 * 43000, SDS-PAGE [24]; <5> 1 * 36500, SDS-PAGE [27]; <1> 1 * 35800, SDS-PAGE [28]) [1, 5, 7, 9, 10, 14, 15, 17, 24, 27, 28]
Additional information <1> (<1> ? * 38000, SDS-PAGE [25]; <1> ? * 36000, SDS-PAGE [26]) [25, 26]

Posttranslational modification

no glycoprotein <4-7> [9]

5 Isolation/Preparation/Mutation/Application

Source/tissue

adrenal gland <5> [27]

aorta <1> [3]
brain <1, 7> (<7> ox [15,16]) [3-6, 10, 15, 16]
erythrocyte <1> [1, 3]
fruit <3, 8> [2]
heart <1> [4]
kidney <1> (<1> not [3]) [4, 17]
leaf <2, 3, 8> [2]
lens <1, 5-7> [4, 7, 9, 11, 14, 20, 25]
liver <7> [12]
lung <1> [4]
muscle <1> [3, 26]
pancreas <1> [4]
placenta <1> (<1> not [3]) [4, 8, 25]
retina <1> [26]
skeletal muscle <5, 6> [9]
testis <1> [4, 21]
Additional information <1> (<1> no activity in human liver [3, 4]) [3, 4]

Purification

<1> (partial [3, 6]; wild type and expressed in Escherichia coli [25]; expressed in baculovirus system [26]) [1, 3-6, 8, 17, 21, 25, 26]
<2> [2]
<4> [9]
<5> (wild type and expressed in Escherichia coli [25]) [25, 27]
<7> (calf [10-12]; 2 enzymes: AR I and AR II [10]; ox [15]; ALR 2 [20]) [7, 10-12, 14, 15, 20]
<9> [13]
<10> [19]
<11> [24]
<12> (ALR 1 und ALR 2 [23]; wild type and recombinant from Escherichia coli [23]) [23]

Crystallization

<1> [18]
<1> (structural model [28]) [28]

Cloning

<1> (overexpression [18]; expression in Escherichia coli [25]; expression in baculovirus system [26]) [18, 25, 26]
<5> (expression in Escherichia coli [25]) [25]
<12> (overexpression in Escherichia coli [23]) [23]

Engineering

H110A <1> (<1> mutant enzyme [18]) [18]
H110Q <1> (<1> mutant enzyme [18]) [18]
K77M <1> (<1> mutant enzyme [18]) [18]
Y48F <1> (<1> mutant enzyme [18]) [18]
Y48H <1> (<1> changing substrate selectivity [18]) [18]
Y48S <1> (<1> mutant enzyme [18]) [18]

6 Stability

pH-Stability

4.8-8.6 <7> (<7> 2 h, stable [10]) [10]
5 <1> (<1> 4°C, unstable below [4]) [4]
5.1 <1> (<1> 90 min, 50% loss of activity [5]) [5]
6 <1> (<1> unstable below [5]) [5]
6-9 <1> (<1> 4°C, several days, stable [4]) [4]

Temperature stability

30-35 <11> (<11> after 48 h incubation activity decreases [24]) [24]
50 <7> (<7> 10 min, inactivation [12]) [12]
53 <9> (<9> after 60 min 40-50% activity left [13]) [13]
55 <7> (<7> thermal aggregation and precipitation is prevented by specific, thermal stress induced complexing of α-crystallin, stable at least for 2 h at 55°C and for hours at room temperature, complexing can be hindered by Tris [20]) [20]
62 <9> (<9> total inactivation within 10 min [13]) [13]

General stability information

<1>, freezing and thawing at temperatures above 50°C results in loss of activity [5]
<1>, freezing of purified enzyme results in complete loss of activity [1]
<7>, -55°C, unstable to prolonged storage [10]
<7>, dithiothreitol, 2 mM stabilizes, other sulfhydryl compounds such as β-mercaptoethanol, glutathione and cysteine are less effective [12]
<7>, extremely labile to freezing and thawing unless Carbowax 100 is included, 10 mg/ml [7]
<7>, freezing and thawing: more than twice, unstable [10]
<11>, 25°C, 0.3 M phosphate or Tris, pH 6.0, more than 2 months [24]
<11>, needs Tween for stability during purification [22]
<12>, thiol stabilizing reagents are necessary [23]

Storage stability

<1>, -20°C, 4 weeks, little loss of activity [6]
<1>, 4°C, 2 weeks in 1% ampholines [8]
<1>, 4°C, 3 weeks, 50% loss of activity [6]
<1>, 4°C, several weeks [5]
<7>, -45°C, stable at any stage of purification [12]
<7>, -60°C, indefinitely [7]

References

[1] Das, B.; Srivastava, S.K.: Purification and properties of aldose reductase and aldehyde reductase II from human erythrocyte. Arch. Biochem. Biophys., **238**, 670-679 (1985)

[2] Negm, F.B.: Purification and properties of an NADPH-aldose reductase (aldehyde reductase) from Euonymus japonica leaves. Plant Physiol., **80**, 972-977 (1986)
[3] Srivastava, S.K.; Ansari, N.H.; Hair, G.A.; Das, B.: Aldose and aldehyde reductases in human tissues. Biochim. Biophys. Acta, **800**, 220-227 (1984)
[4] Wermuth, B.; von Wartburg, J.-P.: Aldose reductase from human tissues. Methods Enzymol., **89**, 181-186 (1982)
[5] Wermuth, B.; Bürgisser, H.; Bohren, K.; von Wartburg, J.-P.: Purification and characterization of human-brain aldose reductase. Eur. J. Biochem., **127**, 279-284 (1982)
[6] O'Brien, M.M.; Schofield, P.J.: Polyol-pathway enzymes of human brain. Partial purification and properties of aldose reductase and hexonate dehydrogenase. Biochem. J., **187**, 21-30 (1980)
[7] Sheaff, C.M.; Doughty, C.C.: Physical and kinetic properties of homogenous bovine lens aldose reductase. J. Biol. Chem., **251**, 2696-2702 (1976)
[8] Clements, R.S.; Winegrad, A.I.: Purification of alditol: NADP oxidoreductase from human placenta. Biochem. Biophys. Res. Commun., **47**, 1473-1479 (1972)
[9] Cromlish, J.A.; Flynn, T.G.: Purification and characterization of two aldose reductase isoenzymes from rabbit muscle. J. Biol. Chem., **258**, 3416-3424 (1983)
[10] Dons, R.F.; Doughty, C.C.: Isolation and characterization of aldose reductase from calf brain. Biochim. Biophys. Acta, **452**, 1-12 (1976)
[11] Hayman, S.; Kinoshita, J.H.: Isolation and properties of lens aldose reductase. J. Biol. Chem., **240**, 877-882 (1965)
[12] Attwood, M.A.; Doughty, C.C.: Purification and properties of calf liver aldose reductase. Biochim. Biophys. Acta, **370**, 358-368 (1974)
[13] Scher, B.M.; Horecker, B.I.: Pentose metabolism in Candida utilis. Arch. Biochem. Biophys., **116**, 117-128 (1966)
[14] Halder, A.B.; Crabbe, M.J.C.: Bovine lens aldehyde reductase (aldose reductase). Biochem. J., **219**, 33-39 (1984)
[15] Ryle, C.M.; Dowling, T. G.; Tipton, K.F.: Purification and properties of low-K_M aldehyde reductase from ox brain. Biochim. Biophys. Acta, **791**, 155-163 (1984)
[16] Ryle, C.M.; Tipton, K.F.: Kinetic studies with low-K_M aldehyde reductase from ox brain. Biochem. J., **227**, 621-627 (1985)
[17] Ansari, N.H.; Bhatnagar, A.; Liu, S.; Srivastava, S.K.: Purification and characterization of aldose reductase and aldehyde reductase from human kidney. Biochem. Int., **25**, 755-765 (1991)
[18] Bohren, K.M.; Grimshaw, C.E.; Lai, C.-J.; Harrison, D.H.; Ringe, D.; Petsko, G.A.; Gabbay, K.H.: Tyrosine-48 is the proton donor and histidine-110 directs substrate stereochemical selectivity in the reduction reaction of human aldose reductase: enzyme kinetics and crystal structure of the Y48H mutant enzyme. Biochemistry, **33**, 2021-2032 (1994)
[19] Gross, W.; Seipold, P.; Schnarrenberger, C.: Characterization and purification of an aldose from the acidophilic and thermophilic red algae Galdiera sulphuraria. Plant Physiol., **114**, 231-236 (1997)

[20] Marini, I.; Bucchioni, L.; Voltarelli, M.; Del Corso, A.; Mura, U.: α-Crystallin-like molecular chaperone against the thermal denaturation of lens aldose reductase: the effect of divalent metal ions. Biochem. Biophys. Res. Commun., **212**, 413-420 (1995)

[21] Tanimoto, T.; Ohta, M.; Tanaka, A.; Ikemoto, I.; Machida, T.: Purification and characterization of human testis aldose and aldehyde reductase. Int. J. Biochem., **23**, 421-428 (1991)

[22] Nidetzky, B.; Mayr, P.; Neuhauser, W.; Puchberger, M.: Structural and functional properties of alsode xylose reductase from the D-xylose-metabolizing Candida tenuis. Chem.-Biol. Interact., **130-132**, 583-595 (2001)

[23] Mayr, P.; Brüggler, K.; Kulbe, K.D.; Nidetzky, B.: D-Xylose metabolism by Candida intermedia: isolation and characterization of two forms of aldose reductase with different coenzyme specificities. J. Chromatogr., **737**, 195-202 (2000)

[24] Neuhauser, W.; Haltrich, D.; Kulbe, K.D.; Nidetzky, B.: NAD(P)H-dependent aldose reductase from the xylose-assimilating yeast Candida tenuis. Biochem. J., **326**, 683-692 (1997)

[25] Sato, S.; Old, S.; Carper, D.; Kador, P.F.: Purification and characterization of recombinant human placental and rat lens aldose reductases expressed in Escherichia coli. Enzymology and molecular biology of carbonyl metabolism, **5**, 259-268 (1995)

[26] Nishimura, C.; Yamaoka, T.; Mizutani, M.; Yamashita, K.; Akera, T.; Tanimoto, T.: Purification and characterization of the recombinant human aldose reductase expressed in baculovirus system. Biochim. Biophys. Acta, **1078**, 171-178 (1991)

[27] Inazu, N.; Nagashima, Y.; Satoh, T.; Fujii, T.: Purification and properties of six aldo-keto reductases from rat adrenal gland. J. Biochem., **115**, 991-999 (1994)

[28] Petrash, J.M.; Tarle, I.; Wilson, D.K.; Quiocho, F.A.: Aldose reductase catalysis and crystallography. Diabetes, **43**, 955-959 (1994)

UDPglucose 6-dehydrogenase 1.1.1.22

1 Nomenclature

EC number
1.1.1.22

Systematic name
UDPglucose:NAD^+ 6-oxidoreductase

Recommended name
UDPglucose 6-dehydrogenase

Synonyms
dehydrogenase, uridine diphosphoglucose
sugarless protein [Swissprot]
UDP-D-glucose dehydrogenase
UDP-Glc DH
UDP-Glc dehydrogenase [Swissprot]
UDP-GlcDH [Swissprot]
UDP-α-D-glucose:NAD oxidoreductase
UDPG dehydrogenase
UDPG:NAD oxidoreductase
UDPGDH [Swissprot]
UDPGlc dehydrogenase
UDPglucose dehydrogenase
UDPglucose:NAD^+ oxidoreductase
uridine diphosphate D-glucose dehydrogenase
uridine diphosphate glucose dehydrogenase
uridine diphosphoglucose dehydrogenase

CAS registry number
9028-26-6

2 Source Organism

<1> *Bos taurus* (calf [2, 9-11]) [1, 2, 9-11, 17]
<2> *Pisum sativum* (pea [10]) [10]
<3> *Escherichia coli* [3, 4, 16]
<4> *Rattus norvegicus* [6, 7, 14]
<5> *Lilium longiflorum* [5]
<6> *Cryptococcus laurentii* [8]
<7> *Gallus gallus* [10]

<8> *Streptococcus pyogenes* (group A [12]) [12]
<9> *Phaseolus vulgaris* (two-dimensional IEF SDS-PAGE showed several isoforms of the purified enzyme [13]) [13]
<10> *Cavia porcellus* [14]
<11> *Homo sapiens* [15]
<12> *Glycine max* [18]

3 Reaction and Specificity

Catalyzed reaction

UDPglucose + 2 NAD^+ + H_2O = UDPglucuronate + 2 NADH (<1,9> also shows alcohol dehydrogenase activity [13,17]; <1> also shows aldehyde dehydrogenase activity [17]; <8> ping-pong reaction mechanism [12]; <1> reaction mechanism, C275 provides SH-group in the catalytic centre [17]; <12> two-step reaction, reaction mechanism [18])

Reaction type

oxidation
reduction

Natural substrates and products

S UDPglucose + NAD^+ + H_2O <1-12> (<5> first step of a branched pathway leading to plant cell-wall polysaccharides which contain glucuronic and galacturonic acids and the pentoses xylose, arabinose and apiose [5]; <8> delivers glucuronic acid for the formation of a antiphagocytic polysaccharide capsule that is required for virulence of pathogenic bacteria [12]) (Reversibility: ? <1-12> [1-18]) [1-18]
P UDPglucuronate + NADH

Substrates and products

S 3-OH-benzopyrene + NAD^+ <4> (Reversibility: ? <4> [14]) [14]
P ?
S 5-fluorouracil + NAD^+ <1> (Reversibility: ? <1> [9]) [9]
P ?
S 6-azauracil + NAD^+ <1> (Reversibility: ? <1> [9]) [9]
P ?
S CDPglucose + NAD^+ + H_2O <3, 5> (<3> reaction rate is 5.5% of that with UDPglucose [4]; <5> 17% [5]) (Reversibility: ? <3, 5> [4, 5]) [4, 5]
P CDPglucuronate + NADH
S TDPglucose + NAD^+ + H_2O <3> (<3> reaction rate is 17% of that with UDPglucose [4]) (Reversibility: ? <3> [4]) [4]
P TDPglucuronate + NADH
S UDP-2-deoxyglucose + NAD^+ + H_2O <1> (Reversibility: ? <1> [2]) [2]
P UDP-2-deoxyglucuronate + NADH
S UDPglucose + NAD^+ + H_2O <1-12> (<1> reversal of reaction cannot be demonstrated [11]) (Reversibility: ? <1-12> [1-18]) [1-18]
P UDPglucuronate + NADH

S dTDPglucose + NAD^+ + H_2O <3> (<3> reaction rate is 16.7% of that with UDPglucose [4]) (Reversibility: ? <3> [4]) [4]

P dTDPglucuronate + NADH

S Additional information <1, 3-6> (<3> no activity with GTPglucose [4]; <3> ADPglucose [4]; <1> guanosine diphosphomannose [11]; <1> uridine diphosphoacetylglucosamine [11]; <1> uridine diphosphoacetylgalactosamine [11]; <1> α-D-glucose-1-phosphate [11]; <1> glucose [11]; <1> ethyl alcohol [11]; <1, 3, 6> cosubstrates which can replace NAD^+: 3-acetylpyridine adenine dinucleotide [4, 8, 11]; <3> 3-pyridinealdehyde adenine dinucleotide [4]; <3, 6> thionicotinamide adenine dinucleotide [4, 8]; <1, 3> deamino adenine dinucleotide [4, 11]; <4> nicotinamide hypoxanthine dinucleotide [7]; <3> no reaction with: 3-pyridinealdehyde deamino adenosine dinucleotide [4]; <4, 6> α-NAD^+ [7, 8]; <1,5> $NADP^+$ [5, 11]; <6> deamino-NAD^+ [8]; <6> 3-formylpyridine adenine dinucleotide [8]; <6> 3-propionylpyridine adenine dinucleotide [8]; <6> ethylnicotinate adenine dinucleotide [8]) [4, 5, 7, 8, 11]

P ?

Inhibitors

ADP <6> [8]
AMP <6> [8]
NADH <3, 4> [4, 7]
NH_2OH <4> (<4> deactivation [7]) [7]
UDP-D-galactose <5> (<5> slight [5]) [5]
UDP-D-xylose <1, 2, 7, 8, 9, 11, 12> (<8, 12> competitive with UDPglucose [12, 18]) [10, 12, 13, 15, 18]
UDParabinose <6> [8]
UDPgalacturonic acid <5> [5]
UDPglucuronate <3-5, 12> (<11, 12> product feed-back inhibition competitive with UDPglucose [15,18]) [4, 5, 7, 15, 18]
UDPxylose <3, 6> [4, 8]
p-chloromercuribenzoate <12> [18]
piperine <4, 10> (<4> non-competitive, reversible, non-pH-dependent, 90% inhibition at concentration 0.05 mM and 10% inhibition at concentration 0.01 mM [14]; <10> stronger inhibition with intestinal cell enzyme than with rat liver enzyme [14]) [14]
thiol group modifiying reagents <8> [12]
uridine 5'-diphosphate chloroacetol <8> (<8> alkylates a thiol group of cysteine in the catalytic centre via being specifically bound instead of the substrate [12]) [12]

Cofactors/prosthetic groups

3-acetylpyridine adenine dinucleotide <1, 3, 6> (<1, 3, 6> can replace NAD^+ [4, 8, 11]) [4, 8, 11]
NAD^+ <5> (<5> specific for [5]) [5]
β-NAD^+ <4> [7]
Additional information <8> (<8> no constantly bound chromophoric cofactors, i.e. NAD^+, required [12]) [12]

Activating compounds

3-pyridinealdehyde adenine dinucleotide <3> (<3> can replace NAD^+ [4]) [4]
deamino adenine dinucleotide <1, 3> (<1, 3> can replace NAD^+ [4, 11]) [4, 11]
nicotinamide hypoxanthine dinucleotide <4> (<4> can replace NAD^+ [7]) [7]
thionicotinamide adenine dinucleotide <3, 6> (<3, 6> can replace NAD^+ [4, 8]) [4, 8]
Additional information <11> (<11> fetal calf serum stimulates enzyme activity [15]) [15]

Metals, ions

Additional information <1, 4> (<1, 4> no metal ion requirement [6, 17]) [6, 17]

Turnover number (min^{-1})

7 <8> (UDPglucose) [12]

Specific activity (U/mg)

1.5 <8> (<8> recombinant and purified [12]) [12]
27.17 <9> (<9> purified 390fold [13]) [13]
Additional information <1, 4, 6> (<6> 14.0, 1 unit is defined as the amount of enzyme required to produce 0.0002 mM of NADH per min at 30°C [8]) [6-8, 11]

K_m-Value (mM)

0.015 <4> (UDPglucose, <4> pH 9.4 [14]) [14]
0.02 <9> (NAD^+) [13]
0.02 <8> (UDPglucose) [12]
0.035 <4> (UDPglucose, <4> pH 8.6 [14]) [14]
0.05 <3> (NAD^+) [4]
0.05 <12> (UDPglucose) [18]
0.06 <8> (NAD^+) [12]
0.0756 <4> (UDPglucose, <4> native enzyme [6]) [6]
0.108 <11> (UDPglucose) [15]
0.17 <11> (NAD^+) [15]
0.3 <5> (UDPglucose) [5]
0.4 <5> (NAD^+) [5]
1 <3, 4> (UDPglucose, <4> dissociated enzyme [6]) [4, 6]
5.5 <9> (UDPglucose) [13]
6.6 <3> (UDPglucose) [16]
8.4 <3> (UDPglucose, <3> recombinant glutathione-S-transferase fusion protein [16]) [16]
Additional information <1, 4, 6, 12> (<1> pH-dependence of K_M [9]; <12> K_M increase at pH higher than 9.0 [18]) [7-9, 18]

pH-Optimum

7.3-7.8 <6> [8]
8.4-8.8 <5> [5]

8.6 <4> (<4> two pH optima [14]) [14]
8.7 <1> [11]
8.9 <4> [7]
9 <3> [4]
9.4 <4> [6]
9.4 <4> (<4> two pH optima [14]) [14]

pH-Range

6.3-8.6 <6> (<6> at pH 6.3 and 8.6 about 50% of activity maximum [8]) [8]
7-9.5 <12> [18]
7.5-9.4 <3> (<3> at pH 7.5 and 9.4 about 50% of activity maximum [4]) [4]
7.8-9.4 <1> (<1> at pH 7.8 and 9.4 about 50% of activity maximum [11]) [11]

Temperature optimum (°C)

30 <3> (<3> assay at [4]) [4]

4 Enzyme Structure

Molecular weight

45500 <8> (<8> electrospray mass spectrometry [12]) [12]
52000 <8> (<8> gel filtation [12]) [12]
72000 <3> (<3> gel filtration [16]) [16]
86000 <3> (<3> gel filtration [4]) [4]
300000 <4> (<4> gel filtration [6]) [6]
305000 <1> (<1> equilibrium measurement under native conditions [1]) [1]

Subunits

? <9> (<9> ? * 40000, SDS-PAGE [13]) [13]
dimer <3> (<3> 2 * 47000, SDS-PAGE [3]; <3> 2 * 44000, SDS-PAGE [16]) [3, 16]
hexamer <1> (<1> 6 * 52000, at pH 5.5-7.8, equilibrium measurement under native and denaturing conditions [1]; <1> 6 * 52000, gel filtration [17]) [1, 17]
monomer <8> (<8> 1 * 45500, SDS-PAGE [12]) [12]
tetramer <4> (<4> 4 * 70000, gel filtration after treatment with SDS [6]) [6]

5 Isolation/Preparation/Mutation/Application

Source/tissue

cartilage <7> [10]
cell suspension culture <9> [13]
fibroblast cell line <11> (<11> enzyme activity depends on growth phase of the culture [15]) [15]
liver <1, 4, 10> [2, 5-7, 10, 11, 14, 17]
pollen <5> [5]

root nodule <12> [18]
skin <4> [7]
small intestine <10> [14]

Localization
cytosol <3, 12> (<3> recombinant fusion protein [16]) [16, 18]
microsome <4> [14]

Purification
<1> [1, 17]
<1> (calf [11]) [11]
<3> (recombinant glutathione-S-transferase fusion protein [16]) [3, 4, 16]
<4> (partial [14]) [6, 7, 14]
<5> (partial [5]) [5]
<6> (partial [8]) [8]
<8> (recombinant [12]) [12]
<9> [13]
<12> [18]

Cloning
<3> (overexpression as glutathione-S-transferase fusion protein in Escherichia coli [16]) [16]
<8> (overexpression in Escherichia coli JM109 [12]) [12]

Engineering
Additional information <1> (<1> comparison of sequence homologies with bacterial enzymes [17]) [17]

Application
pharmacology <8> (<8> target for inhibitor design [12]) [12]

6 Stability

Temperature stability
30 <8> (<8> after incubation for 1.6 h loss of 40% activity [12]) [12]
50 <5> (<5> without substrate, complete inactivation [5]) [5]

General stability information
<3>, 2-mercaptoethanol stabilizes [4]
<3>, UDPglucose stabilizes [4]
<5>, NAD^+ protects against heat inactivation [5]
<5>, UDPglucose protects against heat inactivation [5]
<8>, fairly stable with 2 mM dithiothreitol [12]
<8>, stable during repeated freezing and thawing cycles with 10% glycerol [12]

Storage stability
<1>, -10°C, little loss of activity after several weeks [11]
<1>, -70°C, 5 mM UDPglucose, 1 mM dithiothreitol [2]

<3>, -4°C, frozen in presence of UDPglucose, stable for 2 months [4]
<5>, -12°C, 2 weeks [5]
<5>, 0°C, 24 h [5]
<8>, 5°C, 50 mM Tris-HCl, pH 8.7, 2 mM dithiothreitol, after 24 h 90% activity left [12]

References

[1] Jaenicke, R.; Rudolph, R.: Dissociation and in vitro reconstitution of bovine liver uridine diphosphoglucose dehydrogenase. The paired subunit nature of the enzyme. Biochemistry, **25**, 7283-7287 (1986)

[2] Druzhinina, T.N.; Kusov, Y.Y.; Shibaev, V.N.; Kochetkov, N.K.; Biely, P.: Uridine diphosphate 2-deoxyglucose. Chemical synthesis, enzymic oxidation and epimerization. Biochim. Biophys. Acta, **381**, 301-307 (1975)

[3] Schiller, J.G.; Lamy, F.; Frazier, R.; Feingold, D.S.: UDP-glucose dehydrogenase from Escherichia coli. Purification and subunit structure. Biochim. Biophys. Acta, **453**, 418-425 (1976)

[4] Schiller, J.G.; Bowser, A.M.; Feingold, D.S.: Partial purification and properties of UDPG dehydrogenase from Escherichia coli. Biochim. Biophys. Acta, **293**, 1-10 (1973)

[5] Davies, M.D.; Dickinson, D.B.: Properties of uridine diphosphoglucose dehydrogenase from pollen of Lilium longiflorum. Arch. Biochem. Biophys., **152**, 53-61 (1972)

[6] Sivaswami, A.; Kelkar, S.M.; Nadkarni, G.B.: Uridine diphosphoglucose dehydrogenase from rat liver: purification and effect of pH on regulatory properties. Biochim. Biophys. Acta, **276**, 43-52 (1972)

[7] Molz, R.J.; Danishefsky, I.: Uridine diphosphate glucose dehydrogenase of rat tissue. Biochim. Biophys. Acta, **250**, 6-13 (1971)

[8] Ankel, H.; Ankel, E.; Feingold, D.S.: Biosynthesis of uridine diphosphate D-xylose. 3. Uridine diphosphate D-glucose dehydrogenase of Cryptococcus laurentii. Biochemistry, **5**, 1864-1869 (1966)

[9] Goldberg, N.D.; Dahl, J.L.; Parks, R.E.: pH dependence of the reactions with the 5-fluorouracil and 6-azauracil analogues of uridine diphosphate glucose. J. Biol. Chem., **238**, 3109-3114 (1963)

[10] Neufeld, E.F.; Hall, C.W.: Inhibition of the UDP-D-glucose dehydrogenase by UDP-D-xylose: a possible regulatory mechanism. Biochem. Biophys. Res. Commun., **19**, 456-461 (1965)

[11] Strominger, J.L.; Maxwell, E.S.; Axelrod, J.; Kalckar, H.M.: Enzymatic formation of uridine diphosphoglucuronic acid. J. Biol. Chem., **224**, 79-90 (1957)

[12] Campbell, R.E.; Sala, R.F.; van de Rijn, I.; Tanner, M.E.: Properties and kinetic analysis of UDP-glucose dehydrogenase from group A streptococci. J. Biol. Chem., **272**, 3416-3422 (1997)

[13] Robertson, D.; Smith, C; Bolwell, G.P.: Inducible UDP-glucose dehydrogenase from french bean (Phaseolus vulgaris L.) locates to vascular tissue and has alcohol dehydrogenase activity. Biochem. J., **313**, 311-317 (1996)

[14] Reen, R.K.; Jamwal, D.S.; Taneja, S.C.; Koul, J.L.; Dubey, R.K.; Wiebel, F.J.; Singh, J.: Impairment of UDP-glucose dehydrogenase and glucuronidation activities in liver and small intestine of rat and guinea pig in vitro by piperine. Biochem. Pharmacol., **46**, 229-238 (1993)
[15] Grubb, M.F.; Kasofsky, J.; Strong, J.; Anderson, L.W.; Cysyk, R.L.: Serum stimulation of UDP-glucose dehydrogenase activity in swiss 3T3 fibroblasts. Biochem. Mol. Biol. Int., **30**, 819-827 (1993)
[16] Sieberth, V.; Rigg, G.P.; Roberts, I.S.; Jann, K.: Expression and characterization of UDPGlc dehydrogenase (Kfid), which is encoded in the type-specific region 2 of the Escherichia coli K5 capsule genes. J. Bacteriol., **177**, 4562-4565 (1995)
[17] Hempel, J.; Perozich, J.; Romovacek, H.; Hinich, A.; Kuo, I.; Feingold, D.S.: UDP-glucose dehydrogenase from bovine liver: primary structure and relationship to other dehydrogenases. Protein Sci., **3**, 1074-1080 (1994)
[18] Stewart, D.C.; Copeland, L.: Kinetic properties of UDP-glucose dehydrogenase from soybean nodules. Plant Sci., **147**, 119-125 (1999)

Histidinol dehydrogenase 1.1.1.23

1 Nomenclature

EC number
1.1.1.23

Systematic name
L-histidinol:NAD^+ oxidoreductase

Recommended name
histidinol dehydrogenase

Synonyms
HDH [Swissprot]
L-histidinol dehydrogenase

CAS registry number
9028-27-7

2 Source Organism

<1> *Salmonella typhimurium* [1, 4, 9, 10, 12, 13, 14, 15, 17, 18]
<2> *Brassica oleracea* (cabbage [2]; contains probably multiple L-histidinol isoenzymes encoded by different genes [7]) [2, 3, 5, 6, 7, 8]
<3> *Asparagus officinalis* (asparagus shoots [8]) [8]
<4> *Lactuca sativa* (lettuce shoots [8]) [8]
<5> *Cucumis sativus* (cucumber shoots [8]) [8]
<6> *Brassica olearacea L. var capitata L.* (spring cabbage [8]) [8]
<7> *Triticum aestivum* (wheat germ [8]) [8]
<8> *Rosa sp.* ("Paul's scarlet", rose cell culture [8]) [8]
<9> *Solanum melongena* (eggplant shoots [8]) [8]
<10> *Escherichia coli* [11, 15, 19]
<11> *Bacillus psychrophilus* [16]
<12> *Bacillus subtilis* [16]
<13> *Bacillus stearothermophilus* [16]
<14> *Bacillus subtilis HT1* (obtained by transformation [16]) [16]
<15> *Bacillus caldolyticus* [16]
<16> *Arthrobacter histidinolovorans* [19]

3 Reaction and Specificity

Catalyzed reaction

L-histidinol + 2 NAD^+ = L-histidine + 2 NADH (<1> fast reversible oxidation of L-histidanol, followed by slow irreversible hydride transfer from the intermediate L-histidinal [1]; <2> Bi Uni Uni BI Ping Pong mechanism, recombinant histidinol dehydrogenase [3])

Reaction type

redox reaction

Natural substrates and products

S L-histidinal + NAD^+ <1-16> (<1-16> reaction intermediate [1-19]) (Reversibility: ir <1> [1]; ?<2-16> [2-19]) [1-19]
P L-histidine + NADH <1-16> [1-19]
S L-histidinol + NAD^+ <1-16> (Reversibility: ir <1> [1]; ? <2-16> [2-19]) [1-19]
P L-histidine + NADH <1-16> [1-19]

Substrates and products

S L-histidinal + NAD^+ <1-16> (Reversibility: ir <1> [1]; ? <2-16> [2-19]) [1-19]
P L-histidine + NADH <1-16> [1-19]
S L-histidinol + NAD^+ <1-16> (Reversibility: ir <1> [1]; ? <2-16> [2-19]) [1-19]
P L-histidine + NADH <1-16> [1-19]

Inhibitors

7-chloro-4-nitro-2,1,3-benzoxadiazole <1> (<1> 0.1 mM, pseudo first order inactivation, binds to Cys-116 in active site [9]) [9]
Ba^{2+} <2> (<2> 0.5 mM, 35% inhibition [8]) [8]
Ca^{2+} <2> (<2> 0.5 mM, 64% inhibition [8]) [6, 8, 18]
Cd^{2+} <1> (<1> 0.1 mM, 33% inhibition [18]) [18]
Cu^{2+} <2> (<2> 0.5 mM, 76% inhibition [8]) [6, 8, 18]
EDTA <16> (<16> 30 mM, 50% inhibition at pH 7.5 [19]) [19]
$HgCl_2$ <1, 10> (<10> 0.5 mM, 100% inhibition, 0.05 mM 79% inhibition [11]; <1> 0.1 mM, 80% inhibition, pH 9.4 [18]) [11, 18]
Mg^{2+} <2> (<2> 0.5 mM, 46% inhibition [8]) [6, 8, 18]
N-(4-dimethylamino-3,5-dinitrophenyl)-maleimide <1> (<1> 75% inhibition with one equivalent per subunit [12]) [12, 15]
Ni^{2+} <2> (<2> 0.5 mM, 54% inhibition [8]) [8, 18]
Zn^{2+} <2> (<2> 0.5 mM, 66% inhibition [8]) [6, 8, 18]
borate <16> (<16> 10 mM, 90% inhibition [19]) [19]
histamine <1> (<1> competitive inhibition [9]) [9]
imidazole <1> (<1> competitive inhibition [9]) [9]
p-chloromercuribenzoate <1, 16> (<1> 75% inhibition with two equivalents per subunit [12]; <1> 2.5 mM, 60% inhibition [18]; <16> 0.02 mM, 50% inhibition [19]) [12, 18, 19]

pyridoxal 5'-phosphate <1> (<1> 100 mM, 62.5% inhibition, concentration dependent, almost completely reversible [13]) [13]
urea <10> (<10> 2.2 mM, 50% inhibition, reversible [11]) [11]

Cofactors/prosthetic groups

NAD^+ <1-16> [1-19]

Metals, ions

Cd^{2+} <2> (<2> 0.02-0.2 mM, 75% increase [2]) [2]
Mn^{2+} <2, 10> (<2> 0.5 mM, 26% increase [8]; <10> 0.5 mM, 165% increase [11]) [4, 6, 8, 11]
Zn^{2+} <1, 10> (<1> essential for activity, one Zn atom per subunit [4]; <10> 0.5 mM, 20% increase [11]) [4, 11]

Turnover number (min^{-1})

9 <1> (L-histidinal, <1> H216N mutation [4]) [4]
66 <2> (L-histidinol) [3]
216 <1> (L-histidinal) [4]
560.4 <1> (L-histidinal) [10]
2100 <2> (L-histidinol, <2> recombinant enzyme without transit peptide [5]) [5]

Specific activity (U/mg)

0.0003 <5> (<5> enriched extracts [8]) [8]
0.0004 <4> (<4> enriched extracts [8]) [8]
0.0004 <9> (<9> enriched extracts [8]) [8]
0.0009 <3> (<3> enriched extracts [8]) [8]
0.0016 <2> (<2> enriched extracts [8]) [8]
0.0021 <7> (<7> enriched extracts [8]) [8]
0.0061 <6> (<6> enriched extracts [8]) [8]
0.0062 <8> (<8> enriched extracts [8]) [8]
3.11 <11> [16]
3.56 <12> [16]
3.86 <13> [16]
3.89 <14, 15> [16]
10 <2> (<2> recombinant enzym [6]) [6]
10.16 <2> [8]
12.3 <1> (<1> oxidation of histidinal [10]) [10]
12.5 <1> [15]
12.71 <10> [15]
14.2 <1> [10]
14.3 <1> [18]
15.32 <10> [11]

K_m-Value (mM)

0.0027 <1> (L-histidinal) [4]
0.0027 <2> (L-histidinol) [3]
0.0037 <1> (L-histidinal, <1> H216N mutation [4]) [4]
0.0072 <1> (NAD^+) [4]

0.009 <2> (L-histidinol) [5]
0.01 <2> (L-histidinol, <2> recombinant enzyme [6]) [6]
0.012 <2> (L-histidinol) [6]
0.014 <1> (L-histidinol) [11, 12]
0.016 <2> (L-histidinol) [8]
0.042 <2> (NAD^+) [6]
0.042 <2> (NAD^+) [8]
0.048 <11-15> (L-histidinol) [16]
0.05 <2> (NAD^+, <2> recombinant enzyme [6]) [6]
0.051 <2> (NAD^+) [3]
0.053 <2> (NAD^+) [5]
0.08 <16> (NAD^+) [19]
0.2 <11-15> (NAD^+) [16]
0.24 <1> (NAD^+, <1> H216N mutation [4]) [4]
0.57 <10> (NAD^+) [11]
0.7 <1> (NAD^+) [12]

pH-Optimum
8.6 <10> (<10> crude extracts [19]) [19]
8.7 <16> (<16> partly purified enzyme [19]) [19]
9.2 <1, 2> (<2> recombinant enzyme [6]; <2> purified from cabbage heads [8]) [6, 8, 18]
9.5 <10> [11]
9.6 <11-15> [16]

Temperature optimum (°C)
25 <11> [16]
41 <12> [16]
74 <13> [16]
87 <15> [16]
92 <14> [16]

4 Enzyme Structure

Molecular weight
29000 <11-15> (<11-15> sedimentation equilibrium [16]) [16]
82000 <1> (<1> gel filtration [14]) [14]
91000 <10> (<10> gel filtration [11]) [11]
100000 <2> (<2> recombinant enzyme expressed in SF9 cells, native PAGE [6]) [6]
103000 <2> (<2> gel filtration [8]) [8]

Subunits
homodimer <1> (<1> α,α, 2 * 43000, SDS-PAGE [14]) [14]
homodimer <2, 10> (<2, 10> α,α, 2 * 52000, SDS-PAGE [6, 8, 11]) [6, 8, 11]

5 Isolation/Preparation/Mutation/Application

Source/tissue

germ <7> [8]

Localization

chloroplast <2> (<2> enzyme contains chloroplast transit peptide sequence [5]; <2> encoded as proenzyme with amino terminal extension with composition of chloroplast transit peptides [7]) [5, 7]

Purification

<1> (heating, pH 4.8, ammonium sulfate, DEAE-cellulose [18]) [18]
<2> (ammonium sulfate precipitation, DEAE-chromatography, L-histidinol-Sepharose 4B affinity chromatography [8]) [8]
<2> (enzyme expressed in SF9 cells, cell homogenate, Mono Q FPLC, [6]) [6]
<10> (60°C, pH 5 treatment, ammonium sulfate precipitation, gel filtration, DEAE-cellulose, gel filtration [11]) [11]
<13> (<13> ammonium sulfate precipitation, heating, affinity chromatography, preparative gel [16]) [16]
<1, 10> (55°C, 40% ammonium sulfate, Sephadex G-50, DEAE-cellulose, crystallization [15]) [15]

Crystallization

<1> (1.6-2.0 mM ammonium sulfate solution at 4°C, pH 6.5, space group C2, monoclinic [17]) [17]

Cloning

<1> (expressed in Spodoptera frugiperda, SF9 cells [4]) [4]
<2> (expressed in Spodoptera frugiperda, SF9 cells [3, 5, 6]) [3, 5, 6]

Engineering

C112A <2> (<2> conserved Cys112 is not involved in catalysis, Cys112 corresponds to Cys116 in Salmonella typhimurium [5]) [5]
H261N <1> (<1> mutation causes complete loss of Zn and activity [4]) [4]

6 Stability

pH-Stability

9.2-10 <2> (<2> rapid loss of activity above pH 10 [8]) [8]

Temperature stability

65 <10> (<10> almost no loss of activity after 5 min at pH 7.5, histidinol, 2 mM and $MnCl_2$, 1 mM, protect enzyme from heat inactivation [11]) [11]
70 <1> (<1> 50% activity lost after 2 min [18]) [18]

Storage stability

<1>, 0°C, 0.05 M Tris-maleate, pH 6.5, 0.1 mM $MnCl_2$, 3 months, 45% loss of activity [18]
<2>, 4°C, 1 week, 10% loss of activity [8]
<2>, 4°C, 20% glycerol, 1 week, 0% loss of activity [8]

References

[1] Grubmeyer, C.; Teng, H.: Mechanism of Salmonella typhimurium histidinol dehydrogenase: kinetic isotope effects and pH profiles. Biochemistry, **38**, 7355-7362 (1999)

[2] Kanaori, K.; Ohta, D.; Nosaka, A.Y.: Effect of excess cadmium ion on the metal binding site of cabbage histidinol dehydrogenase studie by 113Cd-NMR spectroscopy. FEBS Lett., **412**, 301-304 (1997)

[3] Kheirolomoom, A.; Mano, J.; Nagai, A.; Ogawa, A.; Iwasaki, G.; Ohta, D.: Steady-state kinetics of cabbage histidinol dehydrogenase. Arch. Biochem. Biophys., **312**, 493-500 (1994)

[4] Nagai, A.; Ohta, D.: Histidinol dehydrogenase loses ist catalytic function through the mutation of His261-Asn due to its inability to ligate the essential Zn. J. Biochem., **115**, 22-25 (1994)

[5] Nagai, A.; Kheirolomoom, A.; Ohta, D.: Site-directed mutagenesis shows that the conserved cysteine residues of histidinol dehydrogenase are not essential for catalysis. J. Biochem., **114**, 856-861 (1993)

[6] Nagai, A.; Suzuku, K.; Ward, E.; Moyer, M.; Hashimoto, M.; Mano, J.; Ohta, D.; Scheidegger, A.: Overexpression of plant histidinol dehydrogenase using baculovirus expression vector system. Arch. Biochem. Biophys., **295**, 235-239 (1992)

[7] Nagai, A.; Ward, E.; Beck, J.; Tada, S.; Chang, J.; Scheidegger, A.: Structural and functional conservation of histidinol dehydrogenase between plants and microbes. Proc. Natl. Acad. Sci. USA, **88**, 4133-4137 (1991)

[8] Nagai, A.; Scheidegger, A.: Purification and characterization of histidinol dehydrogenase from cabbage. Arch. Biochem. Biophys., **284**, 127-132 (1991)

[9] Grubmeyer, C.T.; Gray, W.R.: A cysteine residue (cysteine-116) in the histidinol binding site of histidinoldehydrogenase. Biochemistry, **25**, 4778-4784 (1986)

[10] Görisch, H.; Hölke, W.: Binding of histidinal to histidinol dehydrogenase. Eur. J. Biochem., **150**, 305-308 (1985)

[11] Andorn, N.; Aronovitch, J.: Purification and properties of histidinol dehydrogenase from Escherichia coli B. J. Gen. Microbiol., **128**, 579-584 (1982)

[12] Feingold, D.S.; Franzen, J.S.: Pyridine nucleotide-linked four-electron transfer dehydrogenases. Trends Biochem. Sci., **6**, 103-105 (1981)

[13] Bürger, E.; Görisch, H.: Evidence for an essential lysine at the active site of L-histidinol:NAD^+ oxidoreductase; a bifunctional dehydrogenase. Eur. J. Biochem., **118**, 125-130 (1981)

[14] Bürger, E.; Görisch, H.; Lingens, F.: The catalytically active form of histidinol dehydrogenase from Salmonella typhimurium. Biochem. J., **181**, 771-774 (1979)

[15] Bitar, K.G.; Firca, J.R.; Loper, J.C.: Histidinol dehydrogenase from Salmonella typhimurium and Escherichia coli. Purification, some characteristics and the amino acid sequence around a reactive thiol group. Biochim. Biophys. Acta, **493**, 429-440 (1977)

[16] Lindsay, J.A.; Creaser, E.H.: Purification and properties of histidinol dehydrogenases from psychrophilic, mesophilic and thermophilic bacilli. Biochem. J., **165**, 247-253 (1977)
[17] Yang, H.J.; Lee, B.; Haslam, J.L.: Studies on histidinol dehydrogenase preliminary crystallographic data. J. Mol. Biol., **81**, 517-519 (1973)
[18] Loper, J.C.; Adams, E.: Purification and properties of histidinol dehydrogenase from Salmonella typhimurium. J. Biol. Chem., **240**, 788-795 (1965)
[19] Adams, E.: The enzymatic synthesis of histidine from histidinol. J. Biol. Chem., **209**, 829-846 (1954)

Quinate 5-dehydrogenase 1.1.1.24

1 Nomenclature

EC number
1.1.1.24

Systematic name
quinate:NAD^+ 5-oxidoreductase

Recommended name
quinate 5-dehydrogenase

Synonyms
dehydrogenase, quinate
quinate dehydrogenase
quinate: oxidoreductase
quinate:NAD oxidoreductase
quinic dehydrogenase

CAS registry number
9028-28-8

2 Source Organism

<1> *Neurospora crassa* (<1> bifunctional enzyme: quinate (shikimate) dehydrogenase [1]) [1]
<2> *Aspergillus niger* (<2> bifunctional enzyme: quinate (shikimate) dehydrogenase, different from EC 1.1.1.25 [2]) [2]
<3> *Phaseolus aureus* (mung bean [3]) [3, 7, 10]
<4> *Daucus carota* (carrot) [4, 9]
<5> *Aerobacter aerogenes* [5]
<6> *Aspergillus nidulans* [6]
<7> *Pinus taeda* [9]
<8> *Streptomyces sp.* (diverse number of strains [8]) [8]
<9> *Pseudonocardia sp.* [8]
<10> *Rhodococcus rhodochrous* [8]

3 Reaction and Specificity

Catalyzed reaction
quinate + NAD^+ = 5-dehydroquinate + NADH

Reaction type

oxidation
reduction

Natural substrates and products

- **S** 5-dehydroquinate + NADH <3, 4> (Reversibility: r <3, 4> [4, 9, 10]) [4, 9, 10]
- **P** quinate + NAD^+
- **S** *p*-hydroxybenzoate + O2 <8, 9> (<8, 9> overview substrate specificity, some strains are using the substrate, some are not [8]) [8]
- **P** protocatechuic acid + H_2O
- **S** quinate + NAD^+ <1, 3, 4, 8-10> (<1> first reaction in inducible quinic acid catabolic pathway [1]) (Reversibility: r <3, 4> [4, 9, 10]; ? <8-10> [8]) [1, 4, 8-10]
- **P** 5-dehydroquinate + NADH

Substrates and products

- **S** 5-dehydroquinate + NADH <3> (Reversibility: r <3> [10]) [10]
- **P** quinate + NAD^+
- **S** *p*-hydroxybenzoate + O2 <8, 9> (<8, 9> overview substrate specificity, some strains are using the substrate, some are not [8]) [8]
- **P** protocatechuic acid + H_2O
- **S** quinate + NAD^+ <1-5, 8-10> (<8> overview substrate specificity, some strains are using the substrate, some are not [8]) (Reversibility: r <3> [3, 10]; ? <8-10> [8]) [1-5, 8, 10]
- **P** 5-dehydroquinate + NADH
- **S** shikimate + NAD^+ <1, 2> [1, 2]
- **P** ?

Inhibitors

3-hydroxybenzoic acid <3> [3]
5-dehydroquinic acid <3> [3]
$CuSO_4$ <3> [3]
aromatic amino acids <3> (<3> inhibit reduction of dehydroquinate [10]) [10]
borate <3> [3]
molybdate <3> [3]
p-chloromercuribenzoate <1, 3> [1, 3]
shikimic acid <3> (<3> competitive inhibition [10]) [10]
sulfhydryl inhibitors <3> [3]

Cofactors/prosthetic groups

NAD^+ <1, 3, 4> (<1, 3> specific for NAD^+ [1, 3]; <3> can also utilize $NADP^+$/NADPH [10]; <4> utilizes like all young tissues of coniferous preferably $NADP^+$/NADPH [9]) [1, 3, 9, 10]
NADH <1, 3, 4> (<1, 3> specific for NAD^+ [1, 3]; <3> can also utilize $NADP^+$/NADPH [10]; <4> utilizes like all young tissues of coniferous preferably $NADP^+$/NADPH [9]) [1, 3, 9, 10]

Activating compounds

aromatic amino acids <3> (<3> slightly increased quinate oxidation [10]) [10]
dithiothreitol <1> [1]
erythrose 4-phosphate <3> (<3> erythrose 4-phosphate plus phosphoenolpyruvate increase quinate oxidizing activity by 25% [10]) [10]
phosphoenolpyruvate <3> (<3> erythrose 4-phosphate plus phosphoenolpyruvate increase quinate oxidizing activity by 25% [10]) [10]

Metals, ions

Additional information <3> (<3> no metal ion requirement [3]) [3]

Specific activity (U/mg)

2.15 <5> [5]
29.99 <3> [10]
86.87 <1> [1]
3453 <7> [9]
3700 <9> [8]
3700-6500 <8> [8]
4800 <10> [8]

K_m-Value (mM)

0.005 <7> (NADPH) [9]
0.007 <7> ($NADP^+$) [9]
0.009 <3> (NADH) [10]
0.031 <2> (NAD^+) [2]
0.047 <3> (3-dehydroquinate) [10]
0.05 <3> (NAD^+) [10]
0.12 <3> (NADPH) [10]
0.26 <2> (NAD^+, <2> + quinate [2]) [2]
0.32-0.37 <1, 2> (quinate) [1, 2]
0.37 <3> (quinate) [10]
0.42 <2> (NAD^+, <2> + shikimate [2]) [2]
0.45 <3> (3-dehydroquinate) [10]
0.7 <3> ($NADP^+$) [10]
1 <7> (dehydroquinate) [9]
1.18 <1> (shikimate) [1]
3 <3> (quinate) [10]
3.6 <7> (quinate) [9]
6.25 <2> (quinate) [2]
11.5 <2> (shikimate) [2]

pH-Optimum

9.6 <3> [3]
9.8 <5> [5]

pH-Range

8.5-9.8 <5> (<5> at pH 8.5 and 9.8 about 50% of activity maximum [5]) [5]

Temperature optimum (°C)

25 <5> (<5> assay at [5]) [5]
30 <3> (<3> assay at [3]) [3]

4 Enzyme Structure

Molecular weight

23000 <2> (<2> gel filtration [2]) [2]
35000 <7> (<7> gel filtration [9]) [9]
42000 <1> (<1> sucrose density gradient centrifugation [1]) [1]
110000 <4> (<4> gel filtration [4]) [4]
160000 <3> (<3> gel filtration [10]) [10]
Additional information <4> (<4> transfer of carrot enzyme from dark to light conditions shifts MW from 42000 Da to 110000 Da, probably due to association of a regulatory subunit which may be a calciprotein [4]) [4]

Subunits

dimer <4> (<4> 1 * 40000 + 1 * 60000-63000, SDS-PAGE [4]) [4]
monomer <1> (<1> 1 * 41000, SDS-PAGE in presence of urea or electrophoresis in presence of cetyltrimethyl ammonium bromide [1]) [1]
trimer <3> (<3> 3 * 55000, SDS-PAGE [10]) [10]

5 Isolation/Preparation/Mutation/Application

Source/tissue

cell suspension culture <4> [4]
mycelium <1> [1]
needle <7> [9]
seedling <3> [10]

Purification

<1> (<1> bifunctional enzyme: quinate (shikimate) dehydrogenase [1]) [1]
<3> (<3> to homogenity [10]) [3, 10]
<4> [4]
<5> [5]
<7> [9]

6 Stability

pH-Stability

7.5 <1> (<1> -20°C, highest stability [1]) [1]

Temperature stability

40 <1> (<1> half-life of unprotected enzyme: 20 min, 180 min in presence of NaCl, completely stable in presence of quinate, shikimate or NADH [1]) [1]

General stability information

<1>, high ionic strength, such as 0.1 M $(NH_4)_2SO_4$, NaCl or phosphate buffer or the presence of 0.1 M quinate or shikimate protects against thermal inactivation [1]

Storage stability

<1>, -20°C, 0.1 M NaCl, half-life 40 days [1]
<3>, -20°C, pH 7.5, several weeks [3]
<5>, -15°C, stable for several weeks, 30% loss of activity in 6 months [5]

References

[1] Barea, J.L.; Giles, N.H.: Purification and characterization of quinate (shikimate) dehydrogenase, an enzyme in the inducible quinic acid catabolic pathway of Neurospora crassa. Biochim. Biophys. Acta, **524**, 1-14 (1978)

[2] Cain, R.B.: The identity of shikimate dehydrogenase and quinate dehydrogenase in Aspergillus niger. Biochem. J., **127**, 15P (1972)

[3] Gamborg, O.L.: Aromatic metabolism in plants III. quinate dehydrogenase from mung bean cell suspension cultures. Biochim. Biophys. Acta, **128**, 483-491 (1966)

[4] Graziana, A.; Ranjeva, R.; Salimath, B.P.; Boudet, A. M.: The reversible association of quinate:NAD^+ oxidoreductase from carrot cells with a putative regulatory subunit depends on light conditions. FEBS Lett., **163**, 306-311 (1983)

[5] Davies, B.D.; Gilvarg, C.; Mitsuhyshi, S.: Enzymes of aromatic biosynthesis C. quinic dehydrogenase from Aerobacter aerogenes. Methods Enzymol., **2**, 307-311 (1955)

[6] Hawkins, A.R.; Moore, J.D.; Adeokun, A.M.: Characterization of the 3-dehydroquinase domain of the pentafunctional AROM protein, and the quinate dehydrogenase from Aspergillus nidulans, and the overproduction of the type II 3-dehydroquinase from Neurospora crassa. Biochem. J., **296**, 451-457 (1993)

[7] Kang, X.; Neuhaus, E.; Scheibe, R.: Subcellular localization of quinate: oxidoreductase from Phaseolus mungo L. sprouts. Z. Naturforsch. C, **49c**, 415-420 (1994)

[8] Grund, E.; Kutzner, H.J.: Utilization of quinate and *p*-hydroxybenzoate by actinomycetes: key enzymes and taxonomic relevance. J. Basic Microbiol., **38**, 241-255 (1998)

[9] Ossipov, V.; Bonner, C.; Ossipova, S.; Jensen, R.: Broad specificity quinate (shikimate) dehydrogenase from Pinus taeda needles. Plant Physiol., **38**, 923-928 (2000)

[10] Kang, X.; Scheibe, R.: Purification and characterization of the quinate: oxidoreductase from Phaseolus mungo sprouts. Phytochemistry, **33**, 769-773 (1993)

Shikimate 5-dehydrogenase 1.1.1.25

1 Nomenclature

EC number
1.1.1.25

Systematic name
shikimate:$NADP^+$ 5-oxidoreductase

Recommended name
shikimate 5-dehydrogenase

Synonyms
5-dehydroshikimate reductase
dehydroshikimic reductase
shikimate dehydrogenase
shikimate oxidoreductase
shikimate:NADP oxidoreductase

CAS registry number
9026-87-3

2 Source Organism

<1> *Phaseolus mungo* [3]
<2> *Euterpe oleracea* [9]
<3> *Camellia sinensis* (tea plant [4]) [4]
<4> *Pisum sativum* (pea) [8]
<5> *Escherichia coli* [1, 10, 11]
<6> *Lycopersicon esculentum* (tomato) [2]
<7> *Bambusa sp.* (bamboo) [5]
<8> *Physcomitrella patens* (enzyme complex of EC 4.2.1.10 and EC 1.1.1.25) [6]
<9> *Neurospora crassa* (multienzyme complex) [7]
<10> *Cucumis sativus* [12]
<11> *Capsicum annuum* [13]

3 Reaction and Specificity

Catalyzed reaction
shikimate + $NADP^+$ = 5-dehydroshikimate + NADPH

Reaction type

oxidation
reduction

Natural substrates and products

S 3-dehydroshikimate + NADPH <11> (Reversibility: r <11> [13]) [13]
P shikimate + $NADP^+$
S 5-dehydroshikimate + NADPH <4> (<4> pathway of biosynthesis of aromatic amino acids [8]; <4> enzyme of shikimic acid biosynthesis [8]) (Reversibility: r <4> [8]) [8]
P ?
S shikimate + $NADP^+$ <1-11> (Reversibility: r <4> [8]) [1-13]
P 5-dehydroshikimate + NADPH

Substrates and products

S 3-dehydroshikimate + NADPH <11> (Reversibility: r <11> [13]) [13]
P shikimate + $NADP^+$
S shikimate + $NADP^+$ <1-11> (Reversibility: r <4> [8]) [1-13]
P 5-dehydroshikimate + NADPH

Inhibitors

2,4-dichlorophenoxyacetic acid <11> [13]
3,5-dihydroxybenzoate <1> (<1> moderate [3]) [3]
Cd^{2+} <10> (<10> 500 nM [12]) [12]
$CdSO_4$ <2> [9]
Cu^{2+} <6, 11> [2, 13]
$CuSO_4$ <2> [9]
Hg^{2+} <2, 6> [2, 9]
$HgCl_2$ <10> (<10> complete inhibition at concentration 0.05 mM [12]) [12]
N-ethylmaleimide <6> [2]
NH_4^+ <6> [2]
Zn^{2+} <6, 10, 11> (<10> 0.005 mM [12]) [2, 12, 13]
$ZnCl_2$ <2> [9]
arsenite <3> [4]
borate <6> [2]
guaiacol <11> [13]
iodoacetate <3, 4> (<4> slight [8]) [4, 8]
metal ions <2> [9]
p-chloromercuribenzoate <1-4, 6, 10> (<10> biphasic inhibition: first rapid inhibtion leading to loss of 70% activity, then within 5 min loss of the remaining 30% activity, inhibition can be partially hindered by thiols and chloride [12]) [2, 3, 4, 8, 9, 12]
p-hydroxymercuribenzoate <1> (<1> moderate [3]) [3]
protocatechuic acid <1, 2, 6, 10, 11> (<1> moderate [3]; <11> competitive inhibition [13]) [2, 3, 9, 12, 13]

Cofactors/prosthetic groups

$NADP^+$ <3-6> (<3, 5, 6> specific for [2, 4, 10]) [1, 2, 4, 8, 10, 11]

NADPH <3-6> (<3, 5, 6> specific for [2, 4, 10]) [1, 2, 4, 8, 10, 11]
Additional information <4, 6> (<4,6> no activity with NAD^+ [2, 8]) [2, 8]

Activating compounds

Ca^{2+} <11> [13]
Mg^{2+} <11> [13]
Mn^{2+} <11> [13]
NH_4^+ <11> [13]

Metals, ions

Additional information <4, 6> (<6> no requirement for divalent ions [2]; <4> no metal ion requirement [8]) [2, 8]

Specific activity (U/mg)

0.28 <6> [2]
3.5 <10> (<10> crude extract [12]) [12]
25.4 <10> (<10> purified enzyme [12]) [12]
167 <1> [3]
1100 <5> [1]

K_m-Value (mM)

0.007 <4> ($NADP^+$) [8]
0.01 <10> ($NADP^+$) [12]
0.01 <6> ($NADP^+$) [2]
0.017 <11> ($NADP^+$) [13]
0.025 <1> ($NADP^+$) [3]
0.032 <3> ($NADP^+$) [4]
0.038 <6> (shikimate) [2]
0.05 <10> (shikimic acid, <10> $NADP^+$ [12]) [12]
0.087 <11> (shikimic acid, <11> $NADP^+$ [13]) [13]
0.19-0.28 <4> (shikimate) [8]
0.22 <1> (shikimate, <1> 2 K_M-values [3]) [3]
0.43 <3> (shikimate) [4]
0.91 <1> (shikimate, <1> 2 K_M-values [3]) [3]
Additional information <2, 5> [9, 10]

pH-Optimum

7.7 <3> (<3> Tris-HCl acid buffer [4]) [4]
8.5 <5> [10]
8.7 <10> [12]
9.1 <6> [2]
10 <1, 4> [3, 8]
10.1 <3> (<3> glycine-sodium hydroxide buffer [4]) [4]
10.5 <11> [13]

pH-Range

6-10 <11> [13]
6-11 <6> (<6> no activity below and above [2]) [2]
8.5-10.5 <4> (<4> pH 8.5: about 50% of activity maximum, pH 10.5: about 90% of activity maximum [8]) [8]

Temperature optimum (°C)

22 <3> (<3> assay at [4]) [4]
25 <1> (<1> assay at [3]) [3]
35 <8> (<8> assay at [6]) [6]
50 <10, 11> [12, 13]

Temperature range (°C)

20-55 <10> [12]

4 Enzyme Structure

Molecular weight

29400 <5> (<5> mass spectometry and calculation from DNA sequence [11]) [11]
32000 <5> (<5> gel filtration, SDS-PAGE [1]) [1]
45000 <10> (<10> gel filtration, 2 different isoenzymes with molecular weight about 45000 kDa [12]) [12]
48000 <8> (<8> gel filtration [6]) [6]
50000 <2> (<2> gel filtration [9]) [9]
57000 <1, 4> (<1> disc gel electrophoresis [3]) [3, 8]
67000 <11> (<11> gel filtration, 4 different isoenzymes [13]) [13]
73000 <6> (<6> gel filtration [2]) [2]
73000 <6> (<6> gel filtration [2]) [2]

Subunits

monomer <5> (<5> 1 * 29400, mass spectrometry and gel filtration [11]; <5> 1 * 32000, SDS-PAGE [1]) [1, 11]

5 Isolation/Preparation/Mutation/Application

Source/tissue

epicotyl <4> (<4> etiolated [8]) [8]
fruit <6> [2]
seedling <1> [3]
shoot <3, 7> [4, 5]

Purification

<1> [3]
<2> [9]
<4> [8]
<5> (partial [1, 10]) [1, 10, 11]
<8> (enzyme complex of EC 4.2.1.10 and EC 1.1.1.25 [6]) [6]
<9> (multienzyme complex [7]) [7]
<10> [12]
<11> (separation of isoenzymes by regaining the protein from PA-gel slices [13]) [13]

Crystallization
<5> [11]

Cloning
<5> (overexpression in Escherichia coli K12) [11]

6 Stability

pH-Stability
6-10 <11> (<11> stable for 45 min [13]) [13]

Temperature stability
40 <11> (<11> at temperatures higher than 40°C the enzyme is quickly inactivated [13]) [13]
50 <6> (<6> 5 min, 85% loss of activity [2]) [2]
60 <10> (<10> complete inactivation after 10 min, at 50-55°C biphasic, inactivation can be minimized with bound substrate to about 60% [12]) [12]

General stability information
<10>, stable in crude extract [12]

Storage stability
<5>, -20°C, 50% glycerol, 1 mM benzamidine, 0.4 mM dithiothreitol [1]
<5>, -20°C, Tris-HCl, pH 7.5, 0.4 mM DTT, 50% glycerol, long-term [11]

References

[1] Chaudhuri, S.; Coggins, J.R.: The purification of shikimate dehydrogenase from Escherichia coli. Biochem. J., **226**, 217-223 (1985)
[2] Lourenco, E.J.; Neves, V.A.: Partial purification and some properties of shikimate dehydrogenase from tomatoes. Phytochemistry, **23**, 497-499 (1984)
[3] Koshiba, T.: Purification of two forms of the associated 3-dehydroquinate hydro-lyase and shikimate:NADP$^+$ oxidoreductase in Phaseolus mungo seedlings. Biochim. Biophys. Acta, **522**, 10-18 (1978)
[4] Sanderson, G.W.: 5-Dehydroshikimate reductase in the tea plant (Camellia sinensis L.). Properties and distribution. Biochem. J., **98**, 248-252 (1966)
[5] Higuchi, T.; Shimada, M.: Changes in activity of shikimate: NADP oxidoreductase in relation to lignification of bamboo. Plant Cell Physiol., **8**, 61-69 (1967)
[6] Polley, L.D.: Purification and characterization of 3-dehydroquinate hydrolase and shikmate oxidoreductase. Evidence for a bifunctional enzyme. Biochim. Biophys. Acta, **526**, 259-266 (1978)
[7] Jacobson, J.W.; Hart, B.A.; Doy, C.H.; Giles, N.H.: Purification and stability of the multienzyme complex encoded in the arom gene cluster of Neurospora crassa. Biochim. Biophys. Acta, **289**, 1-12 (1972)

[8] Balinsky, D.; Davies, D.D.: Aromatic biosynthesis in higher plants I. Preparation and properties of dehydroshikimic reductase. Biochem. J., **80**, 292-296 (1961)
[9] Lemos Silva, G.M.; Lourenco, E.J.; Neves, V.A.: Inhibition of shikimate dehydrogenase from heart-of-palm (Euterpe oleracea Mart.). J. Food Biochem., **9**, 105-116 (1985)
[10] Yaniv, H.; Gilvarg, C.: Aromatic biosynthesis XIV. 5-Dehydroshikimic reductase. J. Biol. Chem., **213**, 787-795 (1955)
[11] Maclean, J.; Campbell, S.A.; Pollock, K.; Chackrewarthy, S.; Coggins, J.R.; Lapthorn, A.J.: Crystallization and preliminary X-ray analysis of shikimate dehydrogenase from Escherichia coli. Acta Crystallogr. Sect. D, **56**, 512-515 (2000)
[12] Lourenco, E.J.; Silva, G.M.L.; Neves, V.A.: Purification and properties of shikimate dehydrogenase from cucumber (Cucumis sativa L.). J. Agric. Food Chem., **39**, 458-462 (1991)
[13] Diaz, J.; Merino, F.: Shikimate dehydrogenase from pepper (Capsicum annuum) seedlings. Purification and properties. Physiol. Plant., **100**, 147-152 (1997)

Glyoxylate reductase 1.1.1.26

1 Nomenclature

EC number
1.1.1.26

Systematic name
glycolate:NAD^+ oxidoreductase

Recommended name
glyoxylate reductase

Synonyms
NADH-dependent glyoxylate reductase
NADH-glyoxylate reductase
glycolate oxidase
glyoxylic acid reductase
reductase, glyoxylate

CAS registry number
9028-32-4

2 Source Organism

<1> *Pseudomonas fluorescens* [1]
<2> *Nicotiana tabacum* (tobacco) [4, 5, 11]
<3> *Sinapis alba* [6]
<4> *Euglena gracilis* (non-photosynthetic mutant) [2]
<5> *Rattus norvegicus* [3]
<6> *Spinacia oleracea* (spinach) [4, 7, 8, 11]
<7> *Pisum sativum* [11, 12]
<8> *Cyanophora paradoxa* [12]
<9> *Petroselinum hortense* (parsley) [11]
<10> *Hordeum vulgare* (barley) [11]
<11> *Zea mays* (maize) [11]
<12> *Homo sapiens* [9, 10]

3 Reaction and Specificity

Catalyzed reaction

glycolate + NAD^+ = glyoxylate + NADH (reduces glyoxylate to glycolate or hydroxypyruvate to D-glycerate; <7> can convert L-lactate [12])

Reaction type

oxidation
reduction

Natural substrates and products

S glycolate + NAD^+ <6, 7, 8> (Reversibility: r <6, 7, 8> [11, 12]) [11, 12]
P glyoxylate + NADH
S glyoxylate + NADH <6, 7, 8> (Reversibility: r <6, 7, 8> [11, 12]) [11, 12]
P glycolate + NAD^+
S hydroxypyruvate + NADH <6, 7-9> (<6, 9> located exclusively in the peroxisomes, prefers NAD^+/NADH [11]) (Reversibility: r <6, 7-9> [11, 12]) [11, 12]
P D-glycerate + NAD^+

Substrates and products

S glycolate + NAD^+ <6, 7, 8> (Reversibility: r <6, 7, 8> [11, 12]) [11, 12]
P glyoxylate + NADH
S glyoxylate + NADH <1-8> (<1-8> equilibrium is very far in direction of glycolate formation [1-8, 11, 12]) (Reversibility: r <1-8> [1-8, 11, 12]) [1-8, 11, 12]
P glycolate + NAD^+
S hydroxypyruvate + NADH <1, 6, 7-9> (Reversibility: r <1, 6, 7-9> [1, 7, 11, 12]) [1, 7, 11, 12]
P D-glycerate + NAD^+
S Additional information <6> (<6> no activity with: pyruvate, 2-oxobutyrate, acetaldehyde, oxaloacetate, 2-oxoglutarate [7]) [7]
P ?

Inhibitors

MOPS buffer <8> (<8> 50 mM at acidic and alkaline pH [12]) [12]
acetohydroxamate <6, 7> (<6, 7> inhibits only the cytosolic enzyme type with preference for glyoxylate as substrate [11, 12]) [11, 12]
oxalate <6, 7> (<6, 7> inhibits only the cytosolic enzyme type with preference for hydroxypyruvate as substrate [11, 12]) [11, 12]
oxo acids <1> [1]
p-chloromercuribenzoate <1, 2> [1, 4]
phenylhydrazine <2> [4]
semicarbazide <2> [4]
tartronate <6, 7> (<6, 7> inhibits only the cytosolic enzyme type with preference for hydroxypyruvate as substrate [11, 12]) [11, 12]

Cofactors/prosthetic groups

NAD^+ <1-8> (<8> can also utilize $NADP^+$ or O_2 (reduction to H_2O_2) as cofactors [12]; <6> enzyme of peroxisomes and chloroplasts prefer NAD^+, enzymes of the cytosol prefer $NADP^+$ [11]) [1-6, 11, 12]
NADH <1-8> (<8> can also utilize NADPH or O_2 (reduction to H_2O_2) as cofactors [12]; <6> enzyme of peroxisomes and chloroplasts prefer NADH, enzymes of the cytosol prefer NADPH [11]) [1-6, 11, 12]
Additional information <6> (<6> NADPH cannot replace NADH [7]) [7]

Specific activity (U/mg)

0.032 <7> (<7> 0.5 mM NADPH [12]) [12]
0.033 <6> [7]
0.044 <8> (<8> 75 mM NADH [12]) [12]
0.05-0.23 <6> (<6> NADPH, cytosolic enzyme with preference for hydroxypyruvate as substrate [11]) [11]
0.077 <7> (<7> 0.5 mM NADH [12]) [12]
0.095 <8> (<8> 0.5 mM NADPH [12]) [12]
0.096 <8> (<8> 0.5 mM NADH [12]) [12]
0.11 <8> (<8> 0.5 mM NADPH [12]) [12]
0.134 <8> (<8> 0.5 mM NADH [12]) [12]
0.145 <12> [9]
0.16 <7> (<7> 0.5 mM NADPH [12]) [12]
0.205 <7> (<7> NADH [12]) [12]
0.3-1.5 <6> (<6> NADH, peroxisomal enzyme type with preference for hydroxypyruvate and NADH as substrates [11]) [11]
0.738 <7> (<7> 0.5 mM, NADH [12]) [12]
5 <1> [1]
5.6 <2> [4]
7 <1> [1]
9.1 <2> [5]
Additional information <1, 2> [1, 4]

K_m-Value (mM)

0.07-0.11 <6> (glyoxylate, <6> NADPH, cytosolic enzyme [11]) [11]
0.07-0.12 <6> (hydroxypyruvate, <6> NADH, peroxisomal enzyme type with preference for hydroxypyruvate and NADH as substrates [11]) [11]
0.8 <6> (hydroxypyruvate, <6> NADPH, cytosolic enzyme type with preference for hydroxypyruvate and NADPH as substrates [11]) [11]
1.1 <6> (glyoxylate, <6> NADPH, cytosolic enzyme type with preference for hydroxypyruvate as substrate [11]) [11]
1.25 <12> (glyoxylate, <12> NADPH [9]) [9]
10-25 <6> (glyoxylate, <6> NADH, peroxisomal enzyme type with preference for hydroxypyruvate and NADH as substrates [11]) [11]

pH-Optimum

6-6.8 <1> [1]
6.3-6.6 <2> [5]
6.4 <2> [4]

7.6 <12> (<12> substrate glyoxylate [9]) [9]
7.9 <8> (<8> TES/K^+, substrate glyoxylate [12]) [12]
8.2 <8> (<8> Tris/HCl, substrate glyoxylate [12]) [12]

pH-Range

5.1-7.8 <2> (<2> pH 5.1: about 60% of activity maximum, pH 7.8: about 50% of activity maximum [5]) [5]
5.8-7.4 <6,7> (<6, 7> the optimal pH for hydroxypyruvate reduction is more acidic than that for glyoxylate reduction [11]) [11]
6.5-11 <8> [12]

4 Enzyme Structure

Molecular weight

36500-40000 <12> (<12> calculation from DNA sequence and non-denaturing PAGE [10]) [10]
70000 <6> (<6> cytosolic enzyme with preference for hydroxypyruvate as substrate [11]) [11]
90000 <6> (<6> peroxisomal enzyme with preference for hydroxypyruvate as substrate [11]) [11]
125000 <6> (<6> cytosolic enzyme [11]) [11]
180000 <1> (<1> thin-layer gel filtration [1]) [1]

Subunits

dimer <4, 6> (<6> 2 * 38000, cytosolic enzyme with preference for hydroxypyruvate as substrate [11]; <4> 2 * 41000-44000, peroxisomal enzyme with preference for hydroxypyruvate as substrate [2]) [2, 11]
monomer <12> (<12> 1 * 40000, SDS-PAGE [10]) [10]
tetramer <6> (<6> 4 * 33000, cytosolic enzyme [11]) [11]

5 Isolation/Preparation/Mutation/Application

Source/tissue

cotyledon <3> [6]
kidney <12> [9]
leaf <2, 6, 7, 10, 11> (<10> mutant without cytosolic hydroxypyruvate-metabolizing activity [11]) [4, 5, 8, 11, 12]
liver <5, 12> [3, 9, 10]

Localization

chloroplast <6, 7> (<6, 7> utilizes NADH and NADPH equally [11, 12]) [11, 12]
cyanelle <8> [12]
cytosol <6, 7, 10> (<6, 7> $NADP^+$/NADPH preferring type [11, 12]) [11, 12]
peroxisome <6, 7, 9, 10> (<6, 7, 9> NAD^+/NADH preferring type [11, 12]) [11, 12]

Purification

<1> [1]
<2> [4, 5]
<6> [11]
<6> (partial) [8]
<7, 8> [12]

Crystallization

<2> [5]

Cloning

<12> (expression in Epicurean coli BL21 (DE3) as fusion protein from a His-tag plasmid [10]) [10]

Engineering

Additional information <10> (<10> mutant lacking cytosolic hydroxypyruvate-metabolizing activity (HPR-1) [11]) [11]

6 Stability

Storage stability

<6>, 4°C, crystalline suspension in ammonium sulfate solution, 3.2 mol/l, stable for several months [7]

References

[1] Hullin, R.P.: Glyoxylate reductase, two forms from Pseudomonas. Methods Enzymol., **41 B**, 343-348 (1975)
[2] Yokota, A.; Kitaoka, S.: Occurence and subcellular distribution of enzymes involved in the glycolyte pathway and their physiological function in a bleached mutant of Euglena gracilis. Agric. Biol. Chem., **45**, 15-22 (1981)
[3] Suzuki, S.; Suga, T.; Ninobe, S.: Studies on peroxisomes. IV. Intracellular localization of $NADH_2$-glyoxylate reductase in rat liver. J. Biochem., **73**, 1033-1038 (1973)
[4] Zelitch, I.: Glycolic acid oxidase and glyoxylic acid reductase. Methods Enzymol., **1**, 528-535 (1955)
[5] Zelitch, I.: The isolation and action of crystalline glyoxylic acid reductase from tobacco leaves. J. Biol. Chem., **216**, 553-575 (1955)
[6] Cerff, R.: Glyceraldehyde 3-phosphate dehydrogenases and glyoxylate reductase. Plant Physiol., **51**, 76-81 (1973)
[7] Bergmeyer, H.U.; Graßl, M.; Walter, H.E.: Glyoxylate reductase. Methods Enzym. Anal., 3rd Ed. (Bergmeyer, H.U., ed.), **2**, 126-328 (1974)
[8] Zelitch, I.: Oxidation and reduction of glycolic and glyoxylic acids in plants II. glyoxylic acid reductase. J. Biol. Chem., **201**, 719-726 (1953)
[9] Giafi, C.F.; Rumsby, G.: Kinetic analysis and tissue distribution of human D-glycerate dehydrogenase/glyoxylate reductase and its relevance to the diag-

nosis of primary hyperoxaluria type 2. Ann. Clin. Biochem., **35**, 104-109 (1998)
[10] Rumsby, G.; Cregeen, D.P.: Identification and expression of a cDNA for human hydroxypyruvate/glyoxylate reductase. Biochim. Biophys. Acta, **1446**, 383-388 (1999)
[11] Givan, C.V.; Kleczkowski, L.A.: The enzymic reduction of glyoxylate and hydroxypyruvate in leaves of higher plants. Plant Physiol., **100**, 552-556 (1992)
[12] Betsche, T.; Schaller, D.; Melkonian, M.: Identification and characteriaztion of glycolate oxidase and related enzymes from the endocyanotic alga Cyanophora paradoxa and from pea leaves. Plant Physiol., **98**, 887-893 (1992)

L-Lactate dehydrogenase 1.1.1.27

1 Nomenclature

EC number
1.1.1.27

Systematic name
(S)-lactate:NAD^+ oxidoreductase

Recommended name
L-lactate dehydrogenase

Synonyms
ε crystallin [Swissprot]
Immunogenic protein p36 [Swissprot]
L(+)-nLDH
L-(+)-lactate dehydrogenase
L-LDH [Swissprot]
L-lactic acid dehydrogenase
L-lactic dehydrogenase
LDH [Swissprot]
NAD-lactate dehydrogenase
dehydrogenase, lactate
ε-crystallin
lactate dehydrogenase
lactate dehydrogenase NAD-dependent
lactic acid dehydrogenase
lactic dehydrogenase
proteins, specific or class, anoxic stress response, p34

CAS registry number
9001-60-9

2 Source Organism

<-3> no activity in *Taraxacum officinale* (leaf [17]) [17]
<-2> no activity in *Zea mays* (leaf [17]) [17]
<-1> no activity in *Rubus idaeus* (leaf [17]) [17]
<1> *Lactobacillus acidophilus* [1, 24]
<2> *Lactobacillus helveticus* [1, 41]
<3> *Lactobacillus salivarius* [1]
<4> *Lactobacillus casei* (ATCC 393 [5]; ssp. casein [19]) [1, 5, 19, 24]

<5> *Lactobacillus curvatus* [1, 19, 24]
<6> *Lactobacillus plantarum* [1, 24]
<7> *Lactobacillus brevis* [1]
<8> *Lactobacillus buchnerii* [1]
<9> *Lactobacillus cellobiosus* [1]
<10> *Lactobacillus confusus* [1]
<11> *Lactobacillus fermentum* [1]
<12> *Lactobacillus vermeforme* [1]
<13> *Lactobacillus viridescens* [1]
<14> *Lactobacillus mesenteroides* [1]
<15> *Streptococcus agalactiae* [1]
<16> *Streptococcus uberis* (fructose 1,6-diphosphate dependent enzyme [2]) [1, 2]
<17> *Streptococcus bovis* [1]
<18> *Streptococcus faecalis* [1, 20]
<19> *Streptococcus faecium* [1]
<20> *Streptococcus cremonis* [1, 4]
<21> *Streptococcus raffinolacis* [1]
<22> *Streptococcus acidominimus* [1]
<23> *Streptococcus milleri* [1]
<24> *Streptococcus mutans* [1]
<25> *Streptococcus mitior* [1]
<26> *Streptococcus salivarius* [1]
<27> *Streptococcus sanguis* [1]
<28> *Streptococcus thermophilus* [1]
<29> *Pediococcus acidilactici* [1]
<30> *Pediococcus damnosus* [1]
<31> *Pediococcus dextrinicus* [1]
<32> *Pediococcus halophilus* [1]
<33> *Pediococcus inopinatus* [1]
<34> *Pediococcus pentosaceus* [1]
<35> *Bifidobacterium bifidum* [1]
<36> *Staphylococcus aureus* [1]
<37> *Staphylococcus epidermidis* (ATCC 14990 [6]) [1, 6, 23]
<38> *Staphylococcus urea* [1]
<39> *Acholeplasma laidlawii* [1]
<40> *Mycoplasma sp.* [1]
<41> *Bacillus caldolyticus* [1]
<42> *Bacillus subtilis* [1]
<43> *Butyrivibrio vibrisolvens* [1]
<44> *Selenomonas ruminantium* [1]
<45> *Actinomyces viscosus* [1]
<46> *Rothia dentocariosa* [1]
<47> *Lactococcus lactis* (strain 760 [7]) [1, 7, 22, 25]
<48> *Fundulus heteroclitus* [3]
<49> *Thermoanaerobacter ethanolicus* [39]
<50> *Homarus americanus* [8]

<51> *Mus musculus* [9]
<52> *Ipomoea batatas* [10, 16]
<53> *Alcaligenes eutrophus* [11]
<54> *Solanum tuberosum* [12]
<55> *Homo sapiens* [13, 40]
<56> *Thermus caldophilus* (GK24 [14]) [14]
<57> *Thermotoga maritima* [15, 28]
<58> *Capsella bursa-pastoris* (L. Med [17]) [17]
<59> *Lactuca sativa* [17, 18]
<60> *Thlaspi arvense* [17]
<61> *Cardamine hirsuta* [17]
<62> *Spinacia oleracea* [17]
<63> *Raphanus sativus* [17]
<64> *Acinetobacter calcoaceticus* [21]
<65> *Bifidobacterium longum* [26, 38]
<66> *Methanococcus jannaschii* [27]
<67> *Plasmodium falciparum* [29]
<68> *Bacillus stearothermophilus* [37]
<69> *Molinema dessetaw* [30]
<70> *Rana perezi* [31]
<71> *Rana calamita* [31]
<72> *Saduria entomon* [32]
<73> *Salmo salar* [33]
<74> *Oncorhynchus mykiss* [33]
<75> *Coregonus sardinella* [33]
<76> *Acipenser baerri* [33]
<77> *Acipenser ruthenus* [33]
<78> *Misgurnus fossilis* [33]
<79> *Cyprinus carpio* [33]
<80> *Carassius auratus* [33]
<81> *Danio rerio* [33]
<82> *Coregonus autumnalis* [33]
<83> *Sphyraena argentea* [34]
<84> *Sphyraena lucasana* [34]
<85> *Sphyraena idiastes* [34]
<86> *Sphyraena ensis* [34]
<87> *Sphyraena helleri* [34]
<88> *Sphyraena barracuda* [34]
<89> *Gallus gallus* [35, 42]
<90> *duck* (ε-crystallin [35, 42]) [35, 42]
<91> *Nocardia asteroides* [36]

3 Reaction and Specificity

Catalyzed reaction

(S)-lactate + NAD^+ = pyruvate + NADH (<20>, reduction of pyruvate follows an Theorell-Chance mechanism [4]; <90>, mechanism [42])

Reaction type

oxidation
redox reaction
reduction

Natural substrates and products

S ? <47, 59> (<59>, the enzyme is a component of the system regulating the cellular pH and/or controlling the concentration of reducing equivalents in the cytoplasm of leaf cells [18]; <47>, the enzyme is unlikely to catalyze lactate oxidation at an appreciable rate under physiological conditions [25]) [18, 25]

P ?

S L-lactate + NAD^+ <70, 71> (<70, 71>, the enzyme plays two important roles in heart metabolism, it catalyzes pyruvate reduction, mainly at the beginning of effort, or during hypoxia, and also catalyzes the oxidation of lactate released into the blood by other tissues, such as skeletal muscle, which would be used as substrate fuel by the heart mainly during steady-state exercise or during recuperation [31]) (Reversibility: r <70, 71> [31]) [31]

P pyruvate + NADH

S pyruvate + NADH <52, 56, 57, 66, 70, 71> (<57>, last step in anaerobic glycolysis [15]; <52>, enzyme activity increases in response to infection by black-rot fungus but decreases in response to cutting [16]; <66>, key enzyme in anaerobic metabolism which converts pyruvate to lactate [27]) (Reversibility: ? <52, 56, 57, 66> [14, 15, 16, 27]; r <70, 71> [31]) [14, 15, 16, 27, 31]

P L-lactate + NAD^+

Substrates and products

S (S)-lactate + NAD^+ <1, 4, 5, 6, 16, 37, 47, 50, 51, 52, 56, 58, 59, 64, 69, 70, 71, 72> (<37>, poor reaction [6]; <4, 5>, the maximal velocity of lactate oxidation is about 10% of pyruvate reduction [19]; <47>, inactive towards D-lactate [25]; <69>, pyruvate reduction is the favored reaction [30]) (Reversibility: r <1, 4, 5, 6, 16, 37, 47, 50, 51, 52, 53, 56, 58, 59, 69, 70, 71, 72> [2, 6, 7, 8, 9, 10, 11, 14, 16, 17, 18, 19, 24, 25, 30, 31, 32]; ? <64> [21]) [2, 6, 7, 8, 9, 10, 14, 16, 17, 18, 21, 24, 25, 30, 31, 32]

P pyruvate + NADH <16, 37, 47, 50, 51, 52, 56, 58, 59, 64, 69, 70, 71, 72> [2, 6, 7, 8, 9, 10, 14, 16, 17, 18, 21, 24, 25, 30, 31, 32]

S 2-oxobutyrate + NADH <2, 49, 53, 54> (<53>, 107% of the activity with pyruvate [11]; <2>, 3% of the activity with pyruvate [41]) (Reversibility: ? <2, 49, 53, 54> [11, 12, 39, 41]) [11, 12, 39, 41]

P 2-hydroxybutyrate + NAD^+

S 2-oxoglutarate + NADH <49, 53> (<53>, 79.6% of the activity with pyruvate [11]) (Reversibility: ? <49, 53> [11, 39]) [11, 39]
P 2-hydroxyglutarate + NAD^+
S 2-oxopentanoate + NADH <53> (<53>, 125% of the activity with pyruvate [11]) (Reversibility: ? <53> [11]) [11]
P 2-hydroxypentanoate + NAD^+
S 3-fluoropyruvate + NADH <53, 90> (<53>, 33.6% of the activity with pyruvate [11]) (Reversibility: ? <53, 90> [11, 42]) [11, 42]
P ?
S 3-methyl-2-oxobutanoate + NADH <53> (<53>, 28.5% of the activity with pyruvate [11]) (Reversibility: ? <53> [11]) [11]
P 2-hydroxy-3-methylbutanoate + NAD^+
S 3-methyl-2-oxopentanoate + NADH <53> (<53>, 5.3% of the activity with pyruvate [11]) (Reversibility: ? <53> [11]) [11]
P 2-hydroxy-3-methylpentanoate + NAD^+
S 4-methyl-2-oxopentanoate + NADH <53, 67> (<53>, 39% of the activity with pyruvate [11]) (Reversibility: ? <53, 67> [11, 29]) [11, 29]
P 2-hydroxy-4-methylpentanoate + NAD^+
S bromopyruvate + NADH <90> (Reversibility: ? <90> [42]) [42]
P 5-bromo-2-hydroxypropanoate + NAD^+
S glyoxylate + NADH <54, 59> (Reversibility: ? <54, 59> [12, 18]) [12, 18]
P glycolate + NAD^+
S hydroxypyruvate + NADPH <54, 59> (Reversibility: ? <54, 59> [12, 18]) [12, 18]
P 2,3-dihydroxypropanoate + NAD^+
S phenylpyruvate + NADH <2, 67> (<2>, 28% of the activity with pyruvate [41]) (Reversibility: ? <2, 67> [29, 41]) [29, 41]
P 2-hydroxy-3-phenylpropanoate + NAD^+
S pyruvate + NADH <1-47, 48, 49, 50, 51, 52, 53, 54, 55, 56, 57, 58, 59, 66, 67, 69, 70, 71, 72, 73, 74, 75, 76, 77, 78, 79, 80, 81, 82, 83, 84, 85, 86, 87, 88, 89, 91> (<4, 5>, the maximal velocity of lactate oxidation is only 10% of pyruvate reduction [19]; <69>, pyruvate reduction is the favored reaction [30]) (Reversibility: r <1, 4, 5, 6, 16, 37, 47, 50, 51, 52, 53, 58, 59, 69, 70, 71, 72> [2, 6, 7, 8, 9, 10, 11, 16, 17, 18, 19, 24, 25, 30, 31, 32]; ir <37> [23]; ? <1-47, 49, 54, 55, 57, 66, 67, 73, 74, 75, 76, 77, 78, 79, 80, 81, 82, 83, 94, 85, 86, 87, 88, 91> [1, 4, 5, 12, 13, 15, 20, 22, 27, 28, 29, 33, 34, 36, 39, 40, 41]) [1, 2, 3, 4, 5, 6, 7, 8, 9, 10, 11, 12, 13, 14, 15, 16, 17, 18, 19, 20, 22, 23, 24, 25, 27, 28, 29, 30, 31, 32, 33, 34, 35, 36, 39, 40, 41, 42]
P (S)-lactate + NAD^+ <1-47, 48, 49, 50, 51, 52, 53, 54, 55, 56, 57, 58, 59, 66, 67, 69, 70, 71, 72, 73, 74, 75, 76, 77, 78, 79, 80, 81, 82, 83, 84, 85, 86, 87, 88, 89, 91> [1, 2, 3, 4, 5, 6, 7, 8, 9, 10, 11, 12, 13, 14, 15, 16, 17, 18, 19, 20, 22, 23, 24, 25, 27, 28, 29, 30, 31, 32, 33, 34, 35, 36, 39, 40, 41, 42]
S pyruvate + NADPH <4, 5, 49, 54> (<4>, 15% of the activity with NADH [19]; <5>, 20% of the activity with NADH [19]; <49>, 34% of the activity with NADH [39]) (Reversibility: ? <4, 5, 49, 54> [12, 19, 39]) [12, 19, 39]
P (S)-lactate + $NADP^+$
S pyruvate ethyl ester + NADH <90> (Reversibility: ? <90> [42]) [42]

P 2-hydroxypropanoate ethyl ester

S pyruvate methyl ester + NADH <90> (Reversibility: ? <90> [42]) [42]

P 2-hydroxypropanoate methyl ester + NAD^+

S Additional information <1-47, 90> (<4, 5, 15, 16, 17, 18, 19, 20, 21, 22, 23, 24, 25, 26, 27, 28, 31, 35, 37, 43, 45, 46, 47>, fructose 1,6-diphosphate activated enzymes are virtually nonreversible, enzymes which do not require fructose 1,6-diphosphate catalyze reversible reactions [1]; <90>, in addition of lactate dehydrogenase activity, the ε-crystallin also possesses the enzymatic activity of malate dehydrogenase [35]) [1, 35]

P ?

Inhibitors

3-aminopyridine adenine dinucleotide <90> (<90>, competitive versus NAD^+ and noncompetitive versus L-lactate [42]) [42]

3-fluoropyruvate <53> [11]

3-phosphoglycerate <59> [18]

6-phosphogluconate <4> [5]

ADP <4, 6, 24, 37, 41, 44, 52, 54, 59> (<37>, competitive with NADH [6]; <52>, isoenzyme I and II, inhibition is reduced by $MgCl_2$ [16]) [1, 5, 6, 12, 16, 18]

AMP <37, 52, 59> (<52>, isoenzyme I and II, inhibition is reduced by $MgCl_2$ [16]) [6, 16, 18]

ATP <4, 6, 20, 34, 37, 39, 44, 45, 46, 52, 54, 59> (<54>, at neutral or alkaline pH ATP behaves as a weak competitive inhibitor, potent inhibitor at acid pH values [12]; <52>, isoenzyme I and II, inhibition is reduced by $MgCl_2$ [16]; <59>, competitive with respect to NADH at pH 7.0 and at pH 6.2 [18]) [1, 4, 5, 6, 12, 16, 18]

Ag^+ <42, 53> [1, 11]

Ala <59> [18]

Cd^{2+} <2, 4, 53> (<53>, partial [11]; <4>, at high concentration [1]; <2>, 0.1 mM and 1.0 mM, weak inhibition [41]) [1, 11, 41]

Co^{2+} <2> (<2>, 0.1 mM and 1.0 mM, weak inhibition [41]) [41]

Cu^{2+} <2, 4, 34, 53, 91> (<4, 34>, $CuSO_4$ [1, 5]; <2>, 0.1 mM and 1.0 mM, weak inhibition [41]; <91>, $CuCl_2$ [36]) [1, 5, 11, 36, 41]

D-lactate <90> (<90>, dead-end inhibitor, competitive inhibitor versus L-lactate [42]) [42]

DTNB <50> [8]

GTP <4, 37, 54> [5, 6, 12]

Gly <59> [18]

HEPES buffer <2> [41]

Hg^{2+} <4, 34, 53, 64, 69> (<4, 34, 64>, $HgCl_2$ [1, 5, 21]) [1, 5, 11, 21, 30]

L-lactate <90> (<90>, substrate inhibition is uncompetitive [42]) [42]

MES buffer <2> [41]

Mn^{2+} <4> (<4>, at low pH-values [19]) [19]

NAD^+ <4, 34, 44, 52, 53, 54> (<52>, competitive with respect to NADH, reduction of pyruvate [16]; <54>, product inhibition [12]) [1, 5, 11, 12, 16]

NADH <54> (<54>, competitive with respect to NAD^+ [12]) [12]

NADPH <49> [39]
Ni^{2+} <53> (<53>, partial [11]) [11]
Ser <59> [18]
Tris-HCl <2> [41]
Tris/maleate buffer <4> (<4>, at pH 5.4 [5]) [5]
Zn^{2+} <53> [11]
α-NAD^+ <90> (<90>, noncompetitive inhibitor versus β-NAD^+ [42]) [42]
citrate <59> [18]
citrate/phosphate buffer <4> (<4>, at pH 5.4 [5]) [5]
dihydroxyacetone phosphate <4> [5]
fructose 1,6-diphosphate <56, 59> (<56>, activation at low concentrations, inhibition at high concentrations [14]) [14, 18]
glucose 6-phosphate <59> [18]
iodoacetamide <1, 4> (<1, 4>, slight [1]) [1]
iodoacetate <1, 4, 69> (<5>, no effect [24]) [24, 30]
isocitrate <59> [18]
lactate analogs <42> [1]
nicotinic acid adenine dinucleotide <90> (<90>, competitive versus NAD^+ and noncompetitive versus L-lactate [42]) [42]
oxalate <34, 53, 54, 59, 64, 69> (<69>, noncompetitive with pyruvate, competitive with lactate [30]) [1, 11, 12, 18, 21, 30]
oxaloacetate <42, 53> [1, 11]
oxamate <1, 6, 20, 24, 34, 37, 53, 69, 90> (<69>, competitive with pyruvate, noncompetitive with lactate [30]; <90>, dead-end inhibitor, competitive inhibitor versus pyruvate [42]) [1, 4, 6, 11, 30, 42]
p-chloromercuribenzoate <20, 42, 53, 69> [1, 4, 11, 30]
p-hydroxymercuribenzoate <1, 2, 4, 34, 50> (<1>, slight [24]) [1, 5, 8, 24, 41]
phenylglyoxal <48> [3]
phosphate <47, 56> (<56>, slight activation of non-activated enzyme, inhibition of fructose 1,6-diphosphate activated enzyme [14]) [14, 25]
phosphoenolpyruvate <52, 53, 54> [11, 12, 16]
pyruvate <47, 55, 56, 67, 72, 90> (<47>, above 20 mM [25]; <67>, substrate inhibition in wild type enzyme, lower substrate inhibition in mutant S163L [29]; <72>, at high concentrations [32]; <55>, above 0.3 mM [40]; <90>, substrate inhibition is uncompetitive [42]; <56>, strong substrate inhibition at high concentrations, fructose 1,6-diphosphate activated enzyme [14]) [14, 25, 29, 32, 40, 42]
sodium phosphate <4> [5]
tartronate <90> (<90>, dead-end inhibitor, competitive inhibitor versus L-lactate [42]) [42]
thionicotinamide adenine dinucleotide <90> (<90>, competitive versus NAD^+ and noncompetitive versus L-lactate [42]) [42]
Additional information <67> (<67>, lack of substrate inhibition [29]) [29]

Cofactors/prosthetic groups

3-acetylpyridine adenine dinucleotide <67, 90> (<67>, wild-type and S163 L mutant protein bind 3-acetylpyridine adenine dinucleotide more tightly than they bind NADH [29]; <90>, alternative coenzyme [42]) [29, 42]
FMN <64> (<64>, contains non-covalently bound FMN as prosthetic group [21]) [21]
NAD$^+$ <1, 4, 5, 6, 16, 37, 47, 50, 51, 52, 53, 56, 58, 59, 64, 69, 70, 71, 72> (<1, 4, 5, 6, 16, 37, 47, 50, 51, 52, 53, 56, 58, 59, 64, 69, 70, 71, 72> coenzyme [2, 6, 7, 8, 9, 10, 11, 14, 16, 17, 18, 19, 21, 24, 25, 30, 31, 32]) [2, 6, 7, 8, 9, 10, 11, 14, 16, 17, 18, 19, 21, 24, 25, 30, 31, 32]
NADH <1-47, 48, 49, 50, 51, 52, 53, 54, 55, 56, 57, 58, 59, 66, 67, 69, 70, 71, 72, 73, 74, 75, 76, 77, 78, 79, 80, 81, 82, 83, 84, 85, 86, 87, 88, 89, 91> (<1-47, 48, 49, 50, 51, 52, 53, 54, 55, 56, 57, 58, 59, 66, 67, 69, 70, 71, 72, 73, 74, 75, 76, 77, 78, 79, 80, 81, 82, 83, 84, 85, 86, 87, 88, 89, 91> coenzyme [1, 2 , 3, 4, 5, 6, 7, 8, 9, 10, 11, 12, 13, 14, 15, 16, 17, 18, 19, 20, 22, 23, 24, 25, 27, 28, 29, 30, 31, 32, 33, 34, 35, 36, 39, 40, 41, 42]) [1, 2, 3, 4, 5, 6, 7, 8, 9, 10, 11, 12, 13, 14, 15, 16, 17, 18, 19, 20, 22, 23, 24, 25, 27, 28, 29, 30, 31, 32, 33, 34, 35, 36, 39, 40, 41, 42]
NADPH <4, 5, 49, 54> (<4, 5, 49, 54> coenzyme [12, 19, 39]) [12, 19, 39]
nicotinamide hypoxanthine dinucleotide <90> (<90>, alternative coenzyme [42]) [42]
thio-NAD$^+$ <90> (<90>, alternative coenzyme [42]) [42]

Activating compounds

2-glycerophosphate <54> (<54>, activation [12]) [12]
3-phosphoglycerate <54> (<54>, activation [12]) [12]
5-phosphoribosyl 1-diphosphate <56> (<56>, activates [14]) [14]
fructose 1,6-diphosphate <4, 5, 15, 16, 17, 18, 19, 20, 21, 22, 23, 24, 25, 26, 27, 28, 31, 35, 37, 43, 45, 46, 47, 49, 56, 65> (<4, 5, 15, 16, 17, 18, 19, 20, 21, 22, 23, 24, 25, 26, 27, 28, 31, 35, 37, 43, 45, 46, 47, 49, 56, 65>, L-lactate dehydrogenase form which is activated by fructose 1,6-diphosphate [1, 4, 5, 6, 14, 19, 20, 22, 23, 25, 38, 39]; <1, 2, 3, 4, 6, 7, 8, 9, 10, 11, 12, 13, 14, 15, 17, 24, 28, 29, 30, 32, 33, 34, 35, 36, 40, 41, 42, 44>, L-lactate dehydrogenase form which is not activated by fructose 1,6-diphosphate [1]; <53>, not dependent on fructose 1,6-diphosphate [11]; <16>, dependent on [2]; <37>, concentration required for 50% maximal activity is about 0.15 mM [6]; <56>, maximal activation is about: 0.03 mM at pH 6.0, 0.2 mM at pH 7.1, 2 mM at pH 8.0 and 7 mM at pH 9.0 [14]; <18>, binding of fructose 1,6-diphosphate induces a conformational change in the enzyme which leads to increased activity, without association of enzyme subunits or dimers [20]; <47>, binding of fructose 1,6-diphosphate induces a conformational change in the enzyme, producing a form with reduced protein fluorescence and increased activity towards pyruvate reduction [22]; <49>, at pH 7.5 lactate dehydrogenase activity is absent without fructose 1,6-diphosphate, from pH 5.0 to pH 7.0 lactate dehydrogenase is present without fructose 1,6-diphosphate but is greater in the presence of the activator [39]) [1, 2, 4, 5, 6, 11, 14, 19, 20, 22, 23, 25, 38, 39]
fructose 2,6-diphosphate <56> (<56>, activates [14]) [14]
glucose 1,6-diphosphate <20, 56> (<20, 56>, activates [4, 14]) [4, 14]

phosphate <56> (<56>, slight activation of non-activated enzyme, inhibition of fructose 1,6-diphosphate activated enzyme [14]) [14]
tagatose 1,6-diphosphate <20> (<20>, stimulates [4]) [4]

Metals, ions

Ca^{2+} <2, 4, 5, 91> (<4>, stimulation is as effective as with Mn^{2+}, Co^{2+} and Cd^{2+} [19]; <5>, 50% of the stimulation with Mn^{2+}, Co^{2+} or Cd^{2+}, higher concentrations required [19]; <91>, increases activity [36]; <2>, 1 mM, almost 2fold stimulation [41]) [19, 36, 41]
Cd^{2+} <4, 5> (<4, 5>, stimulation [19]) [19]
Co^{2+} <4, 5, 18, 91> (<4,5,18>, stimulation [19,20]; <18>, activation involves association of enzyme dimers, followed by ligand binding [20]; <91>, increases activity [36]) [19, 20, 36]
Fe^{2+} <91> (<91>, increases activity [36]) [36]
Mn^{2+} <4, 5, 18, 91> (<4, 5, 18>, stimulation [19, 20]; <47>, no effect [25]; <91>, increases activity [36]) [19, 20, 36]
Zn^{2+} <2> (<2>, 1 mM, almost 2fold stimulation [41]) [41]

Turnover number (min^{-1})

19.8 <67> (4-methyl-2-oxopentanoate, <67>, wild-type enzyme [29]) [29]
42 <90> (ε-NAD^+) [42]
66 <90> (nicotinamide hypoxanthine dinucleotide, <89>, enzyme from muscle [42]) [42]
84 <90> (3-acetylpyridine adenine dinucleotide) [42]
198 <89> (NAD^+, <89>, enzyme from muscle [42]) [42]
318 <90> (ε-NAD^+) [42]
330 <90> (nicotinamide hypoxanthine dinucleotide) [42]
372 <90> (ε-NAD^+, <90>, enzyme from heart [42]) [42]
414 <90> (NAD^+) [42]
960 <89> (3-acetylpyridine adenine dinucleotide, <89>, enzyme form heart and muscle [42]) [42]
1128 <90> (nicotinamide hypoxanthine dinucleotide, <90>, enzyme from heart [42]) [42]
2370 <67> (phenylpyruvate, <67>, wild-type enzyme [29]) [29]
3600 <89> (NAD^+, <89>, enzyme from heart [42]) [42]
5880 <89> (fluoropyruvate) [42]
10560 <90> (pyruvate ethyl ester) [42]
10800 <67> (pyruvate, <67>, wild-type enzyme [29]) [29]
12240 <90> (pyruvate) [42]
13620 <90> (pyruvate methyl ester) [42]
16140 <90> (bromopyruvate) [42]
38580 <2> (pyruvate) [41]
140000 <50> (NADH) [8]
Additional information <55, 67> (<67>, turnover-numbers for mutant enzymes with enlarged loop [29]) [29, 40]

Specific activity (U/mg)

0.714 <52> (<52>, isoenzyme I [10]) [10]

1.87 <52> (<52>, isoenzyme II [10]) [10]
18.5 <54> [12]
27.6 <64> [21]
80 <51> (<51>, isoenzyme X [9]) [9]
80.27 <69> [30]
106 <55> [40]
109 <47> [25]
216.9 <91> [36]
419 <48> [3]
458.5 <55> [13]
510 <72> [32]
589 <49> [39]
824.6 <53> [11]
1422 <16> [2]
2030 <5> [19]
2119 <2> (<2>, recombinant enzyme [41]) [41]
2220 <37> [6]
2292 <4> [5]
2320 <4> [19]
2350 <6> [24]
2460 <1> [24]
Additional information <1, 4, 5, 18, 20, 24, 39, 42, 45, 47, 50, 52, 57, 59> [1, 7, 8, 16, 18, 28]

K_m-Value (mM)

0.01 <4> (NADH, <4>, fructose 1,6-diphosphate has no effect [5]) [5]
0.0125 <52> (NADH, <52>, pH 7.3, isoenzyme I and isoenzyme II [16]) [16]
0.0142 <86> (NADH) [34]
0.0143 <88> (NADH) [34]
0.015 <56> (NAD^+) [14]
0.016 <71, 87, 91> (pyruvate, <71>, 20°C, pH 7 [31]) [31, 34, 36]
0.018 <67> (NADH, <67>, wild-type enzyme [29]) [29]
0.018 <57, 71> (pyruvate, <57>, octameric enzyme form [28]; <71>, 15°C, pH 7 [31]) [28, 31]
0.0186 <84> (NADH) [34]
0.0189 <85> (NADH) [34]
0.019 <57> (pyruvate, <57>, tetrameric enzyme form [28]) [28]
0.0196 <83> (NADH) [34]
0.02 <6, 34> (NADH, <34>, pH 5.4 [1]) [1]
0.024 <16> (NADH) [2]
0.025 <71> (pyruvate, <71>, 15°C, pH 7.4 [31]) [31]
0.027 <70> (pyruvate, <70>, 15°C, pH 7 [31]) [31]
0.03 <55> (pyruvate) [40]
0.032 <24> (NADH) [1]
0.033 <5> (NADH) [1]
0.033 <70> (pyruvate, <70>, 20°C, pH 7 [31]) [31]
0.039 <71> (pyruvate, <71>, 20°C, pH 7.4 [31]) [31]

0.04 <34> (NADH, <34>, pH 7.4 [1]) [1]
0.044 <70> (pyruvate, <70>, 15°C, pH 7.4 [31]) [31]
0.045 <4> (NADH) [1]
0.045 <41> (NADH) [1]
0.054 <70> (pyruvate, <70>, 20°C, pH 7.4 [31]) [31]
0.057 <71> (pyruvate, <71>, 15°C, pH 8 [31]) [31]
0.061 <50> (NADH) [8]
0.061 <70, 71> (pyruvate, <70,71>, 30°C, pH 7 [31]) [31]
0.065 <56> (NAD^+) [14]
0.065 <67> (NADH, <67>, mutant enzyme S163L [29]) [29]
0.069 <1> (NADH) [1]
0.078 <71> (pyruvate, <71>, 20°C, pH 8 [31]) [31]
0.08 <47, 49> (NADH, <47>, pH 6.9, 90 mM Tris-maleate buffer, pH 6.9, 0.5 mM fructose 1,6-diphosphate [25]) [1, 25, 39]
0.08 <72> (pyruvate, <72>, at 0°C and at 5°C [32]) [32]
0.083 <64> ((S)-lactate) [21]
0.086 <71> (pyruvate, <71>, 30°C, pH 7.4 [31]) [31]
0.11 <70> (pyruvate, <70>, 15°C, pH 8 [31]) [31]
0.117 <70> (pyruvate, <70>, 30°C, pH 7.4 [31]) [31]
0.12 <72> (pyruvate, <72>, 10°C [32]) [32]
0.13 <72> (pyruvate, <72>, 15°C [32]) [32]
0.14 <72> (pyruvate, <72>, 20°C [32]) [32]
0.145 <71> (pyruvate, <71>, 30°C, pH 8 [31]) [31]
0.15 <72> (pyruvate, <72>, 25°C [32]) [32]
0.171 <70> (pyruvate, <70>, 20°C, pH 8 [31]) [31]
0.18 <69> (NAD^+) [30]
0.18 <72> (pyruvate, <72>, 30°C [32]) [32]
0.19 <88> (pyruvate) [34]
0.2 <54> (pyruvate) [12]
0.22 <86> (pyruvate) [34]
0.23 <16> (NAD^+) [2]
0.24 <50, 87> (pyruvate) [8, 34]
0.25 <69> (NADH) [30]
0.25 <2, 49, 58> (pyruvate, <49>, with 1.0 mM fructose 1,6-diphosphate [39]) [17, 39, 41]
0.26 <84> (pyruvate) [34]
0.312 <70> (pyruvate, <70>, 30°C, pH 7.4 [31]) [31]
0.33 <85> (pyruvate) [34]
0.34 <52, 69, 83> (pyruvate, <52>, pH 7.3, isoenzyme I [16]) [16, 30, 34]
0.35 <59> (pyruvate, <59>, pH 6.2 [18]) [18]
0.398 <55> (pyruvate) [13]
0.4 <20> (NADH, <20>, in presence of fructose 1,6-diphosphate [4]) [1, 4]
0.42 <52> (pyruvate, <52>, pH 7.3, isoenzyme II [16]) [16]
0.5 <52> (pyruvate, <52>, pH 6.5, isoenzyme II [16]) [16]
0.515 <71> ((S)-lactate, <71>, 15°C, pH 8 [31]) [31]
0.517 <70> ((S)-lactate, <70>, 20°C, pH 8 [31]) [31]
0.537 <70> ((S)-lactate, <70>, 20°C, pH 7.4 [31]) [31]

0.541 <70> ((S)-lactate, <70>, 15°C, pH 8 [31]) [31]
0.567 <71> ((S)-lactate, <71>, 15°C, pH 7.4 [31]) [31]
0.616 <71> ((S)-lactate, <71>, 20°C, pH 8 [31]) [31]
0.62 <70> ((S)-lactate, <70>, 15°C, pH 7.4 [31]) [31]
0.641 <71> ((S)-lactate, <71>, 20°C, pH 7.4 [31]) [31]
0.735 <70> ((S)-lactate, <70>, 15°C, pH 7 [31]) [31]
0.803 <71> ((S)-lactate, <71>, 20°C, pH 7 [31]) [31]
0.822 <70> ((S)-lactate, <70>, 30°C, pH 7 [31]) [31]
0.862 <71> ((S)-lactate, <71>, 15°C, pH 7 [31]) [31]
0.9 <42> (NAD^+) [1]
0.909 <71> ((S)-lactate, <71>, 30°C, pH 7 [31]) [31]
0.94 <71> ((S)-lactate, <71>, 30°C, pH 7.4 [31]) [31]
0.95 <52> (NAD^+, <52>, pH 8.2, isoenzyme II [16]) [16]
0.966 <70> ((S)-lactate, <70>, 20°C, pH 7 [31]) [31]
1 <4> (pyruvate, <4>, in presence of 0.02 mM fructose 1,6-diphosphate [5]) [5]
1.016 <70> ((S)-lactate, <70>, 30°C, pH 7.4 [31]) [31]
1.1 <50> (NAD^+) [8]
1.15 <20> (pyruvate, <20>, in absence of fructose 1,6-diphosphate [4]) [4]
1.191 <71> ((S)-lactate, <71>, 30°C, pH 8 [31]) [31]
1.33 <70> ((S)-lactate, <70>, 30°C, pH 7 [31]) [31]
1.5 <47, 52> (pyruvate, <52>, pH 6.5 [16]; <47>, pH 6.9, 90 mM Tris-maleate buffer, pH 6.9, 0.5 mM fructose 1,6-diphosphate [25]) [16, 25]
1.75 <16> (pyruvate) [2]
2.2 <49> (pyruvate, <49>, in absence of fructose 1,6-diphosphate [39]) [39]
2.4 <47> (NAD^+, <47>, in presence of 1 mM fructose 1,6-diphosphate [25]) [25]
2.4 <47> (NAD^+, <47>, in presence of 1.0 mM fructose 1,6-diphosphate [1]) [1]
2.5 <69> ((S)-lactate) [30]
3 <41> (pyruvate) [1]
5 <34> (NAD^+) [1]
7.7 <20> (pyruvate, <20>, in presence of fructose 1,6-diphosphate [4]) [4]
8 <47> (NADH, <47>, without fructose 1,6-diphosphate [1, 25]) [1, 25]
8 <54> (hydroxypyruvate) [12]
13 <58> ((S)-lactate) [17]
15 <47> (pyruvate) [1]
17 <52> ((S)-lactate, <52>, pH 8.2, isoenzyme II [16]) [16]
22 <52> ((S)-lactate, <52>, pH 8.8, isoenzyme I [16]) [16]
25 <54> (glyoxylate) [12]
30 <42> ((S)-lactate) [1]
44 <20> (NADH, <20>, in absence of fructose 1,6-diphosphate [4]) [4]
55 <6> ((S)-lactate) [1]
67 <34> ((S)-lactate) [1]
77 <54> (2-oxobutanoate) [12]
77 <20> (pyruvate) [1]
90.04 <72> ((S)-lactate) [32]

100 <47> ((S)-lactate, <47>, in presence of 1 mM fructose 1,6-diphosphate [25]) [1, 25]
109 <16> ((S)-lactate) [2]
280 <5> ((S)-lactate) [19]
360 <1> ((S)-lactate) [1]
420 <4> ((S)-lactate) [19]
1100 <50> ((S)-lactate) [8]
Additional information <4, 5, 67, 73-82> (<4, 5>, influence of pH [19]; <67>, K_M-values for mutant enzymes with enlarged loop [29]; <73-82>, K_M-values at position of temperatur minimum [33]) [19, 29, 33]

pH-Optimum

4.5 <4, 5> (<5>, reduction of pyruvate [1]; <4>, reduction of pyruvate, with addition of fructose 1,6-diphosphate and Mn^{2+} [1]) [1, 19]
4.5-6 <34> (<34>, reduction of pyruvate [1]) [1]
5 <2> [41]
5-6.2 <45> (<45>, reduction of pyruvate [1]) [1]
5-7 <47> (<47>, reduction of pyruvate, in presence of fructose 1,6-diphosphate [7]) [7]
5.3 <1> (<1>, reduction of pyruvate [1]) [1]
5.4 <4> (<4>, sodium acetate buffer [5]) [5]
5.5 <4, 16> (<4>, reduction of pyruvate [1]) [1, 2]
5.5-7 <24> (<24>, reduction of pyruvate [1]) [1]
5.6 <37> (<37>, reduction of pyruvate [1, 6]) [1, 6]
6 <6, 42> (<6, 42>, reduction of pyruvate [1]) [1]
6.5 <39, 49, 54, 72> (<39>, reduction of pyruvate, without fructose 1,6-diphosphate [1]; <54>, reduction of oxo acids with NADH or NADPH [12]; <72>, reduction of pyruvate [32]) [1, 12, 32, 39]
6.8 <69> (<69>, reduction of pyruvate [30]) [30]
6.9 <47> (<47>, in presence of fructose 1,6-diphosphate [25]) [25]
7 <24, 53, 58> (<24>, reduction of pyruvate, addition of fructose 1,6-diphosphate [1]; <53>, reduction of pyruvate [11]; <58>, reduction of pyruvate [17]) [1, 11, 17]
7.2 <42> (<42>, oxidation of lactate [1]) [1]
7.3 <52> (<52>, reduction of pyruvate, isoenzyme I and II [16]) [16]
7.5 <6, 50, 64> (<6>, oxidation of lactate [1]; <50>, reduction of pyruvate and oxidation of lactate [8]) [1, 8, 21]
7.6 <34> (<34>, oxidation of lactate [1]) [1]
7.8 <1> (<1>, oxidation of lactate [1]) [1]
7.8-8.8 <52> (<52>, oxidation of lactate, isoenzyme I and II [16]) [16]
8 <47> (<47>, reduction of pyruvate, in absence of fructose 1,6-diphosphate [7]) [7]
8.2 <36, 47> (<36>, oxidation of lactate [1]; <47>, in absence of fructose 1,6-diphosphate [25]) [1, 25]
8.5 <45> (<45>, oxidation of lactate [1]) [1]
8.6 <69> (<69>, oxidation of lactate [30]) [30]
9 <58> (<58>, oxidation of lactate [17]) [17]

9.1 <72> (<72>, oxidation of lactate [32]) [32]
9.5 <91> (<91>, glycine-NaOH buffer [36]) [36]
9.8 <53> (<53>, oxidation of lactate [11]) [11]

pH-Range

4.3-8 <16> (<16>, no activity below pH 4.3 and above pH 8.0 [2]) [2]
5-7 <49> (<49>, about 55% of maximal activity at pH 5.0 and at pH 7.0 [39]) [39]
5.2-7.2 <37> (<37>, pH 5.2: about 35% of maximal activity, pH 7.2: about 50% of maximal activity [6]) [6]
5.3-8 <53> (<53>, pH 5.3: about 50% of maximal activity, pH 8.0: about 60% of maximal activity, reduction of pyruvate [11]) [11]
6-8 <69> (<69>, pH 6.0: 85% of maximal activity, pH 8: 45% of maximal activity, reduction of pyruvate [30]) [30]
7.7-9.5 <58> (<58>, pH 7.7: about 60% of maximal activity, pH 9.5: about 30% of maximal activity, reduction of pyruvate [17]) [17]
8-9.6 <69> (<69>, pH 8.0: 60% of maximal activity, pH 9.6: 45% of maximal activity, oxidation of lactate [30]) [30]
8.5-10.7 <53> (<53>, pH 8.5: about 60% of maximal activity, pH 10.7: about 70% of maximal activity, oxidation of lactate [11]) [11]

Temperature optimum (°C)

39 <37> [6]
42-45 <53> [11]
50 <91> [36]

Temperature range (°C)

37-60 <91> (<91>, 37°C: about 85% of maximal activity, 60°C: about 90% of maximal activity [36]) [36]

4 Enzyme Structure

Molecular weight

65000-67000 <53> (<53>, gel filtration [11]) [11]
113000 <69> (<69>, gel filtration [30]) [30]
125000-138000 <2> (<2>, gel filtration [41]) [41]
130000 <37> (<37>, sucrose density gradient centrifugation [6]; <37>, tetramer, sucrose density gradient centrifugation in presence of fructose 1,6-diphosphate and NADH [23]) [6, 23]
132000 <4> (<4>, non-denaturing gel electrophoresis [5]) [5]
134000 <49> (<49>, gel filtration [39]) [39]
135000 <4> (<4>, gel filtration [5]) [5]
139000 <48> (<48>, gel filtration [3]) [3]
140000 <20, 57, 72> (<20>, ultracentrifugal studies [4]; <57>, tetramer, gel filtration [28]; <72>, gel filtration [32]) [4, 28, 32]
144000 <16> (<16>, equilibrium sedimentation [2]) [2]
150000 <52> (<52>, disc gel electrophoresis [16]) [16]

187000 <48> (<48>, non-denaturing gel electrophoresis [3]) [3]
269300 <57> (<57>, meniscus depletion experiments [28]) [28]
280000 <57> (<57>, octamer, gel filtration [28]) [28]
280900 <57> (<57>, equilibrium sedimentation [28]) [28]
670000 <64> (<64>, above, gel filtration [21]) [21]

Subunits

? <48, 64> (<48>, x * 33700, SDS-PAGE [3]; <64>, x * 40000, oligomer, SDS-PAGE [21]) [3, 21]
dimer <53, 69> (<53>, 2 * 37000, SDS-PAGE [11]; <69>, 2 * 58000, isoenzyme LDH1 and LDH2, SDS-PAGE [30]) [11, 30]
octamer <57> (<57>, 8 * 35000, enzyme also exists in an tetrameric enzyme form, SDS-PAGE, meniscus depletion experiments in 4 M guanidinium chloride [28]) [28]
tetramer <4, 16, 37, 49, 55, 72> (<49>, 4 * 33500, SDS-PAGE [39]; <4>, 4 * 34000, SDS-PAGE [5]; <55>, 4 * 35000, SDS-PAGE [40]; <16>, 4 * 35900, SDS-PAGE [2]; <37, 72>, 4 * 36000, SDS-PAGE [6,23,32]) [2, 5, 6, 23, 32, 39, 40]
Additional information <91> (<91>, three bands detected after SDS-PAGE with MW of 60000 Da, 66000 Da and 74000 Da [36]) [36]

5 Isolation/Preparation/Mutation/Application

Source/tissue

embryo <73-82> [33]
heart <71, 89> [31, 35, 42]
leaf <58, 59, 60, 61, 62, 63> [17, 18]
lens <89> [35]
liver <55> [13]
muscle <89> [35, 42]
muscle <50> (<50>, tail muscle [8]) [8]
pectoral muscle <72> [32]
root <52> [10, 16]
testis <51, 55> [9, 40]
tuber <54> [12]

Localization

cytosol <58> [17]
integral to plasma membrane <64> [21]

Purification

<1> [1, 24]
<2> [41]
<4> (fructose 1,6-diphosphate-activated enzyme form [5]) [1, 5, 19, 24]
<5> [1, 19, 24]
<6> [24]
<16> [2]
<18> [1]

<20> [1, 4]
<24> [1]
<37> [6]
<39> [1]
<42> [1]
<45> [1]
<47> [1, 7, 25]
<48> (allozymes: LDH-Ba4, LDH-Ba/Bb and LDH-Bb4 [3]) [3]
<49> [39]
<50> [8]
<51> (isoenzyme A4, B4 and X [9]) [9]
<52> (isoenzyme I and II [10,16]; partial [16]) [10, 16]
<53> [11]
<54> [12]
<55> [13, 40]
<57> [28]
<58> (partial [17]) [17]
<59> [18]
<64> [21]
<67> (wild-type and enlarged loop mutants ELM1 and ELM2 [29]) [29]
<69> (two major isoenzymes LDH1 and LDH2 [30]) [30]
<72> [32]
<91> [36]

Crystallization

<57> (2.1 A resolution as a quarternary complex with the cofactor NADH, the allosteric activator fructose-1,6-bisphosphate and the substrate analog oxamate [15]) [15, 28]
<65> (mutant C210S [38]) [26, 38]
<66> (overexpression in E. coli [27]) [27]
<68> (crystal structure of a mutant into which an additional loop has been engineered in order to prevent tetramerization [37]) [37]

Cloning

<2> [41]
<55> (expression in Escherichia coli [40]) [40]
<57> (expression in Escherichia coli [28]) [28]
<84> [34]

Engineering

D8G <84> (<84>, no measurable effect on the K_m value for pyruvate, increased thermal stability [34]) [34]
S163L <67> (<67>, decrease in substrate inhibition, the K_m value for pyruvate is increased substantially and the turnover number is 60% that of wild type [29]) [29]
Additional information <68, 84> (<68>, a mutant into which an additional loop has been engineered in order to prevent tetramerization [37]) [34, 37]

6 Stability

pH-Stability

5-7.5 <47> (<47>, rapid loss of activity below pH 5.0 and above pH 7.5 [7]) [7]
6.5 <37> (<37>, highest stability at [6]) [6]

Temperature stability

50 <2, 37, 49, 69> (<37>, 2 min, 6% loss of activity in presence of fructose 1,6-diphosphate, NADH and pyruvate, complete loss of activity without protectant [6]; <69>, 15 min, 2% loss of activity [30]; <49>, 900 min, stable [39]; <2>, 10 min, 90% loss of activity [41]) [6, 30, 39, 41]
55 <69, 55> (<69>, stable below [30]; <55>, stable up to [40]) [30, 40]
57.5 <48> (<48>, 20 min, 50% loss of activity, allozyme LDH-Ba4 [3]) [3]
60 <4, 5> (<4>, relatively stable [19]; <5>, 5 min, complete inactivation without stabilizer, Mn^{2+} in combination with fructose 1,6-diphosphate stabilizes the enzyme completely [19]) [19]
61 <48> (<48>, 20 min, 50% loss of activity, allozyme LDH-Ba/Bb [3]) [3]
62.5 <48> (<48>, 20 min, 50% loss of activity, allozyme LDH-Bb4 [3]) [3]
65 <55, 72> (<72>, 20 min, stable up to [32]; <55>, stable for 10 min, 30% loss of activity after 1 h [40]) [32, 40]
69 <73, 74> (<73, 74>, 30 min, 50% loss of activity [33]) [33]
70 <4> (<4>, 3 min, complete inactivation [5]) [5]
71 <76, 77> (<76, 77>, 30 min, 50% loss of activity [33]) [33]
71.5 <72> (<72>, 20 min, 50% loss of activity [32]) [32]
73.5 <81> (<81>, 30 min, 50% loss of activity [33]) [33]
75 <78> (<78>, 30 min, 50% loss of activity [33]) [33]
Additional information <20> (<20>, protection against heat inactivation by 1 mM fructose 1,6-diphosphate, 1 mM ATP, 1 mM glucose 6-phosphate and increased levels of phosphate [4]) [4]

General stability information

<4>, stable to dilution from 0.2-0.01 mg protein per ml [19]
<4>, very stable during 70 h dialysis in acetate buffer pH 5.5, in imidazol buffer, pH 6.5, in phosphate buffer pH 6.5 with dithioerythritol and in Tris-HCl buffer pH 7.5 with Mn^{2+} and fructose 1,6-diphosphate [19]
<5>, Mn^{2+} and fructose 1,6-diphosphate are required during dialysis at pH 5.5, very unstable during dialysis at pH 7.5 although Mn^{2+} and fructose 1,6-diphosphate are added [19]
<5>, irreversible loss of activity after several hours when the concentration is below 0.1 mg protein per ml [19]
<20>, NADH and fructose 1,6-diphosphate partially stabilize the 140000 Da molecular weight form against dissociation in triethanolamine-hydrochloride buffer at pH 8.0 [4]
<37>, fructose-1,6-diphosphate is necessary to stabilize the tetrameric enzyme form [23]
<48>, addition of cofactor stabilizes all allozymes to urea inactivation [3]
<48>, inactivation by pronase, trypsin and chymotrypsin [3]

Storage stability

<16>, -5°C, precipitated from phosphate buffer with $(NH_4)_2SO_4$ and storage in the supernatant solution, stable for several weeks [2]

<20>, -20°C, 50 mM sodium phosphate buffer, pH 7.0, stable for several months [4]

<48>, 4°C, as ammonium sulfate suspension, stable for more than 8 months [3]

<53>, -20°C, stable for months [11]

<53>, stabilized for storage at 4°C by addition of 0.2-1 M potassium chloride or 3.2 M ammonium sulfate [11]

References

[1] Garvie, E.I.: Bacterial lactate dehydrogenases. Microbiol. Rev., **44**, 106-139 (1980)

[2] Williams, R.A.; Andrews, P.: Purification of the fructose 1,6-bisphosphate-dependent lactate dehydrogenase from Streptococcus uberis and an investigation of its existence in different forms. Biochem. J., **236**, 721-727 (1986)

[3] Place, A.R.; Powers, D.A.: Purification and characterization of the lactate dehydrogenase (LDH-B4) allozymes of Fundulus heteroclitus. J. Biol. Chem., **259**, 1299-1308 (1984)

[4] Hillier, A.J.; Jago, G.R.: L-Lactate dehydrogenase, FDP-activated, from Streptococcus cremoris. Methods Enzymol., **89**, 362-367 (1982)

[5] Gordon, G.L.; Doelle, H.W.: Purification, properties and immunological relationship of L(+)-lactate dehydrogenase from Lactobacillus casei. Eur. J. Biochem., **67**, 543-555 (1976)

[6] Goetz, F.; Schleifer, K.H.: Purification and properties of a fructose-1,6-diphosphate activated L-lactate dehydrogenase from Staphylococcus epidermidis. Arch. Microbiol., **105**, 303-312 (1975)

[7] Mou, L.; Mulvena, D.P.; Jonas, H.A.; Jago, G.R.: Purification and properties of nicotinamide adenine dinucleotide-dependent D- and L-lactate dehydrogenases in a group N streptococcus. J. Bacteriol., **111**, 392-396 (1972)

[8] Eichner, R.D.: L(+)-lactate dehydrogenase from Homarus americanus. Methods Enzymol., **89**, 359-362 (1982)

[9] Lee, C.-Y.; Yuan, J.H.; Goldberg, E.: Lactate dehydrogenase isozymes from mouse. Methods Enzymol., **89**, 351-358 (1982)

[10] Oba, K.; Uritani, I.: L-Lactate dehydrogenase isozymes from sweet potato roots. Methods Enzymol., **89**, 345-351 (1982)

[11] Steinbuechel, A.; Schlegel, H.G.: NAD-linked L(+)-lactate dehydrogenase from the strict aerobe Alcaligenes eutrophus. 1. Purification and properties. Eur. J. Biochem., **130**, 321-328 (1983)

[12] Davies, D.D.; Davies, S.: Purification and properties of L(+)-lactate dehydrogenase from potato tubers. Biochem. J., **129**, 831-839 (1972)

[13] Pettit, S.M.; Nealon, D.A.; Henderson, A.R.: Purification of lactate dehydrogenase isoenzyme-5 from human liver. Clin. Chem., **27**, 88-93 (1981)

[14] Taguchi, H.; Machida, M.; Matsuzawa, H.; Ohta, T.: Allosteric and kinetic properties of L-lactate dehydrogenase from Thermus caldophilus GK24, an extremely thermophilic bacterium. Agric. Biol. Chem., **49**, 359-365 (1985)

[15] Auerbach, G.; Ostendorp, R.; Prade, L.; Korndorfer, I.; Dams, T.; Huber, R.; Jaenicke, R.: Lactate dehydrogenase from the hyperthermophilic bacterium Thermotoga maritima: the crystal structure at 2.1 A resolution reveals strategies for intrinsic protein stabilization. Structure, **15**, 769-781 (1998)

[16] Oba, K.; Murakami, S.; Uritani, I.: Partial purification and characterization of L-lactate dehydrogenase isozymes from sweet potato roots. J. Biochem., **81**, 1193-1201 (1977)

[17] Betsche, T.; Bosach, K.; Gerhardt, B.: L-Lactate dehydrogenase from leaves of Capsella bursa-pastoris (L.) Med. I. Identification and partial characterization. Planta, **146**, 567-574 (1979)

[18] Betsche, T.: L-Lactate dehydrogenase from leaves of higher plants. Kinetics and regulation of the enzyme from lettuce (Lactuca sativa L). Biochem. J., **195**, 615-622 (1981)

[19] Hensel, R.; Mayr, U.; Stetter, K.O.; Kandler, O.: Comparative studies of lactic acid dehydrogenases in lactic acid bacteria. I. Purification and kinetics of the allosteric L-lactic acid dehydrogenase from Lactobacillus casei ssp. casei and Lactobacillus curvatus. Arch. Microbiol., **112**, 81-93 (1977)

[20] Hardman, M.J.; Pritchard, G.G.: Kinetics of activation of L-lactate dehydrogenase from Streptococcus faecalis by fructose 1,6-bisphosphate and by metal ions. Biochim. Biophys. Acta, **912**, 185-190 (1987)

[21] Allison, N.; Fewson, C.A.: Purification and properties of L-lactate dehydrogenase from Acinetobacter calcoaceticus. FEMS Microbiol. Lett., **36**, 183-186 (1986)

[22] Hardman, M.J.; Crow, V.L.; Cruickshank, D.S.; Pritchard, G.G.: Kinetics of activation of L-lactate dehydrogenase from Streptococcus lactis by fructose 1,6-bisphosphate. Eur. J. Biochem., **146**, 179-183 (1985)

[23] Goetz, F.; Schleifer, K.H.: Biochemical properties and the physiological role of the fructose-1,6-bisphosphate activated L-lactate dehydrogenase from Staphylococcus epidermidis. Eur. J. Biochem., **90**, 555-561 (1978)

[24] Hensel, R.; Mayr, U.; Fujiki, H.; Kandler, O.: Comparative studies of lactate dehydrogenases in lactic acid bacteria. Amino-acid composition of an active-site region and chemical properties of the L-lactate dehydrogenase of Lactobacillus casei, Lactobacillus curvatus, Lactobacillus plantarum, and Lactobacillus acidophilus. Eur. J. Biochem., **80**, 83-92 (1977)

[25] Crow, V.L.; Pritchard, G.G.: Fructose 1,6-diphosphate-activated L-lactate dehydrogenase from Streptococcus lactis: kinetic properties and factors affecting activation. J. Bacteriol., **131**, 82-91 (1977)

[26] Iwata, S.; Yoshida, S.; Ohta, T.: A regular 1:1 complex of two allosteric states in the single crystal of L-lactate dehydrogenase from Bifidobacterium longum. J. Mol. Biol., **236**, 958-959 (1994)

[27] Lee, B.I.; Chang, C.; Cho, S.J.; Han, G.W.; Yu, Y.G.; Eom, S.H.; Suh, S.W.: Lactate dehydrogenase from hyperthermophilic archaeon Methanococcus

jannaschii: overexpression, crystallization and preliminary X-ray analysis. Acta Crystallogr. Sect. D, **56**, 81-83 (2000)
[28] Ostendorp, R.; Auerbach, G.; Jaenicke, R.: Extremely thermostable L(+)-lactate dehydrogenase from Thermotoga maritima: cloning, characterization, and crystallization of the recombinant enzyme in its tetrameric and octameric state. Protein Sci., **5**, 862-873 (1996)
[29] Hewitt, C.O.; Sessions, R.B.; R.; Dafforn, T.R.; Holbrook, J.J.: Protein engineering tests of a homology model of Plasmodium falciparum lactate dehydrogenase. Protein Eng., **10**, 39-44 (1997)
[30] Marchat, L.; Loiseau, P.M.; Petek, F.: Purification and characterization of lactate dehydrogenase isoenzymes 1 and 2 from Molinema dessetae (Nematoda:Filarioidea). Parasitol. Res., **82**, 672-680 (1996)
[31] Mendiola, P.; De Costa, J.: The effects of temperature and pH on the kinetic properties of heart muscle lactate dehydrogenase from anuran amphibians. Comp. Biochem. Physiol. B, **98B**, 529-534 (1991)
[32] Mulkiewicz, E.; Zietara, M.S.; Stachowiak, K.; Skorkowski, E.F.: Properties of lactate dehydrogenase from the isopod, Saduria entomon. Comp. Biochem. Physiol. B, **126**, 337-346 (2000)
[33] Klyachko, O.S.; Ozernyuk, N.D.: Functional and structural properties of lactate dehydrogenase from embryos of different fishes. Comp. Biochem. Physiol. B, **119B**, 77-80 (1998)
[34] Holland, L.Z.; McFall-Ngai, M.; Somero, G.N.: Evolution of lactate dehydrogenase-A homologs of barracuda fishes (genus Sphyraena) from different thermal environments: differences in kinetic properties and thermal stability are due to amino acid substitutions outside the active site. Biochemistry, **36**, 3207-3215 (1997)
[35] Wu, C.Y.; Chen, S.T.; Chiou, S.H.; Wang, K.T.: Kinetic analysis of duck ε-crystallin with L-lactate dehydrogenase activity: determination of kinetic constants and comparison of substrate specificity. Biochem. Biophys. Res. Commun., **186**, 874-880 (1992)
[36] Ike, J.; Sangan, P.; Gunasekaran, M.: Purification and properties of lactate dehydrogenase from Nocardia asteroides. Microbios, **69**, 119-127 (1992)
[37] Cameron, A.D.; Roper, D.I.; Moreton, K.M.; Muirhead, H.; Holbrook, J.J.; Wigley, D.B.: Allosteric activation in Bacillus stearothermophilus lactate dehydrogenase investigated by an X-ray crystallographic analysis of a mutant designed to prevent tetramerization of the enzyme. J. Mol. Biol., **238**, 615-625 (1994)
[38] Iwata, S.; Kamata, K.; Yoshida, S.; Minowa, T.; Ohta, T.: T and R states in the crystals of bacterial L-lactate dehydrogenase reveal the mechanism for allosteric control. Nat. Struct. Biol., **1**, 176-185 (1994)
[39] Bryant, F.O.: Characterization of the fructose 1,6-bisphosphate-activated, L(+)-lactate dehydrogenase from Thermoanaerobacter ethanolicus. J. Enzyme Inhib., **5**, 235-248 (1991)
[40] LeVan, K.M.; Goldberg, E.: Properties of human testis-specific lactate dehydrogenase expressed from Escherichia coli. Biochem. J., **273**, 587-592 (1991)
[41] Savijoki, K.; Palva, A.: Molecular genetic characterization of the L-lactate dehydrogenase gene (ldhL) of Lactobacillus helveticus and biochemical

characterization of the enzyme. Appl. Environ. Microbiol., **63**, 2850-2856 (1997)

[42] Chang, G.G.; Huang, S.M.; Chiou, S.H.: Kinetic mechanism of the endogenous lactate dehydrogenase activity of duck ε-crystallin. Arch. Biochem. Biophys., **284**, 285-291 (1991)

D-Lactate dehydrogenase 1.1.1.28

1 Nomenclature

EC number
1.1.1.28

Systematic name
(R)-lactate:NAD$^+$ oxidoreductase

Recommended name
D-lactate dehydrogenase

Synonyms
D-(-)-lactate dehydrogenase (NAD)
D-LDH [Swissprot]
D-lactic acid dehydrogenase
D-lactic dehydrogenase
D-specific lactic dehydrogenase
Fermentative lactate dehydrogenase [Swissprot]
LDH
Respiratory D-lactate dehydrogenase [Swissprot]
dehydrogenase, D-lactate
lactic acid dehydrogenase

CAS registry number
9028-36-8

2 Source Organism

<1> *Cardium edule* (bivalve mollusc [1]) [1]
<2> *Pythium debaryanum* [2]
<3> *Sapromyces elongatus* [2]
<4> *Pythium undulatum* [2]
<5> *Polysphondylium pallidum* (slime mold [3]) [3]
<6> *Streptococcus bovis* [4]
<7> *Helix aspersa* (land snail [6,7]) [6, 7]
<8> *Haliotis cracherodii* (black abalone [7]) [7]
<9> *Butyribacterium rettgeri* [8, 12]
<10> *Limulus polyphemus* (horseshoe crab [9,10]) [9, 10]
<11> *Nereis virens* (seaworm [10]) [10]
<12> *Allomyces sp.* [11]
<13> *Lactobacillus bulgaricus* [12, 13, 14, 15, 19, 21, 22]

<14> *Lactobacillus acidophilus* [12]
<15> *Lactobacillus delbrueckii* [12]
<16> *Lactobacillus jensenii* [12]
<17> *Lactobacillus jugurt-helveticus* [12]
<18> *Lactobacillus lactis* [12]
<19> *Lactobacillus leichmannii* [12]
<20> *Lactobacillus casei* [12]
<21> *Lactobacillus plantarum* [12]
<22> *Lactobacillus brevis* [12]
<23> *Lactobacillus buchneri* [12]
<24> *Lactobacillus cellobiosus* [12]
<25> *Lactobacillus confusus* [12]
<26> *Lactobacillus fermentum* [12]
<27> *Lactobacillus vermeforme* [12]
<28> *Lactobacillus viridescens* [12]
<29> *Leuconostoc cremoris* [12]
<30> *Leuconostoc dextranicum* [12]
<31> *Leuconostoc lactis* [12]
<32> *Leuconostoc mesenteroides* [12]
<33> *Leuconostoc paramesenteroides* [12]
<34> *Leuconostoc venosus* [12]
<35> *Streptococcus lactis* [12]
<36> *Pediococcus acidilactici* [12]
<37> *Pediococcus damnosus* [12]
<38> *Pediococcus inopinatus* [12]
<39> *Pediococcus pentosaceus* [12]
<40> *Pediococcus aerogenes* [12]
<41> *Escherichia coli* [12, 20]
<42> *Staphylococcus sp.* [12]
<43> *Staphylococcus aureus* [12]
<44> *Mycoplasma sp.* [12]
<45> *Staphylococcus rettgeri* [12]
<46> *Selenomonas ruminantium* [12]
<47> *Staphylococcus haemolyticus* [12]
<48> *Staphylococcus hominis* [12]
<49> *Staphylococcus warneri* [12]
<50> *mammalia* [12]
<51> *Neisseria meningitidis* [16]
<52> *Leuconostoc mesenteroides subsp. cremoris* [17]
<53> *Enterococcus faecium* [18]
<54> *Aerobacter aerogenes* [5]

3 Reaction and Specificity

Catalyzed reaction

(R)-lactate + NAD$^+$ = pyruvate + NADH (<2-4> binary-binary mechanism with either NAD$^+$ or NADH as the first substrate [2])

Reaction type

oxidation
redox reaction
reduction

Natural substrates and products

S (R)-lactate + NAD$^+$ <1-53> (Reversibility: r <2-4, 7, 8> [2, 6, 7] ; ? <1, 5, 6, 9-53> [1, 3-4, 8-22]) [1-22]
P pyruvate + NADH

Substrates and products

S (R)-lactate + NAD$^+$ <1-53> (Reversibility: r <2-4, 7, 8> [2, 6, 7] ; ? <1, 5, 6, 9-53> [1, 3-4, 8-22]) [1-22]
P pyruvate + NADH
S α-ketobutyrate + NADH <7, 8> (Reversibility: ? <7, 8> [7]) [7]
P 2-hydroxybutyrate + NAD$^+$
S bromopyruvate + NADH <13> (Reversibility: ? <13> [22]) [22]
P 3-bromo-2-hydroxypropanoate + NAD$^+$
S glyoxylate + NADH <13> (Reversibility: ? <13> [22]) [22]
P hydroxyacetic acid + NAD$^+$
S oxalacetate + NADH <13> (Reversibility: ? <13> [15, 22]) [15, 22]
P malate + NAD$^+$

Inhibitors

ADP <13-50> [12]
ATP <9, 12-50> [8, 11, 12]
$CoCl_2$ <51> [16]
Cu^{2+} <31, 39, 51> [12, 16]
D-malate <51> [16]
D-tartrate <51> [16]
Hg^{2+} <9, 31, 39> [8, 12]
$MnCl_2$ <51> [16]
NAD$^+$ <12, 31, 39> [11, 12]
$NiCl_2$ <51> [16]
$ZnCl_2$ <51> [16]
iodoacetamide <10, 13-50> [9, 12]
iodoacetate <10, 13-50> [9, 12]
oxalate <18> [12]
oxamate <9, 10, 13-51> [8, 9, 12, 16]
p-chloromercuribenzoate <31> [12]
p-chloromercuriphenyl sulfonic acid <9> [8, 12]
p-hydroxymercuribenzoate <10, 11, 41> [9, 10, 12]

pyruvate <5, 7, 8, 51> (<5> some salt solutions influence the susceptibility to inhibition by pyruvate [3]; <7> no inhibition [7]) [3, 6, 7, 16]
sodium arsenite <41> [12]

Cofactors/prosthetic groups
FAD <41> [20]
NAD^+ <1-53> [1-12, 16, 17, 18, 21]
NADH <1-53> [1-12, 16, 17, 18]

Activating compounds
2-mercaptoethanol <53> (<53> strongly reducing enviroment required [18]) [18]
acetate <53> [18]
fructose 1,6-diphosphate <2-4, 6> [2, 4]
phosphate <53> [18]

Metals, ions
Ca^{2+} <51> [16]
Mg^{2+} <51> [16]

Turnover number (min^{-1})
38000 <10> (NADH) [9]

Specific activity (U/mg)
2.2 <13> [13]
1950-2100 <13> [15]

K_m-Value (mM)
0.00021 <10> (NADH) [9]
0.00044 <10> (NAD^+) [9]
0.001 <20> (NADH) [12]
0.0035 <53> (NADPH) [18]
0.01 <7> (NADH) [7]
0.015 <7> (NADH) [6]
0.019 <53> (NADH) [18]
0.02 <7> (NAD^+) [6]
0.02 <8> (NADH) [7]
0.03 <52> (NADH) [17]
0.05 <13> (NADH) [13]
0.07 <10> (pyruvate) [9]
0.071 <21> (NADH) [12]
0.1 <8> (NAD^+) [7]
0.12 <7> (NAD^+) [7]
0.16 <52> (NAD) [17]
0.16 <14> (NADH) [12]
0.16 <8> (pyruvate) [7]
0.3 <52> (pyruvate) [17]
0.33 <7> (pyruvate) [7]
0.42 <7> (pyruvate) [6]
0.59 <51> ((R)-lactate) [16]

1 <13> (pyruvate) [13]
1.05 <39> (NAD^+) [12]
1.5 <53, 13> (pyruvate) [18, 22]
1.5-1.6 <13> (pyruvate, <13> recombinant and wild-type enzyme [15]) [15]
1.9 <8> ((R)-lactate) [7]
2.2 <32> (NAD^+) [12]
2.9 <31> (NAD^+) [12]
3.2 <7> ((R)-lactate) [6]
3.8 <10> ((R)-lactate) [9]
4.4 <8> (α-ketobutyrate) [7]
4.8 <7> ((R)-lactate) [7]
7 <13> (oxalacetate) [22]
7.5-9 <13> (oxalacetate, <13> recombinant and wild-type enzyme [15]) [15]
10 <7> (α-ketobutyrate) [7]
10 <13> (bromopyruvate) [22]
19 <52> ((R)-lactate) [17]
20 <18, 26> ((R)-lactate) [12]
70 <16, 19> ((R)-lactate) [12]
100 <13> (glyoxylate) [22]
1250 <14> ((R)-lactate) [12]

pH-Optimum

5.8-6.5 <9> [8]
6 <7, 8> (<7, 8> pyruvate reduction [7]) [7]
6.4 <7> (<7> reduction of pyruvate [6]) [6]
6.4-7.5 <41> [12]
6.6-8 <31> [12]
7-7.5 <13> [15]
7.3 <14> [12]
7.5 <13> [22]
7.5-8 <13> [14]
7.6 <19> [12]
7.8 <16> [12]
8 <51> [16]
8 <39> (<39> 3.6 or 8.0, changes with pyruvate concentration [12]) [12]
8.2-9.7 <7, 8> (<7, 8> (R)-lactate oxidation [7]) [7]
8.6 <26> [12]
9 <7> (<7> oxidation of (R)-lactate [6]) [6]

4 Enzyme Structure

Molecular weight

68000-70000 <1> (<1> gel filtration, analytical ultracentrifugation [1]) [1]
70000 <51> (<51> SDS-PAGE [16]) [16]
70000-80000 <10> (<10> gel filtration, equilibrium sedimentation [9, 10]) [9, 10]

75000 <13> (<13> gel filtration [13]) [13]
98000 <2> (<2> sucrose density gradient sedimentation [2]) [2]
115000 <2> (<2> gel filtration [2]) [2]
140000 <7> (<7> gel filtration [6]) [6]
145000 <9> (<9> gel filtration [8]) [8]

Subunits

dimer <1> (<1> 2 * 35000, analytical ultracentrifugation [1]; <7, 8> 2 * 35000-40000, analytical ultracentrifugation [7]; <10> 2 * 35000, gel filtration [9,10]; <13> 2 * 37000, SDS-PAGE [13-15]; <52> 2 * 40000, SDS-PAGE [17]) [1, 7, 9, 10, 13-15, 17]
tetramer <7> (<7> 4 * 34000, SDS-PAGE [6]) [6]

5 Isolation/Preparation/Mutation/Application

Source/tissue

brain <7> [6]
foot <1, 7, 8> (<1> plus adductor muscle [1]) [1, 6, 7]
gill <1> [1]
heart <7> [6]
hepatopancreas <7> [6]
kidney <7> [6]
mycelium <2-4> [2]
skeletal muscle <10> [9]
vegetative <5> [3]

Localization

inner membrane <41> (<41> cytoplasmic side [20]) [20]
plasma membrane <51> (spheroplast membrane) [16]

Purification

<9> [8]
<10> [9, 10]
<11> [10]
<12> (partial [11]) [11]
<13> [13, 14, 15, 21, 22]
<51> [16]
<52> [17]
<2-4> [2]
<7, 8> [7]

Crystallization

<13> (crystals belong to the orthorhombic space group, diffract beyond 3.0 A resolution [21]) [21]
<41> (1.9 A resolution [20]) [20]
<53> (hexagonal and tetragonal crystal forms, tetragonal form diffracted to 3.0 A resolution [18]) [18]

Cloning

<13> (<13> exprcssion in Escherichia coli [15, 22]) [15, 22]
<51> (<51> expression in Escherichia coli [16]) [16]
<52> (<52> expression in Escherichia coli [17]) [17]
<53> (<53> expression in Escherichia coli as thioredoxin fusion protein [18]) [18]

Engineering

D259N <13> (<13> 56-fold reduction in K_{cat}, fivefold lowering of K_m, shifting of the enzymatic actvity profile towards the acidic range by two units [19]) [19]
E264G <13> (<13> shift of 2 units in optimal pH toward the acidic range [22]) [22]
H205K <13> (<13> 125-fold reduction in K_{cat} [19]) [19]
H296K <13> (<13> no significant change [19]) [19]
H303K <13> (<13> K_m for pyruvate increased fourfold [19]) [19]

6 Stability

Temperature stability

35 <32> (<32> stable below, Tris-HCl, pH 8.5 [12]) [12]
40 <32> (<32> stable below, phosphate buffer, pH 7.5 [12]) [12]
45 <14, 21, 31> (<31, 14> 5 min, 98% inactivation [12]; <21> 3 min, inactivation [12]) [12]
50 <10, 16, 19, 21, 26, 47-49> (<10> 3 min, 5% loss of activity [9]; <16> 5 min, 25% loss of activity [12]; <19> 5 min, 50% loss of activity [12]; <47-49> 10 min, inactivation [12]; <21> 3 min, inactivation [12]; <26> 5 min, 98% inactivation [12]) [9, 12]
55 <39> (<39> 3 min, inactivation [12]) [12]
60 <10> (<10> 3 min, 90% loss of activity [9]) [9]
80 <16, 19> (<16, 19> 5 min, 100% loss of activity [12]) [12]

General stability information

<9>, repeated freezing and thawing, low ionic strength or pH above 6.5 cause loss of activity [8]
<10>, 2-mercaptoethanol and EDTA stabilize [9]
<12>, glycerol and reducing agents stabilize [11]

Storage stability

<9>, 0°C or -20°C, 0.5 M potassium phosphate buffer pH 6.2, two weeks [8]
<12>, -20°C, 50% glycerol, 2 months [11]
<2-4>, -20°C, for at least 2 years [2]
<2-4>, 4°C, several weeks [2]

References

[1] Jaenicke, R.; Müller, K.; Gäde, G.: Dimeric lactic dehydrogenase from the bivalve mollusc Cardium edule. Naturwissenschaften, **68**, 205-206 (1981)

[2] LeJohn, H.B.; Stevenson, R.M.: D(-)-lactate dehydrogenase from fungi. Methods Enzymol., **41 B**, 293-298 (1975)

[3] Garland, R.C.; Kaplan, N.O.: Salt-induced alteration of D(-) lactate dehydrogenase from Polyspondylium pallidum. Biochem. Biophys. Res. Commun., **26**, 679-685 (1967)

[4] Wolin, M.J.: Fructose-1,6-diphosphate requirement of streptococcal lactic dehydrogenases. Science, **146**, 775-776 (1964)

[5] Pascal, M.-C.; Pichinoty, F.: Sur la configuration du lactate produit au cours de la fermentation du glucose par quelques entérobactéries. Ann. Inst. Pasteur, **107**, 55-62 (1964)

[6] Storey, K.B.: Lactate dehydrogenase in tissue extracts of the land snail, Helix aspersa: unique adaptation of LDH subunits in a facultative anaerobe. Comp. Biochem. Physiol. B, **56**, 181-187 (1977)

[7] Long, G.L.; Ellington, W.R.; Duda, T.F.: Comparative enzymology and physiological role of D-lactate dehydrogenase from the foot muscle of two gastropod molluscs. J. Exp. Zool., **207**, 237-248 (1979)

[8] Wittenberger, C.L.: D(-)-lactate dehydrogenase from butyribacterium rettgeri. Methods Enzymol., **41B**, 299-303 (1975)

[9] Long, G.L.: D-lactate dehydrogenase from the horseshoe crab. Methods Enzymol., **41 B**, 313-323 (1975)

[10] Long, G.L.; Kaplan, N.O.: Diphosphopyridine nucleotide-linked D-lactate dehydrogenases from the horseshoe crab, Limulus polyphemus and the seaworm, Nereis virens. I. Physical and chemical properties. Arch. Biochem. Biophys., **154**, 696-710 (1973)

[11] Purohit, K.; Turian, G.: D(-)-lactate dehydrogenase from Allomyces. Partial purification and allosteric properties. Arch. Mikrobiol., **84**, 287-300 (1972)

[12] Garvie, E.L.: Bacterial lactate dehydrogenases. Microbiol. Rev., **44**, 106-139 (1980)

[13] Le Bras, G.; Garel, J.R.: Properties of D-lactate dehydrogenase from Lactobacillus bulgaricus: a possible different evolutionary origin for the D- and L-lactate dehydrogenases. FEMS Microbiol. Lett., **79**, 89-93 (1991)

[14] Kochhar, S.; Hunziker, P.E.; Leong-Morgenthaler, P.; Hottinger, H.: Primary structure, physicochemical properties, and chemical modification of NAD^+-dependent D-lactate dehydrogenase. Evidence for the presence of Arg-235, His-303, Tyr-101, and Trp-19 at or near the active site. J. Biol. Chem., **267**, 8499-8513 (1992)

[15] Kochhar, S.; Chuard, N.; Hottinger, H.: Cloning and overexpression of the Lactobacillus bulgaricus NAD(+)-dependent D-lactate dehydrogenase gene in Escherichia coli: purification and characterization of the recombinant enzyme. Biochem. Biophys. Res. Commun., **185**, 705-712 (1992)

[16] Erwin, A.L.; Gotschlich, E.C.: Oxidation of D-lactate and L-lactate by Neisseria meningitidis: Purification and cloning of meningococcal D-lactate dehydrogenase. J. Bacteriol., **175**, 6382-6391 (1993)
[17] Dartois, V.; Phalip, V.; Schmitt, P.; Divies, C.: Purification, properties and DNA sequence of the D-lactate dehydrogenase from Leuconostoc mesenteroides subsp. cremoris. Res. Microbiol., **146**, 291-302 (1995)
[18] Stoll, V.S.; Manohar, A.V.; Gillon, W.; MacFarlane, E.L.; Hynes, R.C.; Pai, E.F.: A thioredoxin fusion protein of VanH, a D-lactate dehydrogenase from Enterococcus faecium: cloning, expression, purification, kinetic analysis, and crystallization. Protein Sci., **7**, 1147-1155 (1998)
[19] Kochhar, S.; Lamzin, V.S.; Razeto, A.; Delley, M.; Hottinger, H.; Germond, J.E.: Roles of His205, His296, His303 and Asp259 in catalysis by NAD^+-specific D-lactate dehydrogenase. Eur. J. Biochem., **267**, 1633-1639 (2000)
[20] Dym, O.; Pratt, E.A.; Ho, C.; Eisenberg, D.: The crystal structure of D-lactate dehydrogenase, a peripheral membrane respiratory enzyme. Proc. Natl. Acad. Sci. USA, **97**, 9413-9418 (2000)
[21] Nessler, S.; Le Bras, G.; Le Bras, G.; Garel, J.R.: Crystallization of D-lactate dehydrogenase from Lactobacillus bulgaricus. J. Mol. Biol., **235**, 370-371 (1994)
[22] Kochhar, S.; Chuard, N.; Hottinger, H.: Glutamate 264 modulates the pH dependence of the NAD(+)-dependent D-lactate dehydrogenase. J. Biol. Chem., **267**, 20298-20301 (1992)

Glycerate dehydrogenase 1.1.1.29

1 Nomenclature

EC number
1.1.1.29

Systematic name
(R)-glycerate:NAD^+ oxidoreductase

Recommended name
glycerate dehydrogenase

Synonyms
D-glycerate dehydrogenase
GDH [Swissprot]
Glyoxylate reductase [Swissprot]
HPR [Swissprot]
hydroxypyruvate dehydrogenase [Swissprot]
NADH-dependent hydroxypyruvate reductase [Swissprot]
dehydrogenase, glycerate
hydroxypyruvate reductase

CAS registry number
9028-37-9

2 Source Organism

<1> *Zea mays* [1]
<2> *Hyphomicrobium methylovorum* [2, 6]
<3> *Pisum sativum* [3]
<4> *Beta vulgaris* [3]
<5> *Lycopersicon esculentum* [3]
<6> *Raphanus sativus* [3]
<7> *Spinacia oleracea* [3]
<8> *Petraselinum hortense* (parsley) [3, 5]
<9> *lettuce* [3]
<10> *Brassica oleracea* (kohlrabi) [3]
<11> *Daucus carota* (carrot) [3]
<12> *Methylobacterium extorquens* [4]
<13> *Hordeum vulgare* (barley) [7]

3 Reaction and Specificity

Catalyzed reaction

(R)-glycerate + NAD^+ = hydroxypyruvate + NADH

Reaction type

redox reaction

Natural substrates and products

S hydroxypyruvate + NADH <1-11> (Reversibility: ir <2> [2]; r <3-13> [3, 4, 7]; ? <8, 2> [5, 6]) [1-7]
P D-glycerate + NAD^+

Substrates and products

S glyoxylate + NADH <2> (Reversibility: ? <2> [2]; ir <12> [4]) [2, 4]
P glycolate + NAD^+
S hydroxypyruvate + NADH <1-11> (Reversibility: ir <2> [2]; r <3-13> [3, 4, 7]; ? <8, 2> [5, 6]) [1-7]
P D-glycerate + NAD^+

Inhibitors

$AgNO_3$ <2> [2]
$CuSO_4$ <2> (<2> slight [2]) [2]
$HgCl_2$ <2> [2]
NADH <12> (<12> concentration above 0.5 mM [4]) [4]
NADPH <12> (<12> concentration above 0.5 mM [4]) [4]
$ZnCl_2$ <2> (<2> slight [2]) [2]
dihydroxyfumarate <2> [2]
hydroxypyruvate <12> (<12> 99% inhibition of the reverse reaction at 0.5 mM [4]) [4]
hydroxypyruvate <12> (<12> concentration over 4 mM [4]) [4]
p-chloromercuribenzoate <2> [2]
phenylhydrazine <2> [2]
Additional information <2> (<2> not inhibitory: citrate, $NADP^+$, EDTA, 1,10-phenanthroline, iodoacetate, iodoacetamide, N-ethylmaleimide, hydroxylamine, semicarbazide, NaN_3, NaF, KCN, L-ascorbate, dithiothreitol, pyridoxal 5'-phosphate [2]) [2]

Cofactors/prosthetic groups

NAD^+ <3-11> [3]
NADH <1-12> (<2-11> specific for, not replaceable by NADPH [2, 3]) [1-4]
NADPH <12> [4]

Specific activity (U/mg)

1086 <2> [2]
Additional information <2> (<2> activities of several Hypomicrobium and Methylobacterium strains listed [6]) [6]

K_m-Value (mM)

0.0058 <8> (NADH) [5]

0.04 <12> (NADH) [4]
0.055 <2> (NADH) [2]
0.06 <12> (NADPH) [4]
0.062 <8> (hydroxypyruvate) [5]
0.09 <1> (hydroxypyruvate) [1]
0.1 <12> (hydroxypyruvate) [4]
0.12 <12> (hydroxypyruvate) [4]
0.175 <2> (hydroxypyruvate) [2]
1.5 <12> (glyoxylate) [4]
2.6 <12> (glycerate) [4]
10.8 <2> (glyoxylate) [2]
20 <12> (glyoxylate) [4]

pH-Optimum
4-6.5 <12> (<12> reduction of hydroxypyruvate [4]) [4]
6.8 <2> [2]
9-11 <12> (<12> glycerate oxidation [4]) [4]

4 Enzyme Structure

Molecular weight
69000-70000 <2> (<2> gel filtration [2]) [2]
71000 <12> (<12> SDS-PAGE [4]) [4]
90000 <8> (<8> gel filtration [5]) [5]

Subunits
dimer <12> (<12> 2 * 37000, SDS-PAGE [4]) [4]
dimer <2> (<2> 2 * 38000, SDS-PAGE [2, 6]) [2, 6]
dimer <8> (<8> 2 * 45200, SDS-PAGE [5]) [5]

5 Isolation/Preparation/Mutation/Application

Source/tissue
leaf <1, 3-11> [1, 3, 5]

Localization
peroxisome <8> [5]

Purification
<2> [2, 6]
<8> [5]
<12> [4]

6 Stability

pH-Stability

5-9 <2> [2]

Temperature stability

25 <2> (<2> up to [2]) [2]
30 <2> (<2> 30 min, 13.3% loss of activity [2]) [2]
35 <2> (<2> 30 min, 30.8% loss of activity [2]) [2]
40 <2> (<2> 30 min, 100% loss of activity [2]) [2]

Storage stability

<2>, -20°C, 10 mM potassium phosphate buffer, pH 7.0, 45% v/v glycerol, 1.0 M DTT, 18 months [2]
<12>, 4°C, 50 mM Tris-HCl buffer, pH 7.5, containing 10% glycerol and 1 mM dithiothreitol, 2 months, at least 90% of inital activity [4]

References

[1] Kleczkowski, L.A.; Randall, D.D.; Edwards, G.E.: Oxalate as a potent and selective inhibitor of spinach (Spinacia oleracea) leaf NADPH-dependent hydroxypyruvate reductase. Biochem. J., **276**, 125-127 (1991)

[2] Izumi, Y.; Yoshida, T.; Kanzaki, H.; Toki, S.-i.; Miyazaki, S.S.; Yamada, H.: Purification and characterization of hydroxypyruvate reductase from a serine-producing methylotroph, Hyphomicrobium methylovorum GM2. Eur. J. Biochem., **190**, 279-284 (1990)

[3] Stafford, H.A.; Magaldi, A.; Vennesland, B.: The enzymatic reduction of hydroxypyruvic acid to D-glyceric acid in higher plants. J. Biol. Chem., **207**, 621-629 (1954)

[4] Chistoserdova, L.V.; Lidstrom, M.E.: Purification and characterization of hydroxypyruvate reductase from the facultative methylotroph Methylobacterium extorquens AM1. J. Bacteriol., **173**, 7228-7232 (1991)

[5] Julliard, J.H.; Breton-Gilet, A.: Identification of hydroxypyruvate reductase from parsley by peptide sequence comparison after a two-step purification. Protein Expr. Purif., **9**, 10-14 (1997)

[6] Hagishita, T.; Yoshida, T.; Izumi, Y.; Mitsunaga, T.: Immunological characterization of serine-glyoxylate aminotransferase and hydroxypyruvate reductase from a methylotrophic bacterium, Hyphomicrobium methylovorum GM2. FEMS Microbiol. Lett., **142**, 49-52 (1996)

[7] Kleczkowski, L.A.; Randall, D.D.; Edwards, G.E.: Oxalate as a potent and selective inhibitor of spinach (Spinacia oleracea) leaf NADPH-dependent hydroxypyruvate reductase. Biochem. J., **276**, 125-127 (1991)

3-Hydroxybutyrate dehydrogenase 1.1.1.30

1 Nomenclature

EC number
1.1.1.30

Systematic name
(R)-3-hydroxybutanoate:NAD^+ oxidoreductase

Recommended name
3-hydroxybutyrate dehydrogenase

Synonyms
3-D-hydroxybutyrate dehydrogenase
3-HBDH [Swissprot]
3-hydroxybutyrate dehydrogenase [Swissprot]
BDH [Swissprot]
D-(-)-3-hydroxybutyrate dehydrogenase
D-3-hydroxybutyrate dehydrogenase
D-β-hydroxybutyrate dehydrogenase
NAD-β-hydroxybutyrate dehydrogenase
β-hydroxybutyrate dehydrogenase
β-hydroxybutyric acid dehydrogenase
β-hydroxybutyric dehydrogenase
hydroxybutyrate oxidoreductase

CAS registry number
9028-38-0

2 Source Organism

<1> *Homo sapiens* [1, 2, 8]
<2> *Bos taurus* [2, 3, 5, 6, 10, 11, 14, 16, 17, 19, 24, 27, 29, 31, 32]
<3> *Staphylococcus xylosus* [4]
<4> *Rattus norvegicus* [5, 7, 9, 10, 11, 18, 22, 25, 29]
<5> *Rhodopseudomonas spheroides* [12, 20]
<6> *Paracoccus denitrificans* (enzyme shows hysteresis, lag phase in progress curve [13,15]) [13, 15]
<7> *Pseudomonas lemoignei* [21, 26]
<8> *Zoogloea ramigera I-16-M* [23]
<9> *Mycobacterium phlei* [28]
<10> *Cavia porcellus* [29]
<11> *Bacillus cereus* (strain T) [30]

3 Reaction and Specificity

Catalyzed reaction

(R)-3-hydroxybutanoate + NAD^+ = acetoacetate + NADH (<2> enzyme is completely dependent on the presence of phospholipid for activity [31]; ordered Bi Bi mechanism, NAD^+ first substrate, NADH last product to leave [32])

Reaction type

redox reaction

Natural substrates and products

S (R)-3-hydroxybutanoate + NAD^+ <1-11> (Reversibility: r <1-11> [1-32]) [1-32]
P acetoacetate + NADH <1-11> [1-32]

Substrates and products

S (R)-3-hydroxybutanoate + 3-acetylpyridine adenine dinucleotide <2> (Reversibility: r<2> [17]) [17]
P acetoacetate + 3-acetylpyridine adenine dinucleotideH_2 <2> [17]
S (R)-3-hydroxybutanoate + NAD^+ <1-11> (Reversibility: r <1-11> [1-32]) [1-32]
P acetoacetate + NADH <1-11> [1-32]
S 2-hydroxy-propansulfonate + NAD^+ <7> (Reversibility: r <7> [26]) [26]
P acetonylsulfonate + NADH <7> [26]
S 2-methyl-3-hydroxybutyrate + NAD^+ <7> (Reversibility: r <7> [26]) [26]
P 2-methylacetoacetate + NADH <7> [26]

Inhibitors

(DL)-2-hydroxybutyrate <6> (<6> 5 mM, pH 8.5, 25°, 69% inhibition [13]) [13]
(DL)-lactate <6> (<6> 5 mM, pH 8.5, 25°C, 65% inhibition [13]) [13]
2-hydroxybutyrate <8> (<8> 10 mM, 57% inhibition [23]) [23]
2-oxobutyrate <8> (<8> 10 mM, 29% inhibition [23]) [23]
4-mercapto-3-nitrobenzoate <2> (<2> derivatization of enzyme sulfhydryl groups, less than 1% residual activity [14]) [14]
5,5'-dithiobis(2-nitrobenzoic acid) <8> (<8> 0.15 mM, 63% inhibition [23]) [23]
ADP <7> (<7> 2.85 mM, 0.025 mM NADH, 63% inhibition, competitive inhibitor vs. NADH and NAD^+, pH 7-7.5, noncompetitive vs. acetoacetate [21]) [21]
ADP-ribose <6> (<6> competitive inhibition vs. coenzyme, noncompetitive vs. substrate [13]) [13]
AMP <7> (<7> 2.85 mM, 0.025 mM NADH, 55% inhibition, competitive inhibitor vs. NADH and NAD^+, pH 7-7.5 [21]) [21]
ATP <7> (<7> 2.85 mM, 0.025 mM NADH, 79% inhibition, competitive inhibitor vs. NADH and NAD^+, pH 7-7.5, noncompetitive vs. acetoacetate [21]) [21]

$CuSO_4$ <8> (<8> 1 mM, 81% inhibition [23]) [23]
D-lactate <6> (<6> 5 mM, pH 8.5, 25°, 85% inhibition, uncompetitive vs. coenzyme, competitive vs. substrate [13]) [13]
$HgCl_2$ <8, 9> (<8> 0.1 mM, 100% inhibition [23]; <9> enzyme extremely sensitive, NADH and Ca^{2+} protect [28]) [23, 28]
KCl <2> (<2> 80-100 mM, 50% inhibition [32]) [32]
L-3-hydroxybutyrate <6, 6, 8> (<6> 5 mM, pH 8.5, 25°C, 32% inhibition [13] <6> 5 mM, pH 8.5, 25°C, 23% inhibition [13]; <8> 5 mM, 73% inhibition [23]) [13, 13, 23]
LiCl <2> (<2> 80-100 mM, 50% inhibition [32]) [32]
$MnCl_2$ <8> (<8> 1 mM, 50% inhibition [23]) [23]
N-ethylmaleimide <2> (<2> irreversible inhibition [17]; <2> NAD protects against inhibition [24]) [17, 24]
NAD^+ <7> (<7> substrate inhibition, high concentrations, noncompetitive vs. β-hydroxybutyrate [21]) [21]
NADH <2, 7, 8> (<2> competitive vs. NAD^+ [32]; <7> substrate inhibition, above 0.1 mM, competitive vs. acetoacetate [21]; <8> 0.15 mM, 48% inhibition [23]) [21, 23, 32]
NH_4Cl <2> (<2> 80-100 mM, 50% inhibition [32]) [32]
NaCl <2> (<2> 80-100 mM, 50% inhibition [32]) [32]
acetoacetate <7> (<7> substrate inhibition, above 5 mM, noncompetitive vs. NADH [21]; <8> inhibition above 1 mM, 2 mM, 89% inhibition [23]) [21, 23]
acetoacetyl-CoA <8> (<8> 1 mM, 29% inhibition [23]) [23]
acetyl-CoA <8> (<8> 1 mM, 82% inhibition [23]) [23]
acetylphosphonate <7> (<7> 16.7 mM, pH 7, 28% inhibition [26]) [26]
β-hydroxybutyrate <2> (<2> substrate inhibition at concentrations above 20 mM, 50% inhibition at 200 mM, cyanylated enzyme is 3fold more active at high substrate concentration [14]) [14]
bromomalonate <2, 4, 10> [29]
butanedione <2> [17]
butyrate <6> (<6> 5 mM, pH 8.5, 25°C, 22% inhibition [13]) [13]
butyrate <8> (<8> 10 mM, 20% inhibition [23]) [23]
chloromalonate <2, 4, 10> [29]
crotonate <8> (<8> 10 mM, 20% inhibition [23]) [23]
diazenedicarboxylic acid bis(dimethylamide) <2> (<2> 0.20 mM, activity can be recovered with dithiothreitol, enzyme contains a vicinal dithiol that is oxidized by diamide [17]; <2> NAD^+ protects against inhibition [24]) [17, 24]
dimethyl phosphate <7> (<7> 16.7 mM, pH 7, 12% inhibition [26]) [26]
dimethylmalonate <2, 10> (<2> 15.3 mM, 50% inhibition, oxidation, 11.3% reduction [29]; <10> 7.6 mM, 50% inhibition, oxidation, 8.5 mM, reduction [29]) [29]
dimethyloxyphosphinylacetate <7> (<7> competitive inhibitor vs. acetoacetate [21]) [21]
hydroxymalonate <2, 4, 10> (<2> 4.8 mM, 50% inhibition, oxidation, 7.8 reduction [29]; <10> 1.8 mM, 50% inhibition, oxidation, 6.0 mM, reduction [29]) [29]

malonate <2, 4, 8, 10> (<2> 9.5 mM, 50% inhibition, oxidation, 8.2 mM [29]; <8> 10 mM, 66% inhibition, oxidation [23]; <10> 3.7 mM, 50% inhibition, oxidation [29]) [23, 29]
mesoxalate <2, 4, 10> [29]
methanephosphonic acid <7> (<7> 3.3 mM, pH 7, 11% inhibition [26]) [26]
methyl methylphosphonate <7> (<7> 3.3 mM, pH 7, 13% inhibition [26]) [26]
methyl phosphate <7> (<7> 16.7 mM, pH 7, 21% inhibition [26]) [26]
methyl-2-methoxyphosphinylacetate <7> (<7> competitive inhibitor vs. acetoacetate [21]) [21]
methylmalonate <2, 4, 10> (<2> 2.1 mM, 50% inhibition, oxidation [29]; <4> brain 3-hydroxybutyrate dehydrogenase, 0.5 mM 69% inhibition, 0.75 mM 83%, 1 mM 87%, 1.5 mM 3-hydroxybutyrate, liver 3-hydroxybutyrate dehydrogenase, 0.1 mM 37% inhibition, 0.25 mM 41%, 0.5 mM 45%, 1 mM 3-hydroxybutyrate, competitive inhibition [7]; <10> 1.4 mM, 50% inhibition, oxidation, 0.8 mM reduction [29]) [7, 29]
oxalate <7> (<7> 16.7 mM, pH 7, 9% inhibition [26]) [26]
p-chloromercuribenzoate <2, 8, 9> (<8> 0.1 mM, 100% inhibition [23]; <2> 0.001 mM, extremely rapid inhibition, NAD^+ and NADH protect against inhibition [24]; <9> enzyme extremely sensitive, NADH and Ca^{2+} protect [28]) [23, 24, 28]
phenylarsine oxide <2> (<2> NAD^+ and NADH protect against inhibition [24]) [24]
phenylglyoxal <2> (<2> extremely rapid inhibition [17]) [17]
propionate <6> (<6> 5 mM, pH 8.5, 25°C, 9% inhibition [13]) [13]
pyruvate <8> (<8> 4 mM, 5% inhibition [23]; <7> 16.7 mM, pH 7.0, 22% inhibition [26]) [23, 26]
sodium acetonylphosphonate <7> (<7> 3.3 mM, pH 7, 21% inhibition [26]) [26]
sodium monomethyl acetylphosphonate <7> (<7> 16.7 mM, pH 7, 60% inhibition [26]) [26]
sodium monomethylacetonylphosphate <7> (<7> 16.7 mM, pH 7, 30% inhibition [26]) [26]
sodium sulfide <2> (<2> competitive vs. β-hydroxybutyrate [24]) [24]
sodium sulfite <2> (<2> uncompetitive vs. NAD^+ and acetoacetate, competitive vs. β-hydroxybutyrate [24]) [24]
succinate <8> (<8> 10 mM, 12% inhibition [23]) [23]

Cofactors/prosthetic groups

NAD^+ <1-11> [1-32]

Activating compounds

2-mercaptoethanol <8> (<8> 10 mM, 41% increase [23]) [23]
EDTA <8> (<8> 0.5 mM, 21% increase [23]) [23]
KCN <8> (<8> 0.1 mM, 37% increase [23]) [23]
dithiothreitol <8> (<8> 10 mM, 31% increase [23]) [23]
phosphatidylcholine <1-11> (<2> required for formation of tight and functional complexes of enzyme with NAD^+, coenzyme binding strengthens the

interaction of the enzyme with phosphatidylcholine [3]; <2, 4> state of phosphatidylcholine determines nature of activation, noncooperative for soluble phosphatidylcholine, cooperative for bilayer phosphatidylcholine [11]) [1-32]
reduced glutathione <8> (<8> 10 mM, 38% increase [23]) [23]

Metals, ions

Ca^{2+} <3, 8> (<3> 10 mM, more than 10fold increase in activity [4]; <8> 10 mM 20% increase in activity [23]) [4, 23]
Cd^{2+} <3> (<3> 10 mM, 2fold increase in activity [4]) [4]
Cu^{2+} <3> (<3> 10 mM, 2fold increase in activity [4]) [4]
Mg^{2+} <3, 8, 11> (<3> 10 mM, more than 10fold increase [4]; <8> 10 mM 41% increase in activity [23]; <11> 0.1 mM, 2fold increase in activity, crude extracts [30]) [4, 23, 30]
Mn^{2+} <3> (<3> 10 mM, more than 10fold increase in activity [4]) [4]
Zn^{2+} <3> (<3> 10 mM, 2fold increase in activity [4]) [4]

Specific activity (U/mg)

0.02 <2> (<2> cofactor 3-acetylpyridine adenine dinucleotide [17]) [17]
0.038 <4> (<4> brain 3-hydroxybutyrate dehydrogenase, mitochondrial extracts [9]) [9]
0.07 <4> (<4> heart 3-hydroxybutyrate dehydrogenase, mitochondrial extracts [9]) [9]
0.07 <4> (<4> kidney 3-hydroxybutyrate dehydrogenase, mitochondrial extracts [9]) [9]
0.17 <2> (<2> solubilized apo-3-hydroxybutyrate dehydrogenase, reconstituted with phosphatidylcholine/phosphatidylethanolamine/diphosphatidylglycerol, 5/4/1 [2]) [2]
0.22 <1> (<1> apo-3-hydroxybutyrate dehydrogenase, reconstituted with phosphatidylcholine/phosphatidylethanolamine/diphosphatidylglycerol, 5/4/1 [2]) [2]
0.36 <4> (<4> liver 3-hydroxybutyrate dehydrogenase, mitochondrial extracts [7]) [7, 9]
0.5 <1> (<1> enzyme expressed in SF9 cells, S24T mutant [2]) [2]
0.52 <2> (<2> submitochondrial vesicles, membrane bound enzyme [2]) [2]
0.54 <2> (<2> submitochondrial vesicles [27]) [27]
0.54 <2> (<2> submitochondrial vesicles [5]) [5]
0.6 <1> (<1> enzyme expressed in SF9 cells, M92V mutatant [2]) [2]
0.63 <1> (<1> enzyme expressed in SF9 cells, membrane bound [2]) [2]
0.74 <3> [4]
0.8 <1> (<1> enzyme expressed in SF9 cells, C242S mutatant [2]) [2]
1.15 <4> (<4> liver 3-hydroxybutyrate dehydrogenase, reconstituted with microsomal membranes [22]) [22]
1.26 <4> (<4> brain 3-hydroxybutyrate dehydrogenase, mitochondrial extracts [7]) [7]
1.275 <4> (<4> liver 3-hydroxybutyrate dehydrogenase, reconstituted with mitochondrial inner membranes [22]) [22]
2.7 <4> (<4> submitochondrial vesicles [5]) [5]
14 <9> [28]

19.2 <5> [20]
21 <11> [30]
23.8 <2> [31]
41 <4> (<4> purified enzyme, reconstituted with mitochondrial phospholipids [5]) [5]
86 <2> (<2> purified enzyme, reconstituted with mitochondrial phospholipids [5]) [5]
91 <4> (<4> liver 3-hydroxybutyrate dehydrogenase, reconstituted with mitochondrial phospholipids [10]) [10]
97 <6> [15]
100 <2> [16]
123 <8> [23]
145 <2> (<2> heart 3-hydroxybutyrate dehydrogenase, reconstituted with mitochondrial phospholipids [10]) [10]
Additional information <4> (<4> 0.0018 mM/min/g wet weight [18]) [18]

K_m-Value (mM)

0.000445 <3> (NAD^+) [4]
0.00102 <3> (NADH) [4]
0.00547 <3> (acetoacetate) [4]
0.014 <4> (NADH, <4> liver, soluble enzyme [25]) [25]
0.015 <8> (NADH) [23]
0.029 <4> (NADH, <4> heart, soluble enzyme [25]) [25]
0.0391 <3> (β-hydroxybutyrate) [4]
0.057 <8> (NAD^+) [23]
0.067 <4> (NAD^+, <4> heart, soluble enzyme [25]) [25]
0.071 <4> (NAD^+, <4> liver, soluble enzyme [25]) [25]
0.08 <2> (NADH) [24]
0.091 <2> (NADH) [14]
0.149 <4> (acetoacetate, <4> liver, soluble enzyme [25]) [25]
0.15 <8> (acetoacetate) [23]
0.164 <2> (NADH, <2> cyanylated enzyme [14]) [14]
0.175 <2> (NAD^+) [24]
0.24 <4> (NAD^+, <4> liver 3-hydroxybutyrate dehydrogenase [10]) [10]
0.26 <2> (acetoacetate) [24]
0.27 <2> (NAD^+, <2> 240 mM Na^+, purified enzyme, reconstituted with total mitochondrial phospholipids [6]) [6]
0.28 <2> (3-acetylpyridine adenine dinucleotide) [17]
0.312 <4> (acetoacetate, <4> heart, soluble enzyme [25]) [25]
0.32 <8> (β-hydroxybutyrate) [23]
0.36 <2> (NAD^+, <2> heart 3-hydroxybutyrate dehydrogenase [10]) [10]
0.53 <1> (NAD^+, <1> enzyme expressed in SF9 cells [2]) [2]
0.55 <2> (NAD^+) [14]
0.59 <2> ((R)-3-hydroxybutanoate, <2> submitochondrial vesicles [2]) [2]
0.59 <2> (acetoacetate) [14]
0.66 <9> (NAD^+) [28]

0.71 <1> ((R)-3-hydroxybutanoate, <1> enzyme expressed in SF9 cells [2]) [2]
0.8 <1> (NAD^+, <1> enzyme expressed in SF9 cells, C242S mutant [2]) [2]
0.8 <2> (β-hydroxybutyrate) [24]
0.866 <4> (β-hydroxybutyrate, <4> liver, soluble enzyme [25]) [25]
0.902 <2> (NAD^+, <2> cyanylated enzyme [14]) [14]
0.95 <11> (NAD^+) [30]
0.95 <2> (NAD^+, <2> submitochondrial vesicles [2]) [2]
1.05 <2> (β-hydroxybutyrate, <2> heart 3-hydroxybutyrate dehydrogenase [10]) [10]
1.32 <4> (β-hydroxybutyrate, <4> liver 3-hydroxybutyrate dehydrogenase [10]) [10]
1.586 <4> (β-hydroxybutyrate, <4> heart, soluble enzyme [25]) [25]
2.2 <2> (β-hydroxybutyrate, <2> 240 mM Na^+, purified enzyme, reconstituted with total mitochondrial phospholipids [6]) [6]
2.3 <2> (β-hydroxybutyrate) [14]
2.3 <11> (β-hydroxybutyrate) [30]
7.4 <9> (β hydroxybutyrate) [28]
8.7 <2> (acetoacetate, <2> cyanylated enzyme [14]) [14]
10 <1> ((R)-3-hydroxybutanoate, <1> enzyme expressed in SF9 cells, C242S mutant, C242 important for substrate binding [2]) [2]
21 <2> (β-hydroxybutyrate, <2> cyanylated enzyme [14]) [14]
200 <2> (β-hydroxybutyrate, <2> 240 mM Na^+, purified enzyme, thiol group 1 modified with 1,1'-azobis(NN'-dimethylformamide), reconstituted with total mitochondrial phospholipids [6]) [6]
260 <2> (β-hydroxybutyrate, <2> 240 mM Na^+, purified enzyme, thiol group 2 modified with 1,1'-azobis(NN'-dimethylformamide), reconstituted with total mitochondrial phospholipids [6]) [6]

pH-Optimum

6-6.5 <7> (<7> reduction of acetoacetate [26]) [26]
6.4-7 <9> (<9> reduction of acetoacetate [28]) [28]
6.8 <3> (<3> acetoacetate reduction [4]) [4]
7 <6> (<6> reduction of acetoacetate [13]) [13]
8 <11> [30]
8 <8> (<8> oxidation of β-hydroxybutyrate [23]) [23]
8.4 <9> (<9> oxidation of β-hydroxybutyrate [28]) [28]
8.5 <6> (<6> oxidation of β-hydroxybutyrate [13]) [13]
8.6 <3> (<3> oxidation of (R)-3-hydroxybutanoate [4]) [4]
9 <5> [12]

pH-Range

5.5-7.5 <8> (<8> acetoacetate reduction [23]) [23]
5.9-8.6 <3> (<3> acetoacetate reduction, 50% activity at pH 5.0, 66% at pH 8.6 [4]) [4]
7-9 <11> [30]
7.6-9 <9> (<9> oxidation of β-hydroxybutyrate [28]) [28]

Temperature optimum (°C)

35-37 <3> [4]

Temperature range (°C)

25-49 <3> (<3> 80% activity at 25°C and 49°C [4]) [4]

4 Enzyme Structure

Molecular weight

68000 <3> (<3> nondenaturing PAGE [4]) [4]
89000 <5> (<5> gel filtration [20]) [20]
106000 <2> (<2> submitochondrial vesicles, radiation inactivation [19]) [19]
112000 <8> (<8> gel filtration [23]) [23]
119000 <2> (<2> reconstituted enzyme, radiation inactivation, irradiation leads to loss of activity and fragmentation [19]) [19]
132000 <6> (<6> gel filtration [15]) [15]
140000 <3> (<3> gel filtration [4]) [4]

Subunits

tetramer <1> (<1> α_4, 4 * 33117, mature protein, calculated from deduced amino acid sequence [8]) [8]
tetramer <1> (<1> α_4, 4 * 38000, preprotein, calculated from deduced amino acid sequence [8]) [8]
tetramer <2> (<2> α_4, 4 * 31500, SDS-PAGE [16]) [16]
tetramer <2, 4> (<2,4> α_4, 4 * 27000, sedimentation equilibrium [10]) [10]
tetramer <2, 4> (<2,4> α_4, 4 * 31000, SDS-PAGE [10]) [10]
tetramer <3> (<3> α_4, 4 * 39000, SDS-PAGE [4]) [4]
tetramer <5> (<5> α_4, 4 * 23000, SDS-PAGE [20]) [20]
tetramer <6> (<6> α_4, 4 * 32000, SDS-PAGE [15]) [15]
tetramer <8> (<8> α_4, 4 * 28000, SDS-PAGE [23]) [23]

5 Isolation/Preparation/Mutation/Application

Source/tissue

brain <4> [7, 9]
heart <1, 2, 4, 10> [1, 2, 3, 5, 6, 8, 9, 10, 14, 17, 19, 24, 25, 27, 29, 31]
kidney <4> [9]
liver <4> [5, 7, 9, 10, 16, 22, 25]
muscle <1> [8]
stomach <4> (<4> gastric glandular mucosa [18]) [18]

Localization

cytoplasm <1-3> [1-5]
mitochondrion <3> [4]

mitochondrion <1, 2, 4> (<1, 2, 4> located on the matrix face of the inner membrane [1, 2, 22, 27]) [1, 2, 3, 5, 6, 7, 8, 10, 11, 14, 16, 17, 18, 19, 22, 24, 27, 31]

Purification

<2> (ammonium sulfate, acid precipitation, DEAE-Sepharose, glass beads [16]) [16]
<3> (protamine sulfate precipitation, Sephacryl S-400, S200, DEAE-Sepharose [4]) [4]
<5> (affinity chromatography on two triazine dyes [20]) [20]
<6> (DEAE-cellulose, Matrex gel blue A, Sephadex G-200, chromatofocussing [15]) [15]
<8> (ammonium sulfate, DEAE-Sepharose, 60°C, 10 min, hydroxylapatite, octyl-Sepharose, Sephadex G-200, DEAE-Sepharose [23]) [23]
<9> (ammonium sulfate precipitation, DEAE-cellulose [28]) [28]
<11> (ammonium sulfate precipitation, DEAE-Sephadex, Sephadex G-200 [30]) [30]
<2, 4> (phospholipase A_2 treatment, controlled-pore glass beads [10]) [10]

Cloning

<1> (His-tagged 3-hydroxybutyrate dehydrogenase, expressed in Escherichia coli [1]) [1]
<1> (mature form, expressed in Spodoptera frugiperda, SF9 cells [2]) [2]

Engineering

C242S <1> (<1> PCR derived cDNA clone [2]) [2]
M92V <1> (<1> PCR derived cDNA clone [2]) [2]
S24T <1> (<1> PCR derived cDNA clone [2]) [2]

6 Stability

Temperature stability

0-45 <3> (<3> stable at 0°C and 25°C for 90 min, 18% activity after 90 min at 45°C, activity restored after 60 min at 37°C, loss of activity at 60°C [4]) [4]
37 <5> (<5> 75% activity after 30 min in buffer containing 1 mM Ca^{2+}, 46% with 1 mM $MnCl_2$ [12]) [12]
37 <9> (<9> loss of 70% activity after 15 min, 0.24 mg protein/ml, Ca^{2+} and Mn^{2+} stabilize enzyme [28]) [28]

Storage stability

<3>, -20°C, crude extract, partially purified enzyme, several weeks, 0% inactivation [4]
<6>, -20°C, several months, 0%, 4°C, lyophilized, 2 month, 0% inactivation, 4°C, 0.1 M sodium phosphate buffer pH 7-8, 5 days, 50% inactivation, stability increases in the presence of 1-10 mM EDTA [15]
<8>, -20°C, 1 mg enzyme/ml, 0.2 mM potassium phosphate pH 8.0, 10% glycerol, 10 mM 2-mercaptoethanol, 6 months, 0% inactivation, 0°C, 0.1 mg enzyme/ml, 1 day, 50% inactivation [23]

<11>, 4°C, Tris-HCl buffer, 50-100 mM Mg^{2+}, Mn^{2+}, Ba^{2+}, or Ca^{2+}, enzyme looses all activity in the presence of phosphate buffer or EDTA within 2 min [30]

References

[1] Loeb-Hennard, C.; McIntyre, J.O.: (R)-3-Hydroxybutyrate dehydrogenase: selective phosphatidylcholine binding by the C-terminal domain. Biochemistry, **39**, 11929-11938 (2000)

[2] Green, D.; Marks, a.R.; Fleischer, S.; McIntyre, J.O.: Wild type and mutant human heart (R)-3-hydroxybutyrate dehydrogenase expressed in insect cells. Biochemistry, **35**, 8158-8165 (1996)

[3] Klein, K.; Rudy, B.; McIntyre, J.O.; Fleischer, S.; Trommer, W.E.: Specific interaction of(R)-3-hydroxybutyrate dehydrogenase with membrane phosphatidylcholine as studied by ESR spectroscopy in oriented phospholipid multilayers: coenzyme binding enhances the interaction with phosphatidylcholine. Biochemistry, **35**, 3044-3049 (1996)

[4] Szewczyk, E.; Rozolska, M.: Occurence, purification and properties of the staphylococcal β-hydroxybutyrate dehydrogenase. Acta Microbiol. Pol., **43**, 33-45 (1994)

[5] Adam, Pascal; Duncan, T.M.; McIntyre, J.O.; Carter, C.E.; Fu, C.; Melin, M.; Latruffe, N.; Fleischer, S.: Monoclonal antibodies for structure-function studies of (R)-3-hydroxybutyrate dehydrogenase, a lipid-dependent membrane-bound enzyme. Biochem. J., **292**, 863-872 (1993)

[6] Dalton, L.A.; McIntyre, J.O.; Fleischer, S.: Effect of selective thiol-group derivatization on enzyme kinetics of (R)3-hydroxybutyrate dehydrogenase. Biochem. J., **296**, 563-569 (1993)

[7] Dutra, J.C.; Dutra-Filho, C.S.; Cardozo, S.E.C.; Wannmacher, C.M.D.; Sarkis, J.J.F.; Wajner, M.: Inhibition of succinate dehydrogenase and β-hydroxybutyrate dehydrogenase activities by methylmalonate in brain and liver of developing rats. J. Inher. Metab. Dis., **16**, 147-153 (1993)

[8] Marks, A.R.; McIntyre, J.O.; Duncan, T.M.; Erdjument-Bromage, H.; Tempst, P.; Fleischer, S.: Molecular cloning and characterization of (R)3-hydroxybutyrate dehydrogenase from human heart. J. Biol. Chem., **267**, 15459-15463 (1992)

[9] Bailly, A.; Lone, Y.C.; Latruffe, N.: Post-transcriptional analysis of rat mitochondrial D-3-hydroxybutyrate dehydrogenase control through development and physiological stages. Biol. Cell., **73**, 121-129 (1991)

[10] McIntyre, J.O.; Latruffe, N.; Brenner, S.C.; Fleischer, S.: Comparison of 3-hydroxybutyrate dehydrogenase from bovine heart and rat liver mitochondria. Arch. Biochem. Biophys., **262**, 85-98 (1988)

[11] Cortese, J.D.; Fleischer, S.: Noncooperative vs. cooperative reactivation of D-β-hydroxybutyrate dehydrogenase: multiple equilibria for lecithin binding are determined by the physical state (soluble vs. bilayer) and composition of the phospholipids. Biochemistry, **26**, 5283-5293 (1987)

[12] Worral, E.B.; Gassain, S.; Cox, D.J.; Sugden, M.C.; Palmer, T.N.: 3-Hydroxyisobutyrate dehydrogenase, an impurity in commercial 3-hydroxybutyrate dehydrogenase. Biochem. J., **241**, 297-300 (1987)
[13] Kovar, J.; Matyskova, I.; Matyska, L.: Kinetics of D-3-hydroxybutyrate dehydrogenase from Paracoccus denitrificans. Biochim. Biophys. Acta, **871**, 302-309 (1986)
[14] Dubois, H.; Fritsche, T.M.; Trommer, W.E.; McIntyre, J.O.; Fleischer, S.: Cyanylation of 3-hydroxybutyrate dehydrogenase. Biol. Chem. Hoppe-Seyler, **367**, 343-353 (1986)
[15] Matyskova I.; Kovar, J.; Racek, P.: Purification and properties of D-3-hydroxybutyrate dehydrogenase from Paracoccus denitrificans. Biochim. Biophys. Acta, **839**, 300-307 (1985)
[16] Burnett, B.K.; Khorana, H.G.: A rapid and efficient procedure for the purification of mitochondrial β-hydroxybutyrate dehydrogenase. Biochim. Biophys. Acta, **815**, 51-56 (1985)
[17] Phelps, D.C.; Hatefi, Y.: Inhibition of D(-)-β-hydroxybutyrate dehydrogenase by butanedione, phenylglyoxal, and diethyl pyrocarbonate. Biochemistry, **20**, 459-463 (1981)
[18] MacGill, A.K.; Anderson, N.G.; Trotman, C.N.A.; Carrington, J.M.; Hanson, P.J.: D-3-Hydroxybutyrate dehydrogenase and metabolism of ketone bodies by rat stomach. Biochem. Soc. Trans., **607**, 788 (1984)
[19] McIntyre, J.O.; Churchill, P.; Maurer, A.; Berenski, C.J.; Jung, C.Y.; Fleischer, S.: Target size of D-β-hydroxybutyrate dehydrogenase. J. Biol. Chem., **258**, 953-959 (1983)
[20] Scawen, M.D.; Darbyshire, J.; Harvey, M.J.; Atkinson, T.: The rapid purification of 3-hydroxybutyrate dehydrogenase and malate dehydrogenase on triazine dye affinity matrices. Biochem. J., **203**, 699-705 (1982)
[21] Kluger, R.; Tsui, W.-C.: Inhibition of bacterial D-3-hydroxybutyrate dehydrogenase by substrates and substrate analgues. Can. J. Biochem., **59**, 810-815 (1981)
[22] Miyahara, M.; Utsumi, K.; Deamer, D.W.: Selective interaction of D β hydroxybutyrate dehydrogenase with intracellular membranes. Biochim. Biophys. Acta, **641**, 222-231 (1981)
[23] Nakada T.; Fukui, T.; Saito, T.; Miki, K.; Oji, C.; Matsuda, S.; Ushijima, A.; Tomita, K.: Purification and properties of D-β-hydroxybutyrate dehydrogenase from Zoogloea ramiger I-16-M. J. Biochem., **89**, 625-635 (1981)
[24] Phelps, D.C.; Hatefi, Y.: Inhibition of D(-)-β-hydroxybutyrate dehydrogenase by modifiers of disulfides, thiols, and vicinal dithiols. Biochemistry, **20**, 453-458 (1981)
[25] Tucker, G.A.; Dawson, A.P.: The kinetics of rat liver and heart mitochondria β-hydroxybutyrate dehydrogenase. Biochem. J., **179**, 579-581 (1979)
[26] Kluger, R.; Nakaoka, K.; Tsui, W.-C.: Substrate analogue studies of the specificity and catalytic mechanism of D-3-hydroxybutyrate dehydrogenase. J. Am. Chem. Soc., **100**, 7388-7392 (1978)
[27] McIntyre, J.O.; Bock, H.-G.O.; Fleischer, S.: The orientation of D-β-hydroxybutyrate dehydrogenase in the mitochondrial inner membrane. Biochim. Biophys. Acta, **513**, 255-267 (1978)

[28] Dhariwal, K.R.; Venkitasubramanian, T.A.: Purification and properties of β-hydroxybutyrate dehydrogenase from Mycobacterium phlei ATCC354. J. Gen. Microbiol., **104**, 123-126 (1978)
[29] Tan, A.W.H.; Smith, C.M.; Aogaichi, T.; Plaut, G.W.E.: Inhibition of D(-)-3-hydroxybutyrate dehydrogenase by malonate analoges. Arch. Biochem. Biophys., **166**, 164-173 (1975)
[30] Thompson, E.D.; Nakata, H.M.: Characterization and partial purification of β-hydroxybutyrate dehydrogenase from sporulating cells of Bacillus cereus T. Can. J. Microbiol., **19**, 673-677 (1973)
[31] Nielsen, N.C.; Fleischer, S.: Mitochondrial D-β-hydroxybutyrate dehydrogenase. III. Isolation and characterization. J. Biol. Chem., **248**, 2549-2555 (1973)
[32] Nielsen, N.C.; Zahler, W.L.; Fleischer, S.: Mitochondrial D-β-hydroxybutyrate dehydrogenase. IV. Kinetic analysis of reaction mechanism. J. Biol. Chem., **248**, 2556-2562 (1973)

3-Hydroxyisobutyrate dehydrogenase 1.1.1.31

1 Nomenclature

EC number
1.1.1.31

Systematic name
3-hydroxy-2-methylpropanoate:NAD^+ oxidoreductase

Recommended name
3-hydroxyisobutyrate dehydrogenase

Synonyms
HIBADH [Swissprot]
β-hydroxyisobutyrate dehydrogenase

CAS registry number
9028-39-1

2 Source Organism

<1> *Oryctolagus cuniculus* [1, 3, 7, 8]
<2> *Rhodopseudomonas spheroides* [2]
<3> *Candida rugosa* (IFO 1542 [3]; IFO 0750 [3, 4]) [3, 4, 8]
<4> *Candida parapsilosis* [3]
<5> *Corynebacterium hydrocarbons* [3]
<6> *Corynebacterium paurometabolom* [3]
<7> *Endomyces reessii* [3]
<8> *Micrococcus flavus* [3]
<9> *Micrococcus luteus* [3]
<10> *Nocardia lyena* [3]
<11> *Pseudomonas aeruginosa* [3]
<12> *Pseudomonas dacunhae* [3]
<13> *Rhodotorula rubra* [3]
<14> *Trichosporum aculeatum* [3]
<15> *Sus scrofa* [3, 6]
<16> *Bos taurus* [3]
<17> *Pseudomonas putida* (E23 [8]) [5, 8]
<18> *Rattus norvegicus* [7, 9]
<19> *Homo sapiens* [7]

3 Reaction and Specificity

Catalyzed reaction

3-hydroxy-2-methylpropanoate + NAD^+ = 2-methyl-3-oxopropanoate + NADH2

Reaction type

oxidation
redox reaction
reduction

Natural substrates and products

S D,L-3-hydroxyisobutyrate + NAD^+ <2> (Reversibility: ? <2> [2]) [2]
P 2-methyl-3-oxopropionate + NADH
S D,L-3-hydroxyisobutyrate + NAD^+ <3, 4> (<3> strain IFO 0750 [3]) (Reversibility: ? <3, 4> [3]) [3]
P D-2-methyl-3-oxopropionate + NADH
S D,L-3-hydroxyisobutyrate + NAD^+ <3, 7, 10-12> (<3> strain IFO 1542 [3]) (Reversibility: ? <3, 7, 10-12> [3]) [3]
P L-2-methyl-3-oxopropionate + NADH
S L-3-hydroxyisobutyrate + NAD^+ <1, 5, 6, 8, 9, 13-18> (Reversibility: r <1> [1, 8]; ? <5, 6, 8, 9, 13-18> [3, 8, 9]) [1, 3, 8, 9]
P L-2-methyl-3-oxopropionate + NADH

Substrates and products

S 2-methyl-D,L-serine + NAD^+ <17> (Reversibility: ? <17> [8]) [8]
P ?
S 3-hydroxypropionate + NAD^+ <1, 17> (Reversibility: ? <1, 17> [7, 8]) [7, 8]
P ?
S D,L-3-hydroxyisobutyrate + NAD^+ <2> (Reversibility: ? <2> [2]) [2]
P 2-methyl-3-oxopropionate + NADH
S D,L-3-hydroxyisobutyrate + NAD^+ <3, 4> (<3> strain IFO 0750 [3]) (Reversibility: ? <3, 4> [3]) [3]
P D-2-methyl-3-oxopropionate + NADH
S D,L-3-hydroxyisobutyrate + NAD^+ <3, 7, 10-12> (<3> strain IFO 1542 [3]) (Reversibility: ? <3, 7, 10-12> [3]) [3]
P L-2-methyl-3-oxopropionate + NADH
S L-3-hydroxyisobutyrate + NAD^+ <1, 5, 6, 8, 9, 13-18> (Reversibility: r <1> [1, 8]; ? <5, 6, 8, 9, 13-18> [3, 8, 9]) [1, 3, 8, 9]
P L-2-methyl-3-oxopropionate + NADH
S L-serine + NAD^+ <17> (Reversibility: ? <17> [8]) [8]
P ?

Inhibitors

(R)-3-hydroxy-2-methylpropanoate <1> (<1> competitive inhibitor of the S-isomer [1]) [1]
$AgNO_3$ <1, 3, 17> [8]

$HgCl_2$ <15, 17> [6, 8]
NADH <1> (<1> product inhibition [1]) [1]
iodoacetate <15> [6]
p-chloromercuribenzoate <1, 3, 15, 17> [6, 8]

Cofactors/prosthetic groups
NAD^+ <1-3, 15, 17> [1-4, 6, 8]

Turnover number (min^{-1})
29.4 <1> ((R)-3-hydroxyisobutyrate) [1]
426 <1> ((S)-3-hydroxyisobutyrate) [1]

Specific activity (U/mg)
10.7 <1> [1]
20.9 <15> [6]
36.1 <3> [8]
294 <17> [8]
294 <17> [8]

K_m-Value (mM)
0.022 <2, 18> (NAD^+, <18> wild type [9]) [2, 9]
0.023 <1> (NAD^+) [7, 8]
0.05 <3> (NAD^+) [4, 8]
0.054 <15> (NAD^+) [6]
0.056 <18> (L-3-hydroxyisobutyrate, <18> wild type, NAD as cofactor [9]) [9]
0.061 <1> (L-3-hydroxyisobutyrate) [7, 8]
0.1 <2> (D,L-3-hydroxyisobutyrate) [2]
0.104 <18> (L-3-hydroxyisobutyrate, <18> wild type, NADP as cofactor [9]) [9]
0.11 <18> (L-3-hydroxyisobutyrate, <18> mutant D33R, NADP as cofactor [9]) [9]
0.112 <18> (L-3-hydroxyisobutyrate, <18> mutant D33R, NAD as cofactor [9]) [9]
0.12 <15> (D,L-3-hydroxyisobutyrate) [6]
0.12 <17> (L-3-hydroxyisobutyrate) [8]
0.24 <18> ($NADP^+$, <18> mutant D33R [9]) [9]
0.37 <3> (L-3-hydroxyisobutyrate) [4, 8]
0.44 <17> (NAD^+) [8]
0.96 <18> ($NADP^+$, <18> wild type [9]) [9]
1.24 <18> (NAD^+, <18> mutant D33R [9]) [9]
1.25 <3> (D-3-hydroxyisobutyrate) [4, 8]
18 <17> (3-hydroxypropionate) [8]
44 <17> (L-serine) [8]
83 <17> (2-methyl-D,L-serine) [8]

pH-Optimum
8.6 <2> [2]
9 <15> [6]

9-10.5 <3> [8]
9 11 <1> [1, 8]
10 <17> [8]

pH-Range
7-11 <15> [6]
7-11.5 <1> [1]

4 Enzyme Structure

Molecular weight
74000 <1> (<1> gel filtration [1]) [1, 8]
80000 <3> (<3> gel filtration [4]) [4, 8]
120000 <17> (<17> gel filtration [8]) [8]

Subunits
homodimer <1, 3> (<1> α_2, 2 * 34000, SDS-PAGE [1]; <3> α_2, 2 * 40000, SDS-PAGE [4]) [1, 4, 8]
tetramer <17> (<17> α_4, 4 * 30000, SDS-PAGE [8]) [8]
Additional information <2> (<2> hydroxybutyrate dehydrogenase and 3-hydroxyisobutyrate dehydrogenase activity [2]) [2]

5 Isolation/Preparation/Mutation/Application

Source/tissue
heart <15> [6]
kidney <15> [3, 6]
liver <1, 15, 16> [1, 3, 6, 7, 8]

Purification
<1> [1, 7]
<3> [3, 4]
<15> [6]
<17> [5, 8]
<18> [7]
<19> [7]

Crystallization
<3> (powdered ammonium sulfate is added to the enzyme solution until it becomes slightly turbid, after 4 days crystallization is complete [4]) [4]

Cloning
<18> (expression in Escherichia coli [7]) [7]
<19> (expression in Escherichia coli [7]) [7]

Engineering
D33R <18> (<18> higher affinity to cofactor $NADP^+$ than to natural cofactor NAD^+ [9]) [9]

6 Stability

pH-Stability

4.5-7.3 <3> [8]
5.7-9.3 <17> [8]
7-11.5 <1> [8]

References

[1] Rougraff, P.M.; Paxton, R.; Kuntz, M.J.; Crabb, D.W.; Harris, R.A: Purification and characterization of 3-hydroxyisobutyrate dehydrogenase from rat liver. J. Biol. Chem., **263**, 327-331 (1988)
[2] Worall, E.B.; Gassain, S.; Cox, D.J.; Sugden, M.C.; Palmer, T.N.: 3-Hydroxyisobutyrate dehydrogenase, an impurity in commercial 3-hydroxybutyrate dehydrogenase. Biochem. J., **241**, 297-300 (1987)
[3] Hasegawa, J: Distribution in organisms and stereospecificity of β-hydroxyisobutyrate dehydrogenase. Agric. Biol. Chem., **45**, 2899-2901 (1981)
[4] Hasegawa, J: Purification, crystallization and some properties of β-hydroxyisobutyrate dehydrogenase from Candida rugosa IFO 0750. Agric. Biol. Chem., **45**, 2805-2814 (1981)
[5] de Paul Marshall, V.; Sokatch, J.R.: Regulation of valine catabolism in Pseudomonas putida. J. Bacteriol., **110**, 1073-1081 (1972)
[6] Robinson, W.G.; Coon, M.J.: The purification and properties of β-hydroxyisobutyrate dehydrogenase. J. Biol. Chem., **225**, 511-521 (1956)
[7] Hawes, J.W.; Crabb, D.W.; Chan, R.J.; Rougraff, P.M.; Paxton, R.; Harris, R.A.: Mammalien 3-hydroxyisobutyrate dehydrogenase. Methods Enzymol., **324**, 218-228 (2000)
[8] Chowdhury, E.K.; Nagata, S.; Misono, H.: 3-Hydroxyisobutyrate dehydrogenase from Pseudomonas putida E23: purification and characterization. Biosci. Biotechnol. Biochem., **60**, 2043-2047 (1996)
[9] Hawes, J.W.; Harper, E.T.; Crabb, D.W.; Harris, R.A.: Structural and mechanistic aspects of a new family of dehydrogenases, the β-hydroxyacid dehydrogenases. Adv. Exp. Med. Biol., **414**, 395-402 (1996)

Mevaldate reductase 1.1.1.32

1 Nomenclature

EC number
1.1.1.32

Systematic name
(R)-mevalonate: NAD^+ oxidoreductase

Recommended name
mevaldate reductase

Synonyms
mevalonic dehydrogenase

CAS registry number
9028-33-5

2 Source Organism

<1> *Sus scrofa* [1-3]
<2> *Rattus norvegicus* [2]

3 Reaction and Specificity

Catalyzed reaction
(R)-mevalonate + NAD^+ = mevaldate + NADH (<1> formation of mevalonic acid is strongly favoured [2, 3]; <1> stereospecific insertion of hydride into C-5, hydride is added from the re-face of carbonyl group in 3R and 3S-isomers [1])

Reaction type
redox reaction

Natural substrates and products
- **S** (R)-3-hydroxy-3-methyl-5-oxovaleric acid + NADH <1> (Reversibility: ir <1> [2, 3]) [1, 2, 3]
- **P** (R)-3-hydroxy-3-methyl-5-hydroxyvaleric acid + NAD^+ <1> [1, 2, 3]
- **S** (S)-3-hydroxy-3-methyl-5-oxovaleric acid + NADH <1> (Reversibility: ir <1> [2, 3]) [1, 2, 3]
- **P** (S)-3-hydroxy-3-methyl-5-hydroxyvaleric acid + NAD^+ <1> [1, 2, 3]

Substrates and products

- **S** (R)-3-hydroxy-3-methyl-5-oxovaleric acid + NADH <1> (Reversibility: ir <1> [2, 3]) [1, 2, 3]
- **P** (R)-3-hydroxy-3-methyl-5-hydroxyvaleric acid + NAD^+ <1> [1, 2, 3]
- **S** (R)-3-hydroxy-3-methyl-5-oxovaleric acid + NADPH <1> (Reversibility: ir <1> [2, 3]) [1, 2, 3]
- **P** (R)-3-hydroxy-3-methyl-5-hydroxyvaleric acid + $NADP^+$ <1> [1, 2, 3]
- **S** (S)-3-hydroxy-3-methyl-5-oxovaleric acid + NADH <1> (Reversibility: ir <1> [2, 3]) [1, 2, 3]
- **P** (S)-3-hydroxy-3-methyl-5-hydroxyvaleric acid + NAD^+ <1> [1, 2, 3]

Inhibitors

(R,S)-mevaldate <1> (<1> 20 mM or higher [2]) [2]

Cofactors/prosthetic groups

NADH <1> [1, 2, 3]
NADPH <2> [2, 3]

Specific activity (U/mg)

0.2 <1> (<1> partially purified liver enzyme [1]) [1]
0.63 <1> (<1> partially purified liver enzyme [2]) [2]

K_m-Value (mM)

0.003 <1> (NADH) [2]
0.09 <1> (NADPH) [2]
4 <1> ((R)-mevaldic acid) [2]

pH-Optimum

5.4 <1> [2]

pH-Range

4.6-6.9 <1> [2]

5 Isolation/Preparation/Mutation/Application

Source/tissue

heart <1> [3]
kidney <1> [3]
liver <1, 2> [1-3]

Purification

<1> (ammonium sulfate precipitation, calcium phosphate gel, DEAE-cellulose, bentonite precipitation [2]) [2]

References

[1] Ngan, H.-L.; Popjak, G.: Stereochemistry of the reaction catalyzed by mevaldate reductase. Bioorg. Chem., **4**, 166-180 (1975)

[2] Schlesinger, M.J.; Coon, M.J.: Reduction of mevaldic acid to mevalonic acid by a partially purified enzyme from liver. J. Biol. Chem., **236**, 2421-2424 (1961)
[3] Coon, M.J.; Kupiecki, F.P.; Dekker, E.E.; Schlesinger, M.J.; del Campillo, E.: The enzymic synthesis of branched-chain acids. CIBA Symposium on the biosynthesis of terpenes and sterols (Wolstenholme, G.E.W.; O. Conner, M., eds.), 62-72 (1954)

Mevaldate reductase (NADPH) 1.1.1.33

1 Nomenclature

EC number
1.1.1.33

Systematic name
(R)-mevalonate:NADP$^+$ oxidoreductase

Recommended name
mevaldate reductase (NADPH)

Synonyms
mevaldate (reduced nicotinamide adenine dinucleotide phosphate) reductase

CAS registry number
9028-34-6

2 Source Organism

<1> *Sus scrofa* [1]

3 Reaction and Specificity

Catalyzed reaction
(R)-mevalonate + NADP$^+$ = mevaldate + NADPH

Reaction type
redox reaction

Natural substrates and products
- **S** (R)-3-hydroxy-3-methyl-5-oxovaleric acid + NADPH <1> (Reversibility: ir <1> [1]) [1]
- **P** (R)-3-hydroxy-3-methyl-5-hydroxyvaleric acid + NADP$^+$ <1> [1]

Substrates and products
- **S** (R)-3-hydroxy-3-methyl-5-oxovaleric acid + NADPH <1> (Reversibility: ir <1> [1]) [1]
- **P** (R)-3-hydroxy-3-methyl-5-hydroxyvaleric acid + NADP$^+$ <1> [1]
- **S** (S)-3-hydroxy-3-methyl-5-oxovaleric acid + NADH <1> (Reversibility: ir <1> [1]) [1]
- **P** (S)-3-hydroxy-3-methyl-5-hydroxyvaleric acid + NAD$^+$ <1> [1]
- **S** 4-cyanobenzaldehyde + NADPH <1> (Reversibility: ir <1> [1]) [1]

P 4-cyanobenzylalcohol + $NADP^+$ <1> [1]
S pyridine-3-carboxaldehyde + NADPH <1> (Reversibility: ir <1> [1]) [1]
P pyridine-3-methanol + $NADP^+$ <1> [1]

Inhibitors
sodium aminobarbitone <1> (<1> 1 mM, 48% inhibition, noncompetitive [1]) [1]
thiobarbituric acid <1> (<1> 1 mM, 22% inhibition, noncompetitive [1]) [1]

Cofactors/prosthetic groups
NADH <1> (<1> giving one-fifth of the rate obtained with NADPH [1]) [1]
NADPH <1> [1]

Specific activity (U/mg)
0.573 <1> [1]

K_m-Value (mM)
0.09 <1> (NADPH) [1]
0.15 <1> (4-cyanobenzaldehyde) [1]
0.26 <1> ((3RS)-mevaldate) [1]

4 Enzyme Structure

Molecular weight
30000 <1> (<1> gel filtration [1]) [1]

5 Isolation/Preparation/Mutation/Application

Source/tissue
liver <1> [1]

Purification
<1> (ammonium sulfate precipitation, calcium phosphate gel, gel filtration [1]) [1]

6 Stability

Storage stability
<1>, -20°C, 3 months, no loss of activity [1]

References

[1] Beedle, A.S.; Rees, H.H.; Goodwin, T.W.: Some properties and a suggested reclassification of mevaldate reductase. Biochem. J., **139**, 205-209 (1974)

Hydroxymethylglutaryl-CoA reductase (NADPH) 1.1.1.34

1 Nomenclature

EC number
1.1.1.34

Systematic name
(R)-mevalonate:$NADP^+$ oxidoreductase (CoA-acylating)

Recommended name
hydroxymethylglutaryl-CoA reductase (NADPH)

Synonyms
3-hydroxy-3-methylglutaryl-CoA reductase
3-hydroxy-3-methylglutaryl-CoA reductase (NADPH)
HMG-CoA reductase [Swissprot]
HMG2.2 [Swissprot]
HMG3.3 [Swissprot]
HMGCoA reductase-mevalonate:NADP-oxidoreductase (acetylating CoA)
HMGR [Swissprot]
HMGR1 [Swissprot]
HMGR2 [Swissprot]
NADPH-hydroxymethylglutaryl-CoA reductase
S-3-hydroxy-3-methylglutaryl-CoA reductase
β-hydroxy-β-methylglutaryl coenzyme A reductase
β-hydroxy-β-methylglutaryl-Co A reductase
hydroxymethylglutaryl CoA reductase (NADPH)
hydroxymethylglutaryl-coenzyme A reductase (reduced nicotinamide adenine dinucleotide phosphate)
mevalonate:$NADP^+$ oxidoreductase (acetylating CoA)

CAS registry number
9028-35-7

2 Source Organism

<1> *Rattus norvegicus* [1, 3-5, 9, 11-13, 16-18]
<2> *Saccharomyces cerevisiae* [2, 4, 10]
<3> *Fusarium oxysporum* [6, 7]
<4> *hamster* [8]
<5> *Hevea brasiliensis* [12]
<6> *Pisum sativum* [12]

<7> *Nepeta cataria* [14]
<8> *Homo sapiens* [15, 20]
<9> *Schistosoma mansoni* [19]
<10> *Brassica oleracia* (cauliflower) [21]
<11> *Haloferax volcanii* [22]
<12> *Pseudomonas mevalonii* [23]
<13> *Trypanosoma cruzi* (Schizotrypanum) [24]
<14> *Streptomyces sp.* (CL190 [25]) [25]
<15> *Mus musculus* [26]
<16> *Leishmania major* [27]
<17> *Sulfolobus solfataricus* [28-30]
<18> *Staphylococcus aureus* [31]

3 Reaction and Specificity

Catalyzed reaction

(S)-3-hydroxy-3-methylglutaryl-CoA + 2 $NADPH_2$ = (R)-mevalonate + CoA + 2 NADP

Reaction type

oxidation
redox reaction
reduction

Natural substrates and products

S 3-hydroxy-3-methylglutaryl-CoA + NADPH <1-18> (Reversibility: ? <1, 3-10, 12-16> [1, 3, 5-9, 11-21, 23-27]; ir <2> [2]; r <11, 17, 18> [22, 29, 30, 31]) [1-31]
P (R)-mevalonate + CoA + $NADP^+$

Substrates and products

S 3-hydroxy-3-methylglutaryl-CoA + NADPH <1-18> (Reversibility: ? <1, 3-10, 12-16> [1, 3, 5-9, 11-21, 23-27]; ir <2> [2]; r <11, 17, 18> [22, 29, 30, 31]) [1-31]
P (R)-mevalonate + CoA + $NADP^+$
S (R,S)-mevaldehyde + NADPH <11> (Reversibility: <11> [22]) [22]
P ?
S (R,S)-mevaldehyde + acetate <11> (Reversibility: <11> [22]) [22]
P ?

Inhibitors

1,10-phenanthroline <1> [3]
3,3-dimethylglutaryl-CoA <8> (competitive inhibitor [20]) [20]
3-hydroxy-3-methylglutaryl-CoA <2> (<2> 0.05 mM [10]) [10]
3-hydroxyglutaryl-CoA <8> (<8> competitive inhibitor [20]) [20]
3-methylglutaryl-CoA <8> (<8> competitive inhibitor [20]) [20]
CoA disulfide <2> (<2> no inactivation in presence of NADPH 100 μM or HMG-CoA 50 μM [10]) [10]

CoASH <1, 2, 6> (<2> 0.05 mM [10]) [2, 3, 10, 12]
DTNB <2> [10]
Fe^{2+} <1> [3]
Fe^{3+} <1> [3]
$MgATP^{2-}$ <1, 5> (<5> 4 mM [12]) [12]
(R)-3-hydroxy-3-methylglutaryl-CoA <11> (<11> competitive inhibitor [22]) [22]
(S)-4-carboxy-3-hydroxy-3-methylbutyryl-CoA <8> (<8> competitive inhibitor [20]) [20]
(S)-4-carboxy-3-hydroxybutyryl-CoA <8> (<8> competitive inhibitor [20]) [20]
SDS <1> [3]
acetoacetyl-CoA <1> [3]
acetyl-CoA <1> [3]
adenosine-2'-monophospho-5'-diphosphoribose <1> (<1> competitive inhibitor for NADPH binding site [17]) [17]
cholate <1> [1]
compactin <1> (<1> competitive inhibitor for HMG-CoA binding site [17]) [17, 18]
deoxycholate <1> [3]
digitonine <1> (2% digitonin, 80% inhibition [3]) [3]
fluvastatin <18> (<18> competitive inhibitor [31]) [31]
glutathione <2> [10]
hydroxymethylglutarate <5, 6> (<5> 2 mM, slightly inhibitory [12]) [12]
lovastatin <1, 9, 11, 13, 17> (<1, 17> competitive inhibitor for HMG-CoA binding site [17, 28]; <13> competitive inhibitor for HMG-CoA binding site and noncompetitive inhibitor for NADPH binding site [24]) [17-19, 22, 24, 28]
p-chloromercuribenzoate <5> (<5> 1 mM, complete inhibition [12]) [12]
p-hydroxymercuribenzoate <1, 2> (<1> 0.01 mM inhibited 97% of reduction [3]; <2> 3.7 mM, complete inhibition [2]) [2, 3]

Cofactors/prosthetic groups
NADH <3, 5, 18> [7, 12, 31]
NADPH <1-18> [1-31]

Activating compounds
dithiothreitol <5> (<5> 2 mM [12]) [12]
thiol <5> [12]

Specific activity (U/mg)
0.003 <7> [14]
0.0035 <2> [2]
0.0036 <9> [19]
0.0147 <1> [5]
0.0164 <13> [24]
0.03 <1> [3]
0.0335 <16> [27]

0.145 <3> [6, 7]
1.6 <14> [25]
17500 <17> [28]
24000 <11> [22]

K_m-Value (mM)
0.003 <9> (3-hydroxy-3-methylglutaryl-CoA) [19]
0.004 <1> (S-3-hydroxy-3-methylglutaryl-CoA) [11]
0.006 <1> (3-hydroxy-3-methylglutaryl-CoA, <1> only one enantiomer [3]) [3]
0.008 <14> (3-hydroxy-3-methylglutaryl-CoA) [25]
0.01 <15> (3-hydroxy-3-methylglutaryl-CoA, <15> reductase from tumor [26]) [26]
0.012 <1> (3-hydroxy-3-methylglutaryl-CoA) [3]
0.014 <15> (3-hydroxy-3-methylglutaryl-CoA, <15> reductase from liver [26] and reductase from tumor [26]) [26]
0.015 <15> (3-hydroxy-3-methylglutaryl-CoA, <15> reductase from liver, implanted tumor [26]) [26]
0.017 <17> (3-hydroxy-3-methylglutaryl-CoA) [28]
0.019 <15> (3-hydroxy-3-methylglutaryl-CoA, <15> reductase from liver, implanted tumor [26]) [26]
0.021 <3> (3-hydroxy-3-methylglutaryl-CoA) [7]
0.023 <17> (NADPH) [28]
0.024 <15> (3-hydroxy-3-methylglutaryl-CoA, <15> reductase from liver [26]) [26]
0.028 <13> (3-hydroxy-3-methylglutaryl-CoA) [24]
0.03 <11> (CoA) [22]
0.032 <11> (NADPH, <11> reduction of R,S-mevaldehyde [22]) [22]
0.035 <11> ($NADP^+$) [22]
0.037 <13> (NADPH) [24]
0.04 <16> ((R,S)-3-hydroxy-3-methylglutaryl-CoA) [27]
0.045 <17> (3-hydroxy-3-methylglutaryl-CoA, <17> wild type [30]) [30]
0.052 <9> (NADPH) [19]
0.055 <17> (NADPH, <17> wild type [30]) [30]
0.056 <5> ((R,S)-3-hydroxy-3-methylglutaryl-CoA) [12]
0.06 <11> ((S)-3-hydroxy-3-methylglutaryl-CoA) [22]
0.062 <14> (NADPH) [25]
0.066 <11> (NADPH) [22]
0.076 <17> (3-hydroxy-3-methylglutaryl-CoA, <17> mutant [30]) [30]
0.081 <16> (NADPH) [27]
0.083 <17> (NADPH, <17> mutant [30]) [30]
0.47 <11> ((R,S)-mevaldehyde, <11> oxidative acylation of (R,S)-mevaldehyde [22]) [22]
0.55 <11> ((R,S)-mevaldehyde, <11> reduction of (R,S)-mevaldehyde [22]) [22]

pH-Optimum
5.5 <17> [28, 29]
6 <11> (<11> reduction of (R,S)-mevaldehyde [22]) [22]

6.5-7.5 <1> [12]
6.6-6.9 <5> [12]
6.8 <6> [12]
7 <1> [3]
7.2 <14> [25]
7.3 <11> [22]
7.4 <3> [7]
8 <11> (<11> oxidative acylation of (R,S)-mevaldehyde [22]) [22]
9.5 <17> (<17> oxidative acylation of (R,S)-mevaldehyde or mevalonate [29]) [29]

pH-Range

6-9 <3> [7]

Temperature optimum (°C)

30 <5> (<5> assay at [12]) [12]
35-60 <14> [25]
37 <1, 3, 7> (<1,7> assay at [13, 14]) [7, 13, 14]
85 <17> [28]

Temperature range (°C)

15-50 <3> [7]
15-60 <14> [25]

4 Enzyme Structure

Molecular weight

11000 <3> (gel filtration [6]) [6, 7]
66000 <9> (<9> gel filtration [19]) [19]
100000 <1, 14> (<1> PAGE [16, 18]; <14> gel filtration [25]) [16, 18, 25]
184000 <16> (<16> gel filtration [27]) [27]
200000 <1> (<1> PAGE [16]) [16]
217000-226000 <1> (<1> gel filtration, ultracentrifugation [3]) [3]

Subunits

dimer <14> (<14> α_2, 2 * 41000, SDS-PAGE [25]) [25]
monomer <3, 9> (<3> 1 * 11000, SDS-PAGE [6, 7]; <9> 1 * 66000, SDS-PAGE [19]) [6, 7, 19]
tetramer <16> (<16> α_4, 4 * 46000, SDS-PAGE [27]) [27]

Posttranslational modification

phosphoprotein <1, 8, 10> (<1, 8, 10> activity is regulated by phosphorylation and dephosphorylation [9, 15, 17, 21]) [9, 15, 17, 21]

5 Isolation/Preparation/Mutation/Application

Source/tissue

adrenal gland <4> [8]
latex <5> [12]
leaf <7> [14]
leukocyte <8> [15]
liver <1, 15> [1, 3, 9, 13, 16-18, 26]
lymphocyte <8> [15]
mucosa <1> [11]
small intestine <1> [11]

Localization

chloroplast <7> [14]
cytosol <2, 3, 7, 13> [4, 7, 14, 24]
endoplasmic reticulum <1, 6> [12, 18]
glycosome <13> [24]
membrane <5> (lutoid) [12]
microsome <1, 8, 15> [1, 3-5, 9, 13, 15, 16, 18, 26]
mitochondrion <1, 4> [1, 4, 8]

Purification

<1> [1, 3, 4, 9, 11, 13]
<2> [2, 4]
<3> [6, 7]
<4> [8]
<7> [14]
<8> [15]
<9> [19]
<11> [22]
<13> [24]
<14> [25]
<16> [27]
<17> [28-30]
<18> [31]

Cloning

<11> (expression in Escherichia coli [22]) [22]
<12> (expression in Escherichia coli [23]) [23]
<14> (overexpression in Escherichia coli [25]) [25]
<16> (overexpression in Escherichia coli [27]) [27]
<17> (expression in Escherichia coli [28-30]) [28-30]
<18> (expression in Escherichia coli [31]) [31]

Engineering

L403R/G404R/A406S <17> (<17> regulation of activity by phosphorylation [30]) [30]
R387S <12> (<12> regulation of activity by phosphorylation [23]) [23]

6 Stability

Temperature stability

19 <1> (<1> below 19°C reversible inactivation of enzyme activity, prevented by NADPH and $NADP^+$ [5]) [5]
90 <17> (<17> half-life at 90°C is 3.2 h [28]) [28]

Organic solvent stability

dithiothreitol <2> (<2> 10 mM at pH 7 reverses CoA disulfide inactivation of enzyme [10]) [10]

General stability information

<1>, freezing and thawing leads to a doubling of the activity [1]
<1, 5, 6>, presence of reduced thiol compounds is important for enzyme stability [12]

Storage stability

<1>, -10°C, crude microsomal form, 10-20% loss of activity per week [3]
<1>, -20°C, after quick freezing in acetone/solid CO_2 at -80°C, one week, 10% loss of activity [4]
<3>, 0°C, 24 h, 25% loss of activity [7]
<9>, -20°C, crude enzyme, 30 days, no loss in activity [19]
<9>, -20°C, partially purified enzyme, 24-48 h, loss of most of activity [19]

References

[1] Bucher, N.L.R.; Overath, P.; Lynen, F.: β-Hydroxy-β-methylglutaryl coenzyme A reductase, cleavage and condensing enzymes in relation to cholesterol formation in rat liver. Biochim. Biophys. Acta, **40**, 491-501 (1960)
[2] Durr, I.F.; Rudney, H.: The reduction of β-hydroxy-β-methylglutaryl coenzyme A to mevalonic acid. J. Biol. Chem., **235**, 2572-2578 (1960)
[3] Kawachi, T.; Rudney, H.: Solubilization and purification of β-hydroxy-β-methylglutaryl coenzyme A reductase from rat liver. Biochemistry, **9**, 1700-1705 (1970)
[4] Madhosingh, C.; Migicovski, B.B.; Wood, I.M.: Inhibition of yeast hydroxymethylglutaryl-CoA reductase by a rat liver mitochondrial preperation. FEBS Lett., **46**, 20-22 (1974)
[5] Heller, R.A.; Gould, R.G.: Prevention of cold inactivation of 3-hydroxy-3-methylglutaryl coenzyme A reductase by NADPH. Biochim. Biophys. Acta, **388**, 254-259 (1975)
[6] Madhosingh, C.; Orr, W.: Hydroxymethylglutaryl coenzyme A reductase in Fusarium oxysporum. Agric. Biol. Chem., **41**, 1519-1521 (1977)
[7] Madhosingh, C.; Orr, W.: Hydroxymethylglutaryl coenzyme A reductase purification and properties of the enzyme from Fusarium oxysporum. Biochim. Biophys. Acta, **523**, 283-296 (1978)

[8] Preiss, B.; Lehoux, J.-G.: Hydroxymethylglutaryl coenzyme A reductase activity in hamster adrenal mitochondria purified on a sucrose gradient. Biochem. Biophys. Res. Commun., **88**, 1140-1146 (1979)
[9] Ingebritsen, T.S.; Parker, R.A.; Gibson, D.M.: Regulation of liver hydroxymethylglutaryl-CoA reductase by a bicyclic phosphorylation system. J. Biol. Chem., **256**, 1138-1144 (1981)
[10] Gilbert, H.F.; Stewart, M.D.: Inactivation of hydroxymethylglutaryl-CoA reductase from yeast by coenzyme A disulfide. J. Biol. Chem., **256**, 1782-1785 (1981)
[11] Young, N.L.; Saudek, C.D.; Crawford, S.A.; Zuckerbrod, S.L.: Recovery and activation from hydroxymethylglutaryl coenzyme A reductase from rat small intestine. J. Lipid Res., **23**, 257-265 (1982)
[12] Sipat, A.P.: Hydroxymethylglutaryl CoA reductase (NADPH) in the latex of Hevea brasiliensis. Phytochemistry, **21**, 2613-2618 (1982)
[13] van Heusden, G.P.H.; Wirtz, K.W.A.: Hydroxymethylglutaryl CoA reductase and the modulation of microsomal cholesterol content by the none specific lipid transfer protein. J. Lipid Res., **25**, 27-32 (1984)
[14] Arebalo, R.E.; Mitchell, Jr., E.D.: Cellular distribution of 3-hydroxy-3-methylglutaryl coenzyme A reductase and mevalonate kinase in leaves of Nepeta cataria. Phytochemistry, **23**, 13-18 (1984)
[15] Harwood, Jr., H.J.; Schneider, M.; Stacpool, P.W.: Regulation of human leukocyte microsomal hydroxymethylglutaryl-CoA reductase activity by a phosphorylation and dephosphorylation mechanism. Biochim. Biophys. Acta, **805**, 245-251 (1984)
[16] Ness, G.C.; Eales, S.J.; Pendleton, L.C.; Smith, M.: Activation of rat liver microsomal 3-hydroxy-3-methylglutaryl coenzyme A reductase by NADPH. J. Biol. Chem., **260**, 12391-12393 (1985)
[17] Feingold, K.R.; Moser, A.H.: The effect of substrates and competitive inhibitors on the phosphatase dependent activation of hepatic hydroxymethylglutaryl CoA reductase. Arch. Biochem. Biophys., **249**, 46-51 (1986)
[18] Gibson, D.M.; Parker, R.A.: Hydroxymethylglutaryl-coenzyme A reductase. The Enzymes, (Boyer, P.D., ed.), **18**, 179-215 (1987)
[19] Chen, G.-Z.; Foster, L.; Bennet, J.L.: Purification and characterization of 3-hydroxymethylglutaryl-coenzyme A reductase of Schistosoma mansoni: Regulation of parasite enzyme activity differs from mammalien host. Exp. Parasitol., **73**, 82-92 (1991)
[20] Hupperich, M.; Thomer, A.; Kratz, K.; Eggerer, H.: Substrate and inhibitor specificity of 3-hydroxy-3-methylglutaryl-CoA reductase determined with substrate-analogue CoA-thioesters and CoA-thioethers. Biol. Chem. Hoppe-Seyler, **372**, 857-863 (1991)
[21] Ball, K.L.; Dale, S.; Weekes, J.; Hardie, D.G.: Biochemical characterization of two forms of 3-hydroxy-3-methylglutaryl-CoA reductase kinase from cauliflower (Brassica oleracia). Eur. J. Biochem., **219**, 743-750 (1994)
[22] Bischoff, K.M.; Rodwell, V.W.: 3-Hydroxy-3-methylglutaryl-coenzyme A reductase from Haloferax volcanii: Purification, characterization and expression in Escherichia coli. J. Bacteriol., **178**, 19-23 (1996)

[23] Friesen, J.A.; Rodwell, V.W.: Identification of elements critical for phosphorylation of 3-hydroxy-3-methylglutaryl coenzyme A reductase by adenosine monophosphate-activated protein kinase: Protein engineering of the naturally nonephosphorylatable 3-hydroxy-3-methylglutaryl coenzyme A reductase from Pseudomonas mevalonii. Biochemistry, **36**, 1157-1162 (1997)
[24] Concepcion, J.L.; Gonzalez-Pakanowska, D.; Urbina, J.A.: 3-Hydroxy-3-methylglutaryl-CoA reductase in Trypanosoma (Schizotrypanum) cruzi: Subcellular localization and kinetic properties. Arch. Biochem. Biophys., **352**, 114-120 (1998)
[25] Takahashi, S.; Kuzuyama, T.; Seto, H.: Purification, characterization, and cloning of a eubacterial 3-hydroxy-3-methylglutaryl coenzyme A reductase, a key enzyme involved in biosynthesis of terpenoids. J. Bacteriol., **181**, 1256-1263 (1999)
[26] Polo, M.; de Bravo, M.G.; Carbone, C.: 3-Hydroxy-3-methylglutaryl coenzyme A reductase activity in liver of athymic mice with or without an implanted human carcinoma. Comp. Biochem. Physiol. B, **122**, 433-437 (1999)
[27] Montalvetti, A.; Pena-Diaz, J.; Hurtado, R.; Ruiz-Pérez; L.M.; Gonzáles-Pacanowska, D.: Characterization and regulation of Leishmania major 3-hydroxy-3-methylglutaryl-CoA reductase. Biochem. J., **349**, 27-34 (2000)
[28] Bochar, D.A.; Brown, J.R.; Doolittle, W.F.; Klenk, H.-P.; Lam, W.; Schenk, M.E.; Stauffacher, C.V.; Rodwell, V.W.: 3-Hydroxy-3-methylglutaryl coenzyme A reductase of Sulfolobus solfataricus: DNA sequence, phylogeny, expression in Escherichia coli of the hmgA gene, and purification and kinetic properties of the gene product. J. Bacteriol., **179**, 3632-3638 (1997)
[29] Kim, D.-Y.; Bochar, D.A.; Stauffacher, C.V.; Rodwell, V.W.: Expression and characterization of the HMG-CoA reductase of the thermophilic archaeon Sulfolobus solfataricus. Protein Expr. Purif., **17**, 435-442 (1999)
[30] Kim, D.-Y.; Bochar, D.A.; Stauffacher, C.V.; Rodwell, V.W.: Engineering of Sulfolobus solfataricus HMG-CoA reductase to a form whose activity is regulated by phosphorylation and dephosphorylation. Biochemistry, **39**, 2269-2275 (2000)
[31] Wilding, E.I.; Kim, D.-Y.; Bryant, A.P.; Gwynn, N.M.; Lunsford, R.D.; McDavitt, D.; Myers, Jr., J.E.; Rosenberg, M.; Sylvester, D.; Stauffacher, C.V.; Rodwell, V.W.: Essentiality, expression, and characterization of the class II 3-hydroxy-3-methylglutaryl coenzyme A reductase of Staphylococcus aureus. J. Bacteriol., **182**, 5147-5152 (2000)

3-Hydroxyacyl-CoA dehydrogenase 1.1.1.35

1 Nomenclature

EC number
1.1.1.35

Systematic name
(S)-3-hydroxyacyl-CoA:NAD$^+$ oxidoreductase

Recommended name
3-hydroxyacyl-CoA dehydrogenase

Synonyms
1-specific DPN-linked β-hydroxybutyric dehydrogenase
3-L-hydroxyacyl-CoA dehydrogenase
3-L-hydroxybutyryl-CoA dehydrogenase
3-hydroxyacetyl-coenzyme A dehydrogenase
3-hydroxyacyl coenzyme A dehydrogenase
3-hydroxyisobutyryl-CoA dehydrogenase
3-keto reductase
3β-hydroxyacyl coenzyme A dehydrogenase
Endoplasmic reticulum-associated amyloid β-peptide binding protein [Swissprot]
HCDH [Swissprot]
L-3-hydroxyacyl CoA dehydrogenase
L-3-hydroxyacylcoenzyme A dehydrogenase
Scully protein [Swissprot]
Type II HADH [Swissprot]
β hydroxyacyl dehydrogenase
β-hydroxy acid dehydrogenase
β-hydroxyacyl CoA dehydrogenase
β-hydroxyacyl-coenzyme A synthetase
β-hydroxybutyrylcoenzyme A dehydrogenase
β-keto-reductase
β-ketoacyl-CoA reductase
βhydroxyacylcoenzyme A dehydrogenase
short chain L-3-hydroxyacyl-CoA dehydrogenase <1> (<1> displays optimal activity using 6-carbon substrates [1]) [1]

CAS registry number
9028-40-4

2 Source Organism

<1> *Homo sapiens* [1, 2, 4, 5]
<2> *Sus scrofa* [3, 6, 7, 8, 9, 11, 19, 20, 21, 22, 23, 24]
<3> *Rattus norvegicus* [10, 14, 16, 17]
<4> *Escherichia coli* [12, 13]
<5> *Mycobacterium smegmatis* [15]
<6> *Clostridium kluyveri* [18]
<7> *Bos taurus* [25]

3 Reaction and Specificity

Catalyzed reaction
(S)-3-hydroxyacyl-CoA + NAD^+ = 3-oxoacyl-CoA + NADH

Reaction type
oxidation
redox reaction
reduction

Natural substrates and products
S (S)-3-hydroxybutyryl-CoA + NAD^+ <1-6> (Reversibility: r <1-6> [1-25]) [1-25]
P acetoacetyl-CoA + NADH <1-6> [1-25]
S 17β-estradiol + NAD^+ <1> (<1> enzyme conatains 17β-hydroxysteroid and 3α-hydroxysteroid dehydrogenase activity [2]) (Reversibility: ir <1> [2]) [2]
P estrone + NADH <1> [2]

Substrates and products
S (S)-3-hydroxybutyryl CoA + NAD^+ <1-7> (<3> highest activity with acetoacetyl-CoA, significant activity with increasing chain length up to C_{16} [10]) (Reversibility: r <1-7> [1-25]) [1-25]
P acetoacetyl-CoA + NADH <1-7> [1-25]
S (S)-3-hydroxybutyryl-CoA + $NADP^+$ <3, 5> (<3> low activity for mitochondrial enzyme [14]) (Reversibility: r <3, 5> [14]) [14, 15]
P acetoacetyl-CoA + NADPH <3, 5> [14, 15]
S (S)-3-hydroxydecanoyl-CoA + NAD^+ <3, 7> (Reversibility: r <3> [14]) [14, 25]
P 3-ketodecanoyl-CoA + NADH <3, 7> [14, 25]
S (S)-3-hydroxyhexanoyl-CoA + NAD^+ <2, 3, 7> (Reversibility: r <2, 3> [14, 24]) [14, 24, 25]
P 3-ketohexanoyl-CoA + NADH <2, 3, 7> [14, 24, 25]
S (S)-3-hydroxyhexenoyl-CoA + NAD^+ <2> (Reversibility: r <2> [24]) [24]
P 3-ketohexenoyl-CoA + NADH <2> [24]
S (S)-3-hydroxylauryl-CoA + NAD^+ <7> (Reversibility: r <7> [25]) [25]
P 3-ketolauryl-CoA + NADH <7> [25]

S 1-propanol + NAD^+ <1> (<1> multifunctional enzyme from brain [2]) (Reversibility: ir <1> [2]) [2]
P 1-propanal + NADH <1> [2]
S 17β-estradiol + NAD^+ <1> (<1> enzyme from brain contains 17β-hydroxysteroid and 3α-hydroxysteroid dehydrogenase activity [2]) (Reversibility: ir <1> [2]) [2]
P estrone + NADH <1> [2]
S 2-propanol + NAD^+ <1> (<1> multifunctional enzyme from brain [2]) (Reversibility: ir <1> [2]) [2]
P acetone + NADH <1> [2]
S 3-ketohexadecanoyl-CoA + $NADH^+$ <1, 2> (Reversibility: r <1, 2> [3, 4]) [3, 4]
P (S)-3-hydroxhexadecanoyl-CoA + NAD^+ <1, 2> [3, 4]
S 3-ketooctanoyl-CoA + $NADH^+$ <1, 2, 3, 7> (Reversibility: r <1, 2, 3> [3, 4]) [3, 4, 14, 25]
P (S)-3-hydroxyoctanoyl-CoA + NAD^+ <1, 2, 3, 7> [3, 4, 14, 25]
S 5α-dihydrotestosterone + NADH <1> (Reversibility: r <1> [2]) [2]
P dihydroandrosterone + NAD^+ <1> [2]
S acetoacetyl-N-acetylcysteamine + NADH <2> (Reversibility: r <2> [6]) [6, 19]
P (S)-3-hydroxybutyryl-N-acetylcysteamine + NAD^+ <2> [6, 19]
S acetoacetyl-N-β-alanylcysteamine + NADH <2> (Reversibility: r <2> [6]) [6]
P (S)-3-hydroxybutyryl-N-β-alanylcysteamine + NAD^+ <2> [6]
S acetoacetyl-S-6,8-dithioloctanoic acid + NADH <2> (Reversibility: r <2> [24]) [24]
P (S)-3-hydroxybutyryl-S-6,8-dithioloctanoic acid + NAD^+ <2> [24]
S acetoacetyl-S-6-ethyl-monothiolooctanoic acid + NADH <2> (Reversibility: r <2> [24]) [24]
P (S)-3-hydroxybutyryl-S-6-ethyl-monothioloctanoic acid + NAD^+ <2> [24]
S acetoacetyl-cysteamine-2,2,5,5-tetramethyl-1-oxy-3-pyrroline-3-carboxylic acid amide + NADH <2> (Reversibility: r <2> [6]) [6]
P (S)-3-hydroxybutyryl-cysteamine-2,2,5,5-tetramethyl-1-oxy-3-pyrroline-3-carboxylic acid amide + NAD^+ <2> [6]
S acetoacetyl-pantetheine + NADH <2> (Reversibility: r <2> [6]) [6, 19]
P (S)-3-hydroxybutyryl-pantetheine <2> [6, 19]
S acetoacetyl-pantetheine-4'-(2,2,5,5-tetramethyl-1-oxy-3-pyrroline-3-carboxylic acid ester) + NADH <2> (Reversibility: r <2> [6]) [6]
P (S)-3-hydroxybutyryl-pantetheine-4'-(2,2,5,5-tetramethyl-1-oxy-3-pyrroline-3-carboxylic acid ester) + NAD^+ <2> [6]
S acetoacetyldecanoate + NADH <2> (Reversibility: r <2> [24]) [24]
P (S)-3-hydroxybutyryldecanoate + NAD^+ <2> [24]
S androsterone + NAD^+ <1> (Reversibility: ir <1> [2]) [2]
P androstanedione + NADH <1> [2]

Inhibitors

5,5'-dithiobis(2-nitrobenzoic acid) <2> [19]
N-bromsuccinimide <2> [19]
N-ethylmaleimide <2> [19]
iodoacetamide <2> [19]
iodoacetic acid <2> [19]
p-chloromercuribenzoate <2> [19]
p-chloromercuriphenyl sulfonic acid <5> (<5> 95% inhibition at 0.1 mM [15]) [15]

Cofactors/prosthetic groups

NAD^+ <1-7> [1-25]
$NADP^+$ <3, 5> [14, 15]

Turnover number (min^{-1})

0.66 <1> (17β-estradiol, <1> enzyme from brain [2]) [2]
0.66 <1> (androsterone, <1> enzyme from brain [2]) [2]
5.58 <1> (dihydroandrosterone, <1> enzyme from brain [2]) [2]
1440 <1> (3-ketohexdecanoyl-CoA, <1> enzyme from brain, pH 7.0 [4]) [4]
1680 <1> (3-ketooctanoyl-CoA, <1> enzyme from brain, pH 7.0 [4]) [4]
2220 <1> (acetoacetyl-CoA, <1> enzyme from brain, pH 7.0 [4]) [4]
2700 <2> (3-ketohexadecanoyl-CoA) [3]
6120 <2> (3-ketooctanoyl-CoA) [3]
6780 <1> (acetoacetyl-CoA, <1> pH 8.0 [1]) [1]
9300 <1> (acetoacetyl-CoA, <1> pH 5.0 [1]) [1]
14760 <1> (acetoacetyl-CoA, <1> pH 6.0 [1]) [1]
15120 <1> (acetoacetyl-CoA, <1> pH 7.0 [1]) [1]
19800 <2> (acetoacetyl-CoA, <2> pH 7.0 [3]) [3]

Specific activity (U/mg)

0.002 <1> (<1> enzyme from brain, substrate 1-propanol [2]) [2]
0.0121 <1> (<1> enzyme from brain, substrate androsterone [2]) [2]
0.0156 <1> (<1> enzyme from brain, substrate 17β-estradiol [2]) [2]
0.033 <3> (<3> enzyme activity in brain mitochondria, substrate acetoacetyl-CoA [10]) [10]
0.087 <1> (<1> enzyme from brain, substrate 5α-dihydrotestosterone [2]) [2]
0.13 <1> (<1> enzyme from brain, substrate dihydroandrosterone [2]) [2]
0.41 <6> (<6> enzyme activity in solubilized sediments of homogenized cells [18]) [18]
0.48 <3> (<3> enzyme activity in inner membrane of liver mitochondria, substrate acetoacetyl-CoA [10]) [10]
0.75 <2> (<2> 72% activivty is bound to the matrix surface of the inner mitochondria membrane, binding is inhibited by increasing ionic strength and pH [8]) [8]
0.78 <3> (<3> enzyme activity in kidney mitochondria, substrate acetoacetyl-CoA [10]) [10]
1.5 <3> (<3> enzyme activity in heart mitochondria, substrate acetoacetyl-CoA [10]) [10]

1.67 <3> (<3> enzyme activity in liver mitochondria, substrate acetoacetyl-CoA [10]) [10]
2.3 <3> (<3> enzyme activity in liver mitoplasts, substrate acetoacetyl-CoA [10]) [10]
3.4 <3> (<3> enzyme activity in matrix of liver mitochondria, substrate acetoacetyl-CoA [10]) [10]
6 <7> [25]
12.6 <3> (<3> peroxisomal enzyme [14]) [14]
17.35 <1> (<1> enzyme from brain [4]) [4]
20 <4> [12, 13]
41.2 <5> [15]
176 <2> [22]
220 <2> [19]
511 <2> (<2> recombinant enzyme [3]) [3]
565 <3> (<3> mitochondrial enzyme [14]) [14]
1200 <1> (<1> His-tagged recombinant enzyme [1]) [1]

K_m-Value (mM)
0.003 <2> (NADH) [3]
0.003 <2> (acetoacetyl-CoA) [3]
0.0054 <3> (NADH) [14]
0.0138 <1> (acetoacetyl-CoA, <1> pH 8.0 [1]) [1]
0.0163 <3> ((S)-3-hydroxyoctanoyl-CoA, <3> mitochondrial enzyme, oxidation of (S)-3-hydroxyoctanoyl-CoA [14]) [14]
0.0169 <3> (acetoacetyl-CoA) [14]
0.0187 <1> (acetoacetyl-CoA, <1> pH 7.0 [1]) [1]
0.02 <1> (NADH, <1> enzyme from brain [4]) [4]
0.0286 <3> ((S)-3-hydroxyhexanoyl-CoA, <3> mitochondrial enzyme, oxidation of (S)-3-hydroxyhexanoyl-CoA [14]) [14]
0.033 <5> (NADH) [15]
0.034 <1> (dihydroandrosterone, <1> enzyme from brain [2]) [2]
0.0345 <1> (acetoacetyl-CoA, <1> pH 5.0 [1]) [1]
0.036 <5> (acetoacetyl-CoA) [15]
0.043 <1> (17β-estradiol, <1> enzyme from brain [2]) [2]
0.045 <1> (acetoacetyl-CoA, <1> pH 6.0 [1]) [1]
0.045 <1> (androsterone, <1> enzyme from brain [2]) [2]
0.05 <1> (NAD^+, <1> enzyme from brain, substrate 17β-estradiol [2]) [2]
0.0585 <3> (NAD^+) [14]
0.06 <2> (acetoacetyl-CoA) [19, 22]
0.066 <4> (acetoacetyl-CoA) [12]
0.0699 <3> ((S)-3-hydroxybutyryl-CoA, <3> mitochondrial enzyme, oxidation of (S)-3-hydroxybutyryl-CoA [14]) [14]
0.08 <2> (acetoacetyl-pantetheine) [6]
0.083 <5> (NAD^+) [15]
0.088 <3> ((S)-3-hydroxydecanoyl-CoA, <3> mitochondrial enzyme, oxidation of (S)-3-hydroxydecanoyl-CoA [14]) [14]
0.089 <1> (acetoacetyl-CoA, <1> enzyme from brain [4]) [4]

0.1 <7> ((S)-3-hydroxylauryl-CoA) [25]
0.21 <1> (NAD^+, <1> enzyme from brain, substrate dihydroandrosterone [2]) [2]
0.242 <1> (NAD^+, <1> enzyme from brain, substrate androsterone [2]) [2]
0.31 <7> ((S)-3-hydroxybutyryl-CoA) [25]
0.34 <7> ((S)-3-hydroxyhexanoyl-CoA) [25]
0.35 <7> ((S)-3-hydroxyoctanoyl-CoA) [25]
0.4 <2> (acetoacetyl-pantetheine) [19, 22]
0.435 <3> (NADPH) [14]
0.56 <3> (acetoacetyl-pantetheine) [14]
1.19 <5> (acetoacetyl-pantetheine) [15]
1.4 <2> (acetoacetyl-N-β-alanylcysteamine) [6]
10 <2> (acetoacetyl-N-acetylcysteamine) [6]
10 <2> (acetoacetyl-Nacetylcysteamine) [19, 22]
11 <3> ($NADP^+$) [14]
44.4 <5> (acetoacetyl-N-acetyl-cysteamine) [15]

pH-Optimum
4.5 <3> (<3> mitochondrial enzyme, reduction of acetoacetyl CoA [14]) [14]
6 <5> (<5> reduction of acetoacetyl-CoA [15]) [15]
6-7 <2> (<2> reduction of acetoacetyl-CoA [24]) [24]
6.2 <3> (<3> peroxisomal enzyme, reduction of acetoacetyl-CoA [14]) [14]
6.5-7 <1> (<1> enzyme from brain, reduction of acetoacetyl-CoA [2]) [2]
9 <5> (<5> oxidation of (S)-3-hydroxybutyryl-CoA [15]) [15]
9.5-10 <1> (<1> enzyme from brain, oxidation of 17β-estradiol [2]) [2]
9.6 <2> (<2> oxidation of (S)-3-hydroxybutyryl-CoA [24]) [24]
9.6 <3> (<3> mitochondrial enzyme, oxidation of (S)-3-hydroxybutyryl-CoA [14]) [14]
9.8 <3> (<3> peroxisomal enzyme, oxidation of (S)-3-hydroxybutyryl-CoA [14]) [14]
10 <7> [25]

4 Enzyme Structure

Molecular weight
40000 <3> (<3> liver enzyme [17]) [17]
50300 <5> (<5> gel filtration [15]) [15]
65000 <2> (<2> sedimentation equilibrium [19]) [19, 20, 23]
65000 <3> (<3> mitochondrial enzyme, gel filtration [14]) [14]
66000 <2> (<2> gel filtration [3]) [3]
70000 <3> (<3> gel filtration [10]) [10]
70000 <3> (<3> peroxisomal enzyme, gel filtration [14]) [14]
75000 <2> (<2> gel filtration [20]) [20, 22]
76000 <2> (<2> gel filtration [23]) [23]
108000 <1> (<1> enzyme from brain, gel filtration [4]) [4]
260000 <4> (<4> gel filtration [13]) [13]
270000 <4> (<4> gel filtration [12]) [12]

Subunits

dimer <1> (<1> 2 * 34000, SDS-PAGE [1]) [1]
dimer <2> (<2> 2 * 33000, SDS-PAGE [3]) [3, 19, 20, 23]
dimer <3> (<3> 2 * 32000, mitochondrial enzyme, SDS-PAGE [14]) [14]
monomer <3> (<3> 1 * 77000, peroxisomal enzyme, SDS-PAGE [14]) [14]
tetramer <1> (<1> 4 * 27000, enzyme from brain, SDS-PAGE [4]) [4]
tetramer <4> (<4> 2 * 78000 + 2 * 42000, SDS-PAGE [13]) [13]

5 Isolation/Preparation/Mutation/Application

Source/tissue

brain <1, 3> [2, 4, 10]
heart <1, 2, 3> [1, 8, 9, 10, 11, 19, 20, 21, 22, 23, 24]
kidney <1, 3> [5, 10]
liver <1, 2, 3, 7> [3, 5, 10, 14, 17, 25]
pancreas <1> [5]
skeletal muscle <1> [5]

Localization

endoplasmic reticulum <1> [2, 4]
mitochondrion <1, 2, 3, 7> (<3> matrix enzyme [16]) [1, 2, 3, 4, 8, 10, 14, 15, 16, 19, 20, 25]
peroxisome <3> [14]

Purification

<1> (overexpressed in Escherichia coli, hydroxylapatite [4]) [4]
<2> (ammonium sulfate, first CM-cellulose, gel filtration, second and third CM-cellulose [19]) [19, 22]
<2> (overexpressed in Escherichia coli, phosphocellulose [3]) [3]
<3> (calcium phosphate gel, ammonium sulfate, CM-Sephadex, Blue Dextran-Sepharose 4B, calcium phosphate gel-cellulose, Blue Dextran-Sepharose 4B, mitochondrial enzyme [14]) [14]
<3> (phosphocellulose, ammonium sulfate, CM-cellulose, ammonium sulfate, peroxisomal enzyme [14]) [14]
<4> (60°C for 10 min, phosphocellulose, Sephacryl S200 [12]) [12]
<5> (ammonium sulfate, Sephadex G-150, hydroxyapatite, NAD-Sepharose 4B [15]) [15]
<7> (ammonium sulfate, 50°C for 20 min, $Zn(OH)_2$-gel [25]) [25]

Renaturation

<2> (after denaturation with 2-3 M urea or guanidinium-HCl, 94% activity can be recovered by dialyzing the enzyme at 4°C against 100 mM sodium pyrophosphate, pH 7.3, 10 mM β-mercaptoethanol [19]) [19, 23]

Crystallization

<1> (50 mM N(2-acetamido)-2-iminodiacetic acid, pH 6.5, polyethylene glycol 4000, 5 mM NAD^+ hanging drop, crystals within 3 to 5 days at 18°C, en-

zyme structure is compromised of two domains, a NAD^+-binding domain and a helical C-terminal domain [1]) [1]
<2> (50% saturation with ammonium sulfate solution, 0.1 M potassium phosphate, pH 6.8, 1 mM EDTA, 2 mM β-mercaptoethanol, 4°C, crystals appear after 2 days [19]) [19]
<2> (dialyisis against 40% saturated ammonium sulfate containing 100 mM phosphate, 2 mM β-mercaptoethanol, 1 mM EDTA, pH 6.9, 7.5 or 8.2, vapor diffusion crystallization, crystals are obtained in the ammonium sulfate saturation range of 41% to 48% [21]) [21]
<2> (polyethylene glycol, pH 8, orthorhombic crystals, 2.7 A resolution, crystallisation at pH 5 leads to trigonal space group [7,9]) [7, 9]
<2> (two dimers of the enzyme in the asymmetric unit of an orthorombic cell, two coenzyme binding sites per dimer [11]) [11]

Cloning

<1> (C-terminal hexameric histidine tag, expressed in Escherichia coli [1]) [1, 4]
<2> (expressed in Escherichia coli [3]) [3]

6 Stability

Storage stability

<2>, 4°C, 90% ammonium sulfate, at least 6 monts, no loss of activity [19]
<3>, -20°C, 120 mM potassium phosphate, pH 7.0, several months without loss of activity [14]
<4>, -76°C, 0.2 M potassium phosphate, pH 6.6, 25% glycerol, 10 mM mercaptoethanol, several months, no loss in activity [12]

References

[1] Barycki, J.J.; O'Brian, L.K.; Bratt, J.M.; Zhang, R.; Sanishvili, R.; Strauss, A.W.; Banaszak, L.J.: Biochemical characterization and crystal structure determination of human heart short chain L-3-hydroxyacyl-CoA dehydrogenase provide insights into catalytic mechanism. Biochemistry, **38**, 5786-5789 (1999)
[2] He, X-Y.; merz, G.; Mehta, P.; Schulz, H.; Yang, S-Y.: Human brain short chain L-3-hydroxyacyl Coenzyme A dehydrogenase is a singl-domain multifunctional enzyme. J. Biol. Chem., **274**, 15014-15019 (1999)
[3] He, X-Y.; Yang, S-Y.: Molecular cloning, expression in Escherichia coli, and characterization of a novel L-3-hydroxyacyl coenzyme A dehydrogenase from pig liver. Biochim. Biophys. Acta, **1392**, 119-126 (1998)
[4] He, X-Y.; Schulz, H.; Yang, S-Y.: A human brain L-3-hydroxyacyl-coenzyme a dehydrogenase is identical to an amyloid β-peptide-binding protein involved in Alzheimer's disease. J. Biol. Chem., **273**, 10741-10746 (1998)

[5] Vredendaal, P.J.C.M.; van den Berg, I.E.T.; Malingre, H.E.M.; Stroobants, A.K.; Olde weghuis, D.E.M.; Berger, R.: Human short-chain L-3-hydroxyacyl-CoA dehydrogenase: cloning and characterization of the coding sequence. Biochem. Biophys. Res. Commun., **223**, 718-723 (1996)
[6] Hartmann, D.; Philipp, R.; Schmadel, K.; Birktoft, J.J.; Banaszak, L.J.; Trommer, W.E.: Spatial arrangement of coenzyme and substrates bound to L-3-hydroxyacyl-CoA dehydrogenase as studied by spin-labeled analogues of NAD^+ and CoA. Biochemistry, **30**, 2782-2790 (1991)
[7] Birktoft, J.J.; holden, H.M.; Hamlin, R.; Xuong, N: H.; Banaszak, L.J.: Structure of L-3-hydroxyacyl-coenzyme A dehydrogenase: preliminary chain tracing at 2.8 A resolution. Proc. Natl. Acad. Sci. USA, **84**, 8262-8266 (1987)
[8] Sumegi, B.; Srere, P.A.: Binding of the enzymes of fatty acid β-oxidation and some related enzymes to pig heart inner mitochondrial membrane. J. Biol. Chem., **259**, 8748-8752 (1984)
[9] Holden, H.M.; Banaszak, L.J.: L-3-Hydroxyacyl coenzyme A dehydrogenase. J. Biol. Chem., **258**, 2383-2389 (1983)
[10] El-Fahkri, M.; Middleton, B.: The existence of an inner-membrane-bound, long acyl-chain-specific 3-hydroxyacyl-CoA dehydrogenase in mammalian mitochondria. Biochim. Biophys. Acta, **713**, 270-279 (1982)
[11] Holden, H.M.; Banaszak, L.J.; Frieden, C.; McLoughlin, D.J.: Differences in the binding of coenzyme to L-3-hydroxyacyl-Coenzyme A dehydrogenase in the crystalline state and in solution. FEBS Lett., **132**, 15-18 (1981)
[12] Binstock, J.F.; Schulz, H.: Fatty acid oxidation complex from Escherichia coli. Methods Enzymol., **71**, 403-411 (1981)
[13] Pawar, S.; Schulz, H.: The structure of the multienzyme complex of fatty acid oxidation from Escherichia coli. J. Biol. Chem., **256**, 3894-3899 (1981)
[14] Osumi, T.; Hashimoto, T.: Purification and properties of mitochondrial and peroxisomal 3-hydroxyacyl-CoA dehydrogenase from rat liver. Arch. Biochem. Biophys., **203**, 372-383 (1980)
[15] Shimakata, T.; Fujita, Y.; Kusaka, T.: Purification and characterization of 3-hydroxyacyl-CoA dehydrogenase of Mycobacterium smegmatis. J. Biochem., **86**, 1191-1198 (1979)
[16] El-Fakhri, M.; Middleton, B.: The existence of two different L-3-hydroxyacyl-coenzyme A dehydrogenases in rat tissues. Biochem. Soc. Trans., **7**, 392-393 (1979)
[17] Middleton, B.: 3-Oxoacyl-coenzyme A thiolases and L-3-hydroxyacyl-coenzymeA dehydrogenases. Biochem. Soc. Trans., **6**, 80-83 (1978)
[18] Hillmer, P.; Gottschalk, G.: Solubilization and partial characterization of particulate dehydrogenases from Clostridium kluyveri. Biochim. Biophys. Acta, **334**, 12-23 (1974)
[19] Bradshaw, R.A.; Noyes, B.E.: L-3-Hydroxylacyl coenzyme A dehydrogenase from pig heart muscle. Methods Enzymol., **35**, 122-128 (1975)
[20] Noyes, B.E.; Glatthaar, B.E.; Garavelli, J.S.; Bradshaw, R.A.: Structural and functional similarities between mitochondrial malate dehydrogenase and L-3-hydroxyacyl CoA dehydrogenase. Proc. Natl. Acad. Sci. USA, **71**, 1334-1338 (1974)

[21] Wininger, M.; Noyes, B.E.; Bradshaw, R.A.; Banaszak, L.J.: L-3-Hydroxacyl coenzyme A dehydrogenase: crystallographic properties of the pig heart enzyme. J. Mol. Biol., **90**, 409-413 (1974)
[22] Noyes, B.E.; Bradshaw, R.A.: L-3-Hydroxyacyl coenzyme A dehydrogenase from pig heart muscle. J. Biol. Chem., **248**, 3052-3059 (1973)
[23] Noyes, B.E.; Bradshaw, R.A.: L-3-Hydroxyacyl coenzyme A dehydrogenase from pig heart muscle. II. Subunit structure. J. Biol. Chem., **248**, 3060-3066 (1973)
[24] Stern, J.R.: Crystalline β-hydroxybutyryl dehydrogenase from pig heart. Biochim. Biophys. Acta, **26**, 448-449 (1957)
[25] Wakil, S.J.; Green, D.E.; Mil, S.; Mahler, H.R.: Studies on the fatty acid oxidizing system of animal tissues. IV. β-hydroxyacyl coenzyme A dehydrogenase. J. Biol. Chem., **207**, 631-638 (1954)

Acetoacetyl-CoA reductase 1.1.1.36

1 Nomenclature

EC number
1.1.1.36

Systematic name
(R)-3-hydroxyacyl-CoA:NADP$^+$ oxidoreductase

Recommended name
acetoacetyl-CoA reductase

Synonyms
(R)-3-hydroxyacyl-CoA dehydrogenase
D(-)-β-hydroxybutyryl CoA-NADP oxidoreductase
D-3-hydroxyacyl-CoA reductase
NADP-linked acetoacetyl CoA reductase
NADPH:acetoacetyl-CoA reductase
acetoacetyl coenzyme A reductase
β-ketoacyl-CoA reductase
hydroxyacyl coenzyme-A dehydrogenase
short chain β-ketoacetyl(acetoacetyl)-CoA reductase

CAS registry number
9028-41-5

2 Source Organism

<1> *pigeon* (low activity [3]) [1, 3, 7]
<2> *Zoogloea ramigera* (I-16-M [7]; gene cloned in Escherichia coli [8]) [2, 7, 8, 9, 10, 11]
<3> *Azotobacter beijerinckii* [2, 9]
<4> *Rattus norvegicus* [3, 5, 6]
<5> *Bos taurus* [3]
<6> *goose* (low activity [3]) [3]
<7> *Gallus gallus* (low activity [3]) [3]
<8> *yeast* (low activity [3]) [3]
<9> *Ralstonia eutropha* (gene expressed in Brassica napus [13]) [13]
<10> *Streptomyces coelicolor* [4]
<11> *Syntrophomonas wolfei* (ssp. Wolfei [9]) [9]
<12> *Methylobacterium extorquens* (15 [10]) [10]
<13> *Methylobacterium rhodesianum* (MB 126 [10,11]) [10, 11]

<14> *Alcaligenes eutrophus* [9, 10, 12]
<15> *Azotobacter vinelandii* [10]
<16> *Rhodospirillum rubrum* [10, 11]
<17> *Rhodococcus ruber* [11]

3 Reaction and Specificity

Catalyzed reaction

(R)-3-hydroxyacyl-CoA + NADP = 3-oxoycyl-CoA + NADPH2

Reaction type

redox reaction

Natural substrates and products

S acetoacetyl-CoA + NADH <11> (<11> PHA-sythesis [9]) (Reversibility: ir <11> [9]) [9]
P D-3-hydroxybutyryl-CoA + NAD^+ <11> [9]
S acetoacetyl-CoA + NADH <2, 16> (<2> low activity [2]) (Reversibility: r <2> [2]; ? <2, 16> [10, 11]) [2, 10, 11]
P L-3-hydroxybutyryl-CoA + NAD^+ <2, 16> [2, 10, 11]
S acetoacetyl-CoA + NADPH <2, 3, 11, 12, 14, 17> (<2> high activity [2]; <12> tricarboxylic acid cycle, poly-3-hydroxybutyrate synthesis, aerobic serine pathway [10]; <2, 3, 14, 17> PHB synthesis [2, 9, 10, 11]) (Reversibility: r <2, 12> [2, 10]; ? <3, 14, 17> [2, 9, 11]) [2, 9, 10, 11]
P D-3-hydroxybutyryl-CoA + $NADP^+$ <2, 3, 12, 14, 17> [2, 10, 11]
S acetoacetyl-CoA + NADPH <2> (biosynthesis of poly(D-3-hydroxybutyrate)) [8]
P ?

Substrates and products

S 2-methylacetoacetyl-CoA + NADPH <2> (Reversibility: ? <2> [8]) [8]
P 2-methyl-3-hydroxybutyryl-CoA + $NADP^+$ <2> [8]
S ? + NADPH <14> (Reversibility: ? <14> [12]) [12]
P (R)-3-hydroxybutyrylphosphopantetheine + $NADP^+$ <14> [12]
S D-3-hydroxybutyryl-CoA + $NADP^+$ <2> (Reversibility: r <2> [7]) [7]
P acetoacetyl-CoA + NADPH <2> [7]
S S-acetoacetyl-N-acetylcysteamine + NADPH <1, 4-8, 10> (Reversibility: ? <1, 4-8, 10> [3, 4]) [3, 4]
P 3-hydroxybutyryl-N-acetylcysteamine + $NADP^+$ <1, 4-8> [3]
S acetoacetyl-CoA + NADH <4> (Reversibility: ? <4> [5, 6]) [5, 6]
P D-β-hydroxybutyryl-CoA + NAD^+ <4> [5]
S acetoacetyl-CoA + NADH <13> (<13> reverse only with L-form [11]) (Reversibility: r <13> [11]) [11]
P L-β-hydroxybutyryl-CoA + NAD^+ <13> [11]
S acetoacetyl-CoA + NADPH <1-5, 10> (Reversibility: r <1-4, 10> [1, 2, 4, 5, 7, 8]; ir <4, 5> [3]) [1-5, 7-9]
P (3R)-D-3-hydroxybutyryl-CoA + $NADP^+$ <2> [8]

S acetoacetyl-CoA + NADPH <1, 13, 9> (<13> reverse only with D-form [11]) (Reversibility: r <13> [11]; ir <1> [1]; ? <9> [13]) [1, 11, 13]
P D-β-hydroxybutyryl-CoA + NADP$^+$ <1, 13, 9> [1, 11, 13]
S acetoacetyl-S-(D-pantetheine)11-pivalate + NADPH <2> (Reversibility: ? <2> [8]) [8]
P 3-hydroxybutyryl-S-(D-pantetheine) 11-pivalate + NADP$^+$ <2> [8]
S β-ketopalmitoyl-CoA + ? <4> (Reversibility: ? <4> [6]) [6]
P ? <4> [6]
S butyrylacetyl-CoA + NADPH <2> (Reversibility: ? <2> [8]) [8]
P 3-hydroxyhexanoyl-CoA + NADP$^+$ <2> [8]
S pantetheine 11-pivalate + ? <14> (Reversibility: ? <14> [12]) [12]
P ?
S propionylacetyl-CoA + NADPH <2> (Reversibility: ? <2> [8]) [8]
P 3-hydroxypentanoyl-CoA + NADP$^+$ <2> [8]
S Additional information <1, 2, 10> (<2> poor substrate: D-3-hydroxyvaleryl-CoA [7]; <1,10> not: crotonyl-CoA [1, 4]; <1> octanoyl-CoA [1]; <1> β-hydroxybutyryl-CoA [1]; <2> D-3-hydroxyacyl-CoA's with chain length exceeding 6 [7]; <2> L-3-hydroxyacyl-CoA [7]) [1, 4, 6, 7]
P ?

Inhibitors

2-methylcrotonyl-CoA <1, 4-8> [3]
N-methylmaleimide <1, 4-8> [3]
N-phenylmaleimide <1, 4-8> [3]
NAD$^+$ <12, 13> (<12> 17% inhibition [10]; <13> NADH-linked [11]) [10, 11]
NADP$^+$ <12, 13> (<12> 40% inhibition [10]) [10, 11]
NADPH <12, 13, 15> (<12> concentration > 100 mM [10]) [10]
acetoacetyl-CoA <2, 12, 13, 14, 15> (<2> markedly at concentrations above 0.005 mM and NADPH 0.01 mM, less inhibition when NADPH increases from 0.01 to 1 mM [7]; <2> above 0.01 mM [8]; <2> above 0.01 mM [2]; <12> concentration above 40 mM [10]; <13> NADPH-linked, concentration above 0.025 mM, nearly independent of NADPH concentrations 0.03-0.5 mM [11]; <13> NADH-linked, less sensitive [11]) [2, 7, 10, 11]
acetyl-CoA <1, 4-8> [3]
butyryl-CoA <1, 4-8> [3]
iodoacetamide <1, 4-8> [3]
malonyl-CoA <1, 4-8> [3]
methylmalonyl-CoA <1, 4-8> [3]
propionylacetyl-CoA <2> [8]
sulfhydryl reagents <1, 4-8> [3]
Additional information <1, 10, 13> (<1> not inhibited by *p*-hydroxymercuribenzoate or N-ethylmaleimide from 0.1-10 mM [1]; <10> not inhibited by acetoacetyl-CoA at high concentrations [4]; <13> NADH-linked: not inhibited by NADPH at concentrations of 1 mM, NADPH-linked: not inhibited by NAD$^+$ at concentrations of 1 mM [11]) [1, 4, 11]

Cofactors/prosthetic groups

NADH <2-4, 10, 11, 13, 14, 16> (<3, 14> appreciable activity [9]) [2, 4-6, 9, 11]
NADP$^+$ <1, 2, 3, 4, 5, 6, 7, 8, 11> [1-3, 5-7, 9]
NADPH <1, 2, 3, 4, 5, 6, 7, 8, 10, 11, 12, 13, 14, 15, 17> (<1, 2, 4-8> specific for [3, 7, 8, 9]; <2> B-specific, i.e. transfer of pro-S-hydrogen from NADPH to acetoacetyl-CoA [8]; <2, 12> absolute specificity [8, 10]; <3, 14> preferentially [9]) [1, 2, 3, 4, 5, 6, 7, 8, 9, 10, 11]

Activating compounds

ATP <12> (<12> 32% activation [10]) [10]
CoA <4, 5> (<4, 5> stimulating [3]) [3]
α-ketoglutarate <12> (<12> 20% activation [10]) [10]
citrate <12> (<12> 30% activation [10]) [10]
isocitrate <12> (<12> 25% activation, stimulates NADP$^+$ regeneration [10]) [10]
pantetheine <4, 5> (<4, 5> stimulating [3]) [3]

Turnover number (min^{-1})

540 <2> (butyrylacetyl-CoA) [8]
1620 <2> (2-methylacetoacetyl-CoA) [8]
5400 <2> (D-3-hydroxybutyryl-CoA) [8]
7440 <2> (propionylacetyl-CoA) [8]
16020 <2> (acetoacetyl-S-(D-pantetheine) 11-pivalate) [8]
18000 <2> (acetoacetyl-CoA) [8]

Specific activity (U/mg)

4.82 <11> (<11> NADPH [9]) [9]
7.7 <2> (<2> NADPH-linked [2]) [2]
9 <4> (<4> β-ketopalmitoyl-CoA [6]) [6]
10.6 <11> (<11> NADH [9]) [9]
15.2 <4> (<4> acetoacetyl-CoA [6]) [6]
87.9 <1> (<1> fed pigeons livers [1]) [1]
170.1 <1> (<1> fasted pigeons livers [1]) [1]
184 <12> [10]
250 <1> [1]
400 <13> (<13> NADPH-linked [11]) [11]
412 <2> [7]
608 <2> [8]
771 <13> (<13> NADH-linked, 345 with NADPH [11]) [11]
Additional information <11> (no activity at pH 8.0 [9]) [9]

K_m-Value (mM)

0.002 <2> (acetoacetyl-CoA) [8]
0.002 <2> (propionylacetyl-CoA) [8]
0.005 <14> (acetoacetyl-CoA) [10]
0.005 <4, 5> (acetoacetyl-CoA) [3, 10]
0.006 <4> (β-ketopalmitoyl-CoA) [6]
0.008 <10> (S-acetoacetyl-N-acetylcysteamine, <10> NADH [4]) [4]

0.0083 <2> (acetoacetyl-CoA, <2> NADPH-linked [2]) [2]
0.01 <10> (acetoacetyl-CoA, <10> NADPH [4]) [4]
0.01 <2> (butyrylacetyl-CoA) [8]
0.011 <15> (acetoacetyl-CoA, <15> 100 mM NADPH [10]) [10]
0.0116 <12> (acetoacetyl-CoA) [10]
0.015 <13> (acetoacetyl-CoA, <13> NADPH-linked [11]) [10, 11]
0.018 <4> (NADH, <4> acetoacetyl-Co-A [5]) [5]
0.018 <13> (NADPH, <13> NADPH-linked [11]) [10, 11]
0.018 <10> (acetoacetyl-CoA, <10> NADH [4]) [4]
0.019 <14> (NADPH) [10]
0.02 <15> (NADPH, <15> 32 mM acetoacetyl-CoA [10]) [10]
0.02-0.021 <2> (NADPH, <2> NADPH-linked, acetoacetyl-CoA [2, 8]) [2, 8]
0.021 <4> (acetoacetyl-CoA, <4> NADH [5]) [5]
0.022 <13> (L-β-hydroxybutyryl-CoA) [11]
0.026 <2> (D-3-hydroxybutyryl-CoA) [8]
0.03 <13> (D-β-hydroxybutyryl-CoA) [10, 11]
0.03 <13> (acetoacetyl-CoA, <13> cofactor NADH [11]) [11]
0.031 <14> ($NADP^+$) [10]
0.033 <14> (D-hydroxybutyryl-CoA) [10]
0.037 <4> (acetoacetyl-CoA) [6]
0.04 <5> (S-acetoacetyl-N-acetylcysteamine) [3]
0.041 <12> (NADPH) [10]
0.06 <13> ($NADP^+$, <13> NADPH-linked [11]) [10, 11]
0.14 <13> (NADPH, <13> NADH-linked [11]) [11]
0.25 <14> (NAD^+) [10]
0.4 <14> (NADH) [10]
0.74 <2> (2-methylacetoacetyl-CoA) [8]
0.74 <1> (acetoacetyl-CoA, <1> [1]) [1]
0.82 <5> (acetoacetyl-pantetheine) [3]
0.99 <2> (acetoacetyl-S-(D-pantetheine) 11-pivalate) [8]

pH-Optimum

6.5-8 <4> [5]
7 <13> (<13> NADH-linked, reverse reaction pH 9.5 [11]) [11]
7 <13> (<13> NADPH-linked, reverse reaction pH 10.0 [11]) [11]
7.5 <12> [10]
8.1 <2> (<2> reduction of acetoacetyl-CoA [2, 9]) [2, 9]

pH-Range

6.5-8.5 <1> (<1> 70-35% of initial acetoacetyl-CoA reduced [1]) [1]

Temperature optimum (°C)

25 <2> [2]
30 <2, 12> [7, 10]
36 <4> [6]
37 <1, 4> [1, 5]

4 Enzyme Structure

Molecular weight

76000 <4> (<4> SDS-PAGE [6]) [6]
77000 <15> [10]
84000 <14> [10]
92000 <2> (<2> gel electrophoresis [7]) [7, 10]
100000 <2> (<2> gel filtration [8]) [8]
140000 <12> (<12> gel-filtration [10]) [10]
141000 <12> (<12> gradient PAGE [10]) [10]
250000 <10, 13> (<10> gel filtration [4]; <13> NADPH-linked: gel filtration [11]) [4, 10, 11]
340000 <13> (<13> NADH-linked: gel filtration [11]) [11]

Subunits

? <13> (<13> NADH-linked: 55000, SDS-PAGE, NADPH-linked: 42000 [11]) [11]
tetramer <2, 10, 12> (<2> 4 * 25500, SDS-PAGE [7]; <2> 4 * 24000 SDS-PAGE, 4 * 25200 gene sequence [8]; <10> 2 * 55000 + 2 * 65000, SDS-PAGE [4]; <12> 4 * 31000, 12% PAGE [10]) [4, 7, 8, 10]

5 Isolation/Preparation/Mutation/Application

Source/tissue

adipose tissue <4> [3]
liver <1, 2, 4, 7> [1, 3, 5, 6, 7]
mammary gland <4, 5> [3]
uropygial gland <6> [3]

Localization

cytosol <4> [5]
microsome <4> [5, 6]
soluble <2, 11, 12> [8, 9, 10]

Purification

<1> (fractionated with ammonium sulfate) [1]
<2> (ammonium sulfate fractionation, hydroxylapatite, Sephadex G-200 chromatography [2]; ammonium sulfate fractionation, DEAE-Sephasrose CL-6B, hydroxyapatite, Red-Sepharose CL-6B chromatography [7]; DEAE-Sepharose chromatography, Blue gel column [8]) [2, 7, 8]
<4> (KCl extraction [5, 6]) [5, 6]
<10> (gel-filtration Sephadex G-200, DEAE-cellulose chromatography, sucrose-gradient centrifugation [4]) [4]
<11> (ammonium sulfate fractionation, ion exchange, hydroxyapatite, affinity chromatography [9]) [9]
<12> (PEG, DEAE-Toyopearl column [10]) [10]

<13> (NADH-linked: DEAE-Sepharose CL-6B, Blue Sepharose CL-6B, botyl Sepharose 4B, Mono Q 5/5 [11]) [11]
<13> (NADPH-linked: DEAE-Sepharose CL-6B, Red 120-agarose, phenyle Sepharose [11]) [11]
<4, 5> [3]

Cloning
<2> (expressed in Escherichia coli JM155 [8]) [8]
<9> (expressed in Brassica napus, transported with fusion peptide to seed leukoplast [13]) [13]

Application
synthesis <14> (<14> PHB-synthesis for thermoplastics [12]) [12]

6 Stability

pH-Stability
6-8.5 <4> [5]
Additional information <1> (<1> above pH 8.5 acetoacetyl-CoA not stable [1]) [1]

Organic solvent stability
glycerol <2> (<2> during purification [7]) [7]

General stability information
<2>, glycerol stabilizes during purification [7]

Storage stability
<2>, -20°C, 10 mM Tris/HCl buffer, pH 7.5, 10 mM 2-mercaptoethanol, 50% glycerol, several months [7]
<2>, 0°C and -20°C, 10 mM Tris/HCl buffer, pH 7.5, 10 mM 2-mercaptoethanol, rather unstable [7]
<4>, -70°C [6]
<12>, 0°C, 25 mM Tris-HCl, pH 8.0 [10]

References

[1] Wakil, S.J.; Bressler, R.: Studies on the mechanism of fatty acid synthesis. J. Biol. Chem., **237**, 687-693 (1962)

[2] Saito, T.; Fukui, T.; Ikeda, F.; Tanaka, Y.; Tomita, K.: An NADP-linked acetoacetyl CoA reductase from Zoogloea ramigera. Arch. Microbiol., **114**, 211-217 (1977)

[3] Dodds, P.F.; Guzman, M.G.F.; Chalberg, S.C.; Anderson, G.J.; Kumar, S.: Acetoacetyl-CoA reductase activity of lactating bovine mammary fatty acid synthase. J. Biol. Chem., **256**, 6282-6290 (1981)

[4] Packter, N.M.; Flatman, S.: Characterization of acetoacetyl-CoA reductase (3-oxoreductase) from Streptomyces coelicolor: Its possible role in polyhydroxybutyrate biosynthesis. Biochem. Soc. Trans., **11**, 598-599 (1973)
[5] Prasad, M.R.; Cook, L.; Vieth, R.; Cinti, D.L.: Rat hepatic microsomal acetoacetyl-CoA reductase. A β-ketoacyl-CoA reductase distinct from the long chain β-ketoacyl-CoA reductase component of the microsomal fatty acid chain elongation system. J. Biol. Chem., **259**, 7460-7467 (1984)
[6] Cook, L.; Prasad, M.R.; Cook, W.R.; Cinti, D.L.: Isolation of rat liver microsomal short-chain β-ketoacyl-coenzyme A reductase and trans-2-enoyl-coenzyme A hydratase: evidence for more than one hydratase. Arch. Biochem. Biophys., **246**, 206-216 (1986)
[7] Fukui, T.; Ito, M.; Saito, T.; Tomita, K.: Purification and characterization of NADP-linked acetoacetyl-CoA reductase from Zoogloea ramigera I-16-M. Biochim. Biophys. Acta, **917**, 365-371 (1987)
[8] Ploux, O.; Masamune, S.; Walsh, C.T.: The NADPH-linked acetoacetyl-CoA reductase from Zoogloea ramigera. Characterization and mechanistic studies of the cloned enzyme over-produced in Escherichia coli. Eur. J. Biochem., **174**, 177-182 (1988)
[9] Amos, D.A.; McInerney, M.J.: Formation of D-3-hydroxybutyryl-coenzyme A by an acetoacetyl A reductase in Syntrophomonas wolfei subsp. wolfei. Arch. Microbiol., **159**, 16-20 (1993)
[10] Belova, L.L.; Sokolov, A.P.; Sidorov, I.A.; Trotsenko, Y.A.: Purification and characterization of NADPH-dependent acetoacetyl-CoA reductase from Methylobacterium extorquens. FEMS Microbiol. Lett., **156**, 275-279 (1997)
[11] Mothes, G.; Babel, W.: Methylobacterium rhodesianum MB 126 possesses two acetoacetyl-CoA reductases. Arch. Microbiol., **161**, 277-280 (1994)
[12] Gerngross, T.U.; Martin, D.P.: Enzyme-catalyzed synthesis of poly[(R)-(-)-3-hydroxybutyrate]: Formation of macroscopic granules in vitro. Proc. Natl. Acad. Sci. USA, **92**, 6279-6283 (1995)
[13] Houmiel, K.L.; Slater, S.; Broyles, D.; Casagrande, L.; Colburn, S.; Gonzalez, K.; Mitsky, T.A.; Reiser, S.E.; Shah, D.; Taylor, N.B.; Tran, M.; Valentin, H.E.; Gruys, K.J.: Poly(β-hydroxybutyrate) production in oilseed leukoplasts of Brassica napus. Planta, **209**, 547-550 (1999)

Malate dehydrogenase 1.1.1.37

1 Nomenclature

EC number
1.1.1.37

Systematic name
(S)-malate:NAD$^+$ oxidoreductase

Recommended name
malate dehydrogenase

Synonyms
(R)-2-hydroxyacid dehydrogenase [Swissprot]
L-malate dehydrogenase
MDH <10> [42]
MalDH <81> [50]
NAD-L-malate dehydrogenase
NAD-dependent malate dehydrogenase
NAD-dependent malic dehydrogenase
NAD-linked malate dehydrogenase
NAD-malate dehydrogenase
NAD-malic dehydrogenase
NAD-specific malate dehydrogenase
VEG69 [Swissprot]
Vegetative protein 69 [Swissprot]
m-MDH <4, 63> [4, 29]
mNAD-MDH [Swissprot]
malate (NAD) dehydrogenase
malate dehydrogenase (NAD)
malic acid dehydrogenase
malic dehydrogenase
mbNAD-MDH [Swissprot]
s-MDH <9, 63> [2, 29]

CAS registry number
9001-64-3

2 Source Organism

<-4> no activity in *Chromatium sp.* [36]
<-3> no activity in *Mycobacterium fortuitum* [10]
<-2> no activity in *Mycobacterium smegmatis 607* [10]

<-1> no activity in *Mycobacterium smegmatis* [10]
<1> *Drosophila virilis* (Texmelucan [1]) [1, 2, 4, 33]
<2> *Drosophila melanogaster* (Oregon K [1]) [1]
<3> *Bos taurus* (ox [1, 4, 5, 36]; bovine [10]) [1, 2, 4, 5, 7, 8, 10, 20, 21, 36]
<4> *Sus scrofa* (pig [1, 4, 5, 7, 8, 11, 12-16, 20-22, 25, 26, 31, 33, 35-37, 39, 40, 43, 47-50]) [1, 4, 5, 7, 8, 11, 12-16, 20-22, 25, 26, 31, 33, 35-37, 39, 40, 43, 47-50]
<5> *Equus sp.* (horse [1]) [1, 4, 5, 20, 33]
<6> *Gallus sp.* (chicken [1, 2, 4, 7, 12, 20-22, 26, 33]) [1, 2, 4, 7, 12, 20-22, 26, 33]
<7> *Thunnus albacares* (tuna [1, 2, 4, 20]) [1, 2, 4, 20]
<8> *Locusta migratoria* [1]
<9> *Physarum polycephalum* [2, 7, 21]
<10> *Saccharomyces cerevisiae* (yeast [2, 14, 20, 21, 33, 37, 42, 47]) [2, 14, 20, 21, 33, 37, 42, 47]
<11> *Opuntia ficus-indica* (cactus [2, 51]) [2, 51]
<12> *Salmo sp.* (salmonid fish [2, 7]; salmon [4, 20]) [2, 4, 7, 20]
<13> *Ilyanassa obsoleta* (marine snail [2]) [2]
<14> *Strongylocentrotus purpuratus* (sea urchin [2, 27]) [2, 27]
<15> *Phycomyces blakesleeanus* (wild-type strain NRRL 1555 (-) [26]) [2, 4, 26]
<16> *Chlorella pyrenoidosa* [2]
<17> *Neurospora crassa* [2-4, 10, 16, 20, 33]
<18> *Spinacia oleracea* (spinach [2, 3, 43]) [2, 3, 43]
<19> *Ascaris suum* [2]
<20> *Citrus sp.* (lemon fruit [2]) [2]
<21> *Zea mays* (maize [2, 30, 43, 47]; corn [51]) [2, 30, 43, 47, 51]
<22> *Fundulus heteroclitus* (teleost [2]) [2]
<23> *Lemna minor* [3]
<24> *Rattus norvegicus* (rat [4, 20, 22, 25, 40, 47]; Sprague-Dawley [22, 40]) [4, 20, 22, 25, 40, 47]
<25> *Bacillus subtilis* [4, 5, 8, 10, 15, 16, 20, 23, 28, 30, 34, 36, 37, 44]
<26> *Bacillus stearothermophilus* (N.C.A 1503, N.C.A. 1518 Ra2, 2184 [16]) [4, 5, 15, 16, 30, 36, 37, 44]
<27> *Escherichia coli* [4, 5, 8, 10, 14-16, 20, 23, 30, 35-37, 41, 43, 44, 47-49]
<28> *Ovis aries* (lamb [4]) [4]
<29> *Pseudomonas testosteroni* (Stanier strain 78, ATCC 11996 [5]) [5, 8, 15, 20, 23]
<30> *Pseudomonas acidovorans* [5]
<31> *Triticum aestivum* (wheat [5]) [5]
<32> *Solanum tuberosum* (potato [5]) [5]
<33> *Moraxella lwoffi* [6]
<34> *Micrococcus lysodeikticus* [6]
<35> *Acetobacter xylinium* [6]
<36> *Penaeus aztecus* (brown shrimp [7]) [7]
<37> *Halobacterium cutirubrium* [8, 9, 17, 18, 20, 34, 36]
<38> *Thermus aquaticus* (strain YT-1 [16]; strain B [23]) [8, 16, 20, 23, 32, 35, 45]
<39> *Mycobacterium phlei* [10, 20]

Substrates and products

S (-)-tartrate(S,S) + NAD^+ <29, 33> (Reversibility: r <29, 33> [5, 6]) [5, 6]
P ? + NADH
S 1,4,5,6-tetrahydronicotinamide + L-malate <4> (Reversibility: ? <4> [49]) [49]
P ?
S 3-acetylpyridine adenine dinucleotide + L-malate <78> (Reversibility: ? <78> [46]) [46]
P ?
S L-(-)α-hydroxyglutarate + NAD^+ <29> (<29> oxidized at an extremely slow rate, less than 0.001% of the rate of L-malate oxidation [5]) (Reversibility: r <29> [5]) [5]
P ? + NADH
S L-lactate + NAD^+ <4> (<4> cMDH less catalytically efficient than against its natural substrate malate [49]) (Reversibility: ? <4> [49]) [49]
P ? + NADH
S L-malate + NAD^+ <1-43, 45-112> (Reversibility: r <1-43, 45-112> [1-51]) [1-51]
P oxaloacetate + NADH
S L-malate + NAD^+ <77> (Reversibility: ? <77> [45]) [45]
P acetate + CO_2 + H_2 + NADH <77> [45]
S α-keto-malonate + NADH <4, 29, 68> (<4> reduced by m-MDH with half of the turnover number observed with oxaloacetate [4]) (Reversibility: ? <4, 68> [4, 32, 49]; ir <29> [5]) [4, 5, 32, 49]
P ? + NAD^+
S α-ketobutyrate + NADH <68> (Reversibility: ? <68> [32]) [32]
P ? + NAD^+
S citrate + NAD^+ <27> (Reversibility: ? <27> [41]) [41]
P ? + NADH
S hydroxymalonate + NAD^+ <29, 33> (<33> tartronic acid [6]) (Reversibility: ? <29> [5]; r <33> [6]) [5, 6]
P ? + NADH
S meso-tartrate + NAD^+ <29, 33, 68> (<68> slightly oxidized, 0.82 of the rate for L-malate oxidation [32]) (Reversibility: r <29, 33, 68> [5, 6, 32]) [5, 6, 32]
P ? + NADH
S nicotinamide 1-N^6-ethenoadenine dinucleotide + L-malate <78> (Reversibility: ? <78> [46]) [46]
P ?
S nicotinamide hypoxanthine dinucleotide + L-malate <78> (Reversibility: ? <78> [46]) [46]
P ?
S oxaloacetate + NADH <1-43, 45-112> (Reversibility: r <1-43, 45-112> [1-51]; ir <44> [20]) [1-51]
P L-malate + NAD^+
S oxaloglycolate + NAD^+ <29> (Reversibility: r <29> [5]) [5]
P ? + NADH

S pyridine 3-aldehyde adenine dinucleotide + L-malate <78> (Reversibility: ? <78> [46]) [46]

P ?

S pyruvate + NAD$^+$ <4> (<4> cMDH less catalytically efficient than against its natural substrate malate [49]) (Reversibility: ? <4> [49]) [49]

P ? + NADH

S thionicotinamide adenine dinucleotide + L-malate <78> (Reversibility: ? <78> [46]) [46]

P ?

S Additional information <29, 68> (<29> D-malate, (+)tartrate, succinate and L-(+)aspartate are not oxidized [5]; <68> D-malate, L-tartarate, D-tartarate, hydroxymalonate, dihydroxyfumarate, L-lactate, D-lactate DL-α-hydroxybutyrate, pyruvate, α-hydroxypyruvate, α-ketoisovalerate and α-ketoglutarate are no substrates [32]) [5, 32]

P ?

Inhibitors

2',3'-cyclic AMP <44> [14]
2-hydroxy-5-nitrobenzyl bromide <4> [4]
3-aminopyridine adenine dinucleotide <78> [46]
3-methylpyridine adenine dinucleotide <78> [46]
3-pyridylacetonitrile adenine dinucleotide <78> [46]
3-pyridylcarbinol adenine dinucleotide <78> [46]
ADP <44, 62, 78> [14, 19, 27, 46]
AMP <78> [46]
ATP <44, 51, 62, 75, 78, 93> [14, 19, 27, 39, 46]
$AgNO_3$ <33> [6]
Ca^{2+} <25, 33> [6, 10]
Co^{2+} <33, 39> [6, 10]
CoA <75> [39]
Cu^{2+} <39> [10]
D-fructose 1,6-diphosphate <3> (<3> heart s-MDH, inhibits the binding of NADH [4]) [4]
Hg^{2+} <39> [10]
L-malate <4, 6, 9, 24, 43, 44, 70> (<70> product inhibition [35]) [2, 12-14, 19, 22, 25, 35, 40]
Mg^{2+} <25, 33, 39> [6, 10]
Mn^{2+} <39> [10]
NAD$^+$ <15, 70, 78> (<70> product inhibition [35]) [26, 35, 46]
NADH <4, 33, 44, 78> (<4, 78> product inhibition [13, 46]; <33> substrate inhibition [6]) [6, 13, 14, 46]
NADP$^+$ <70> (<70> product inhibition [35]) [35]
Na_2SO_3 <62> [27]
Ni^{2+} <39> [10]
Zn^{2+} <25, 39, 62> [10, 27]
acetyl-CoA <44, 75> [14, 39]
adenosine <78> [46]

adenosine 5'-diphosphoribose <78> [46]
α-ketoglutarate <44, 70> (<44> end-product inhibition [14, 19]) [14, 19, 35]
citrate <75> [39]
dicarbonate <64, 65> [30]
diethyldicarbonate <46> [15]
hydroxymalonate <4, 53, 78> [4, 21, 46]
iodoacetamide <65> [30]
iodoacetate <39> [10]
oxaloacetate <1, 4, 9, 15, 24, 29, 33, 46, 51, 62, 70, 75-77, 105, 106> (<33, 46, 62, 70> substrate inhibition [6, 15, 27, 35]) [1, 2, 5, 6, 12, 13, 15, 19, 26, 27, 35, 39, 40, 44, 45]
p-chloromercuribenzoate <39> [10]
p-chloromercuriphenylsulfonate <62> [27]
phenylmethanesulfonylfluoride <65> [30]
pyridine adenine dinucleotide <78> [46]
sulfite <3, 4, 6, 62> [3, 4, 27]
thyroxine <53> [21]

Cofactors/prosthetic groups

NAD^+ <68-70, 76, 78, 82> (<68> can not be replaced by $NADP^+$ [32]; <69> purified enzyme shows activity with both coenzymes [34]; <78> $NADP^+$ does not serve as a coenzyme [46]; <82> no detectable activity with $NADP^+$ [51]) [32, 34, 35, 44, 46, 51]
$NADP^+$ <69, 70, 76> (<69> purified enzyme shows activity with both coenzymes [34]; <76> low activity with [44]) [34, 35, 44]

Activating compounds

2-mercaptoethanol <39> [10]
L-malate <4, 29> [4, 5]
NaCl <108> (<108> activation of the enzyme obtained by increasing the salt concentration to 3 M NaCl/l [47]) [47]
cysteine <39> [10]
dithiothreitol <62> [27]
glutathione <39> [10]
sodium phosphate <62> [27]

Metals, ions

$CaCl_2$ <62> [27]
$MgCl_2$ <62> [27]
Na_2HAsO_4 <62> [27]
NaF <62> [27]

Turnover number (min^{-1})

108 <69> (oxaloacetate) [34]
157 <70> (oxaloacetate) [35]
4600 <4> (L-malate) [13]

Specific activity (U/mg)

1.24 <44> [14]

1.28 <62> [27]
1.5 <70> (<70> L-malate oxidation in presence of NAD^+ [35]) [35]
1.6 <68> [36]
1.773 <73> [36]
2.324 <71> [36]
2.762 <29> [5]
43.47 <39> [10]
46.5 <43> [25]
62 <44> [20]
68 <24> [40]
83 <47> [23]
100.5 <38> [23]
104 <37> [8]
108 <74> [38]
110 <4> [13]
110.4 <75> [39]
180 <69> [34]
187 <55> [23]
222 <78> [46]
277 <70> [35]
322 <25> [23]
348 <68> [32]
352.7 <77> [45]
550 <53> [21]
774 <9> [2]
810 <9> (<9> cytoplasmic isoenzyme [7]) [7]
868 <72> [36]
1020 <9> (<9> mitochondrial isoenzyme [7]) [7]
1270 <64> [30]
1494 <82> (<82> cytoplasmic isoenzyme [51]) [51]
1670 <1> [1]
1675 <65> [30]
1728 <76> [44]
1920 <27> [23]
2000-2500 <82> (<82> mitochondrial isoenzyme [51]) [51]

K_m-Value (mM)

0.0041 <70> (NADH) [35]
0.0056 <69> (oxaloacetate) [34]
0.006 <46> (oxaloacetate, <46> 30°C [15]) [15]
0.0094 <10> (oxaloacetate, <10> fusion protein of MDH with mitochondrial citrate synthase [42]) [42]
0.0098 <46> (NADH, <46> 30°C [15]) [15]
0.0106 <77> (NADH, <77> Tv 10-02 [45]) [45]
0.011 <106> (NADH) [45]
0.0117 <69> (NADH) [34]
0.013 <105> (NADH) [45]

0.0135 <33> (NADH) [6]
0.0136 <10> (NADH, <10> fusion protein of MDH with mitochondrial citrate synthase [42]) [42]
0.014 <10> (L-malate, <10> fusion protein of MDH with mitochondrial citrate synthase [42]) [42]
0.014 <46> (oxaloacetate, <46> 37°C [48]) [48]
0.0152 <82> (NADH, <82> pH 7.5 [51]) [51]
0.016 <106> (oxaloacetate) [45]
0.016 <80> (oxaloacetate, <80> 4°C [48]) [48]
0.0162 <77> (oxaloacetate, <77> Tv 10-02 [45]) [45]
0.0165 <78> (NADH) [46]
0.0175 <105> (oxaloacetate) [45]
0.019 <46> (L-malate, <46> 30°C) [15]
0.02 <77> (NADH, <77> Tv-1961 [45]) [45]
0.02 <77> (NADH, <77> Tv-1973 [45]) [45]
0.02 <46> (oxaloacetate, <46> 60°C [15]) [15]
0.02 <77> (oxaloacetate, <77> Tv-1961 [45]) [45]
0.022 <77> (NAD^+, <77> Tv-1973 [45]) [45]
0.023 <10> (oxaloacetate) [42]
0.023 <27> (oxaloacetate, <27> 4°C [48]) [48]
0.023 <63> (oxaloacetate, <63> *m*-MDH [29]) [29]
0.023 <80> (oxaloacetate, <80> 10°C [48]) [48]
0.024 <75> (NAD^+) [39]
0.025 <68> (NADH) [32]
0.026 <46> (NAD^+, <46> 30°C) [15]
0.026 <27> (oxaloacetate, <27> 10°C [48]) [48]
0.027 <25> (NADH) [10]
0.027 <44> (oxaloacetate) [14, 20]
0.028 <10> (L-malate) [42]
0.0286 <39> (NADH) [10]
0.029 <63> (NADH, <63> m-MDH [29]) [29]
0.03 <62> (NADH) [27]
0.03 <10> (NADH) [42]
0.03 <76> (NADH) [44]
0.03 <46> (NADH, <46> 60°C [15]) [15]
0.03 <53> (NADH, <53> whole cell MDH [21]) [21]
0.03 <15> (oxaloacetate) [26]
0.03 <4> (oxaloacetate, <4> cytoplasmic isoenzyme) [12]
0.034 <63> (oxaloacetate, <63> s-MDH [29]) [29]
0.036 <4> (oxaloacetate) [13]
0.036 <63> (oxaloacetate) [28]
0.038 <29> (NADH) [5]
0.038 <6> (oxaloacetate) [12]
0.0387 <33> (oxaloacetate) [6]
0.04 <53> (NADH, <53> mitochondrial MDH [21]) [21]
0.04 <4> (oxaloacetate) [12]
0.04 <1> (oxaloacetate, <1> cytoplasmic enzyme [1]) [1]

0.041 <82> (NADH, <82> pH 6.8 [51]) [51]
0.043 <77> (oxaloacetate, <77> Tv-1973 [45]) [45]
0.044 <27> (oxalacetate, <27> mutant L305C [43]) [43]
0.0443 <39> (oxalacetate) [10]
0.045 <80> (oxaloacetate, <80> 37°C [48]) [48]
0.047 <1> (oxaloacetate, <1> mitochondrial enzyme [1]) [1]
0.048 <68> (NAD^+) [32]
0.048 <9> (NADH, <9> S-MDH [2]) [2]
0.049 <27> (oxalacetate, <27> wild-type [43]) [43]
0.05 <76> (oxaloacetate) [44]
0.052 <27> (NADH, <27> mutant N122C-L305C(diamide) [43]) [43]
0.052 <63> (NADH, <63> s-MDH [29]) [29]
0.052 <70> (oxaloacetate) [35]
0.056 <27> (NADH, <27> mutant N122C-L305C(DTT) [43]) [43]
0.056 <27> (oxaloacetate, <27> 37°C [48]) [48]
0.057 <29> (oxaloacetate) [5]
0.06 <9> (oxaloacetate, <9> s-MDH [2]) [2]
0.061 <27> (NADH, <27> wild-type [43]) [43]
0.061 <25> (oxalacetate) [10]
0.063 <25> (NADH) [28]
0.0643 <82> (oxaloacetate, <82> pH 6.8 [51]) [51]
0.0667 <78> (3-acetylpyridine adenine dinucleotide) [46]
0.069 <29> (L-malate, <29> 0.03-0.1 mM substrate [5]) [5]
0.0699 <77> (NAD^+, <77> Tv 10-02 [45]) [45]
0.07 <106> (NAD^+) [45]
0.072 <78> (oxaloacetate) [46]
0.074 <9> (oxaloacetate, <9> M-MDH [2]) [2]
0.08 <44> (NADH) [20]
0.081 <44> (NADH) [14]
0.083 <51> (oxaloacetate, <51> mitochondrial MDH [19]) [19]
0.085 <68> (oxaloacetate) [32]
0.087 <27> (NADH, <27> mutant L305C [43]) [43]
0.089 <29> (3-acetylnicotinamide-NAD^+) [5]
0.09 <51> (oxaloacetate, <51> cyt MDH [19]) [19]
0.095 <46> (L-malate, <46> 60°C) [15]
0.1 <53> (NAD^+, <53> mitochondrial MDH [21]) [21]
0.109 <29> (NAD^+) [5]
0.11 <10> (NAD^+, <10> fusion protein of MDH with mitochondrial citrate synthase [42]) [42]
0.117 <78> (NAD^+) [46]
0.132 <78> (thionicotinamide adenine dinucleotide) [46]
0.14 <10> (L-malate, <10> fusion protein of MDH with mitochondrial citrate synthase [42]) [42]
0.14 <25> (NAD^+) [10]
0.14 <4> (NAD^+) [13]
0.14 <53> (NAD^+) [21]
0.14 <46> (NAD^+, <46> 60°C) [15]

0.14 <51> (NADH, <51> mit MDH [19]) [19]
0.17 <53> (oxaloacetate) [21]
0.174 <33> (L-malate) [6]
0.178 <75> ($NADP^+$) [39]
0.19 <9> (NAD^+, <9> m-MDH [2]) [2]
0.19 <78> (pyridine 3-aldehyde adenine dinucleotide) [46]
0.192 <51> (NADH, <51> cytosolic MDH) [19]
0.2 <106> (L-malate) [45]
0.2 <9> (NADH, <9> m-MDH [2]) [2]
0.2 <70> (NADPH) [35]
0.22 <82> (NAD^+, <82> pH 7.5 [51]) [51]
0.23 <29> (thionicotinamide-NAD^+) [5]
0.25 <78> (nicotinamide hypoxanthine dinucleotide) [46]
0.26 <27> (NAD^+, <27> wild-type [43]) [43]
0.263 <29> (L-malate, <29> 0.2-10 mM substrate [5]) [5]
0.264 <29> (deamino-NAD^+) [5]
0.27 <53> (oxaloacetate, <53> mitochondrial MDH [21]) [21]
0.28 <10> (L-malate) [42]
0.28 <82> (NAD^+, <82> pH 6.8 [51]) [51]
0.287 <82> (oxaloacetate, <82> pH 7.5 [51]) [51]
0.3 <10> (NAD^+) [42]
0.3 <63> (NAD^+, <63> s-MDH [29]) [29]
0.34 <27> (oxalacetate, <27> mutant N122C-L305C(DTT) [43]) [43]
0.38 <27> (NAD^+, <27> mutant N122C-L305C(DTT) [43]) [43]
0.435 <33> (NAD^+) [6]
0.459 <39> (NAD^+) [10]
0.46 <82> (L-malate, <82> pH 7.5 [51]) [51]
0.5 <27> (NAD^+, <27> mutant N122C-L305C(diamide) [43]) [43]
0.55 <78> (malate) [46]
0.58 <27> (oxalacetate, <27> mutant N122C-L305C(diamide) [43]) [43]
0.6 <62> (NAD^+) [27]
0.71 <63> (L-malate, <63> s-MDH [29]) [29]
0.71 <63> (NAD^+, <63> m-MDH [29]) [29]
0.75 <9> (NAD^+, <9> S-MDH [2]) [2]
0.833 <39> (L-malate) [10]
0.9 <25> (L-malate) [10]
0.9 <63> (L-malate, <63> m-MDH [29]) [29]
1 <77> (L-malate, <77> Tv-1961 [45]) [45]
1.1 <76> (NAD^+) [44]
1.22 <77> (L-malate, <77> Tv 10-02 [45]) [45]
1.33 <53> (L-malate, <53> mitochondrial MDH [21]) [21]
1.4 <77> (L-malate, <77> Tv-1973 [45]) [45]
1.464 <78> (nicotinamide 1-N^6-ethenoadenine dinucleotide) [46]
2.1 <68> (L-malate) [32]
2.1 <82> (L-malate, <82> pH 6.8 [51]) [51]
2.2 <68> (ketomalonate) [32]
2.6 <27> (L-malate, <27> wild-type [43]) [43]

4 <76> (L-malate) [44]
5 <53> (L-malate) [21]
5.3 <27> (L-malate, <27> mutant N122C-L305C(DTT) [43]) [43]
5.6 <75> (L-malate) [39]
6.8 <27> (L-malate, <27> mutant N122C-L305C(diamide) [43]) [43]
7 <62> (L-malate) [27]
8 <1> (L-malate, <1> cytoplasmic isoenzyme [1]) [1]
20 <1> (L-malate, <1> mitochondrial isoenzyme [1]) [1]

pH-Optimum

4.3-7 <68> (<68> reduction of ketomalonate [32]) [32]
6 <27> (<27> mutant N122C-L305C [43]) [43]
6.5-7.3 <65> (<65> conversion of oxaloacetate to L-malate [30]) [30]
6.6-7 <82> (<82> oxaloacetate reduction [51]) [51]
6.7-7.5 <75> [39]
6.8-7 <82> [51]
7-7.5 <91> [39]
7.2 <46, 63> (<46> oxaloacetate reduction, 60°C [15]) [15, 28]
7.3-7.5 <92> [39]
7.5 <76> (<76> reduction of oxaloacetate [44]) [44]
7.5-8.3 <77> (<77> Tv 10-02, reduction of oxaloacetate [45]) [45]
7.5-9.5 <63> (<63> s-MDH [29]) [29]
7.6 <3, 6, 9> (<3, 6> oxaloacetate reduction, s-MDH [2]; <9> oxaloacetate reduction, both isoenzymes [2]) [2]
7.75-8.5 <68> (<68> reduction of oxaloacetate [32]) [32]
7.8 <3, 6, 77> (<3, 6> oxaloacetate reduction, m-MDH [2]; <77> strain Tv-1973 [45]) [2, 45]
7.8-7.9 <77> (<77> Tv-1961, reduction of oxaloacetate [45]) [45]
8 <44, 63, 105> (<63> m-MDH [29]; <105> oxaloacetate reduction [45]) [14, 29, 45]
8-8.5 <64> (conversion of oxaloacetate to L-malate [30]) [30]
8.1 <69> (<69> reduction of oxaloacetate [34]) [34]
8.2-8.6 <65> (<65> conversion of L-malate to oxaloacetate [30]) [30]
8.2-9.4 <62> (<62> reduction of oxaloacetate with NADH [27]) [27]
8.5 <1, 2, 76> (<1,2> oxaloacetate reduction [1]) [1, 44]
8.6-9 <64> (<64> conversion of L-malate to oxaloacetate [30]) [30]
8.8 <29> (<29> oxaloacetate reduction [5]) [5]
9 <1, 2, 39, 76, 77, 106> (<1, 2> malate oxidation, cytoplasmic isoenzymes [1]; <39> maximum conversion of malate to oxalacetate [10]; <76> oxidation of L-malate [44]; <77> malate oxidation, above pH 9.0, strain Tv-1973 [45]; <106> oxalacetate reduction [45]) [1, 10, 44, 45]
9.2-9.6 <39> (<39> reduction of oxalacetate to malate [10]) [10]
9.2-10.7 <77> (<77> malate oxidation, strain Tv-1961 [45]) [45]
9.3-9.5 <53> (<53> both isoenzymes [21]) [21]
9.5 <1, 2, 39, 46, 68, 77> (<1, 2> mitochondrial isoenzyme [1]; <46> 30°C [15]; <68> L-malate oxidation [32]; <77> reverse reaction from malate to oxalacetate [45]) [1, 10, 15, 32, 45]

10 <7, 9> (<9> malate oxidation [2]) [2]
10.2 <62> (<62> malate oxidation with NAD^+ [27]) [27]
10.4 <29> (<29> malate oxidation [5]) [5]

pH-Range
4-10 <39> [10]
5.5-9 <15> [26]

Temperature optimum (°C)
40 <63> [28]
60 <76> [44]
65 <69> [34]
70 <38> [20]

Temperature range (°C)
4-10 <80> [48]
8-80 <70> [35]
10-70 <37> [18]

4 Enzyme Structure

Molecular weight
32000 <18> [3]
35000 <18> (<18> K^+ and/or dithioerythritol treated [3]) [3]
39000 <18> [3]
43000 <30> (<30> sedimentation equilibrium centrifugation [5]) [5]
52000 <3> (<3> s-MDH [20]) [20]
53000 <52> [20]
54000 <17> [20]
56000 <4, 27, 73> (<4, 27, 73> gel filtration [36]) [36]
57000 <73> (<73> native PAGE [36]) [36, 37]
60000 <4, 27, 46, 73, 77> (<4, 27, 73> sucrose density gradient centrifugation [36]; <46> gel filtration [15]; <77> gel filtration [45]) [15, 20, 36, 45]
60000-67000 <27> (<27> sedimentation equilibrium centrifugation [5]) [5]
60300 <33> (<33> ultracentrifugation [6]) [6]
61000 <33> (<33> gel filtration [6]) [6]
61700 <44> (<44> gel filtration [14, 20]) [14, 20]
62000 <3> (<3> m-MDH [20]) [20]
64000 <18> (<18> repeatedly frozen and thawed, most stable unit, gel filtration [3]) [3]
64200 <44> (<44> ultracentrifugation equlibrium [20]) [20]
65000 <62> [27]
67000 <4, 6, 7, 17, 28, 46, 50> (<4, 6, 7, 17, 28> gel filtration [4]; <46> sedimentation equilibrium centrifugation [15]; <50> gel filtration [16]) [1, 4, 15, 16, 20]

68000 <1, 2, 4, 10, 12, 51, 76> (<1, 2> both isoenzymes, gel filtration [1]; <4> mitochondrial isoform [4]; <10> m-MDH [20]; <51> gel filtration [19]; <76> gel filtration [44]) [1, 4, 19, 20, 33, 44]
68400 <78> (<78> gel filtration [46]) [46]
69000 <63, 65> (<65> m-MDH [30]) [30]
69400 <65> (<63, 65> m-MDH [30,33]) [30, 33]
69500 <9> (<9> both isoenzymes, sedimentation equilibrium [7]) [7]
70000 <7, 23, 24, 29, 48, 53, 100> (<29> gel filtration, analytical Sephadex [5]; <48> gel filtration [16]; <23> various isoenzymes [3]) [3, 5, 16, 20, 21, 47]
71000 <38> (<38> gel filtration [16]) [16]
71600 <24> (<24> gel filtration [22]) [22]
72000 <4, 43, 46, 71, 106> (<4> cytoplasmic isoform [4]; <43> gel filtration [25]; <71> cross-linking experiment [36]; <106> gel filtration [45]) [4, 20, 25, 36, 45]
72400 <4, 63> (<63> gel filtration [33]) [33]
73500 <38> [20]
74000 <4> [11]
74200 <29> (<29> sedimentation equilibrium centrifugation [5]) [5]
75000 <10, 29> (<29> Svedberg equation [5]; <10> s-MDH [20]) [5, 20]
76000 <74, 90> (<74> gel filtration [38]) [38]
79120 <48> (<48> sedimentation equilibrium method [16]) [16]
80000 <68> (<68> gel filtration [32]) [32, 36]
84000 <37> (<37> gel filtration [8]) [8, 9, 17, 20]
85040 <38> (<38> sedimentation equilibrium method [16]) [16]
86200 <39> (<39> SDS-PAGE [10]) [10]
86860 <39> (<39> gel filtration [10]) [10, 20]
90000 <75> (<75> gel filtration [39]) [39]
97000 <70> (<70> gel filtration [35]) [35]
105000 <69> (<69> gel filtration [34]) [34]
110000 <71> (<71> cross-linking experiment [36]) [36]
117000 <27> [4]
117000-148000 <25> (<25> sedimentation equilibrium centrifugation [5]) [5]
122000 <72> (<72> gel filtration [36]) [36]
124000 <47> (<47> gel filtration [16]) [16]
125000 <49> (<49> gel filtration [16]) [16]
127000 <18> (<18> gel filtration [3]) [3]
128000 <72> [37]
128500 <70> (<70> ultracentrifugation, sedimentation equilibrium studies [35]) [35]
130000 <69, 71, 85> (<69> sedimentation equilibrium experiment [34]; <71,85> gel filtration [36]) [34, 36]
130000-140000 <68, 71, 72> (<68, 71, 72> gel filtration [36]) [36]
132000 <81> [50]
133600 <49> (<49> sedimentation equilibrium method [16]) [16]
134000 <70, 72> (<70> cross-linking data [35]; <72> sucrose density gradient centrifugation [36]) [35, 36]

134900 <47> (<47> sedimentation equlibrium method [16]) [16]
135000 <55> (<55> gel filtration [16]) [16]
137000 <68, 71> [37]
137100 <26> (<26> N.C.A. 1518 Ra2, N.C.A. 1503, sedimentation equilibrium method [16]) [16]
138000 <26, 85> (<26> N.C.A. 1518 Ra2, gel filtration [16]; <85> sucrose gradient centrifugation [36]) [16, 36]
138600 <64> [30]
138600 <25> (<25> sedimentation equilibrium method [16]) [16]
139000 <64> (<64> gel filtration [30]) [30]
140000 <23, 49, 71> (<23> various isoenzymes [3]; <71> native PAGE [36]) [3, 35, 36]
140900 <55> (<55> sedimentation equilibrium method [16]) [16]
142000 <25, 68, 71> (<25> gel filtration [16]; <68,71> sucrose density gradient centrifugation [36]) [16, 36]
144000 <69> (<69> cross-linking data [34]) [34]
146000 <71> (<71> cross-linking experiment [36]) [36]
148000 <25, 68> (<68> enzyme cross-linking [36]) [20, 36]
160000 <82> (<82> native PAGE [51]) [51]
170000 <10> (<10> fusion protein of MDH with mitochondrial citrate synthase, gel filtration [42]) [42]
180000 <18> (<18> K^+ and/or dithioerythritol treated [3]) [3]
244000 <105> (<105> gel filtration [45]) [45]

Subunits

dimer <3, 4, 9, 10, 29, 37, 38, 44, 46, 50, 51, 55, 62-64, 65, 68, 71, 73, 74, 76-78, 80, 82, 89, 90, 109> (<3,4> 2 * 33000-35000 [4]; <9> 2 * 35600, cytoplasmic form, SDS-PAGE [7]; <9> 2 * 37000, mitochondrial form, SDS-PAGE [7]; <10> 2 * 90000-95000, fusion protein of MDH with mitochondrial citrate synthase, SDS-PAGE [42]; <29> 2 * 36000, SDS-PAGE [5]; <37> 2 * 39000, SDS-PAGE [8]; <38> 2 * 36000, SDS-PAGE [16]; <44> 2 * 31350, SDS-PAGE [20]; <46> 2 * 35000, SDS-PAGE [15,47]; <51> 2 * 35000, SDS-PAGE [19]; <55> 2 * 34000, SDS-PAGE [16]; <62> 2 * 35000, SDS-PAGE [27]; <63> 2 * 34000, m-MDH, SDS-PAGE [33]; <63> 2 * 36000, s-MDH, SDS-PAGE [33]; <65> 1 * 39000 + 1 * 37000, SDS-PAGE [30]; <68> 2 * 35000, SDS-PAGE [32]; <73> 2 * 27500, SDS-PAGE [36]; <74> 2 * 40000, SDS-PAGE [38]; <76> 2 * 35000, SDS-PAGE [44]; <77> 2 * 34500, SDS-PAGE [45]; <78> 2 * 34600, SDS-PAGE [46]; <82> 2 * 39500, pH 7.0, SDS-PAGE [51]; <109> 2 * 38000 [47]) [4, 5, 7-9, 11, 13, 15-20, 27, 30, 32, 33, 36, 38, 42, 44-48, 50, 51]
hexamer <105> (<105> 6 * 37000 [45]) [45]
octamer <79> [47]
tetramer <25-27, 39, 47, 49, 55, 64, 68, 69, 70-72, 81, 82, 85> (<25> 4 * 33500, SDS-PAGE [4,5]; <25> 4 * 35000, SDS-PAGE [16]; <26> 4 * 35000, SDS-PAGE [16]; <27> 4 * 33500 [4]; <39> 4 * 21550, SDS-PAGE [10]; <47> 4 * 35300, SDS-PAGE [16]; <49> 4 * 32000, SDS-PAGE [16]; <50> 4 * 34000, SDS-PAGE [16]; <64> 4 * 36000, SDS-PAGE [30]; <68> 4 * 37000, SDS-PAGE [36]; <69> 4 * 36600, SDS-PAGE [34]; <70> 4 * 34000, SDS-PAGE [35]; <71> 4 * 37000,

SDS-PAGE [36]; <72> 4 * 36000, SDS-PAGE [36]; <82> 4 * 39500, pH 8.0, SDS-PAGE [51]) [5, 10, 16, 30, 34-37, 50, 51]
trimer <71> [36]

5 Isolation/Preparation/Mutation/Application

Source/tissue

bacteroid <64, 66> [30]
cotyledon <90, 99> [38, 51]
egg <62> [27]
endosperm <21> [2]
erythrocyte <43> [10]
fruit <74> [38]
germ <31> [5]
heart <3, 4-7, 12> [1, 3-5, 7, 8, 10-16, 20-22, 25, 26, 33, 35-37, 39, 40, 47-49]
hypocotyl <93> [39]
kidney <3> [1, 2, 7]
leaf <18, 23, 56, 66, 82> [2, 3, 24, 30, 47, 51]
liver <6, 24, 43> [4, 20, 22, 25, 40, 47]
muscle <12> [20]
mycelium <15> [26]
plant <82> [51]
root <65, 66> (<65> nodule [30]) [30]
seed <56> [24, 47]
seedling <56> [24]
sporangiophore <15> [26]
spore <15> [26]
tuber <32> [5]

Localization

chloroplast <18, 21, 56, 63, 66> [28, 43, 47, 51]
cytoplasm <1, 3-7, 9, 10, 15, 17-19, 22, 51, 53, 78, 98, 110> [1, 2, 4, 11, 12, 19-21, 26, 44, 46, 47, 49-51]
cytosol <3, 4, 6, 9, 10, 24, 43, 46, 56, 62, 63-65, 74, 77, 82, 98, 110, 111, 112> [2, 22, 24, 25, 27-30, 33, 38, 40, 45, 47, 49, 51]
glyoxysome <63, 82, 99> [29, 47, 51]
hydrogenosome <77> [45]
microbody <63> [28, 29, 47]
microsome <63> [28]
mitochondrion <1-7, 9, 10, 17-19, 22-24, 46, 51, 53, 56, 62, 63-65, 74, 82, 98, 99, 112> (<4> inner or outer membrane, m-MDH located in the inner membrane matrix [4]; <63> outer membrane, intermembrane space, inner membrane, matrix [28, 47]) [1-5, 7, 11-13, 15, 16, 19-22, 24, 27-31, 36-40, 44, 47, 51]
peroxisome <18, 23, 63, 82, 112> [3, 28, 47, 51]
plastid <11, 56> [51]

Purification

<1> [1, 33]
<2> [1]
<4> (both isoenzymes, sMDHase and mMDHase [11]) [11, 13, 33]
<5> [33]
<6> [33]
<8> [1]
<9> (2 isoenzymes, S-MDH and M-MDH [2]) [2, 7]
<10> (fusion protein of MDH with mitochondrial citrate synthase [42]) [42]
<14> [27]
<15> [26]
<17> [33]
<21> [47]
<24> [22, 40]
<25> [4, 5, 14, 23, 36]
<26> [5, 14, 36]
<27> [5, 16, 23, 36, 44]
<29> [5]
<33> [6]
<37> [8, 17]
<38> [23]
<39> [10]
<43> [25]
<44> [14, 20]
<46> [15, 44, 47]
<47> [23]
<51> (2 isoenzymes, cytoplasmic and mitochondrial [19]) [19]
<53> [21]
<62> (M-MDH [27]) [27]
<63> [33]
<64> [30]
<65> [30]
<68> [32, 36, 37]
<69> [34]
<70> [35]
<71> [36, 37]
<72> [36, 37]
<73> [36, 37]
<74> [38]
<75> [39]
<76> [44]
<77> [45]
<78> [46]
<80> [48]
<82> [51]
<83> [32]
<95> [44]

<96> [44]
<97> [44]
<98> [44]
<99> [44]
<102> [44]
<103> [44]
<104> [44]
<108> [47]

Crystallization

<3> (s-MDH [4]) [4]
<4> (cytoplasmic enzyme [4]; recombinant porcine cMDH [49]) [4, 11, 49]
<5> (s-MDH [4]) [4]
<21> [47]
<27> [41, 43, 47]
<28> (s-MDH [4]) [4]
<29> [5]
<30> [5]
<39> [10]
<46> [47]
<69> [34]
<70> [35]
<80> [48]
<81> [50]
<108> [47]
<111> [51]

Cloning

<4> (cMDH cloned and overexpressed in Escherichia coli [49]) [49]
<10> (gene encoding mitochondrial isoenzyme of MDH is present in a single copy in genomic DNA [47]) [47]
<27> (Escherichia coli malate dehydrogenase gene cloned into expression vector pASK40 by PCR [43]) [43]
<46> (gene for tMDH expressed in Escherichia coli and sequence reported [47]) [47]
<80> [48]
<81> (structural gene expressed in Escherichia coli expression vector pET11a, large amounts of soluble but inactive form of the enzyme produced upon its induction [47]; gene encoding the halophilic MalDH cloned into M13mp18 phage [50]) [47, 50]
<109> (cDNA encoding mitochondrial malate dehydrogenase sequenced [47]) [47]

Application

agriculture <78> (<78> swine respiratory pathogen is the etiological agent of Glaesser's disease, swine industry economic losses worldwide [46]) [46]
medicine <77> (<77> therapy development for sexually transmitted disease of humans [45]) [45]

6 Stability

pH-Stability

6.5-9 <44> (<44> no activity remains below pH 5.5 and above pH 11 [14]) [14]
7-9 <70> (<70> quite stable [35]) [35]
8 <81> (<81> Hm MalDH dissociates into dimers at 2 M NaCl, the dimer is as active as the wild type tetramer at pH 8, but loses its activity at pH 7 [50]) [50]

Temperature stability

35 <72> (<72> loses activity above [37]) [37]
38 <15> (<15> retains 100% activity after 45 min incubation, only 50% activity remains after 15 min at 45°C [26]) [26]
40 <68> (<68> loses its activity completely upon incubation for 10 min [32]) [32, 37]
45 <56, 62> (<62> after incubation for 50 min enzyme loses about 90% of its activity [27]) [24, 27]
50 <25, 27, 39> (<25, 27> stable for 20 min, loses more than 95% activity when subjected to 55°C for the same period of time [30]; <39> thermostable up to for 15 min [10]) [10, 30]
51 <9> (<9> m-MDH, activity reduced to 50% [2]) [2]
55 <9, 46, 80> (<9> s-MDH, activity reduced to 50% [2]; <46> retains full activity for more than 2 h [48]; <80> half life of 10 min [48]) [2, 48]
60 <69> (<69> quite heat-stable, stable up to, loses 50% of its original activity after 10 min at 77°C [34]) [34]
65 <26, 69, 71> (<71> extremely heat-stable, withstanding for over 1 h with no loss of activity [36]; <71> completely stable to 2 h [37]) [36, 37]
75 <26> (<26> stable up to, but loses all activity when heated to 80°C for 20 min [30]) [30]
80 <82> (<82> activity increases up to [51]) [51]
90 <46> (<46> remarkably heat stable without losing activity after incubation for 60 min, 50% activity lost after 30 min at 96°C [15]; <46> fully active for 1 h [48]) [15, 48]

General stability information

<18>, both Mg^{2+} and NADH stabilizes the 127000 Da MW form [3]
<46>, remarkably resistant to urea and SDS [15]
<65>, mitochondrial malate dehydrogenase is very unstable [30]
<68>, addition of NAD^+, NADH or oxaloacetate stabilizes the enzyme [32]
<81>, only stable in highly concentrated solutions of certain salts [47]

Storage stability

<6>, -20°C, stable [22]
<9>, 0°C, purified enzymes are stable for up to 2 months in 0.05 M potassium phosphate buffer, pH 7.5 containing 0.001 M 2-mercaptoethanol [2]
<24>, -20°C, completely inactivated [22]
<24>, 4°C, no appreciable loss in activity for up to 3 months [22]

<39>, 20°C, dilute enzyme preparation stable for at least 2 weeks [10]
<39>, 25°C, loses not more than 20% activity within 2 weeks [10]
<43>, 4°C, purified enzyme remains stable for up to 1 month [25]
<44>, 4°C, at pH 8 in 0.05 M Tris/HCl containing 1 M NaCl, several weeks without appreciable loss of activity [20]
<44>, 4°C, storage in Tris-HCl, 0.01 M, pH 8, 0.2 M NaCl results in a 96% loss of activity within 48 h, loss of activity can be prevented completely for at least 4 weeks by storage in the same buffer, containing 1 M NaCl, under the latter conditions the enzyme is inactive, activity is completely restored upon dilution of the salt to 20 mM [14]
<62>, -20°C, can be stored at least several months without any appreciable loss in activity [27]
<68>, -10°C, in 20 mM potassium phosphate buffer, pH 7.0, 0.01% 2-mercaptoethanol/40% glycerol, can be stored for at least 2 months without loss of activity [32]
<70>, -20°C, can be stored over months [35]
<78>, -15°C, in 10 mM potassium phosphate buffer, pH 7.5, containing 0.1 M KCl, stable for at least 3 months with no apparent loss of activity [46]
<82>, -20°C, activity tends to decay rapidly after purification [51]
<82>, -80°C, remains active for several weeks [51]

References

[1] McReynolds, M.S.; Barrie Kitto, G.: Purification and properties of Drosophila malate dehydrogenase. Biochim. Biophys. Acta, **198**, 165-175 (1970)

[2] Teague, W.M.; Henney, H.R.: Purification and properties of cytoplasmic and mitochondria malate dehydrogenase of Physarum polycephalum. J. Bacteriol., **116**, 673-684 (1973)

[3] Ziegler, I.: Malate dehydrogenase of spinach: Substrate kinetics of different forms. Phytochemistry, **13**, 2403-2410 (1974)

[4] Banaszak, L.J.; Bradshaw, R.A.: Malate dehydrogenases. The Enzymes (3rd Ed.), **11**, 369-396 (1975)

[5] You, K.S.; Kaplan, N.O.: Purification and properties of malate dehydrogenase from Pseudomonas testosteroni. J. Bacteriol., **123**, 704-716 (1975)

[6] Fernandes, R.; Jones, M.; King, H.K.: Purification and properties of malate-NAD^+ dehydrogenase of Moraxella lwoffi (N.C.I.B.8250). Biochem. Soc. Trans., **4**, 1080 (1976)

[7] Teague, W.M.; Henney jr., H.R.: Physical properties and chemical compositions of cytoplasmic and mitochondrial malate dehydrogenase from Physarum polycephalum. Biochim. Biophys. Acta, **434**, 118-125 (1976)

[8] Mevarech, M.; Eisenberg, H.; Neumann, E.: Malate dehydrogenase isolated from extremely halophilic bacteria of the dead sea. 1. Purification and molecular characterization. Biochemistry, **16**, 3781-3785 (1977)

[9] Mevarech, M.; Neumann, E.: Malate dehydrogenase isolated from extremely halophilic bacteria of the dead sea. 2. Effect of salt on the catalytic activity and structure. Biochemistry, **16**, 3786-3793 (1977)
[10] Tyagi, A.K.; Siddiqui, F.A.; Venkitasubramanian: Studies on the purification and characterization of malate dehydrogenase from Mycobacterium phlei. Biochim. Biophys. Acta, **485**, 255-267 (1977)
[11] Weininger, M.S.; Banaszak, L.J.: Mitochondrial malate dehydrogenase. Crystallographic properties of the pig heart enzyme. J. Mol. Biol., **119**, 443-449 (1978)
[12] Bernstein, L.H.; Grisham, M.B.; Cole, K.D.; Everse, J.: Substrate inhibition of the mitochondrial and cytoplasmic malate dehydrogenases. J. Biol. Chem., **253**, 8697-8701 (1978)
[13] Mueggler, P.A.; Wolfe, R.G.: Malate dehydrogenase. Kinetic studies of substrate activation of supernatant enzyme by L-malate. Biochemistry, **17**, 4615-4620 (1978)
[14] Sprott, G.D.; McKellar, R.C.; Shaw, K.M.; Giroux, J.; Martin, W.G.: Properties of malate dehydrogenase isolated from Methanospirillum hungatii. Can. J. Microbiol., **25**, 192-200 (1978)
[15] Iijima, S.; Saiki, T.; Beppu, T.: Physicochemical and catalytic properties of thermostable malate dehydrogenase from an extreme thermophile Thermus flavus AT-62. Biochim. Biophys. Acta, **613**, 1-9 (1980)
[16] Sundaram, T.K.; Wright, I.P.; Wilkinson, A.E.: Malate dehydrogenase from thermophilic and mesophilic bacteria. Molecular size, subunit structure, amino acid composition, immunochemical homology, and catalytic activity. Biochemistry, **19**, 2017-2022 (1980)
[17] Pundak, S.; Eisenberg, H.: Structure and activity of malate dehydrogenase from the extreme halophilic bacteria of the dead sea. 1. Conformation and interaction with water and salt between 5 M and 1 M NaCl concentration. Eur. J. Biochem., **118**, 463-470 (1981)
[18] Pundak, S.; Aloni, H.; Eisenberg, H.: Structure and activity of malate dehydrogenase from the extreme halophilic bacteria of the dead sea. 2. Inactivation, dissociation and unfolding at NaCl concentrations below 2 M. Salt, salt concentration and temperature dependence of enzyme stability. Eur. J. Biochem., **118**, 471-477 (1981)
[19] Ma, H.; Kubicek, C.P.; Roehr, M.: Malate dehydrogenase isoenzymes in Aspergillus niger. FEMS Microbiol. Lett., **12**, 147-151 (1981)
[20] Storer, A.C.; Sprott, G.D.; Martin, W.G.: Kinetic and physical properties of the L-malate-NAD^+ oxidoreductase from Methanospirillum hungatii and comparison with the enzyme from other sources. Biochem. J., **193**, 235-244 (1981)
[21] Emyanitoff, R.G.; Kelly, P.J.: Kinetic characterization of Mitochondrial malate dehydrogenase from Dictyostelium discoideum. J. Gen. Microbiol., **128**, 1767-1771 (1982)
[22] Crow, K.E.; Braggins, T.J.; Batt, R.D.; Hardman, M.J.: Rat liver cytosolic malate dehydrigenase: Purification, kinetic properties, role in control of free cytosolic NADH concentration. J. Biol. Chem., **257**, 14217-14225 (1982)

[23] Smith, K.; Sundaram, T.K.; Kernick, M.; Wilkinson, A.E.: Purification of bacterial malate dehydrogenases by selective elution from a triazinyl dye affinity column. Biochim. Biophys. Acta, **708**, 17-25 (1982)
[24] Potvin, C.; Simon, J.-P.; Blanchard, M.-H.: Thermal properties of NAD malate dehydrogenase and glutamate oxaloacetate transaminase in two genotypes of Arabidopsis thaliana (Cruciferae) from contrasting environments. Plant Sci. Lett., **31**, 35-47 (1983)
[25] Crow, K.E.; Braggins, T.J.; Hardman, M.J.: Human liver cytosolic malate dehydrogenase: Purification, kinetic properties, and role in ethanol metabolism. Arch. Biochem. Biophys., **225**, 621-629 (1983)
[26] de Arriaga, D.; Teixido, F.; Busto, F.; Soler, J.: The nature of the substrate inhibition of cytoplasmic malate dehydrogenase from Phycomyces blakesleeanus. Biochim. Biophys. Acta, **784**, 158-163 (1984)
[27] Okabayashi, K.; Nakano, E.: Purification and properties of mitochondrial malate dehydrogenase from unfertilized eggs of the sea urchin, Anthocidaris crassispina. J. Biochem., **95**, 1625-1632 (1984)
[28] Isegawa, Y.; Nakano, Y.; Kitaoka, S.: Submitochondrial location and some properties of NAD^+- and $NADP^+$-linked malate dehydrogenase in Euglena. Agric. Biol. Chem., **48**, 549-552 (1984)
[29] Miyatake, K.; Washio, K.; Yokota, A.; Nakano, Y.; Kitaoka, S.: Occurence of two NAD^+-linked malate dehydrogenase isozymes in Euglena gracilis. Agric. Biol. Chem., **49**, 859-860 (1985)
[30] Waters, J.K.; Karr, D.B.; Emerich, D.W.: Malate dehydrogenase from Rhizobium japonicum 3I1b-143 bacteroids and Glycine max root-nodule mitochondria. Biochemistry, **24**, 6479-6486 (1985)
[31] Mueller, J.: Binary and ternary complexes of malate dehydrogenase with substrates and substrate analogs. Biochim. Biophys. Acta, **830**, 95-100 (1985)
[32] Ohshima, T.; Sakuraba, H.: Purification and characterization of malate dehydrogenase from the phototrophic bacterium, Rhodopseudomonas capsulata. Biochim. Biophys. Acta, **869**, 171-177 (1986)
[33] Miyatake, K.; Washio, K.; Yokota, A.; Nakano, Y.; Kitaoka, S.: Purification and immunological properties of NAD^+-linked malate dehydrogenase isozymes from Euglena gracilis. Agric. Biol. Chem., **50**, 2651-2653 (1986)
[34] Grossebueter, W.; Hartl, T.; Goerisch, H.; Stezowski, J.J.: Purification and properties of malate dehydrogenase from the thermophilic archaebacterium Thermoplasma acidophilum. Biol. Chem. Hoppe-Seyler, **367**, 457-463 (1986)
[35] Hartl, T.; Grossebueter, W.; Goerisch, H.; Stezowski, J.J.: Crystalline NAD/NADP-dependent malate dehydrogenase; the enzyme from the thermoacidophilic archaebacterium Sulfolobus acidocaldarius. Biol. Chem. Hoppe-Seyler, **368**, 259-267 (1987)
[36] Tayeh, M.A.; Madigan, M.T.: Malate dehydrogenase in phototrophic purple bacteria: Purification, molecular weight, and quaternary structure. J. Bacteriol., **169**, 4196-4202 (1987)

[37] Tayeh, M.A.; Madigan, M.T.: Malate dehydrogenases in phototrophic purple bacteria. Thermal stability, amino acid composition and immunological properties. Biochem. J., **252**, 595-600 (1988)

[38] Jeffery, D.; Goodenough, P.W.; Weitzman, P.D.J.: Citrate synthase and malate dehydrogenase from tomato fruit. Phytochemistry, **27**, 41-44 (1988)

[39] Norman, E.G.; Colman, B.: Characterization of a malate dehydrogenase in the cyanobacterium Coccochloris peniocystis. Arch. Microbiol., **156**, 28-33 (1991)

[40] Wiseman, M.S.; McKay, D.; Crow, K.E.; Hardman, M.J.: Rat liver mitochondrial malate dehydrogenase: Purification, kinetic properties, and role in ethanol metabolism. Arch. Biochem. Biophys., **290**, 191-196 (1991)

[41] Hall, M.D.; Banaszak, L.J.: Crystal structure of a ternary complex of Escherichia coli malate dehydrogenase citrate and NAD at 1.9 A resolution. J. Mol. Biol., **232**, 213-222 (1993)

[42] Lindbladh, C.; Rault, M.; Hagglund, C.; Small, W.C.; Mosbach, K.; Buelow, L.; Evans, C.; Srere, P.A.: Preparation and kinetic characterization of a fusion protein of yeast mitochondrial citrate synthase and malate dehydrogenase. Biochemistry, **33**, 11692-11698 (1994)

[43] Muslin, E.H.; Dong, L.; Stevens, F.J.; Donnelly, M.; Schiffer, M.; Anderson, L.E.: Engineering a domain-locking disulfide into a bacterial malate dehydrogenase produces a redox-sensitive enzyme. Biophys. J., **68**, 2218-2223 (1995)

[44] van Kuijk, B.L.M.; Stams, A.J.M.: Purification and characterization of malate dehydrogenase from the syntrophic propionate-oxidizing bacterium strain MPOB. FEMS Microbiol. Lett., **144**, 141-144 (1996)

[45] Drmota, T.; Tachezy, J.; Kulda, J.: Isolation and characterization of cytosolic malate dehydrogenase from Trichomonas vaginalis. Folia Parasitol., **44**, 103-108 (1997)

[46] Wise, D.J.; Anderson, C.D.; Anderson, B.M.: Purification and kinetic characterization of Haemophilus parasuis malate dehydrogenase. Arch. Biochem. Biophys., **344**, 176-183 (1997)

[47] Musrati, R.A.; Kollarova, M.; Mernik, N.; Mikulasova, D.: Malate dehydrogenase: Distribution, function and properties. Gen. Physiol. Biophys., **17**, 193-210 (1998)

[48] Kim, S.-Y.; Hwang, K.Y.; Kim, S.-H.; Sung, H.-C.;Han, Y.S.; Cho, Y.: Structural basis for cold adaptation. Sequence, biochemical properties, and crystal structure of malate dehydrogenase from a psychrophile Aquaspirillium arcticum. J. Biol. Chem., **274**, 11761-11767 (1999)

[49] Chapman, A.D.M.; Cortes, A.; Dafforn, T.R.; Clarke, A.R.; Brady, R.L.: Structural basis of substrate specificity in malate dehydrogenases: Crystal structure of a ternary complex of porcine cytoplasmic malate dehydrogenase, α-ketomalonate and tetrahydoNAD. J. Mol. Biol., **285**, 703-712 (1999)

[50] Madern, D.; Ebel, C.; Mevarech, M.; Richard, S.B.; Pfister, C.; Zaccai, G.: Insights into the molecular relationships between malate and lactate dehydrogenases: Structural and biochemical properties of monomeric and dimeric intermediates of a mutant of tetrameric L-[LDH-like] malate dehy-

drogenase from the halophilic archaeon Haloarcula marismortui. Biochemistry, **39**, 1001-1010 (2000)
[51] Cuevas, I.C.; Podesta, F.E.: Purification and physical and kinetic characterization of an NAD^+-dependent malate dehydrogenase from leaves of pineapple (Ananas comosus). Physiol. Plant., **108**, 240-248 (2000)

Malate dehydrogenase (oxaloacetate-decarboxylating) 1.1.1.38

1 Nomenclature

EC number
1.1.1.38

Systematic name
(S)-malate:NAD^+ oxidoreductase (oxaloacetate-decarboxylating)

Recommended name
malate dehydrogenase (oxaloacetate-decarboxylating)

Synonyms
NAD-ME [Swissprot]
NAD-specific malic enzyme
NAD-malic enzyme
malic enzyme
pyruvic-malic carboxylase

CAS registry number
9080-52-8

2 Source Organism

<1> *Rhizobium meliloti* (now Sinorhizobium meliloti [1]) [1, 6]
<2> *Ascaris suum* (parasitic roundwurm [17]) [2, 5, 8, 9, 10, 11, 17]
<3> *Mangifera indica* [3]
<4> *Schizosaccharomyces pombe* [4]
<5> *Atriplex spongiosa* [7]
<6> *Panicum miliaceum* [7, 15]
<7> *Urochloa panicoides* [7]
<8> *Escherichia coli* [12, 13, 14]
<9> *Eleusina indica* [15]
<10> *Digitaria sanguinalis* [15]
<11> *Zea mays* [15]
<12> *Eriochloa borumensis* [15]
<13> *Panicum maximum* [15]
<14> *Escherichia coli W* [16, 18, 21]
<15> *Escherichia coli B* [16]
<16> *Escherichia coli K12* [16]
<17> *Escherichia freundii* [16]
<18> *Aerobacter aerogenes IFO 3319* [16]

<19> *Aerobacter aerogenes IFO 3320* [16]
<20> *Erwinia aroideae* [16]
<21> *Proteus vulgaris* [16]
<22> *Agrobacterium tumefaciens* [16]
<23> *Clostridium tetanomorphum* [16]
<24> *Lactobacillus plantarum B 38* [19]
<25> *Leuconostoc mesenteroides 99* [19]
<26> *Saccharomyces cerevisiae* [20]
<27> *Rattus norvegicus* [22]
<28> *Lactobacillus arabinosus* [23]

3 Reaction and Specificity

Catalyzed reaction
(S)-malate + NAD^+ = pyruvate + NADH (<2> stepwise reaction mechanism in which hydride transfer preceeds decarboxylation [5])

Reaction type
oxidation
oxidative decarboxylation
redox reaction
reduction

Natural substrates and products
S L-malate + NAD^+ <1-28> (Reversibility: r <1-28> [9]) [1-23]
P CO_2 + pyruvate + NADH <1-28> [1-23]
S L-malate + $NADP^+$ <1> (<1> 20% activity compared to cofactor NAD^+ [1]) (Reversibility: ? <1> [1, 2]) [1, 2]
P CO_2 + pyruvate + NADPH <1> [1, 2]
S oxaloacetate <1-28> (Reversibility: ir <1-28> [1-23]) [1-23]
P CO_2 + pyruvate <1-28> [1-23]
S oxaloacetate + NADH <8> (Reversibility: ? <8> [13]) [13]
P L-malate + NAD^+ <8> [13]
S pyruvate + NADH <2, 8> (Reversibility: r <2> [9]; ? <8> [13]) [9, 13]
P L-lactate + NAD^+ <2, 8> [9, 13]

Substrates and products
S L-malate + NAD^+ <1-28> (Reversibility: r <1-28> [1-23]) [1-23]
P CO_2 + pyruvate + NADH <1-28> [1-23]
S L-malate + $NADP^+$ <1> (<1> 20% activity compared to cofactor NAD^+ [1]) (Reversibility: ? <1> [1, 2]) [1, 2]
P CO_2 + pyruvate + NADPH <1> [1, 2]
S oxaloacetate <1-28> (<2> Mg^{2+} or Mn^{2+} are required [9]; <14> NAD^+ inhibits oxaloacetate decarboxylation activity [21]) (Reversibility: ir <1-28> [1-23]) [1-23]
P CO_2 + pyruvate <1-28> [1-23]

S oxaloacetate + NADH <8> (<8> divalent metal ions are required [13]) (Reversibility: ? <8> [13]) [13]
P L-malate + NAD^+ <8> [13]
S pyruvate + NADH <2, 8> (Reversibility: r <2> [9]; ? <8> [13]) [9, 13]
P L-lactate + NAD^+ <2, 8> [9, 13]

Inhibitors

2-methylbutanoate <2> (<2> 5.22 mM, competitive inhibitor, 60% inhibition at pH 7.5, 54% inhibition at pH 7.0 [17]) [17]
2-methylpentanoate <2> (<2> 5.22 mM, competitive inhibitor, 45% inhibition at pH 7.5, 40% inhibition at pH 7.0 [17]) [17]
4,4'-dithiodipyridine <2> (<2> sulfhydryl reagent, rapid and complete loss of activity, presence of Mg^{2+} or Mn^{2+} enhances inactivation rate [11]) [11]
5,5'-dithiobis(2-nitrobenzoic acid) <2> (<2> sulfhydryl reagent, rapid and complete loss of activity, presence of Mg^{2+} or Mn^{2+} enhances inactivation rate [11]; <8> inhibition of decarboxylation activity [13]) [11, 13]
ADP <5, 6, 7> (<5> 57% inhibition, 0.01 mM CoA, 1 mM malate, 2 mM NAD^+, 4 mM Mn^{2+} [7]; <7> 26% inhibition, 0.005 mM CoA, 5 mM malate, 2 mM NAD^+, 4 mM Mn^{2+} [7]) [7]
AMP <5, 6> (<5> 50% inhibition, 0.01 mM CoA, 1 mM malate, 2 mM NAD^+, 4 mM Mn^{2+} [7]; <7> 39% inhibition, 0.005 mM CoA, 5 mM malate, 2 mM NAD^+, 4 mM Mn^{2+} [7]) [7]
ATP <5, 6, 8, 27> (<5> 72% inhibition, 0.01 mM CoA, 1 mM malate, 2 mM NAD^+, 4 mM Mn^{2+} [7]; <8> more effective in the presence of Mn^{2+} than in the presence of Mg^{2+}, 50% inhibition with 1.6 mM Mn^{2+} and 3.9 mM in the presence of Mg^{2+} respectively [14]; <8> allosteric inhibition [12]; <27> competitive inhibition [22]) [7, 12, 14, 22]
$AgNO_3$ <3> (<3> inhibition with 20 mM, complete inhibition with 300 mM [3]) [3]
Ca^{2+} <1> [1]
Cl^- <2> (<2> competitive vs. malate and NAD^+ [10]) [10]
CoA <8, 14> (<8> allosteric inhibition [12]; <8> 50% inhibition with 0.75 mM CoA in the presence of Mg^{2+} and 3.4 mM in the presence of Mn^{2+} [14]; <14> in the absence of L-aspartate 50% inhibition with 0.22 mM CoA, in the presence of 2 mM L-aspartate 50% inhibition with 1.1 mM [18]) [12, 14, 18, 21]
Cu^{2+} <1, 14> (<14> 0.025 mM, 70% activity lost after 5 min [21]) [1, 21]
D-malate <14> (<14> competitive inhibition [21]) [21]
N-ethylamaleimide <2, 8> (<2> sulfhydryl reagent, rapid and complete loss of activity, presence of Mg^{2+} or Mn^{2+} enhances inactivation rate [11]; <8> L-malate and tartronate strongly protect the enzyme in the presence of $MnCl_2$ and NAD^+, 80% inhibition of decarboxylation activity, 10fold increase in reduction activity [13]) [11, 13]
UTP <5, 6> (<5> 25% inhibition, 0.01 mM CoA, 1 mM malate, 2 mM NAD^+, 4 mM Mn+ [7]) [7]
Zn^{2+} <1> (<1> 5 mM $ZnCl_2$ [1]) [1]

acetyl-CoA <1, 8> (<1> 0.005 mM, 18% inhibition, 0.01 mM, 42% inhibition, 0.1 mM, 68% inhibition [1]; <8> allosteric inhibition [12]) [1, 12]
diethyldicarbonate <2> (<2> 80% inactivation after treatment with 5 mM diethyldicarbonate for 25 min, biphasic inactivation, 40-50% inactivation in first phase, 1-2 histidine residues are acylated by diethyldicarbonate, 250 mM malate provides complete protection, 50% protection with 50 mM $MgSO_4$, 55% enzyme activtity is restored with 0.5 M hydroxylamine [8]) [8]
iodoacetate <2> (<2> complete inactivation [11]; <8> inhibition of decarboxylation activity [13]) [11, 13]
malonate <14> (<14> competitive inhibition [21]) [21]
o-phenanthroline <14> (<14> 0.025 mM, 70% activity lost after 5 min [21]) [21]
oxaloacetate <2, 14> (<2> 1 mM, 80% inhibition at pH 7.5, 50% inhibition at pH 7.0 [17]; <14> 50% inhibition with 0.09 mM [21]) [17, 21]
p-(chloromercuri)benzoate <2, 8, 14> (<2> complete inactivation [11]; <14> 0.0025 mM, 80% inhibition within 2 min [21]) [11, 13, 21]
succinate <14-18> (<14-18> 0.005 mM, slight inhibition [16]) [16]
tartronate <2> (<2> competitive inhibition vs. malate, noncompetitve vs. NAD^+ [10]) [10]
tiglic acid <2> (<2> 5.22 mM, competitive inhibitor, 58% inhibition at pH 7.5, 52% inhibition at pH 7.0 [17]) [17]

Cofactors/prosthetic groups

NAD^+ <1-28> [1-23]
$NADP^+$ <1, 14, 26> (<14> 2% activity [21]) [1, 2, 20, 21]

Activating compounds

5,5'-dithiobis(2-nitrobenzoic acid) <8> (<8> 0.5 mM, 2.7fold increase in oxaloacetate reduction activity [13]) [13]
ATP <7> (<7> 68% activation, 0.005 mM CoA, 5 mM malate, 2 mM NAD^+, 4 mM Mn^{2+} [7]) [7]
CoA <5, 7> (<5> sigmoidal response to CoA concentration [7]) [7]
D-aspartate <14> (<14> 1.27fold activation [18]) [18]
L-aspartate <14> (<14> 1.35fold activation [18]) [18, 21]
N-ethylmaleimide <8> (<8> 1 mM, 8.9fold increase in oxaloacetate reduction activity [13]) [13]
NH_4^+ <1> (<1> 30 mM Na^+L-malate instead of K^+L-malate reduces enzyme activity to 18% in the presence of 3 mM $MnCl_2$, activity is restored with 5 or 50 mM NH_4Cl [1]) [1]
UTP <7> (<7> 35% activation, 0.005 mM CoA, 5 mM malate, 2 mM NAD^+, 4 mM Mn^{2+} [7]) [7]
α-methyl-DL-aspartate <14> (<14> 1.35fold activation [18]) [18]
aspartate <8> (<14> 0.005 mM, 15% activation [16]; <15, 16, 20> 0.005 mM, 20% activation [16]; <17> 0.005 mM, 18% activation [16]; <21> 0.005 mM, 35% activation [16]) [12, 16]
β-methyl-DL-aspartate <14> (<14> 1.34fold activation [18]) [18]
fumarate <2> (<2> 0.7 mM, 2.1fold activation [17]) [17]

p-chloromercuribenzoate <8> (<8> 0.02 mM, 2.5fold increase in oxaloacetate reduction activity [13]) [13]
phosphoenolpyruvate <26> (<26> 0.5 mM [20]) [20]
succinate <27> [22]

Metals, ions

Co^{2+} <1, 14> (<1> at 0.002 mM $MgCl_2$ enzyme activity is reduced to 25% without additional divalent cations, activity is restored with 0.5 or 5.0 mM $CoCl_2$ [1]) [1, 21]
Cs^+ <1> (<1> 30 mM Na^+L-malate instead of K^+L-malate reduces enzyme activity to 18% in the presence of 3 mM $MnCl_2$, activity is restored with 5 or 50 mM CsCl [1]) [1]
K^+ <1> (<1> 30 mM Na^+L-malate instead of K^+L-malate reduces enzyme activity to 18% in the presence of 3 mM $MnCl_2$, activity is restored with 50 mM KCl [1]) [1]
Mg^{2+} <1, 14> (<1> at 0.002 mM $MgCl_2$ enzyme activity is reduced to 25% without additional divalent cations, activity is restored with 0.5 or 5.0 mM $MgCl_2$ [1]; <8> stabilizes two distinct conformational states of the enzyme, which differ in response to various substrate and effector concentrations [12]; <8> complex sigmoidal saturation curve of free malate concentration in the presence of 5 mM Mg^{2+} [14]) [1, 12, 14, 21]
Mn^{2+} <1, 8, 14, 26, 28> (<1> at 0.002 mM $MgCl_2$ enzyme activity is reduced to 25% without additional divalent cations, activity is restored with 0.5 or 5.0 mM $MnCl_2$ [1]; <8> stabilizes two distinct conformational states of the enzyme, which differ in response to various substrate and effector concentrations [12]; <8> hyperbolic saturation curve of free malate concentration in the presence of 0.3 mM Mn^{2+} [14]; <28> Mn^{2+} protects enzyme against heat denaturation [23]) [1, 12, 20, 21, 23]
Ni^{2+} <14> [21]
Zn^{2+} <14> [21]

Turnover number (min^{-1})

3.6 <2> (pyruvate, <2> in the presence of NAD^+, 20 mM Mn^{2+} [9]) [9]
5.4 <2> (pyruvate, <2> in the absence of NAD^+, 20 mM Mn^{2+} [9]) [9]
1680 <2> (oxaloacetate, <2> in the absence of NAD^+, 20 mM Mn^{2+} [9]) [9]
1800 <2> (oxaloacetate, <2> in the presence of NAD^+, 20 mM Mn^{2+} [9]) [9]
7800 <2> (L-malate, <2> native and recombinant enzyme [2]) [2]

Specific activity (U/mg)

0.07 <8> (<8> pyruvate reduction [13]) [13]
0.076 <1> [6]
0.2 <26> [20]
0.545 <1> [1]
1.7 <8> (<8> oxaloacetate reduction [13]) [13]
2.3 <15> [16]
5.5 <18> [16]
7 <23> [16]
8.1 <14> [16]

9 <16> [16]
17.5 <25> [19]
30 <20> [16]
32 <19> [16]
32 <2> [5]
35 <2> [9]
38.5 <2> (<2> recombinant enzyme [2]) [2]
65 <22> [16]
70 <21> [16]
93 <17> [16]
170 <24> [19]
177 <14> [21]
3900 <28> [23]

K_m-Value (mM)

0.004 <2> (NAD$^+$, <2> pH 9.0 [10]) [10]
0.011 <2> (NAD$^+$, <2> native enzyme [2]) [2]
0.011 <2> (NAD$^+$, <2> pH 7.3 [10]) [10]
0.015 <2> (NAD$^+$, <2> recombinant enzyme [2]) [2]
0.025 <8> (NADH, <8> reduction of oxaloacetate [13]) [13]
0.046 <14> (NAD$^+$, <14> in the presence of 5 mM L-aspartate [18]) [18]
0.05 <24> (NAD$^+$) [19]
0.055 <14> (NAD$^+$) [21]
0.058 <14> (NAD$^+$, <14> in the presence of 0.5 mM L-aspartate [18]) [18]
0.06 <2> (NAD$^+$, <2> at pH 6.5, 7.0 and 7.5 [17]) [17]
0.063 <8> (NAD$^+$, <8> decarboxylation of L-malate [13]) [13]
0.071 <14> (NAD$^+$, <14> nearly hyperbolic NAD$^+$ saturation curve [18]) [18]
0.12 <24> (Mn^{2+}) [19]
0.142 <1> (NAD$^+$) [1]
0.19 <14> (L-malate, <14> in the presence of 1 mM L-aspartate, pH 7.9 [18]) [18]
0.22 <14> (L-malate, <14> in the presence of 2 mM L-aspartate, pH 5.5 [18]) [18]
0.26 <14> (L-malate, <14> in the absence of L-aspartate, pH 5.5 [18]) [18]
0.32 <14> (Mn^{2+}) [21]
0.39 <9> (NAD$^+$) [15]
0.4 <21> (L-malate) [16]
0.4 <14> (L-malate) [21]
0.4 <14> (L-malate, <14> in the absence of L-aspartate, pH 7.9 [18]) [18]
0.45 <8> (L-malate, <8> decarboxylation of L-malate [13]) [13]
0.45 <6> (NAD$^+$) [15]
0.46 <10> (NAD$^+$) [15]
0.48 <11> (NAD$^+$) [15]
0.49 <13> (NAD$^+$) [15]
0.5 <20> (L-malate) [16]
0.5 <26> (NAD$^+$) [20]
0.51 <12> (NAD$^+$) [15]

0.58 <14> (L-malate, <14> in the presence of 2 mM L-aspartate, pH 8.5 [18]) [18]
0.7 <14> (L-malate) [16]
0.8 <17> (L-malate) [16]
1 <16> (L-malate) [16]
1 <2> (L-malate, <2> native enzyme [2]) [2]
1.1 <27> (L-malate) [22]
1.1 <2> (L-malate, <2> recombinant enzyme [2]) [2]
1.18 <2> (L-malate, <2> pH 7.3 [10]) [10]
1.2 <15> (L-malate) [16]
1.34 <6> (L-malate) [15]
1.4 <14> (L-malate, <14> in the absence of L-aspartate, pH 8.5 [18]) [18]
1.63 <13> (L-malate) [15]
1.76 <9> (L-malate) [15]
1.96 <6> (L-malate) [15]
2.1 <8> (oxaloacetate, <8> reduction of oxaloacetate [13]) [13]
2.15 <10> (L-malate) [15]
2.2 <22> (L-malate) [16]
2.21 <11> (L-malate) [15]
2.3 <14> ($NADP^+$) [21]
3.1 <27> (L-malate, <27> in the presence of 0.5 mM ATP [22]) [22]
4.3 <18> (L-malate) [16]
4.8 <8> (oxaloacetate, <8> decarboxylation of oxaloacetate [13]) [13]
5 <14> (oxaloacetate) [21]
5.2 <2> (L-malate, <2> pH 9.0 [10]) [10]
7.3 <27> (L-malate, <27> in the presence of 1.0 mM ATP [22]) [22]
8 <24> (L-malate) [19]
8.6 <1> (L-malate) [1]
50 <26> (L-malate) [20]

pH-Optimum

4 <8> (<8> oxaloacetate reduction [13]) [13]
5 <8> (<8> pyruvate reduction [13]) [13]
7.2 <14> (<14> 0.3 mM L-malate [18]; <14> with increasing substrate concentrations pH optimum is shifted towards higher pH values [21]) [18, 21]
7.3-7.6 <6, 9-13> [15]
7.4 <8> [14]
7.5 <26> [20]
7.6 <14> (<14> 0.3 mM L-malate, in the presence of 2 mM L-aspartate [18]) [18]
7.8 <1> [1]

4 Enzyme Structure

Molecular weight

84000 <23> (<23> gel filtration [16]) [16]

140000 <19> (<19> gel filtration [16]) [16]
180000 <21> (<21> gel filtration [16]) [16]
190000 <17> (<17> gel filtration [16]) [16]
198000 <14> (<14> low speed sedimentation equilibrium [21]) [21]
200000 <15> (<15> gel filtration [16]) [16]
200000 <20> (<20> gel filtration [16]) [16]
203000 <14> (<14> sucrose density gradient centrifugation [21]) [21]
220000 <14, 16> (<15,16> gel filtration [16]) [16]
240000 <18> (<18> gel filtration [16]) [16]
580000 <22> (<22> gel filtration [16]) [16]

5 Isolation/Preparation/Mutation/Application

Source/tissue

adrenal cortex <27> [22]
bacteroid <1> (<1> enzyme is necessary for symbiotic N_2-fixation in alfalfa nodules [6]) [6]
liver <27> [22]

Localization

mitochondrion <2, 6, 9-13, 27> [2, 5, 15, 17, 22]

Purification

<2> (ammonium sulfate, DEAE-cellulose, Blue-B agarose, Orange-A agarose [5]) [5]
<2> (streptomycin sulfate ammonium sulfate, Phenyl-sepharose, Orange-A-Sepharose [2]) [2]
<14> (streptomycin, ammonium sulfate, heat treatment, acid treatment, calcium phosphate gel, DEAE-sephadex, hydroxylapatite, Sepharose 6B [21]) [21]
<24> ($MnCl_2$ precipitation, protaminsulfate precipitation, Sephadex G-25, Sephadex G-200, hydroxylapatite [19]) [19]
<25> ($MnCl_2$ precipitation, protaminsulfate precipitation, Sephadex G-25, hydroxylapatite [19]) [19]
<26> (protamine sulfate precipitation, ammonium sulfate precipitation, DEAE-cellulose [20]) [20]
<28> (38-40°C treatment, ammonium sulfate, protamine sulfate precipitation, calcium phosphate gel, ammonium sulfate [23]) [23]

Cloning

<1> (overexpressed in Escherichia coli [1]) [1]
<2> (expressed in Escherichia coli [2]) [2]

Engineering

G444A <4> (<4> inactive enzyme [4]) [4]

6 Stability

Temperature stability

50 <14> (<14> 22% activity left after 4 min, 80% activity left in the presence of 10 mM L-aspartate and 1 mM $MnCl_2$, 78% activity left in the presence of 0.7 mM CoA [18]) [18]

Oxidation stability

<14>, enzyme is reversibly inhibited by air oxidation, activity is completely restored in 30 min by the addition of 2 mM dithiotreitol [21]

Storage stability

<2>, -20°C, 15 mM triethanolamine-maleate, pH 7.5, 1 mM EDTA, 10 mM dithiotreitol, 5% glycerol, several months, no loss of activity [9]
<14>, 4°C, 50 mM Tris-HCl, pH 7.4, 1 mM EDTA, 10 mM 2-mercaptoethanol, 10 mM $MgCl_2$, 5 mM L-aspartate, 2 months, 10% activity lost [21]
<14>, 4°C, 50 mM Tris-HCl, pH 7.4, 1 mM EDTA, 10 mM 2-mercaptoethanol, 2 months, 15% activity lost [21]
<14>, 4°C, 50 mM Tris-HCl, pH 7.4, 1 month, 90% activity lost [21]

References

[1] Driscoll, B.T.; Finan, T.M.: Properties of NAD^+ and NADP+-dependent malic enzymes of Rhizobium (Sinorhizobium) meliloti and differential expression of their genes in nitrogen-fixing bacteroids. Microbiology, **143**, 489-498 (1997)

[2] Chooback, L.; Karsten, W.E.; Kulkarni, G.; Nalabolu, S.R.; Harris, B.G.; Cook, P.F.: Expression, purification, and characterization of the recombinant NAD-malic renzyme from Ascaris suum. Protein Expr. Purif., **10**, 51-54 (1997)

[3] Srivatava, G.C.; Yanru, Z.Y.; Pandey, M.; Prasad, N.K.: Effect of silver nitrate on activity of malic enzyme during ripening in mango, Mangifera indica L.. Indian J. Exp. Biol., **34**, 575-576 (1996)

[4] Viljoen, M.; van der Merwe, M.; Subden, R.E.; van Vuuren, H.J.J.: Mutation of Gly-444 inactivates the S. pombe malic enzyme. FEMS Microbiol. Lett., **167**, 157-162 (1998)

[5] Karsten, W.E.; Cook, P.F.: Stepwise versus concerted oxidative decarboxylation catalyzed by malic enzyme: a reinvestigation. Biochemistry, **33**, 2096-2103 (1994)

[6] Driscoll, B.T.; Finan, T.M.: NAD^+-dependent malic enzyme of Rhizobium meliloti is required for symbiotic nitrogen fixation. Mol. Microbiol., **7**, 865-873 (1993)

[7] Furbank, R.T.; Agostino, A.M.; Hatch, M.D.: Regulation of C_4 photosynthess: modulation of mitochondrial NAD-malic enzyme by adenylates. Arch. Biochem. Biophys., **289**, 376-381 (1991)

[8] Rao, J.G.S.; Harris, B.G.; Cook, P.F.: Diethylpyrocarbonate inactivation of NAD-malic enzyme from Ascaris suum. Arch. Biochem. Biophys., **241**, 67-74 (1985)
[9] Park, S-H.; Harris, B.G.; Cook, P.F.: pH dependence of kinetic parameters for oxalacetate decarboxylation and pyruvate reduction reactions catalyzed by malic enzyme. Biochemistry, **25**, 3752-3759 (1986)
[10] Park, S-H.; Kiick, D.M.; Harris, B.G.; Cook, P.F.: Kinetic mechanism in the direction of oxidative decarboxylation for NAD-malic enzyme from Ascaris suum. Biochemistry, **23**, 5446-5453 (1984)
[11] Kiick, D.M.; Allen, B.L.; Rao, J.G.S.; Harris, B.G.; Cook, P.F.: Determination of dissociation constants for enzyme-reactant complexes for NAD-malic enzyme by modulation of the thiol inactivation rate. Biochemistry, **23**, 5454-5459 (1984)
[12] Cook, R.A.: Distinct metal cofactor-induced conformational states in the NAD-specific malic enzyme of Escherichia coli as revealed by proteolysis studies. Biochim. Biophys. Acta, **749**, 198-203 (1983)
[13] Yamaguchi, M.: Studies on regulatory functions of malic enzymes. IV. Effects of sulfhydryl group modification on the catalytic function of NAD-linked malic enzyme from Escherichia coli. J. Biochem., **86**, 325-333 (1979)
[14] Milne, J.A.; Cook, R.A.: Role of metal cofactors in enzyme regulation. Differences in the regulatory properties of the Escherichia coli nicotinamide adenine dinucleotide specific malic enzyme depending on whether Mg^{2+} or Mn^{2+} serves as divalent cation. Biochemistry, **18**, 3605-3610 (1979)
[15] Rhathnam, C.K.M.: Studies with isolated bundle sheath mitochondria. Evidence for NAD-malic enzyme-catalyzed decarboxylation of C_4 acids in species representing the three C_4 metabolic subtypes. FEBS Lett., **96**, 367-372 (1978)
[16] Iwakura, M.; Tokushige, M.; Katsuki, H.; Muramatsu, S.: Studies on regulatory functions of malic enzymes. V. Comparative studies of malic enzymes in bacteria. J. Biochem., **83**, 1387-1394 (1978)
[17] Landsperger, W.J.; Harris, B.G.: NAD^+-malic enzyme. Regulatory properties of the enzyme from Ascaris suum. J. Biol. Chem., **251**, 3599-3602 (1976)
[18] Yamaguchi, M.; Tokushige, M.; Takeo, K.; Katsuki, H.: Studies on the regulatory functions of malic enzymes: III. Regulatory effects of L-aspartate, Coenzyme A, and divalent metal ions on NAD-linked malic enzyme from Escherichia coli. J. Biochem., **76**, 1259-1268 (1974)
[19] Schütz, M.; Radler, F.: Das "Malatenzym" von Lactobacillus plantarum und Leuconostoc mesenteroides. Arch. Mikrobiol., **91**, 183-202 (1973)
[20] Fuck, E.; Stärk, G.; Radler, F.: Äpfelsäurestoffwechsel bei Saccharomyces II. Anreicherung und Eigenschaften eines Malatenzymes. Arch. Mikrobiol., **89**, 223-231 (1973)
[21] Yamaguchi M.; Tokushige, M.; Katsuki, H.: Studies on regulatory functions of malic enzymes II. Purification and molecular properties of nicotinamide adenine dinucleotide-linked malic enzyme from Escherichia coli. J. Biochem., **73**, 169-180 (1973)

[22] Sauer, L.A.: An NAD- and NADP-dependent malic enzyme with regulatory properties in rat liver and adrenal cortex mitochondrial fractions. Biochem. Biophys. Res. Commun., **50**, 524-531 (1973)
[23] Kaufman, S.; Korkes, S.; del Campillo, A.: Biosynthesis of dicarboxylic acids by carbon dioxide filtration V. Further study of the "malic" enzyme of Lactobacillus arabinosus. J. Biol. Chem., **192**, 301-312 (1951)

Malate dehydrogenase (decarboxylating) 1.1.1.39

1 Nomenclature

EC number
1.1.1.39

Systematic name
(S)-malate:NAD^+ oxidoreductase (decarboxylating)

Recommended name
malate dehydrogenase (decarboxylating)

Synonyms
DMA
ME
NAD-ME
NAD-malic enzyme
NAD-specific malic enzyme
dehydrogenase, malate
malic enzyme
pyruvic-malic carboxylase

CAS registry number
9028-46-0

2 Source Organism

<1> *Brassica oleracea* (var. botrytis [6]) [1, 5, 6, 12]
<2> *Solanum tuberosum* (var. urgenta [5]; var. Chieftan [11]) [1, 2, 3, 5, 11]
<3> *Atriplex spongiosa* [1, 14]
<4> *Crassula argentea* [1, 3, 4, 8, 10]
<5> *Rhizobium meliloti* [7]
<6> *Amarathus retroflexus* [9]
<7> *Glossina morsitans* [13]
<8> *Glossina austeni* [13]
<9> *Glossina pallidipes* [13]
<10> *Glossina longipennis* [13]
<11> *Sarcophaga morsitans* [13]
<12> *Schistocerca gregaria* [13]
<13> *Periplaneta americana* [13]
<14> *Odontotermes sp.* [13]
<15> *Spodoptera exempta* [13]

<16> *Apis mellifera* [13]
<17> *Heliocarpus sp.* [13]
<18> *Bradyrhizobium japonicum* [21]
<19> *Amaranthus edulis* [14]
<20> *Panicum miliaceum* [14]
<21> *Homo sapiens* [15]
<22> *Bos taurus* [15]
<23> *salmon* [15]
<24> *cod* [15]
<25> *Oryctolagus cuniculus* [16]
<26> *Pseudomonas aeruginosa* [17]
<27> *Tritrichomonas foetus* [18]
<28> *Pseudomonas diminuta* [19]
<29> *Vigna unguiculata* [20]

3 Reaction and Specificity

Catalyzed reaction

(S)-malate + NAD^+ = pyruvate + CO_2 + NADH (<1>, sequential mechanism, each of the substrate pairs binds randomly to the enzyme [6]; <4>, sequential mechanism [8]; <4>, slow reaction transient in the form of a lag before reaching a steady-state rate in assay [10]; <1>, rapid equilibrium reaction of the intersecting type [12]; <29>, sequential mechanism with each substrate bound randomly [20])

Reaction type

oxidation
oxidative decarboxylation
redox reaction
reduction

Natural substrates and products

S (S)-malate + NAD^+ <1, 2> (<1, 2>, the enzyme plays a special role in the decarboxylation of C_4 acids to pyruvate and CO_2, which are used in subsequent photosynthesis. pH, NAD^+, and coenzyme A levels in the matrix act together to regulate (S)-malate oxidation [5]) (Reversibility: ? <1, 2> [5]) [5]
P ?

Substrates and products

S (S)-malate + $NADP^+$ <2, 3, 4, 7, 25, 26, 27> (<4>, 14% of the activity with NAD^+ [8]; <7>, at 1.5% of the activity with NAD^+ [13]; <3>, 15% of the activity with NAD^+ [14]) (Reversibility: r <26> [17]; ir <25> [16]; ? <2, 3, 4, 7, 27> [8, 11, 13, 14, 18]) [8, 11, 13, 14, 16, 17, 18]
P pyruvate + CO_2 + NADPH <2, 3, 4, 7, 25, 26, 27> [8, 11, 13, 14, 16, 17, 18]
S malate + NAD^+ <1-29> (<1>, no decarboxylation of malate in absence of either Mg^{2+} or NAD^+ [6]; <4>, activity of the reverse reaction is 1.5% of

that of the forward reaction [8]; <1>, ionized malic acid is the true substrate [12]) (Reversibility: ? <1, 2, 3, 4, 6, 8, 9, 10, 11, 12, 13, 14, 15, 16, 17, 18, 19, 20, 21, 22, 23, 24, 27, 28, 29> [1, 2, 3, 4, 5, 6, 9, 10, 11, 12, 14, 15, 18, 19, 20, 21]; r <4, 26> [8, 17]; ir <5, 7, 25> [7, 13, 16]) [1-21]

P pyruvate + CO_2 + NADH <1-29> [1-21]

S pyruvate + CO_2 + NADH <4> (<4>, activity is 1.5% of the decarboxylation of (S)-malate [8]) (Reversibility: r <4> [8]) [8]

P (S)-malate + NAD^+ <4> [8]

S Additional information <5, 25, 28> (<5, 25, 28>, no decarboxylation of oxaloacetate [7, 16, 19]) [7, 16, 19]

P ?

Inhibitors

(S)-malate <3> (<3>, above 10 mM [14]) [14]
2-ketoglutarate <18> [21]
AMP <18> [21]
ATP <18, 25> (<25>, competitive with respect to (S)-malate [16]) [16, 21]
Cl^- <4> [8]
CoA <18> [21]
$CuSO_4$ <28> [19]
D-tartrate <5> [7]
DL-isocitrate <18> [21]
DL-lactate <5> [7]
EDTA <28> [19]
HCO_3^- <3> [14]
$HgCl_2$ <28> [19]
KCNO <1> [6]
L-asparatate <18> [21]
NADH <4, 5> (<4>, product inhibition [8]) [7, 8]
NEM <1, 28> [6, 19]
PCMB <28> [19]
acetyl-CoA <5> (<5>, potent inhibitor [7]) [7]
bicarbonate <3, 5, 19> [7, 14]
bromopyruvate <2> [2]
citrate <4> (<4>, competitive [8]) [8]
fumarate <18> [21]
glyoxylate <5> [7]
hydroquinone <28> [19]
malonate <5, 18> [7, 21]
oxalate <25, 28> [16, 19]
oxaloacetate <1, 2, 5> [5, 7]
phosphoenolpyruvate <4, 5, 18> (<4>, activates at low concentrations, deactivation at high concentrations [8]) [7, 8, 21]
pyruvate <4, 5, 18> (<4>, product inhibition [8]) [7, 8, 21]
tartrate <18> [21]
tartronate <5> [7]

Cofactors/prosthetic groups

NAD$^+$ <1-29> (<1-29>, cofactor [1-21]) [1-21]

NADP$^+$ <2, 3, 4, 5, 7, 25, 27, 28> (<4, 5>, cofactor [7, 8]; <4>, 14% of the activity with NAD$^+$ [8]; <7>, at 1.5% of the activity with NAD$^+$ [13]; <3>, 15% of the activity with NAD$^+$ [14]; <27>, activity is maximal 65% of the activity with NAD$^+$ [18]; <28>, activity with NADP$^+$ is about twice that measured with NAD$^+$ [19]) [7, 8, 11, 13, 14, 16, 18, 19]

Activating compounds

1-N^6-etheno-CoA <4> (<4>, activates [8]) [8]

ADP <5> (<5>, slight stimulstion [7]) [7]

AMP <4> (<4>, competitive activation [8]) [8]

ATP <5> (<5>, slight stimulstion [7]) [7]

D-fructose 6-phosphate <18> (<18>, activation [21]) [21]

D-glucose 6-phosphate <18> (<18>, activation [21]) [21]

SO_4^{2-} <3> (<3>, activates [14]) [14]

adenosine 5'-diphosphoglucose <18> (<18>, activates [21]) [21]

α-D-glucose 1-phosphate <18> (<18>, activation [21]) [21]

aspartate <5> (<5>, stimulates [7]) [7]

coenzyme A <1, 2, 3, 4, 19, 20> (<1, 2>, serves to broaden the pH-optimum, at pH 7.5 approximately 4fold stimulation, the effect is more marked when HCO_3^- is also present, below pH 7 no effect on activity [5]; <4>, activates [8]; <3, 19, 20>, increases activity [14]) [5, 8, 14]

fructose 1,6-bisphosphate <4, 5, 18> (<5>, stimulates [7]; <4, 18>, activates [8, 21]) [7, 8, 21]

fumarate <4, 5> (<5>, stimulates [7]; <4>, fumarate is a strong activator [8]) [7, 8]

γ-amino-n-butyrate <18> (<18>, slight activation [21]) [21]

hydroxy-n-butyrate <18> (<18>, slight activation [21]) [21]

hydroxypyruvate <18> (<18>, slight activation [21]) [21]

phosphoenolpyruvate <4> (<4>, activates at low concentrations, decativation at high concentrations [8]) [8]

succinate <5, 18> (<5, 18>, stimulates [7, 21]) [7, 21]

uridine 5'-diphosphoglucose <18> (<18>, activates [21]) [21]

Metals, ions

Co^{2+} <1> (<1>, can replace Mg^{2+} in activation. K_m: 0.018 mM [6]) [6]

Mg^{2+} <1, 2, 4, 7, 18, 25, 26, 27, 28, 29> (<1>, no decarboxylation of malate in absence of either Mg^{2+} or NAD$^+$. K_m: 0.04 mM [6]; <4>, uses Mg^{2+} or Mn^{2+} as the required divalent cation [8]; <2, 4>, activation by Mg^{2+} or Mn^{2+} [10, 11]; <7, 27>, completely dependent on the presence of Mg^{2+} or Mn^{2+} [13, 18]; <3, 19, 20>, no activation by Mg^{2+} [14]; <25>, Mn^{2+} or Mg^{2+} required [16]; <25>, K_m with NAD$^+$: 4.25 mM, K_m with NADP$^+$: 13.3 mM [16]; <26>, bivalent metal ion required, Mn^{2+} is more effective than Mg^{2+} [17]; <27> activation by Mn^{2+}, at 1 mM, is 10% and 20% higher than activation with 1 mM Mg^{2+} in the presence of NAD$^+$ and NADP$^+$ [18]; <28>, divalent metal ion required, NAD$^+$-linked activity shows maximal activity at 5 mM Mg^{2+} [19]; <29>, preferably used as metal cofactor [20]; <18>, strict requirement for a divalent cation, maximal activation at 2 mM [21]) [6, 8, 10, 11, 13, 16, 17, 18, 19, 20, 21]

Mn^{2+} <1, 2, 3, 4, 7, 18, 19, 20, 25, 26, 27> (<1>, can replace Mg^{2+} in activation [6]; <4>, divalent cation required, Mg^{2+} or Mn^{2+} [8]; <2, 4>, activation by Mn^{2+} or Mg^{2+} [10, 11]; <7, 27>, completely dependent on the presence of Mg^{2+} or Mn^{2+} [13, 18]; <3, 19, 20>, absolute requirement for Mn^{2+}, no activation by Mg^{2+} [14]; <25>, Mn^{2+} or Mg^{2+} required [16]; <25>, K_m with NAD^+: 0.14 mM, K_m with $NADP^+$: 0.81 mM [16]; <26>, bivalent metal ion required, Mn^{2+} is more effective than Mg^{2+} [17]; <18>, strict requirement for a divalent cation, maximal activation at 5 mM [21]) [6, 8, 10, 11, 13, 14, 16, 17, 18, 21]

Ni^{2+} <28> (<28>, divalent metal ion required, $NADP^+$-linked activity exhibits a maximum at 5 mM Ni^{2+} [19]) [19]

Turnover number (min^{-1})

37000 <2> ((S)-malate) [11]

Specific activity (U/mg)

2.74 <28> (<28>, NAD^+-linked activity [19]) [19]
5.45 <28> (<28>, $NADP^+$-linked activity [19]) [19]
20.5 <27> [18]
24 <7> [13]
35.4 <4> [8]
45.2 <5> [7]
72.5 <2> [11]
81.5 <4> [3]
Additional information <6, 18, 29> (<6>, design of an assay procedure to minimize the influence of (S)-malate dehydrogenase and other factors [9]) [9, 20, 21]

K_m-Value (mM)

0.018 <27> (NAD^+) [18]
0.089 <5> (NAD^+) [7]
0.1 <27> ((S)-malate, <27>, with NAD^+ as cofactor [18]) [18]
0.1 <25> (NAD^+, <25>, activation by Mn^{2+} [16]) [16]
0.12 <4> (NADH, <4>, activation by Mn^{2+} [8]) [8]
0.207 <27> ($NADP^+$) [18]
0.21 <7> (NAD^+) [13]
0.3 <3> (NAD^+) [14]
0.3 <2> ($NADP^+$, <2>, activation by Mn^{2+} [11]) [11]
0.458 <27> ((S)-malate, <27>, with $NADP^+$ as cofactor [18]) [18]
0.47 <1> (NAD^+, <1>, with 30 mM Mg^{2+} [6]) [6]
0.48 <1> (NAD^+, <1>, with 8 mM Mg^{2+} [6]) [6]
0.5 <1, 2> (NAD^+, <1>, with 80 mM Mg^{2+} [6]; <2>, activation by Mn^{2+} [11]) [6, 11]
0.52 <26> ((S)-malate) [17]
0.59 <4> ((S)-malate, <4>, activation by Mn^{2+} [8]) [8]
0.76 <2> ((S)-malate, <2>, with NAD^+ and Mn^{2+} [11]) [11]

0.77 <4> (NAD^+, <4>, activation by Mg^{2+} [8]) [8]
0.8 <4> ((S)-malate, <4>, activation by Mn^{2+} [8]) [8]
0.82 <25> (NAD^+, <25>, activation by Mg^{2+} [16]) [16]
0.9 <2> (NAD^+, <2>, activation by Mg^{2+} [11]) [11]
1.32 <25> ($NADP^+$, <25>, activation by Mn^{2+} [16]) [16]
1.56 <5> ($NADP^+$) [7]
1.7 <2> ($NADP^+$, <2>, with $NADP^+$ as coenzyme [11]) [11]
2.5 <7> ((S)-malate) [13]
3.2 <25> ((S)-malate, <25>, with NAD^+ as coenzyme, activation by Mn^{2+} [16]) [16]
4.2 <2> ((S)-malate, <2>, with $NADP^+$ and Mn^{2+} [11]) [11]
5.97 <28> ((S)-malate, <28>, with NAD^+ as coenzyme [19]) [19]
6.03 <4> ((S)-malate, <4>, activation by Mg^{2+} [8]) [8]
6.12 <25> ($NADP^+$, <25>, activation by Mg^{2+} [16]) [16]
8.34 <28> ((S)-malate, <28>, with $NADP^+$ as coenzyme [19]) [19]
9.8 <25> ((S)-malate, <25>, with $NADP^+$ as coenzyme, activation by Mn^{2+} [16]) [16]
12.3 <2> ((S)-malate, <2>, with $NADP^+$ and Mg^{2+} [11]) [11]
12.54 <1> ((S)-malate) [12]
13 <25> ((S)-malate, <25>, with NAD^+ as coenzyme, activation by Mg^{2+} [16]) [16]
13.48 <4> (CO_2, <4>, activation by Mg^{2+} [8]) [8]
14 <2> ((S)-malate, <2>, with NAD^+ and Mg^{2+} [11]) [11]
15.03 <4> (pyruvate, <4>, activation by Mn^{2+} [8]) [8]
22.5 <25> ((S)-malate, <25>, with $NADP^+$ as coenzyme, activation by Mg^{2+} [16]) [16]
Additional information <4, 29> [4, 20]

pH-Optimum

6.4 <2> (<2>, activity with $NADP^+$ and Mg^{2+} [11]) [11]
6.5 <2> (<2>, activity with $NADP^+$ and Mn^{2+} [11]) [11]
6.7 <2> (<2>, enzyme complexed to Mg^{2+} and NAD^+ [2]) [2]
6.8 <2, 29> (<2>, activity with NAD^+ [11]) [11, 20]
6.9 <3> (<3>, with NAD^+, without activator [14]; <3>, activity with $NADP^+$ [14]) [14]
7 <2, 25> (<2>, catalytically competent enzyme-substrate complex formed upon binding malate [2]) [2, 16]
7-8 <28> [19]
7.1 <4> (<4>, activated by Mg^{2+} [10]) [10]
7.2 <3> (<3>, with NAD^+ plus CoA [14]) [14]
7.46 <4> (<4>, activated by Mn^{2+} [10]) [10]
7.6-7.9 <7> [13]

pH-Range

5.8-7.6 <2> (<2>, pH 5.8: about 60% of maximal activity, pH 7.6: about 40% of maximal activity [2]) [2]
6-8.5 <28> (<28>, about 40% of maximal activity at pH 6.0 and at pH 8.5 [19]) [19]

7-8.7 <7> (<7>, pH 7.0: about 30% of maximal activity, pH 8.7: about 45% of maximal activity [13]) [13]

Temperature optimum (°C)

50 <28> [19]

Temperature range (°C)

30-60 <28> (<28>, 30°C: about 35% of maximal activity, 60°C: about 50% of maximal activity [19]) [19]

4 Enzyme Structure

Molecular weight

119000 <4> (<4>, dimer, disc gel electrophoresis [8]) [8]
120000 <4, 29> (<4>, dimer, non-denaturing PAGE [1]; <29>, non-denaturing PAGE [20]) [1, 20]
170000 <2> (<2>, dimer, gel filtration [1]) [1]
200000 <26> (<26>, sucrose density gradient centrifugation [17]) [17]
230000 <4> (<4>, tetramer, non-denaturing PAGE [1]) [1]
231000 <4> (<4>, tetramer, disc gel electrophoresis [8]) [8]
270000 <25, 26> (<25, 26>, gel filtration [16, 17]) [16, 17]
279000 <3> (<3>, tetramer, gel filtration [1]) [1]
300000 <3> (<3>, gel filtration [14]) [14]
308000 <27> (<27>, gel filtration [18]) [18]
350000 <4> (<4>, non-denaturing PAGE [8]) [8]
388000 <18> (<18>, gel filtration [21]) [21]
400000 <1> (<1>, octamer, gel filtration [1]) [1]
490000 <2, 4> (<2>, octamer, gel filtration, enzyme in octameric aggregation state [1]; <4>, octamer, non-denaturing PAGE [1]; <2>, gel filtration in presence of 5 mM dithiothreitol and 2 mM $MgCl_2$ [11]) [1, 11]
491000 <4> (<4>, tetramer, disc gel electrophoresis [8]) [8]
680000 <28> (<28>, gel filtration [19]) [19]

Subunits

? <4, 5> (<4>, x * 55000, α, + x * 61000, β, SDS PAGE [3]; <5>, x * 82000, SDS-PAGE [7]) [3, 7]
decamer <28> (<28>, 10 * 64000, SDS-PAGE [19]) [19]
dimer <29> (<29>, 1 * 61000 + 1 * 64000, SDS-PAGE [20]) [20]
octamer <2> (<2>, $\alpha_4\beta_4$, 4 * 61000 + 4 * 58000, SDS-PAGE [11]) [11]
tetramer <18> (<18>, 4 * 85000, SDS-PAGE [21]) [21]
Additional information <4> (<4>, the enzyme exists in at least three aggregational states: dimer, tetramer and octamer [4]) [4]

5 Isolation/Preparation/Mutation/Application

Source/tissue

bud <1, 2> [1, 5]
flight muscle <7> [13]
floret <1> [6]
heart <25> [16]
hypocotyl <29> (<29>, etiolated [20]) [20]
leaf <3, 4, 6, 19, 20> (<3, 19, 20>, bundle sheath [14]) [4, 8, 9, 10, 14]
spermatozoon <21, 22, 23, 24> [15]
tuber <2> [1, 2, 11]

Localization

cytosol <7> (<7>, 30% of the activity [13]) [13]
hydrogenosome <27> [18]
mitochondrion <1, 2, 6, 7, 25> (<7>, predominantly in the matrix space [13]; <7>, 65% of the activity [13]) [5, 9, 13, 16]

Purification

<1> [6]
<2> [11]
<3> [14]
<4> [3, 4, 8]
<5> [7]
<18> [21]
<25> [16]
<27> [18]
<28> [19]
<29> [20]

6 Stability

pH-Stability

7-9 <28> (<28>, 25°C, 5 min, stable [19]) [19]

Temperature stability

45 <28> (<28>, 15 min, stable at or below [19]) [19]
55 <28> (<28>, 15 min, complete inactivation [19]) [19]

Oxidation stability

<2>, photooxidation with methylene blue [2]

General stability information

<2>, photooxidation with methylene blue [2]
<3>, freezing of crude enzyme extract treated on Sephadex G-25 destroys activity [14]

Storage stability

<3>, crude and partially purified enzyme retains activity for several months when stored as a protein suspension in 75% saturated $(NH_4)_2SO_4$ solution [14]

<5>, -20°C, stable for several weeks [7]

<5>, 4°C, loss of activity upon prolonged storage [7]

References

[1] Artus, N.N.; Edwards, G.E.: NAD-malic enzyme from plants. FEBS Lett., **182**, 225-233 (1985)

[2] Willeford, K.O.; Wedding, R.T.: pH effects on the activity and regulation of the NAD malic enzyme. Plant Physiol., **84**, 1084-1087 (1987)

[3] Willeford, K.O.; Wedding, R.T.: Evidence for a multiple subunit composition of plant NAD malic enzyme. J. Biol. Chem., **262**, 8423-8429 (1987)

[4] Willeford, K.O.; Wedding, R.T.: Regulation of the NAD malic enzyme from Crassula. Plant Physiol., **80**, 792-795 (1986)

[5] Day, D.A.; Neuburger, M.; Douce, R.: Activation of NAD-linked malic enzyme in intact plant mitochondria by exogenous coenzyme A. Arch. Biochem. Biophys., **231**, 233-242 (1984)

[6] Canellas, P.F.; Wedding, R.T.: Kinetic properties of NAD malic enzyme from cauliflower. Arch. Biochem. Biophys., **229**, 414-425 (1984)

[7] Voegele, R.T.; Mitsch, M.J.; Finan, T.M.: Characterization of two members of a novel malic enzyme class. Biochim. Biophys. Acta, **1432**, 275-285 (1999)

[8] Wedding, R.T.; Black, M.K.: Physical and kinetic properties and regulation of the NAD malic enzyme purified from leaves of Crassula argentea. Plant Physiol., **72**, 1021-1028 (1983)

[9] Hatch, M.D.; Tsuzuki, M.; Edwards, G.E.: Determination of NAD malic enzyme in leaves of C_4 plants. Plant Physiol., **69**, 483-491 (1982)

[10] Wedding, R.T.; Canellas, P.F.; Black, M.K.: Slow transients in the activity of the NAD malic enzyme from Crassula. Plant Physiol., **68**, 1416-1423 (1981)

[11] Grover, S.D.; Canellas, P.F.; Wedding, R.T.: Purification of NAD malic enzyme from potato and investigation of some physical and kinetic properties. Arch. Biochem. Biophys., **209**, 396-407 (1981)

[12] Canellas, P.F.; Wedding, R.T.: Substrate and metal ion interactions in the NAD^+ malic enzyme from cauliflower. Arch. Biochem. Biophys., **199**, 259-264 (1980)

[13] Hoek, H.B.; Pearson, D.J.; Olembo, N.K.: Nicotinamide-adenine dinucleotide-linked malic enzyme in flight muscle of the tse-tse fly (Glossina) and other insects. Biochem. J., **160**, 253-262 (1976)

[14] Hatch, M.D.; Mau, S.-L.; Kagawa, T.: Properties of leaf NAD malic enzyme from plants with C_4 pathway photosynthesis. Arch. Biochem. Biophys., **165**, 188-200 (1974)

[15] Mounib, M.S.: NAD- and NADP-malic enzymes in spermatozoa of mammals and fish. FEBS Lett., **48**, 79-84 (1974)

[16] Lin, R.C.; Davis, E.J.: Malic enzymes of rabbit heart mitochondria. Separation and comparison of some characteristics of a nicotinamide adenine dinucleotide-preferring and a nicotinamide adenine dinucleotide phosphate specific enzyme. J. Biol. Chem., **249**, 3867-3875 (1974)

[17] Eyzaguirre, J.; Cornwell, E..; Borie, G.; Ramirez, B.: Two malic enzymes in Pseudomonas aeruginosa. J. Bacteriol., **116**, 215-221 (1973)

[18] Hrdy, I.; Mertens, E.: Purification and partial characterization of malate dehydrogenase (decarboxylating) from Tritrichomonas foetus hydrogenosomes. Parasitology, **107**, 379-385 (1993)

[19] Suye, S.; Okada, Y.; Funada, A.; Kawagoe, M.; Inuta, S.: Purification and properties of malic enzyme from Pseudomonas diminuta IFO-13182. J. Ferment. Bioeng., **73**, 343-347 (1992)

[20] Farias de Aragao, M.E.; Jolivet, Y.; Fernandes de Melo, D.; da Guia Silva Lima, M.; Dizengremel, P.: Purification and partial characterization of NAD-dependent malic enzyme isolated from Vigna unguiculata hypocotyls. Plant Physiol. Biochem., **34(3)**, 363-368 (1996)

[21] Chen, F.; Okabe, Y.; Osano, K.; Tajima, S.: Purification and characterization of an NAD-malic enzyme from Bradyrhizobium japonicum A1017. Appl. Environ. Microbiol., **64**, 4073-4075 (1998)

Malate dehydrogenase (oxaloacetate-decarboxylating, $NADP^+$)

1.1.1.40

1 Nomenclature

EC number

1.1.1.40

Systematic name

(S)-malate:$NADP^+$ oxidoreductase (oxaloacetate-decarboxylating)

Recommended name

malate dehydrogenase (oxaloacetate-decarboxylating, $NADP^+$)

Synonyms

L-malate:NADP oxidoreductase
$NADP^+$ dependent malic enzyme
NADP-ME
NADP-linked decarboxylating malic enzyme
NADP-malic enzyme
NADP-specific malate dehydrogenase
NADP-specific malic enzyme
TME
malate dehydrogenase (NADP, decarboxylating)
malate dehydrogenase (decarboxylating, NADP)
malic enzyme
pyruvic-malic carboxylase

CAS registry number

9028-47-1

2 Source Organism

<1> *Triticum aestivum* [1, 12]
<2> *pigeon* [2, 6, 11, 16]
<3> *duck* [3]
<4> *Gallus gallus* [4, 23]
<5> *Sus scrofa* [5]
<6> *Rattus norvegicus* [6, 10]
<7> *Cavia porcellus* [6]
<8> *Oryctolagus cuniculus* [6, 26]
<9> *seagull* [6]
<10> *Locusta migratoria* [6]
<11> *Pennisetum purpureum* [7]

<12> *Zea mays* [8, 12]
<13> *Streptomyces aureofaciens* [9]
<14> *Haloxylon persicum* [12]
<15> *Flaveria floridana* [12]
<16> *Flaveria bidentis* [12]
<17> *Mesembryanthemun crystallinum* [12]
<18> *Egeria densa* [12]
<19> *Aloe arborescens* [12]
<20> *poplar* [12]
<21> *Apium graveolens* [12]
<22> *Lycopersicon esculentum* [12, 17]
<23> *Magnifera indica* [13]
<24> *Homo sapiens* [14, 15, 18, 20, 21, 25]
<25> *Corynebacterium glutamicum* [19]
<26> *Ricinus communis* [22]
<27> *Rhizobium meliloti* [24]
<28> *Bos taurus* [25]
<29> *salmon* [25]
<30> *cod* [25]
<31> *Pseudomonas aeruginosa* [27]
<32> *Pandalus borealis* [28]

3 Reaction and Specificity

Catalyzed reaction

(S)-malate + NADP$^+$ = pyruvate + CO_2 + NADPH (<24>, sequential ordered bi-ter kinetic mechanism with NADP$^+$ as the leading substrate followed by L-malate. The products are released in the order of CO_2, pyruvate and NADPH [20])

Reaction type

decarboxylation
oxidation
redox reaction
reduction

Natural substrates and products

S L-malate + NADP$^+$ <12, 14, 15, 32> (<12, 14, 15>, enzyme plays a specialized role in bundle sheath chloroplasts, where it provides CO_2 for fixation by EC 4.1.1.39 [12]; <32>, major role of the enzyme is believed to be the supplier of NADPH for the reductive steps of lipogenesis [28]) (Reversibility: ? <12, 14, 15, 32> [12, 14, 28]) [12, 14, 28]

P pyruvate + CO_2 + NADPH <12, 14, 15> [12]

S Additional information <4, 22, 24> (<24>, one may speculate that in vivo the reaction catalyzed by cytosolic enzyme supplies dicarboxylic acids, anaplerotic function, for the formation of neurotransmitters while the mitochondrial enzyme regulates the flux rate via Krebs cycle by disposition of the tricarboxylic acid cycle intermediates, cataplerotic function [14]; <22>, the specific activity falls rapidly as the fruit ripens [17]; <4>, cell-specific regulation and transcription of the malic enzyme gene, cell-specific differences in T3 responsiveness of the malic enzyme gene are mediated in large part by by nonreceptor proteins that augment the transcriptional activity of the nuclear T3 receptor [23]) [14, 17, 23]

P ?

Substrates and products

S L-malate + NAD^+ <24> (<24>, lower activity than with $NADP^+$ [15]; <24>, at 39% of the activity with $NADP^+$ [18]) (Reversibility: ir <24> [18]; ? <24> [15]) [15, 18]

P pyruvate + CO_2 + NADH <24> [15, 18]

S L-malate + $NADP^+$ <1, 2, 3, 4, 5, 8, 11, 12, 13, 14, 15, 17, 18, 23, 24, 25, 26, 27, 28, 29, 30, 31> (<2>, the singly-ionized species is the substrate, doubly-ionized malate is unreactive [2]; <4>, carboxyl is replaced by hydrogen without net change of configuration [4]; <2>, Arg residue is involved in the binding of C-1 carboxyl group of malate [16]; <26>, reaction velocity is 29times higher in the direction of decarboxylation than in the direction of the carboxylation [22]; <24>, the reverse reaction is catalyzed at 84% of the activity [18]) (Reversibility: r <2, 5, 8, 12, 24, 25, 26, 31> [2, 5, 8, 15, 16, 18, 19, 21, 22, 26, 27]; ir <27> [24]; ? <1, 3, 4, 11, 12, 13, 14, 15, 17, 18, 23, 24, 28, 29, 30> [3, 4, 7, 9, 12, 13, 14, 20, 25]) [2, 3, 4, 5, 7, 9, 12, 13, 14, 15, 16, 18, 19, 20, 21, 22, 24, 25, 26, 27]

P pyruvate + CO_2 + NADPH <2, 3, 4, 5, 8, 24, 25> [2, 3, 4, 5, 15, 16, 18, 19, 21, 26]

S oxaloacetate <1, 2, 8, 24> (<24>, at 1.3% of the rate of $NADP^+$-linked oxidative decarboxylation [18]; <27>, no activity [24]) (Reversibility: ? <1, 2, 8, 24> [1, 2, 15, 18, 26]) [1, 2, 15, 18, 26]

P pyruvate + CO_2 <1, 2> [1, 2]

S pyruvate + CO_2 + NADPH <1, 2, 5, 8, 12, 24, 25, 26, 31, 32> (<26>, reaction velocity is 29times higher in the direction of decarboxylation than in the direction of the carboxylation [22]; <24>, at 84% of the activity of the $NADP^+$-linked oxidative decarboxylation of malate [18]; <8>, 6% of the activity of the $NADP^+$-linked oxidative decarboxylation of malate [26]) (Reversibility: r <2, 5, 8, 12, 24, 25, 26, 31> [2, 5, 8, 15, 18, 19, 21, 22, 26, 27]; ? <1, 32> [1, 28]) [1, 2, 5, 8, 15, 18, 19, 21, 22, 26, 27, 28]

P L-malate + $NADP^+$ <1, 2, 5, 8, 12, 24, 25, 26, 31> [1, 2, 5, 8, 15, 19, 22, 26, 27]

S pyruvate + NADPH <24> (<24>, at 5.4% of the activity of NADP-dependent oxidative decarboxylation of malate [18]) (Reversibility: ? <24> [18]) [18]

P ?

Inhibitors

$(NH_4)_2SO_4$ <24> (<24>, at high concentrations [20]) [20]

2,3-butanedione <2> (<2>, pseudo-first-order loss of oxidative decarboxylase activity [16]) [16]

2-mercaptoethanol <6> (<6>, inactivation in absence of Mg^{2+} [10]) [10]
2-oxoglutarate <22> (<22>, strong [17]) [17]
ADP <2> [2]
ATP <24> (<24>, non-competitive versus L-malate [20,21]) [20, 21]
Cu^{2+} <2> (<2>, 3 mM, weak [2]) [2]
D-malate <27> [24]
D-tartrate <27> [24]
EDTA <12> [8]
Hg^{2+} <2> (<2>, 0.0005 mM, almost complete inhibition [2]) [2]
L-tartrate <27> [24]
Mg^{2+} <5> (<5>, mitochondrial enzyme, decarboxylation reaction, above 6 mM [5]) [5]
NAD^+ <2> (<2>, weak [2]) [2]
NaCl <24> (<24>, at high concentrations [20]) [20]
SO_3^{2-} <12> (<12>, in decarboxylation of malate: partially competitive with respect to malate, in carboxylation of pyruvate: fully competitive for CO_2 or HCO_3^- [8]) [8]
acetyl-CoA <5> (<5>, inhibition is much more pronounced in the case of the mitochondrial enzyme than in the cytosolic enzyme and occurs at physiological acetyl-CoA concentrations [5]) [5]
adenosine 2'-phosphate <2> [2]
arsenite <2> (<2>, weak [2]) [2]
diphenyliodonium chloride <2> (<2>, weak [2]) [2]
fumarate <22> (<22>, at high concentration [17]) [17]
glutamate <25> [19]
glutathione <6> (<6>, strongly inactivates in absence of Mg^{2+} [10]) [10]
glyoxylate <27> [24]
iodoacetate <2> (<2>, weak [2]) [2]
malate <2, 12, 14, 24> (<2>, substrate inhibition [11]; <12, 14>, inhibition at pH 7.0 [12]; <1, 15, 17, 18>, no inhibition at pH 7.0 [12]; <24>, excess of malate inhibits the oxidative decarboxylation catalyzed by the cytosolic enzyme at pH 7.0, and below, decarboxylation catalyzed by mitochondrial enzyme is unaffected by the substrate [14]) [11, 12, 14]
malonate <27> [24]
malonyl-CoA <5> (<5>, inhibition of mitochondrial enzyme and cytosolic enzyme to a much lower extent than with acetyl-CoA [5]) [5]
o-iodosobenzoate <2> (<2>, strong [2]) [2]
oxalate <8> [26]
oxaloacetate <22, 25> (<22>, strong [17]) [17, 19]
p-mercuribenzoate <2> (<2>, strong [2]) [2]
phenylglyoxal <2> (<2>, pseudo-first-order loss of oxidative decarboxylase activity [16]) [16]
pyruvate <11, 22> (<11>, competitive [7]; <22>, non-competitive inhibition [17]) [7, 17]
succinate <22> (<22>, slight inhibition at high concentrations [17]) [17]
sulfite <12> [8]

tartronate <24, 27> (<24>, noncompetitive inhibitor with respect to L-malate [20]) [20, 24]

Cofactors/prosthetic groups

$NADP^+$ <1, 2, 3, 4, 5, 12, 13, 14, 15, 17, 18, 22, 23, 24, 25, 26, 27, 31, 32> (<1, 2, 3, 4, 5, 12, 13, 14, 15, 17, 18, 22, 23, 24, 25, 26> cofactor [1, 2, 3, 4, 5, 8, 9, 12, 13, 14, 15, 16, 18, 19, 21, 22]; <2>, stimulation of decarboxylation of oxaloacetate [2]) [1, 2, 3, 4, 5, 8, 9, 12, 13, 14, 15, 16, 18, 19, 21, 22, 24, 27, 28]
NADPH <2, 5, 12, 24, 25, 26> (<2, 5, 12, 24, 25, 26> cofactor [2, 5, 8, 15, 18, 19, 21, 22]) [2, 5, 8, 15, 18, 19, 21, 22]

Activating compounds

CoA <26> (<26>, activates [22]) [22]
NH_4^+ <25> (<25>, activates [19]) [19]
NH_4^+ <31> (<31>, activates [27]) [27]
acetyl-CoA <26> (<26>, activates [22]) [22]
fumarate <22, 24> (<24>, increases activity of the mitochondrial enzyme at low malate concentrations, no effect on the activity of the cytosolic enzyme [14]; <22>, slight activation at low concentration [17]) [14, 17]
palmitoyl-CoA <26> (<26>, activates [22]) [22]
succinate <22, 24, 26> (<24>, increases activity of the mitochondrial enzyme at low malate concentrations, no effect on the activity of the cytosolic enzyme [14]; <22>, slight activation up to 10 mM [17]; <26>, activates [22]) [14, 17, 22]

Metals, ions

Co^{2+} <2, 25> (<2>, less effective activation than Mn^{2+} [2]; <25>, activates [19]) [2, 19]
Cu^{2+} <2> (<2>, weak activation [2]) [2]
K^+ <25, 31> (<25, 31>, activates [19, 27]; <31>, K_m: 0.7 mM [27]) [19, 27]
Mg^{2+} <1, 2, 5, 8, 14, 15, 17, 18, 22, 23, 25, 26, 31> (<2>, promotes activity [2]; <5>, activity of cytosolic enzyme and mitochondrial enzyme is strictly dependent on [5]; <5>, optimal concentration: 6 mM for the decarboxylation of the mitochondrial enzyme, 20 mM for the cytosolic enzyme [5]; <13>, divalent cation required, Mn^{2+} is more effective than Mg^{2+} [9]; <12>, K_m: 0.05-0.23 mM [12]; <14>, K_m: 0.011-3.5 mM [12], <15>, K_m: 0.005 0.16 mM [12]; <1>, K_m: 0.006-0.2 mM [12]; <18>, K_m: 0.00146 mM [12]; <23>, Mg^{2+} and Mn^{2+} stabilize two structurally distinct forms of the enzyme which vary in catalytic and regulatory properties [13]; <22>, K_m: 0.315 mM [17]; <22>, absolute requirement for a divalent cation, Mn^{2+} or Mg^{2+} [17]; <25>, activates [19]; <26>, Mn^{2+} or Mg^{2+} required [22]; <8>, requires low concentration of Mn^{2+} or Mg^{2+} for activity, K_m: 0.24 mM [26]; <31>, divalent metal ion required, Mn^{2+} is more effective than Mg^{2+}, K_m: 0.34 mM [27]) [2, 5, 9, 12, 13, 17, 19, 22, 26, 27]
Mn^{2+} <2, 5, 8, 11, 12, 13, 22, 23, 25, 26, 31> (<8>, requires low concentration of Mn^{2+} or Mg^{2+} for activity [26]; <2>, 0.004 mM, 50% of the maximal activity of the malic activity. 0.2 mM, 50% of the maximal activity of oxaloacetic decarboxylase [2]; <5>, activity of cytosolic enzyme and mitochondrial en-

zyme is strictly dependent on [5]; <5>, optimal concentration is 15 mM for the cytosolic enzyme and 10 mM for the mitochondrial enzyme [5]; <11>, required, K_m: 0.008 mM [7]; <12>, required for decarboxylation of malate and decarboxylation of pyruvate, K_M: 0.0018 mM [8]; <13>, divalent cation required, Mn^{2+} is more effective than Mg^{2+} [9]; <23>, at pH 7.1, K_m: 0.0143 [13]; <23>, Mg^{2+} and Mn^{2+} stabilize two structurally distinct forms of the enzyme which vary in catalytic and regulatory properties [13]; <22>, K_m: 0.210 mM [17]; <22>, absolute requirement for a divalent cation, Mn^{2+} or Mg^{2+} [17]; <25>, dependent on presence of divalent cations. Maximal activity in presence of 5 mM Mn^{2+} [19]; <26>, Mn^{2+} or Mg^{2+} required [22]; <8>, K_m: 0.01 mM [26]; <31>, divalent metal ion required, Mn^{2+} is more effective than Mg^{2+} [27]; <31>, K_m: 0.004 mM [27]) [2, 5, 7, 8, 9, 13, 17, 19, 22, 26, 27]
Ni^{2+} <2, 25> (<2>, weak activation [2]; <25>, activates [19]) [2, 19]
Zn^{2+} <2> (<2>, weak activation [2]) [2]

Turnover number (min^{-1})
9600 <3> (L-malate) [3]
34920 <27> (L-malate, <27>, assuming an octameric oligomerization state [24]) [24]
Additional information <24> [20]

Specific activity (U/mg)
5.04 <24> [21]
10.9 <25> [19]
41 <6> [10]
49 <27> [24]
56.35 <24> [18]
90.9 <24> (<24>, mitochondrial enzyme [15]) [15]
91.4 <24> (<24>, cytosolic enzyme [15]) [15]
7910 <23> [13]
Additional information <2, 5, 22, 26> [2, 5, 17, 22]

K_m-Value (mM)
0.00118 <5> ($NADP^+$, <5>, cytosolic enzyme [5]) [5]
0.00139 <24> ($NADP^+$) [21]
0.00188 <5> ($NADP^+$, <5>, mitochondrial enzyme [5]) [5]
0.002 <11> ($NADP^+$) [7]
0.002 <24> (NADPH) [21]
0.0031 <24> (NADP) [15]
0.00315 <14> ($NADP^+$) [12]
0.0046 <3> ($NADP^+$) [3]
0.0053 <24> (NADPH) [18]
0.0082 <26> ($NADP^+$) [22]
0.0086 <12> ($NADP^+$) [12]
0.0092 <24> ($NADP^+$) [18]
0.0099 <17> ($NADP^+$) [12]
0.012 <15> ($NADP^+$) [12]
0.014 <8> ($NADP^+$, <8>, activation by Mn^{2+} [26]) [26]

0.015 <12> (NADP$^+$) [8]
0.016 <2, 23> (NADP$^+$, <2>, pH 7.5 [2]) [2, 13]
0.025 <8> (NADP$^+$, <8>, activation by Mg^{2+} [26]) [26]
0.037 <1> (NADP$^+$) [12]
0.038 <22> (NADP$^+$) [17]
0.04 <11> (L-malate) [7]
0.041 <2> (L-malate, <2>, pH 6.5 [2]) [2]
0.045 <12> (NADPH) [8]
0.046 <13> (NADP$^+$) [9]
0.0472 <18> (NADP$^+$) [12]
0.06 <25> (NADPH) [19]
0.07 <31> (NADP$^+$) [27]
0.073 <3> (L-malate) [3]
0.08 <14> (L-malate) [12]
0.083 <25> (NADP$^+$) [19]
0.09 <12> (L-malate) [8]
0.12 <24> (L-malate, <24>, with NADP$^+$ as cosubstrate [18]) [18]
0.14 <5> (L-malate, <5>, cytosolic enzyme [5]) [5]
0.19 <12> (L-malate) [12]
0.216 <24> (L-malate) [21]
0.35 <17> (L-malate) [12]
0.39 <2> (L-malate, <2>, pH 7.5 [2]) [2]
0.46 <15> (L-malate) [12]
0.53 <22> (L-malate) [17]
0.6 <24> (L-malate, <24>, cytosolic enzyme [14]) [14]
0.666 <23> (L-malate, <23>, at 1.0 mM Mn^{2+} [13]) [13]
0.68 <26> (L-malate) [22]
0.96 <1, 24> (L-malate, <24>, with NAD$^+$ as cosubstrate [18]) [12, 18]
1.06 <8> (malate, <8>, activation by Mn^{2+} [26]) [26]
1.08 <23> (L-malate, <23>, at 5.0 mM Mn^{2+} [13]) [13]
1.25 <5> (L-malate, <5>, mitochondrial enzyme [5]) [5]
2.1 <13> (L-malate) [9]
2.6 <27> (L-malate) [24]
2.7 <31> (malate) [27]
3 <12> (pyruvate) [8]
3.3 <2> (L-malate, <2>, pH 8.5 [2]) [2]
3.63 <8> (malate, <8>, activation by Mg^{2+} [26]) [26]
3.8 <25> (L-malate) [19]
4.5 <18> (L-malate) [12]
4.6 <24> (NAD$^+$) [18]
4.8 <24> (pyruvate) [21]
5.9 <24> (pyruvate) [18]
7.2 <24> (pyruvate) [15]
8.2 <26> (NADP$^+$) [22]
10 <5> (pyruvate, <5>, cytosolic enzyme [5]) [5]
13.3 <24> (bicarbonate) [21]
13.8 <25> (pyruvate) [19]

23 <24> ($NaHCO_3$) [15]
25 <5> (pyruvate, <5>, mitochondrial enzyme [5]) [5]
27.9 <24> (bicarbonate) [18]

pH-Optimum

5.2 <1> (<1>, decarboxylation of oxaloacetate [1]) [1]
6.6 <5> (<5>, mitochondrial enzyme, carboxylation [5]) [5]
6.8 <23> (<23>, MES-NaOH buffer [13]) [13]
6.9-7.2 <26> [22]
7 <5, 23> (<5>, carboxylation and decarboxylation, cytosolic enzyme [5]; <23>, TES-NaOH buffer [13]) [5, 13]
7-8 <8> [26]
7.1 <23, 24, 26> (<23>, Tris-HCl buffer [13]; <24>, NADP$^+$-linked decarboxylation of malate, at 0.2 mM malate [18]) [13, 18, 22]
7.2 <1, 2, 17, 24> (<2>, in presence of 0.1 mM malate [2]) [2, 12, 21]
7.3 <1, 18, 24> (<1,24>, carboxylation of pyruvate [1,18]; <24>, NADP$^+$-linked decarboxylation of malate, at 1 mM malate [18]) [1, 12, 18]
7.4 <12, 24> (<12>, carboxylation of pyruvate [8]; <24>, cytosolic enzyme, NADPH-dependent reductive carboxylation of pyruvate [15]) [8, 15]
7.5 <15, 23, 25> (<23>, TES-NaOH buffer [13]; <25>, carboxylation of pyruvate [19]) [12, 13, 19]
7.6 <5> (<5>, cytosolic enzyme, decarboxylation [5]) [5]
7.8 <25> (<25>, oxidative decarboxylation of L-malate [19]) [19]
8 <11, 12, 24, 32> (<12>, decarboxylation of malate [8]; <24>, NADP$^+$-dependent decarboxylation of malate, at 10 mM malate, mitochondrial enzyme and cytosolic enzyme [15]; <24>, NADP$^+$-linked decarboxylation of malate, at 10 mM malate [18]) [7, 8, 12, 15, 18, 28]
8.2 <14> [12]
8.3 <13> (<13>, with 0.1 mM L-malate [9]) [9]
8.5-9 <2> (<2>, in presence of 0.1 M malate [2]) [2]
8.8 <13> (<13>, with 1 mM L-malate [9]) [9]

pH-Range

6.5-7.5 <26> (<26>, pH 6.5: about 55% of maximal activity, pH 7.5: about 35% of maximal activity [22]) [22]
6.8-8 <24> (<24>, pH 6.8: about 40% of maximal activity, pH 8.0: about 70% of maximal activity, cytosolic enzyme, NADPH-dependent reductive carboxylation of pyruvate [15]) [15]
7-9 <11> (<11>, pH 7.0: about 60% of maximal activity, pH 9.0: about 55% of maximal activity [7]) [7]
7.1-8.6 <24> (<24>, pH 7.1: about 65% of maximal activity, pH 8.6: about 75% of maximal activity, NADP$^+$-dependent decarboxylation of malate, at 10 mM malate, mitochondrial enzyme and cytosolic enzyme [15]) [15]
7.5-9 <32> (<32>, pH 7.5: about 50% of maximal activity, pH 9.0: about about 85% of maximal activity [28]) [28]

Temperature optimum (°C)

50 <25> [19]

4 Enzyme Structure

Molecular weight

68000 <31> (<31>, gel filtration [27]) [27]
90000 <31> (<31>, sucrose density gradient centrifugation [27]) [27]
105000 <25> [19]
180000 <8> (<8>, gel filtration [26]) [26]
216000 <5> (<5>, cytosolic enzyme, density gradient centrifugation [5]) [5]
227000 <5> (<5>, mitochondrial enzyme, density gradient centrifugation [5]) [5]
230000-240000 <12> (<12>, and a second molecular weight form of 460000-480000, gel filtration [8]) [8]
257000 <24> (<24>, gel filtration [15]) [15]
258000 <23> (<23>, gel filtration [13]) [13]
260000-265000 <22> (<22>, gel filtration [17]) [17]
285000 <26> (<26>, gel filtration [22]) [22]
460000-480000 <12> (<12>, and a second molecular weight form of 230000-240000, gel filtration [8]) [8]

Subunits

? <1, 12, 14, 15, 17, 18, 27> (<12, 15>, x * 62000, SDS-PAGE [12]; <17>, x * 64000, SDS-PAGE [12]; <14>, x * 67000, SDS-PAGE [12]; <1, 18>, x * 72000, SDS-PAGE [12]; <27>, x * 82000, SDS-PAGE [24]) [12, 24]
dimer <25> (<25>, 2 * 42000, SDS-PAGE [19]) [19]
tetramer <22, 23, 24, 26> (<24>, 4 * 63000, SDS-PAGE [18]; <24>, 4 * 64000, SDS-PAGE [15]; <22>, 4 * 64000-65000, SDS-PAGE [17]; <23>, 4 * 64900, SDS-PAGE [13]; <24>, 4 * 65000, SDS-PAGE [21]; <26>, 4 * 72000, SDS-PAGE [22]) [13, 15, 17, 18, 21, 22]

5 Isolation/Preparation/Mutation/Application

Source/tissue

brain <24> [14, 25]
breast cell line <24> (<24>, breast cancer cell line [20, 21]) [20, 21]
bundle sheath <14, 15> [12]
cardiac muscle <5, 6, 7, 8, 9> [5, 6, 26]
cotyledon <26> [22]
embryo <4> (<4>, hepatocytes and fibroblasts [23]) [23]
fruit <22, 23> [13, 17]
germ <1> [1]
kidney <6> (<6>, cortex [6]) [6]
leaf <11, 12, 18> (<12>, etiolated [12]) [7, 12]
liver <2, 3, 4, 6, 24> (<7>, no activity [6]) [2, 3, 4, 6, 10, 11, 16, 18]
muscle <6, 7, 8, 9, 10, 32> (<6, 7>, m. rectus, m. soleus [6]; <8>, m. psoas, m. semitend. [6]; <9>, m. pectoralis [6]; <10>, flight muscle [6]; <32>, abdominal muscle [28]) [6, 28]

mycelium <13> [9]
root <12> [12]
semen <24, 28, 29, 30> [25]

Localization

cytosol <5, 17, 22, 24, 25> (<5>, 2 distinct enzyme forms are present in cytosol and in mitochondria [5]; <25>, represents about 1% of total cytosolic enzyme [19]) [5, 12, 14, 15, 17, 18, 19, 20, 21]
mitochondrion <5, 8, 24> (<5>, 2 distinct enzyme forms are present in cytosol and in mitochondria [5]) [5, 14, 15, 18, 26]
plastid <12, 14, 15> [12]

Purification

<1> [1]
<2> [2]
<3> [3]
<5> (2 distinct enzyme forms [5]) [5]
<6> [10]
<8> [26]
<11> (partial [7]) [7]
<13> [9]
<23> [13]
<24> [15, 21, 18]
<25> [19]
<26> (partial [22]) [22]
<27> [24]
<31> [27]

Crystallization

<6> [10]

Cloning

<3> (<3>, expression in Escherichia coli TH-2 cells as a fusion protein including a 15-residue N-terminal leader from β-galactosidase coded by the lacZ gene [3]) [3]
<25> [19]

Engineering

C99S <3> (<3>, turnover number decreases 3fold and K_m for malate increases 4fold [3]) [3]
R70Q <3> (<3>, kinetic parameters are similar except for a slightly lower turnover number and higher K_m for L-malate [3]) [3]

6 Stability

Temperature stability

4 <25> (<25>, stable for at least 120 min [19]) [19]
30 <24> (<24>, completely stable for at least 20 min [20]) [20]

33 <25> (<25>, 120 min, 24% loss of activity [19]) [19]
40 <25> (<25>, 120 min, 70% loss of activity [19]) [19]
Additional information <2> (<2>, Mg^{2+} or Mn^{2+} increase heat stability [2]; <2>, malic acid decreases heat stability at pH values above 5.0 [2]) [2]

General stability information

<2>, Mg^{2+} or Mn^{2+} increase heat stability [2]
<2>, malic acid decreases heat stability at pH values above 5.0 [2]
<27>, loss of activity upon repeated freezing and thawing [24]

Storage stability

<27>, -20°C, stable for several weeks [24]
<27>, 4°C, loss of activity upon prolonged storage [24]

References

[1] Harary, I.; Korey, S.R.; Ochoa, S.: Biosynthesis of dicarboxylic acids by carbon dioxide fixation. J. Biol. Chem., **203**, 595-604 (1953)

[2] Rutter, W.J.; Lardy, H.A.: Purification and properties of pigeon liver malic enzyme. J. Biol. Chem., **233**, 374-382 (1958)

[3] Hsu, R.Y.; Glynias, M.J.; Satterlee, J.; Feeney, R.; Clarke, A.R.; Emery, D.C.; Roe, B.A.; Wilson, R.K.; Goodridge, A.G.; Holbrook, J.J.: Duck liver 'malic' enzyme. Expression in Escherichia coli and characterization of the wild-type enzyme and site-directed mutants. Biochem. J., **284**, 869-876 (1992)

[4] Clifford, K.H.; Cornforth, J.W.; Donninger, C.; Mallaby, R.: Stereochemical course of decarboxylation of S-malate on malic enzyme. Eur. J. Biochem., **26**, 401-406 (1972)

[5] Bartholome, K.; Brdiczka, D.G.; Pette, D.: Purification and properties of extra- and intramitochondrial malate dehydrogenase (NADP; decarboxylating) from pig heart. Hoppe-Seyler's Z. Physiol. Chem., **353**, 1487-1495 (1972)

[6] Nolte, J.; Brdiczka, D.; Pette, D.: Intracellular distribution of phosphoenolpyruvate carboxylase and (NADP) malate dehydrogenase in different muscle types. Biochim. Biophys. Acta, **284**, 497-507 (1972)

[7] Coombs, J.; Baldry, C.W.; Bucke, C.: The C-4 pathway in Pennisetum purpureum. Planta, **110**, 109-120 (1973)

[8] Ziegler, I.: Malate dehydrogenase in Zea mays: properties and inhibition by sulfite. Biochim. Biophys. Acta, **364**, 28-37 (1974)

[9] Jechova, V.; Hostalek, Z.; Vanek, Z.: Regulation of biosynthesis of secondary metabolites. Folia Microbiol., **20**, 137-141 (1975)

[10] Wada, F.; Numata, M.; Eguchi, Y.; Sakamoto, Y.: Crystallization and properties of rat liver malate dehydrogenase (decarboxylating) (NADP). Biochim. Biophys. Acta, **410**, 237-242 (1975)

[11] Reynolds, C.H.; Hsu, R.Y.; Matthews, B.; Pry, T.A.; Dalziel, K.: Transient kinetic studies of malic enzyme. A conformational change associated with substrate inhibition by malate. Arch. Biochem. Biophys., **189**, 309-316 (1978)

[12] Drincovich, M.F.; Casati, P.; Andreo, C.S.: NADP-malic enzyme from plants: a ubiquitous enzyme involved in different metabolic pathways. FEBS Lett., **490**, 1-6 (2001)

[13] Dubery, I.A.; Schabort, J.C.: Purification and characterization of the NADP-linked malate dehydrogenase (decarboxylating) from Magnifera indica. Biochim. Biophys. Acta, **662**, 102-110 (1981)

[14] Bukato, G.; Kochan, Z.; Swierczynski, J.: Different regulatory properties of the cytosolic and mitochondrial forms of malic enzyme isolated from human brain. Int. J. Biochem. Cell Biol., **27**, 1003-1008 (1995)

[15] Bukato, G.; Kochan, Z.; Swierczynski, J.: Purification and properties of cytosolic and mitochondrial malic enzyme isolated from human brain. Int. J. Biochem. Cell Biol., **27**, 47-54 (1995)

[16] Vernon, C.M.; Hsu, R.Y.: Pigeon liver malic enzyme: involvement of an arginyl residue at the binding site for malate and its analogs. Arch. Biochem. Biophys., **225**, 296-305 (1983)

[17] Goodenough, P.W.; Prosser, I.M.; Young, K.: NADP-linked malic enzyme and malate metabolism in ageing tomato fruit. Phytochemistry, **24**, 1157-1162 (1985)

[18] Zelewski, M.; Swierczynski, J.: Malic enzyme in human liver. Intracellular distribution, purification and properties of cytosolic isozyme. Eur. J. Biochem., **201**, 339-345 (1991)

[19] Gourdon, P.; Baucher, M.-F.; Lindley, N.D.; Guyonvarch, A.: Cloning of the malic enzyme gene from Corynebacterium glutamicum and role of the enzyme in lactate metabolism. Appl. Environ. Microbiol., **66**, 2981-2987 (2000)

[20] Chang, G.G.; Huang, T.M.; Wang, J.K.; Lee, H.J.; Chou, W.Y.; Meng, C.L.: Kinetic mechanism of the cytosolic malic enzyme from human breast cancer cell line. Arch. Biochem. Biophys., **296**, 468-473 (1992)

[21] Chang, G.G.; Wang, J.K.; Huang, T.M.; Lee, H.J.; Chou, W.Y.; Meng, C.L.: Purification and characterization of the cytosolic NADP$^+$-dependent malic enzyme from human breast cancer cell line. Eur. J. Biochem., **202**, 681-688 (1991)

[22] Colombo, S.L.; Andreo, C.S.; Podesta, F.E.: Carbon metabolism in germinating Ricinus communis cotyledons. Purification, characterization and developmental profile of NADP-dependent malic enzyme. Physiol. Plant., **101**, 821-826 (1997)

[23] Fang, X.; Hillgartner, F.B.: Cell-specific regulation of transcription of the malic enzyme gene: characterization of cis-acting elements that modulate nuclear T3 receptor activity. Arch. Biochem. Biophys., **349**, 138-152 (1998)

[24] Voegele, R.T.; Mitsch, M.J.; Finan, T.M.: Characterization of two members of a novel malic enzyme class. Biochim. Biophys. Acta, **1432**, 275-285 (1999)

[25] Mounib, M.S.: NAD- and NADP-malic enzymes in spermatozoa of mammals and fish. FEBS Lett., **48**, 79-84 (1974)

[26] Lin, R.C.; Davis, E.J.: Malic enzymes of rabbit heart mitochondria. Separation and comparison of some characteristics of a nicotinamide adenine dinucleotide-preferring and a nicotinamide adenine dinucleotide phosphate-specific enzyme. J. Biol. Chem., **249**, 3867-3875 (1974)

[27] Eyzaguirre, J.; Cornwell, E..; Borie, G.; Ramirez, B.: Two malic enzymes in Pseudomonas aeruginosa. J. Bacteriol., **116**, 215-221 (1973)
[28] Rouleau, C.; Pellerin-Massicotte, J.; Pelletier, E.: Some properties of malic enzyme and glutamate dehydrogenase activities in the shrimp Pandalus borealis from the Saguenay Fjord (Quebec,Canada). Comp. Biochem. Physiol. B, **99B**, 903-909 (1991)

Isocitrate dehydrogenase (NAD^+) 1.1.1.41

1 Nomenclature

EC number
1.1.1.41

Systematic name
isocitrate:NAD^+ oxidoreductase (decarboxylating)

Recommended name
isocitrate dehydrogenase (NAD^+)

Synonyms
IDH1 [Swissprot]
NAD dependent isocitrate dehydrogenase
NAD isocitrate dehydrogenase
NAD isocitric dehydrogenase
NAD^+-specific ICDH [Swissprot]
NAD-linked isocitrate dehydrogenase
NAD-specific isocitrate dehydrogenase
β-ketoglutaric-isocitric carboxylase
isocitric acid dehydrogenase
isocitric dehydrogenase

CAS registry number
9001-58-5

2 Source Organism

<1> *Cavia porcellus* [1]
<2> *Bos taurus* [1, 4, 9, 10, 12, 14, 16, 17, 20]
<3> *pigeon* [1]
<4> *Homo sapiens* [1, 21]
<5> *Aspergillus niger* [1]
<6> *yeast* [1, 22-24]
<7> *Pisum sativum* (pea [1]) [1, 26]
<8> *Rattus norvegicus* [1, 19]
<9> *Hydrogenomonas eutropha* [2]
<10> *Sus scrofa* [3, 7, 11, 27]
<11> *Lepas antifera* [5]
<12> *waterbug* [5]
<13> *rhinoceros beetle* [5]

<14> *locust* [5]
<15> *Blattera americana* [5]
<16> *Sarcophaga barbata* [5]
<17> *Ovis aries* [5]
<18> *Candida tropicalis* [6]
<19> *Pseudomonas sp.* (W 6 [8]) [8]
<20> *Solanum tuberosum* [13]
<21> *Phormia regina* [15]
<22> *Rhodosporidium toruloides* [18]
<23> *Chlamydomonas reinhardtii* [25]
<24> *Phycomyces blakesleeanus* [28]
<25> *Escherichia coli* [29]

3 Reaction and Specificity

Catalyzed reaction
isocitrate + NAD^+ = 2-oxoglutarate + CO_2 + NADH

Reaction type
oxidation
oxidative decarboxylation
redox reaction
reduction
reductive carboxylation

Natural substrates and products
S isocitrate + NAD^+ <1-25> (Reversibility: ? <1-25> [1-29]) [1-29]
P 2-oxoglutarate + CO_2 + NADH

Substrates and products
S 8-(4-bromo-2,3-dioxobutylthio)-NAD + isocitrate <10> (Reversibility: ? <10> [27]) [27]
P 2-oxoglutarate + CO_2 + 8-(4-bromo-2,3-dioxobutylthio)-NADH
S isocitrate + NAD^+ <1-25> (Reversibility: ? <1-25> [1-29]) [1-29]
P 2-oxoglutarate + CO_2 + NADH

Inhibitors
2-hydroxyethylamine-N,N-diacetate <2> [17]
2-oxoglutarate <9, 22> [2, 18]
2-oxopropane-1,1,3-tricarboxylate <2> (<2> only in presence of ADP [17]) [17]
ADP <23> (<23> 10 mM [25]) [25]
ATP <2, 9, 19, 20, 22, 23> (<23> 10 mM [25]) [1, 2, 8, 10, 13, 16, 18, 25]
CN^- <2, 6, 7> [1]
Ca^{2+} <11-17, 21> [5, 15]
Cl^- <23> (<23> 100 mM [25]) [25]

D,L-hibiscusate <6> (<6> with constant concentrations of isocitrate, NAD$^+$ and Mg^{2+} up to 10 mM and without AMP, D,L-hibiscusate is an activator [22]) [22]
D,L-threo-α-methylisocitrate <6> (<6> inhibition in presence and absence of AMP, competitive inhibition [22]) [22]
D-malate <6> (<6> inhibition only in absence of AMP [22]) [22]
D-tartrate <6> (<6> inhibition only in absence of AMP [22]) [22]
EDTA <2> [17]
EGTA <2> [17]
I^- <23> (<23> 100 mM [25]) [25]
L-garciniate <6> (<6> inhibition in presence and absence of AMP [22]) [22]
L-glutamate <22> [18]
L-hibiscusate <6> (<6> with constant concentrations of isocitrate, NAD$^+$ and Mg^{2+} up to 10 mM and without AMP, L-hibiscusate is an activator [22]) [22]
L-malate <6> (<6> inhibition only in absence of AMP, competitive inhibition [22]) [22]
L-tartrate <6> (<6> inhibition only in absence of AMP [22]) [22]
Mg^{2+} <21> (<21> free form [15]) [15]
N_3^- <2, 6> [1]
NADH <2, 7, 9, 19, 20, 22-24> (<7, 23> competitive inhibitor [25, 26]; <24> inhibits ICDH-2 [28]) [2, 8, 13, 14, 16, 18, 25, 26, 28]
NADPH <2, 23> (<23> noncompetitive inhibitor [25]) [14, 16, 17, 25]
NO_3^- <23> (<23> 100 mM [25]) [25]
β-mercapto-α-ketoglutarate <6> (<6> inhibition in presence and absence of AMP, competitive inhibition [22]) [22]
bilirubin <2> [12]
chicago sky blue <2> [12]
citrate <6, 22, 23> (<6> with constant concentrations of isocitrate, NAD$^+$ and Mg^{2+} up to 10 mM and without AMP, citrate is an activator [22]; <23> 10 mM [25]) [18, 22, 25]
congo red <2> [12]
fluorescein <2> (<2> halogenated derivatives [12]) [12]
fluorocitrate <6> (<6> with constant concentrations of isocitrate, NAD$^+$ and Mg^{2+} up to 10 mM and without AMP, fluorocitrate is an activator [22]) [22]
fumarate <6> (<6> inhibition only in absence of AMP [22]) [22]
glyoxylate <9, 18> (<9> with oxaloacetate, concomitantly added [2]) [2, 6]
homocitrate <6> (<6> inhibition only in absence of AMP [22]) [22]
maleate <6> (<6> inhibition only in absence of AMP [22]) [22]
malonate <6> (<6> inhibition only in absence of AMP [22]) [22]
mitochondrial COX2 UTR RNA <6> [24]
molybdate <2, 6> [1]
nitrilotriacetate <2> [17]
oxaloacetate <22> [18]
p-hydroxymercuribenzoate <2> [1]
pyruvate <9> (<9> with oxaloacetate, concomitantly added, less effective than glyoxylate + oxaloacetate [2]) [2]
succinate <6> (<6> inhibition only in absence of AMP [22]) [22]

sulfobromophthalein <2> [12]
thiocyanate <2> [1]

Cofactors/prosthetic groups

NAD$^+$ <1-25> [1-29]
NADP$^+$ <25> [29]

Activating compounds

ADP <10> [27]
AMP <6, 24> (<24> activates ICDH-1 [28]) [1, 24, 28]

Metals, ions

Ca^{2+} <2> (<2> less effective than Mn^{2+} or Mg^{2+} [10]) [10, 14]
Co^{2+} <2, 10> (<2, 10> less effective than Mg^{2+} or Mn^{2+} [3, 10]) [3, 10]
Mg^{2+} <2, 5-7, 10, 21, 24> (<10, 21, 24> required, active substrate is Mg-DL-isocitrate chelate [3, 15, 28]; <21> free form inhibits [15]) [1, 3, 10, 15, 22, 23, 26-28]
Mn^{2+} <2, 6, 7, 10, 24> (<2, 6, 7, 10, 24> requires divalent cations [1, 3, 26, 28]) [1, 3, 26-28]
Zn^{2+} <10> (<10> less effective than Mn^{2+} or Mg^{2+} [3]) [3]

Specific activity (U/mg)

0.9 <9> [2]
7.3 <22> [18]
16.6 <23> [25]
27-31 <10> [7]
37.5 <24> (<24> ICDH-1 [28]) [28]
48.9 <24> (<24> ICDH-2 [28]) [28]
21300 <18> [6]

K_m-Value (mM)

0.017 <25> (NADP$^+$, <25> wild type [29]) [29]
0.03 <9> (D,L-isocitrate) [2]
0.06-0.09 <2> (NAD$^+$) [12]
0.099 <25> (NAD$^+$, <25> 7-fold mutant [29]) [29]
0.11 <6> (D,L-isocitrate) [1]
0.25 <19> (NAD$^+$) [8]
0.3 <9> (NAD$^+$) [2]
0.33 <7> (NAD$^+$, <7> membrane-associated ICDH [26]) [26]
0.45 <2> (D,L-isocitrate) [1]
4.7 <25> (NAD$^+$, <25> wild type [29]) [29]
5.8 <25> (NADP$^+$, <25> 7-fold mutant [29]) [29]

pH-Optimum

6.5 <2> [1]
7-7.2 <14> [5]
7.2-8.2 <22> [18]
7.5 <6, 7, 11-16> [1, 5, 26]
7.5-9 <24> [28]
8.1 <17> (<17> vertebrate muscle [5]) [5]

8.1-8.3 <18> [6]
8.3 <19> [8]
8.4 <2> [9]

pH-Range

5.5-8.6 <2> [4]
6-8 <10> [3]
6-8.5 <2> [9]
6.5-8.5 <6, 7> [1]

Temperature optimum (°C)

22 <2> [4]
25 <2, 11-17, 20> (<11-17> assay condition [5]) [5, 9, 10, 13]
30 <8, 18, 19> [6, 8, 19]

Temperature range (°C)

12-37 <2> [4]
30-60 <24> [28]

4 Enzyme Structure

Molecular weight

191800 <22> (<22> gel filtration [18]) [18]
224000 <10> (<10> ultracentrifugation [11]) [11]
245000 <10> (<10> sedimentation equilibrium analysis [11]) [11]
300000-340000 <10> (<10> gel filtration [7]) [7]
320000 <7, 23> (<7, 23> gel filtration [25, 26]) [25, 26]
338000 <24> (<24> gel filtration [28]) [28]

Subunits

dimer <10> (<10> $\alpha\beta$, 1 * 39000 + 1 * 41000, SDS-PAGE) [7]
octamer <6, 23, 24> (<6> $\alpha_4\beta_4$, IDH1 and IDH2, 4 * 38001 + 4 * 37755, SDS-PAGE [23]; <23> 8 * 45000, SDS-PAGE [25]; <24> α8, 8 * 40500, SDS PAGE, ICDH-1 and ICDH-2 [28]) [23, 25, 28]
tetramer <4, 10> (<10> $\alpha_2\beta\gamma$, 4 * 39000-41000, SDS-PAGE [11, 27]) [11, 21, 27]

Posttranslational modification

glycoprotein <10> [7]

5 Isolation/Preparation/Mutation/Application

Source/tissue

brain <2> [9, 10]
heart <1, 4, 2, 8, 10> [1, 3, 4, 7, 11, 12, 14, 17, 19, 21, 27]
muscle <3, 4, 11-17, 21> (<11-17, 21> flight muscle [5, 15]; <3> breast [1]; <4> skeletal muscle [21]) [1, 5, 15, 21]
placenta <4> [1]

Localization

mitochondrion <1-8, 18, 20, 22> [1, 6, 13, 14, 18, 19, 26]

Purification

<2> [4, 9, 10, 17]
<6> [24]
<7> [26]
<9> [2]
<10> [3, 7, 11]
<18> [6]
<19> [8]
<20> [13]
<21> [15]
<22> [18]
<23> [25]
<24> [28]

Crystallization

<25> (<25> crystallization in artificial mother liquor supplemented with 100 mM NAD^+ [29]) [29]

Cloning

<4> (expression in Escherichia coli [21]) [21]

Engineering

C201M/C332Y/K344D/Y345I/V351A/Y391K/R395S <25> (<25> converts the cofactor specificity from 7000-fold preference of $NADP^+$ to a 200-fold preference of NAD^+ [29]) [29]

6 Stability

pH-Stability

5.5-8.5 <2> [9]

Temperature stability

0 <22> (<22> full loss in activity in a few hours [18]) [18]
42 <23> (<23> 10 min, 50% loss in activity [25]) [25]
48 <9> [2]

General stability information

<24>, ICDH-2 is more stable to urea than ICDH-1 [28]
<11-17>, 20 h in medium containing ADP [5]

Storage stability

<7>, 4°C, glycerol 20%, 35-40% of original activity after 96 h, membrane-associated ICDH [26]
<2, 10>, -90°C, no loss of activity after 6 weeks [3, 4]

References

[1] Plaut, G.W.E.: Isocitrate dehydrogenase. The Enzymes, 2nd Ed. (Boyer, P.D., Lardy, H., Myrbäck, K., eds.), 7, 105-126 (1963)

[2] Glaeser, H.; Schlegel, H.G.: NADP- und NAD-spezifische dehydrogenase in Hydrogenomas eutropha stamm H 16. Arch. Microbiol., **86**, 327-337 (1972)

[3] Cohen, P.F.; Colman, R.F.: Role of manganous ion in the kinetics of pig-heart NAD-specific isocitrate dehydrogenase. Eur. J. Biochem., **47**, 35-45 (1974)

[4] Silinski, J.M.; Colman, R.F.: The kinetics of NAD-linked isocitrate dehydrogenase from calf heart. Biochim. Biophys. Acta, **370**, 1-25 (1974)

[5] Alp, P.R.; Newsholme, E.A.; Zammit, V.A.: Activities of citrate synthase and NAD$^+$-linked and NADP$^+$-linked isocitrate dehydrogenase in muscle from vertebrates and invertebrates. Biochem. J., **154**, 689-700 (1976)

[6] Nabeshima, S.; Ishiyama, S.; Tanaka, A.; Fukui, S.: Partial purification and some kinetic properties of NAD-linked and NADP-linked isocitrate dehydrogenase from Candida tropicalis. Agric. Biol. Chem., **41**, 509-516 (1977)

[7] Ramachandran, N.; Colman, R.F.: Evidence for the presence of two nonidentical subunits in NAD-dependent isocitrate dehydrogenase of pig heart. Proc. Natl. Acad. Sci. USA, **75**, 252-255 (1978)

[8] Hoffmann, K.H.; Babel, W.: Regulation of NAD$^+$- and NADP$^+$-linked isocitrate dehydrogenase in the obligate methylotrophic bacterium Pseudomonas W6. Z. Allg. Mikrobiol., **20**, 399-404 (1980)

[9] Willson, V.J.C.; Tripton, K.F.: The effect of pH on the allosteric behaviour of ox-brain NAD$^+$-dependent isocitrate dehydrogenase. Eur. J. Biochem., **109**, 411-416 (1980)

[10] Willson, V.J.C.; Tripton, K.F.: The activation of ox-brain NAD$^+$-dependent isocitrate dehydrogenase by magnesium ions. Eur. J. Biochem., **113**, 477-483 (1981)

[11] Ehrlich, R.S.; Hayman, S.; Ramachandran, N.; Colman, R.F.: Re-evaluation of molecular weight of pig heart NAD-specific isocitrate dehydrogenase. J. Biol. Chem., **256**, 10560-10564 (1981)

[12] Gabriel, J.L.; Plaut, G.W.E.: Specific site inhibitors of NAD-specific isocitrate dehydrogenase from bovine heart. J. Biol. Chem., **257**, 8021-8029 (1981)

[13] Tezuka, T.; Laties, G.G.: Isolation and characterization of inner membrane-associated and matrix NAD-specific isocitrate dehydrogenase in potato mitochondria. Plant Physiol., **72**, 959-963 (1982)

[14] Gabriel, J.L.; Plaut, G.W.E.: Inhibition of bovine heart NAD-specific isocitrate dehydrogenase by reduced pyridine nucleotides: modulation of inhibition by ADP, NAD$^+$, Ca^{2+}, citrate, and isocitrate. Biochemistry, **23**, 2773-2778 (1984)

[15] Bulos, B.A.; Thomas, B.J.; Sacktor, B.: Calcium inhibition of the NAD$^+$-linked isocitrate dehydrogenase from blowfly flight muscle mitochondria. J. Biol. Chem., **259**, 10232-10237 (1984)

[16] Gabriel, J.L.; Milner, R.; Plaut, G.W.E.: Inhibition and activation of bovine heart NAD-specific isocitrate dehydrogenase by ATP. Arch. Biochem. Biophys., **240**, 128-134 (1985)

[17] Gabriel, J.L.; Plaut, G.W.E.: NAD-specific isocitrate dehydrogenase from bovine heart. Interaction with Ca^{2+} chelators. Biochem. J., **229**, 817-822 (1985)

[18] Evans, C.T.; Ratledge, C.: The role of the mitochondrial NAD^+:isocitrate dehydrogenase in lipid accumulation by the oleaginous yeast Rhodosporidium toruloides CBS 14. Can. J. Microbiol., **31**, 845-850 (1985)

[19] Rutter, G.A.; Denton, R.M.: Regulation of NAD^+-linked isocitrate dehydrogenase and 2-oxoglutarate dehydrogenase by Ca^{2+} ions within toluene-permeabilized rat heart mitochondria. Interactions with regulation by adenine nucleotides and NADH/NAD^+ ratios [published erratum appears in Biochem J 1988 Aug 1;253(3):935]. Biochem. J., **252**, 181-189 (1988)

[20] Gabriel, J.L.; Plaut, G.W.E.: Citrate activation of NAD-specific isocitrate dehydrogenase from bovine heart. J. Biol. Chem., **259**, 1622-1628 (1984)

[21] Kim, Y.-O.; Koh, H.-J.; Kim, S.-H.; Jo, S.-H.; Huh, J.-W.; Jeong, K.-S.; Lee, I.J.; Song, B.J.; Huh, T.-L.: Identification and functional characterization of a novel tissue-specific NAD^+-dependent isocitrate dehydrogenase β subunit isoform. J. Biol. Chem., **274**, 36866-36875 (1999)

[22] Gabriel, J.L.; Plaut, G.W.E.: Kinetic regulation of yeast NAD-specific isocitrate dehydrogenase by citrate. Biochemistry, **30**, 2594-2599 (1991)

[23] Panisko, E.A.; McAllister-Henn, L.: Subunit interactions of yeast NAD^+-specific isocitrate dehydrogenase. J. Biol. Chem., **276**, 1204-1210 (2001)

[24] Anderson, S.L.; Minard, K.I.; McAllister-Henn, L.: Allosteric inhibition of NAD^+-specific isocitrate dehydrogenase by a mitochondrial mRNA. Biochemistry, **39**, 5623-5629 (2000)

[25] Martinez-Rivas, J.M.; Vega, J.M.: Purification and characterization of NAD-isocitrate dehydrogenase from Chlamydomonas reinhardtii. Plant Physiol., **118**, 249-255 (1998)

[26] McIntosh, C.A.: Partial purification and characteristics of membrane-associated NAD^+-dependent isocitrate dehydrogenase activity from etiolated pea mitochondria. Plant Sci., **129**, 9-20 (1997)

[27] Huang, Y.-C.; Kumar, A.; Colman, R.F.: Identification of the subunits and target peptides of pig heart NAD-specific isocitrate dehydrogenase modified by the affinity label 8-(4-bromo-2,3-dioxobutylthio)NAD. Arch. Biochem. Biophys., **348**, 207-218 (1997)

[28] Alvarez-Villafane, E.; Soler, J.; del Valle, P.; Busto, F.; de Arriaga, D.: Two NAD^+-dependent isocitrate dehydrogenase forms in Phycomyces blakesleeanus. Induction in response to acetate growth and characterization, kinetics, and regulation of both enzyme forms. Biochemistry, **35**, 4741-4752 (1996)

[29] Hurley, J.H.; Chen, R.; Dean, A.M.: Determinants of cofactor specificity in isocitrate dehydrogenase: structure of an engineered $NADP^+$ to NAD^+ specificity-reversal mutant. Biochemistry, **35**, 5670-5678 (1996)

Isocitrate dehydrogenase ($NADP^+$) 1.1.1.42

1 Nomenclature

EC number
1.1.1.42

Systematic name
isocitrate:$NADP^+$ oxidoreductase (decarboxylating)

Recommended name
isocitrate dehydrogenase ($NADP^+$)

Synonyms
CtIDP1 [Swissprot]
CtIDP2 [Swissprot]
IDH [Swissprot]
IDP [Swissprot]
NADP isocitric dehydrogenase
$NADP^+$-linked isocitrate dehydrogenase
$NADP^+$-specific ICDH [Swissprot]
NADP-dependent isocitrate dehydrogenase
NADP-dependent isocitric dehydrogenase
NADP-linked isocitrate dehydrogenase
NADP-specific isocitrate dehydrogenase
PS-NADP-IDH [Swissprot]
isocitrate dehydrogenase (NADP)
isocitrate dehydrogenase (NADP-dependent)
isocitrate dehydrogenase (nicotinamide adenine dinucleotide phosphate)
oxalosuccinate decarboxylase
oxalsuccinic decarboxylase

CAS registry number
9028-48-2

2 Source Organism

<1> *Aspergillus niger* [1, 16]
<2> *Trypanosoma cruzi* [1]
<3> *Escherichia freundii* [1]
<4> *Mycobacterium tuberculosis* [1]
<5> *Saccharomyces cerevisiae* [1]
<6> *Sus scrofa* (pig) [1, 5, 20]
<7> *Thermus aquaticus* [2]

<8> *Bacillus subtilis* [2]
<9> *Chlamydomonas reinhardtii* [2, 29]
<10> *Bacillus stearothermophilus* [2, 5]
<11> *Thermus flavus* [3]
<12> *Acinetobacter calcoaceticus* [4]
<13> *Zea mays* [5]
<14> *Azotobacter sp.* [5]
<15> *Rhodopseudomonas sphaeroides* (i.e. Rhodobacter sphaeroides) [5]
<16> *Bombyx mori* [7]
<17> *Homo sapiens* [9]
<18> *Escherichia coli* [2, 9, 13, 24, 27]
<19> *Vibrio sp.* (strain ABE-1) [10, 14]
<20> *Bos taurus* [6, 8, 11]
<21> *Mytilus edulis* [12]
<22> *Caldococcus noboribetus* [21]
<23> *Halobacterium cutirubrum* [14]
<24> *Fundulus heteroclitus* [15]
<25> *Phycomyces blakesleeanus* [16]
<26> *Blastocladiella emersonii* [16]
<27> *Pisum sativum* [17]
<28> *Mycobacterium phlei* [18]
<29> *Brevibacterium flavum* [18]
<30> *Thiobacillus novellus* [18]
<31> *Crithidia fasciculata* [18]
<32> *Bacillus sp.* [19]
<33> *Azotobacter vinelandii* [2]
<34> *Phormidium laminosum* [22]
<35> *Nicotiana tabacum* [23, 33]
<36> *Rhodomicrobium vannielii* [25]
<37> *Synechocystis sp.* (<37> PCC 6803 [26]) [26]
<38> *Pinus spp.* (scots pine) [28]
<39> *Desulfobacter vibrioformis* [30]
<40> *Archaeoglobus fulgidus* [31]
<41> *Thermoleophilum minutum* (YS-4) [32]
<42> *Corynebacterium glutamicum* [36]
<43> *Vibrio parahaemolyticus* (Y-4) [34]
<44> *Rattus norvegicus* [35]

3 Reaction and Specificity

Catalyzed reaction

isocitrate + $NADP^+$ = 2-oxoglutarate + CO_2 + NADPH

Reaction type

oxidative decarboxylation
redox reaction
reductive carboxylation

Natural substrates and products

- **S** isocitrate + $NADP^+$ <1-44> (Reversibility: r <6> [1]; ? <1-5, 7-44> [1-36]) [1-36]
- **P** 2-oxoglutarate + NADPH + CO_2

Substrates and products

- **S** 3-fluoroisocitrate + $NADP^+$ <6> (Reversibility: ? <6> [20]) [20]
- **P** 3-fluoro-2-oxoglutarate + NADPH + CO_2
- **S** 3-hydroxyisocitrate + $NADP^+$ <6> (Reversibility: ? <6> [20]) [20]
- **P** 3-hydroxy-2-oxoglutarate + NADPH + CO_2
- **S** isocitrate + $NADP^+$ <1-44> (Reversibility: r <6> [1]; ? <1-5, 7-44> [1-36]) [1-36]
- **P** 2-oxoglutarate + NADPH + CO_2
- **S** isopropylmalate + $NADP^+$ <18> (Reversibility: ? <18> [24]) [24]
- **P** ?

Inhibitors

2',5'-ADP <17> [9]
2,3-butanedione <9> [29]
2-mercaptoethanol <28> [18]
2-oxoglutarate <1, 12, 16, 17, 34, 37> (<1, 12, 16, 17, 34> substrate inhibition [4, 7, 9, 16, 22]; <37> non-competitive inhibitor [26]; <38> inhibition by chelating Mg^{2+} [28]) [4, 7, 9, 16, 22, 26, 28]
3-phosphoglycerate <32> [19]
5'-ADP <17, 34, 37, 38> (<37> non-competitive inhibitor [26]; <38> inhibition by chelating Mg^{2+} [28]) [9, 22, 26, 28]
ATP <1, 12, 17, 32, 34, 38> (<38> inhibition by chelating Mg^{2+} [28]) [4, 9, 19, 16, 22, 28]
Ag^+ <6, 16> (<6, 16> decarboxylation of isocitrate and oxalosuccinate [1, 7]) [1, 7]
Cu^{2+} <6, 32> [1, 19]
Hg^{2+} <28, 32> [18, 19]
N-ethylmaleimide <32> [19]
NADPH <1, 32, 34, 37> (<37> competitive inhibitor [26]) [16, 19, 22, 26]
Zn^{2+} <16> [7]
adenosine-3',5'-cyclic monophosphate <16> [7]
cis-aconitate <32> [19]
citrate <1, 16-17> [7, 9, 16]
diphenylchloroarsine <6> [1]
dithiothreitol <28> [18]
glutathione <28> [18]
glyceraldehyde-3-phosphate <32> [19]
glyoxylate <1, 7-9, 11, 12, 19, 20, 28> (<1, 7-9, 11, 12, 19, 20, 28> plus oxaloacetate [2, 3, 4, 10, 11, 16, 18]) [2, 3, 4, 10, 11, 16, 18]
isocitrate <6> (<6> competitive inhibition of decarboxylation of oxalosuccinate [1]) [1]
maleate <16> [7]
monoiodoacetate <16, 28> [7, 18]

nicotinamide mononucleotide <17> [9]
oxaloacetate <12, 16, 32> [4, 7, 19]
oxalomalate <9, 28> [2, 18]
oxalylglycine <6> [20]
p-chloromercuribenzoate <6, 16, 20, 28, 32, 41> [1, 7, 11, 18, 19, 32]
p-hydroxymercuribenzoate <6> (<6> decarboxylation of isocitrate and oxalosuccinate [1]) [1]
phenarsazines <6> [1]
phenylglyoxal <9> [29]
phenylmercuric nitrate <6> [1]
phosphoenolpyruvate <32> [19]
propanetricarboxylate <17> [9]
pyridoxal 5'-phosphate <9> [29]
threo-L-isocitrate <17> [9]
urea <12> (<12> low molecular weight form [4]) [4]

Cofactors/prosthetic groups
NAD$^+$ <36, 40> [25, 31]
NADP$^+$ <1-6, 9, 11, 12, 19, 20, 23, 25-32, 34-44> (<1-6> natural coenzyme [1]) [1, 3, 4, 10, 11, 14, 16-23, 25, 26, 28-36]
NADPH <1-6, 20> [1, 6, 8]

Metals, ions
Cd^{2+} <6, 20, 32> (<6, 20, 32> absolute requirement for divalent cations [11, 19, 20]) [11, 19, 20]
Co^{2+} <16, 20, 32> (<16, 20, 32> absolute requirement for divalent cations [7, 11, 19]) [7, 11, 19]
Mg^{2+} <1, 6, 12, 16, 19, 20, 22, 24, 27-32, 37-41, 44> (<1, 6, 12, 16, 19, 20, 22, 24, 27-32, 37-41, 44> absolute requirement for divalent cations [1, 4, 7, 10-12, 14-21, 26, 28, 30-33, 35]) [1, 4, 7, 10-12, 14-21, 26, 28, 30-33, 35]
Mn^{2+} <1, 6, 9, 12, 16, 19, 20, 22, 24, 27-32, 34, 37, 39-41> (<1, 6, 12, 16, 19, 20, 22, 24, 27-32, 34, 37, 39-41> absolute requirement for divalent cations [1, 4, 7,10-12, 14-22, 26, 30-33]) [1, 4, 7, 10-12, 14-22, 26, 29, 30-33]
Ni^{2+} <6> (<6> absolute requirement for divalent cations [20]) [20]
Zn^{2+} <6, 20, 28, 32> (<6, 20, 28, 32> absolute requirement for divalent cations [11, 18-20]) [11, 18-20]

Turnover number (min^{-1})
3500 <6> (isocitrate, <6> oxidative decarboxylation [1]) [1]

Specific activity (U/mg)
14.4 <1> [16]
15.7 <37> [26]
18.5 <36> (NAD$^+$-linked reaction [25]) [25]
18.6 <35> (<35> ICDH$_2$ [33]) [33]
23 <36> (NADP$^+$-linked reaction [25]) [25]
33.5 <13> [5]
50 <35> (<35> ICDH1 [33]) [33]
109 <28> [18]

141 <40> [31]
144 <22> [21]
156.6 <43> [34]
202.2 <39> [30]
500 <34> [22]
3884 <41> [32]

K_m-Value (mM)

0.0017 <13> (D,L-isocitrate, <13> isozyme 2 [5]) [5]
0.0022 <17> (D,L-isocitrate) [9]
0.0025 <36> (NADP$^+$) [25]
0.0026 <6> (D,L-isocitrate) [1]
0.003 <13> (NADP$^+$, <13> isozyme 2 [5]) [5]
0.0031 <13> (D,L-isocitrate, <13> isozyme 1 [5]) [5]
0.004 <37> (Mn^{2+}, <37> Mn^{2+}-dependent reaction [26]) [26]
0.0057 <37> (D,L-isocitrate, <37> Mn^{2+}-dependent reaction [26]) [26]
0.006 <37> (NADP$^+$, <37> Mn^{2+}-dependent raction [26]) [26]
0.0063 <13> (NADP$^+$, <13> isozyme 1 [5]) [5]
0.007 <35> (NADP$^+$, <35> Mg^{2+}-dependent reaction, ICDH2 [33]) [33]
0.0073 <32> (NADP$^+$) [19]
0.0075 <35> (NADP$^+$, <35> Mg^{2+}-dependent reaction, ICDH1 [33]) [33]
0.0091 <8> (D,L-isocitrate) [2]
0.0092 <6> (NADPH) [1]
0.0097 <7> (D,L-isocitrate, <7> 25°C assay temperature [2]) [2]
0.01 <9, 27, 38> (NADP$^+$) [2, 17, 28]
0.0102 <39> (NADP$^+$, <39> in presence of Mg^{2+} [30]) [30]
0.011 <20> (NADP$^+$) [11]
0.012 <35> (Mg^{2+}, <35> Mg^{2+}-dependent reaction, ICDH2 [33]) [33]
0.012 <37> (NADP$^+$, <37> Mg^{2+}-dependent reaction [26]) [26]
0.0125 <9> (D,L-isocitrate) [2]
0.013 <35> (Mg^{2+}, <35> Mg^{2+}-dependent reaction, ICDH1 [33]) [33]
0.0165 <7> (NADP$^+$, <7> 25°C assay temperature) [2]
0.0181 <7> (D,L-isocitrate, <7> 70°C assay temperature [2]) [2]
0.0204 <39> (D,L-isocitrate, <39> in presence of Mg^{2+} [30]) [30]
0.022 <34> (D,L-isocitrate, <34> Mn^{2+}-dependent reaction [22]) [22]
0.024 <34> (Mn^{2+}, <34> Mn^{2+}-dependent reaction [22]) [22]
0.0274 <8> (NADP$^+$) [2]
0.028 <35> (D-isocitrate, <35> Mg^{2+}-dependent reaction, ICDH2 [33]) [33]
0.03 <40> (NADP$^+$) [31]
0.031 <7> (NADP$^+$, <7> 70°C assay temperature) [2]
0.032 <34> (NADP$^+$, <34> Mg^{2+}-dependent reaction [22]) [22]
0.033 <41> (D,L-isocitrate) [32]
0.041 <35> (D-isocitrate, <35> Mg^{2+}-dependent reaction, ICDH1 [33]) [33]
0.043 <22> (NADP$^+$) [21]
0.048 <41> (NADP$^+$) [32]
0.05 <34> (NADP$^+$, <34> Mn^{2+}-dependent reaction [22]) [22]
0.053 <28> (NADP$^+$) [18]

0.059 <37> (D,L-isocitrate, <37> Mg^{2+}-dependent reaction [26]) [26]
0.07 <34> (D,L-isocitrate, <34> Mg^{2+}-dependent reaction [22]) [22]
0.072 <22> (D,L-isocitrate) [21]
0.074 <28> (D,L-isocitrate) [18]
0.08 <38> (D,L-isocitrate) [28]
0.118 <40> (D,L-isocitrate) [31]
0.125 <37> (Mg^{2+}, <37> Mg^{2+}-dependent reaction [26]) [26]
0.13 <6> (2-oxoglutarate) [1]
0.159 <34> (Mg^{2+}, <34> Mg^{2+}-dependent reaction [22]) [22]
0.56 <6> (oxalosuccinate, <6> reductase [1]) [1]
0.77 <36> (NAD^+) [25]
1.2 <6> (oxalosuccinate, <6> decarboxylase [1]) [1]
5.4 <40> (NAD^+) [31]
25-26 <6> (oxalosuccinate, <6> decarboxylase [1]) [1]

pH-Optimum

6-8 <1> [16]
6.6 <16> (<16> 2-oxoglutarate + CO_2 + NADPH [7]) [7]
7 <22> [21]
7.2 <41> [32]
7.2-8.2 <20> [11]
7.4 <13> [5]
7.5 <12> (<12> high molecular weight form [4]) [4]
7.5-9.5 <27> [17]
7.6-8 <19> (<19> isoenzyme I [10]) [10]
7.8 <16> (<16> isocitrate + $NADP^+$ [7]) [7]
7.9-8.4 <32> [19]
8 <12> (<12> low molecular weight form [4]) [4]
8-8.6 <19> (<19> isoenzyme II [10]) [10]
8.5 <28, 34, 39> [18, 22, 30]
8.6 <40> [31]
9 <37> [26]

pH-Range

5.6-8.5 <6> [1]
6-9.5 <20> [11]
6.3-7.8 <1> [16]
6.5-10 <27, 28> [17, 18]
7.5-10 <37> [26]

Temperature optimum (°C)

20 <19> (<19> isoenzyme II [10, 14]) [10, 14]
23 <16> [7]
25 <7-10> (<7-10> assay condition [2]) [2]
25-30 <28> [18]
28 <1> [16]
37 <8> [2]
40 <19> (<19> isoenzyme I [10, 14]) [10, 14]

45 <39> [30]
55 <34> [22]
70 <7> [2]
75 <32, 41> [19, 32]
90 <40> [31]

Temperature range (°C)

25-65 <34> [22]
30-80 <11> [3]

4 Enzyme Structure

Molecular weight

45000 <21> (<21> PAGE [12]) [12]
53000 <17> (<17> gel filtration [9]) [9]
55000 <20> (<20> PAGE [11]) [11]
60000 <1, 41> (<1> PAGE [16]; <41> PAGE and gel filtration [32]) [16, 32]
60000-70000 <7> (<7> gel filtration [2]) [2]
61000-64000 <6> (ultracentrifugation [1]) [1]
68000 <6> (<6> gel filtration [5]) [5]
70000 <9> (<9> gel filtration [29]) [29]
75000-80000 <8, 36> (<8> gel filtration [2]; <36> PAGE [25]) [2, 25]
80000 <18, 33, 40, 43> (<18, 33, 40, 43> gel filtration [2, 31, 34]) [2, 31, 34]
81000-83000 <13> (<13> gel filtration [5]) [5]
85000 <19> (<19> gel filtration, isoenzyme I: dimer [10, 14]) [10, 14]
86000 <16> (<16> gel filtration [7]) [7]
87000-92600 <10> (<10> gel filtration [2]) [2]
90000 <12, 20, 22, 24, 27, 32> (<12> gel filtration, low molecular weight form [4]; <20> gel filtration, ultracentrifugation, PAGE [6, 8]; <22> gel filtration [21]; <24> gel filtration, 3 isoenzymes [15]; <27> gel filtration [17]; <32> gel filtration [19]) [4, 6, 8, 15, 17, 19, 21]
90000-100000 <9> (<9> gel filtration [2]) [2]
95000 <38> (<38> gel filtration [28]) [28]
105000 <15> (<15> gel filtration [5]) [5]
107000 <34> (<34> gel filtration [22]) [22]
108000 <37> (<37> gel filtration [26]) [26]
117000 <35> (<35> gel filtration, ICDH1 [33]) [33]
136000 <35> (<35> gel filtration, $ICDH_2$ [33]) [33]
360000 <12> (<12> gel filtration, high molecular weight form [4]) [4]

Subunits

dimer <9, 19, 20, 22, 34, 37, 38, 40, 41> (<9> α_2, 2 * 40000, SDS-PAGE [29]; <19> isoenzyme I [10, 14]; <20> α_2, 2 * 45000, SDS-PAGE [8]; <22> α_2, 2 * 48000, SDS-PAGE [21]; <34> α_2, 2 * 53000, SDS-PAGE [22]; <37> α_2, 2 * 57000, SDS-PAGE [26]; <38> α_2, 2 * 46000, SDS-PAGE [28]; <40> α_2, 2 * 42000, SDS-PAGE [31]; <41> α_2, 2 * 30500, SDS-PAGE [32]) [8, 10, 14, 21, 22, 26, 28, 29-32] monomer <19, 36, 39, 43, 42> (<19> isoenzyme II, 1 * 85000, SDS-PAGE [10, 14]; <36,39,43> 1 * 80000, SDS-PAGE [25, 30, 34]) [10, 14, 25, 30, 34, 36]

5 Isolation/Preparation/Mutation/Application

Source/tissue

angiosperm <38> [28]
brain <24> [15]
cotyledon <38> [28]
heart <6, 17, 20> [1, 5, 6, 8, 9, 20]
intestine <24> [15]
leaf <27, 35> [17, 33]
liver <20, 24> [8, 15]
mammary gland <20> [11]
ovary <44> [35]
pupa <16> [7]
seed <13> [5]

Localization

chloroplast <35> (<35> ICDH2 [33]) [33]
cytoplasm <1, 21, 24-26> [12, 15, 16]
cytosol <9, 20, 27, 35, 38, 44> (<35> ICDH1 [33]) [6, 8, 17, 23, 28, 29, 33, 35]
mitochondrion <1, 6, 10, 13-15, 20, 24-26> [5, 6, 8, 15, 16]

Purification

<1> [16]
<7> [2]
<8> [2]
<9> [2, 29]
<11> [3]
<13> [5]
<16> [7]
<17> [9]
<18> [13]
<19> (strain ABE-1) [10, 14]
<20> [6, 8, 11]
<21> [12]
<22> [21]
<24> [15]
<27> [17]
<28> [18]
<32> [19]
<34> [22]
<35> [23, 33]
<36> [25]
<37> [26]
<38> [28]

<39> [30]
<40> [31]
<41> [32]
<43> [34]
<44> [35]

Renaturation

<6, 10, 13-15, 19, 23> (<6, 10, 13-15> overnight dialysis against guanidine-HCl, GDP buffer [5]; <19> with 2-mercaptoethanol after inactivation at 40°C, isoenzyme II, 40% [14]) [5, 14]

Crystallization

<18> (<18> S113E [24]) [24]
<20> [6, 8]
<42> (<42> crystallization with polyethylene glycol 2000 monomethyl ether [36]) [36]

Cloning

<22> (expression in Escherichia coli [21]) [21]
<35> (expression in Escherichia coli [23]) [23]

Engineering

C201M <18> (<18> higher affinity for NAD^+ than for $NADP^+$ [27]) [27]
C201N <18> (<18> higher affinity for NAD^+ than for $NADP^+$ [27]) [27]
C201V <18> (<18> higher affinity for NAD^+ than for $NADP^+$ [27]) [27]
S113E <18> (<18> affinity for isopropylmalate is 37-fold compared to wild-type [24]) [24]

6 Stability

pH-Stability

6 <20> [11]
6-9 <7> [2]
8-10 <32> [19]

Temperature stability

15 <19> (<19> isoenzyme II [14]) [14]
20 <19> (<19> isoenzyme II [10]) [10]
30 <9> (<9> 50% loss of activity in 1 h [2]) [2]
35 <24> (<24> isoenzyme C, T_{50}-value [15]) [15]
40 <22> (<22> nearly no activity at 40°C or lower [21]) [21]
45 <9, 16, 32> (<9> after ten min completely inactivated in absence of substrate [29]) [7, 19, 29]
49 <24> (<24> isoenzyme A, T_{50}-value [15]) [15]
50 <34> (<34> fully stable [22]) [22]
52 <24> (<24> isoenzyme C, T_{50}-value, presence of isocitrate [15]) [15]
54 <8, 24> (<8> 50% loss of activity in 1 h [2]; <24> isoenzyme B, T_{50}-value [15]) [2, 15]

58 <24> (<24> isoenzyme A, T_{50}-value, presence of isocitrate [15]) [15]
69 <24> (<24> isoenzyme B, T_{50}-value, presence of isocitrate [15]) [15]
70 <11, 41> (<41> after 1 h incubation retains 57% of initial activity [32]) [3, 32]
80 <22> (<22> no loss in activity [21]) [21]
88 <7> (<7> 50% loss of activity in 1 h [2]) [2]
90 <40> (<40> half life of 22 min in 50 mM tricine-KOH buffer, pH 8 [31]) [31]

General stability information
<12>, low molecular weight form of Acinetobacter calcoaceticus enzyme more heat stable than high molecular weight form [4]
<20>, stabilization by dithiothreitol [11]
<6, 13>, instable in solutions of low ionic strength, stabilization by ammonium sulfate [1, 5]

Storage stability
<16>, -20°C, 6 months, 10% loss of activity [7]
<20>, 2°C, several weeks [6]
<40>, 4°C, several months, under oxic conditions [31]

References

[1] Plaut, G.W.E.: Isocitrate dehydrogenase. The Enzymes, 2nd Ed. (Boyer, P.D., Lardy, H., Myrbäck, K., eds.), 7, 105-126 (1963)
[2] Ramaley, R.F.; Hudock, M.O.: Purification and properties of isocitrate dehydrogenase (NADP) from Thermus aquaticus YT-1, Bacillus subtilis-168 and Chlamydomonas reinhardti-Y-2. Biochim. Biophys. Acta, **315**, 22-36 (1973)
[3] Saiki, T.; Arima, K.: Studies on NADP$^+$-specific isocitrate dehydrogenase from an extreme thermophile, Thermus flavus AT-62. J. Biochem., **77**, 233-240 (1975)
[4] Kleber, H.-P.: Eigenschaften der NADP$^+$-abhängigen Isocitratdehydrogenasen aus Acinetobacter calcoaceticus. Z. Allg. Mikrobiol., **15**, 431-435 (1975)
[5] Curry, R.A.; Ting, I.P.: Purification, properties, and kinetic observations on the isoenzymes of NADP isocitrate dehydrogenase of maize. Arch. Biochem. Biophys., **176**, 501-509 (1976)
[6] Macfarlane, N.; Mathews, B.; Dalziel, K.: The purification and properties of NADP-dependent isocitrate dehydrogenase from ox-heart mitochondria. Eur. J. Biochem., **74**, 553-559 (1977)
[7] Miake, F.; Torikata, T.; Koga, K.; Hayashi, K.: Isolation and characterization of NADP$^+$-specific isocitrate dehydrogenase from the pupa of Bombyx mori. J. Biochem., **82**, 449-454 (1977)
[8] Macfarlane, N.; Reynolds, H.; Matthews, B.; Dalziel, K.: Purification and molecular properties of NADP-linked isocitrate dehydrogenase from ox heart mitochondria. Biochem. Soc. Trans., **5**, 790-793 (1977)

[9] Seelig, G.F.; Colman, R.F.: Characterization of the physicochemical and catalytic properties of human heart NADP-dependent isocitrate dehydrogenase. Arch. Biochem. Biophys., **188**, 394-409 (1978)
[10] Ochiai, T.; Fukunaga, N.; Sasaki, S.: Purification and some properties of two $NADP^+$-specific isocitrate dehydrogenases from an obligately psychrophilic marine bacterium, Vibrio sp., strain ABE-1. J. Biochem., **86**, 377-384 (1979)
[11] Farrell, H.M.: Purification and properties of $NADP^+$:isocitrate dehydrogenase from lactating bovine mammary gland. Arch. Biochem. Biophys., **204**, 551-559 (1980)
[12] Head, E.J.H.: NADP-dependent isocitrate dehydrogenase from the mussel Mytilus edulis L. 1. Purification and characterisation. Eur. J. Biochem., **111**, 575-579 (1980)
[13] Vasquez, B.; Reeves, H.C.: NADP-specific isocitrate dehydrogenase from Escherichia coli. V. Multiple forms of the enzyme. Biochim. Biophys. Acta, **660**, 16-22 (1981)
[14] Ochiai, T.; Fukunaga, N.; Sasaki, S.: Two structurally different NADP-specific isocitrate dehydrogenases in an obligately psychrophilic bacterium, Vibrio sp. strain ABE-1. J. Gen. Appl. Microbiol., **30**, 479-487 (1984)
[15] Gonzalez-Villasenor, L.I.; Powers, D.A.: A multilocus system for studying tissue and subcellular specialization. The three NADP-dependent isocitrate dehydrogenase isozymes of the fish Fundulus heteroclitus. J. Biol. Chem., **260**, 9106-9113 (1985)
[16] Meixner-Monori, B.; Kubicek, C.P.; Harrer, W.; Schreferl, G.; Rohe, M.: NADP-specific isocitrate dehydrogenase from the citric acid-accumulating fungus Aspergillus niger. Biochem. J., **236**, 549-557 (1986)
[17] Ni, W.; Robertson, E.F.; Reeves, H.C.: Purification and characterization of cytosolic NADP specific isocitrate dehydrogenase from Pisum sativum. Plant Physiol., **83**, 785-788 (1987)
[18] Dhariwal, K.R.; Venkitasubramanian, T.A.: NADP-specific isocitrate dehydrogenase of Mycobacterium phlei ATCC 354: purification and characterization. J. Gen. Microbiol., **133**, 2457-2460 (1987)
[19] Shikata, S.; Ozaki, K.; Kawai, S.; Ito, S.; Okamoto, K.: Purification and characterization of $NADP^+$-linked isocitrate dehydrogenase from an alkalophilic Bacillus. Biochim. Biophys. Acta, **952**, 282-289 (1988)
[20] Grissom, C.B.; Cleland, W.W.: Isotope effect studies of the chemical mechanism of pig heart NADP isocitrate dehydrogenase. Biochemistry, **27**, 2934-2943 (1988)
[21] Aoshima, M.; Oshima, T.: Purification and characterization of isocitrate dehydrogenase from a hyperthermophilic archaebacterium, Caldococcus noboribetus. Biochim. Biophys. Acta, **1340**, 227-234 (1997)
[22] Pardo, M.A.; Llama, M.J.; Serra, J.L.: Purification, properties and enhanced expression under nitrogen starvation of the $NADP^+$-isocitrate dehydrogenase from the cyanobacterium Phormidium laminosum. Biochim. Biophys. Acta, **1431**, 87-96 (1999)
[23] Gálvez, S.; Hodges, M.; Bismuth, E.; Samson, I.; Teller, S.; Gadal, P.: Purification and characterization of a fully active recombinant cytosolic NADP-dependentent isocitrate dehydrogenase in Escherichia coli: evidence for a

role for the N-terminal region in enzyme activity. Arch. Biochem. Biophys., **323**, 164-168 (1995)
[24] Doyle, S.A.; Beernink, B.T.; Koshland, Jr., D.E.: Structural basis for a change in substrate specificity: crystal structur of S113E isocitrate dehydrogenase in a complex with isopropylmalate, Mg^{2+}, and NAPD. Biochemistry, **40**, 4234-4241 (2001)
[25] Leyland, M.L.; Kelly, D.J.: Purification and characterization of a monomeric isocitrate dehydrogenase with dual coenzyme specificity from the photosynthetic bacterium Rhodomicrobium vannielii. Eur. J. Biochem., **202**, 85-93 (1991)
[26] Muro-Pastor, M.I.; Florencio, F.J.: Purification and properties of NADP-isocitrate dehydrogenase from the unicellular cyanobacterium Synechocystis sp. PCC 6803. Eur. J. Biochem., **203**, 99-105 (1992)
[27] Chen, R.; Greer, A.F.; Dean, A.M.: Structural constrains in protein engineering. The coenzyme specificity of Escherichia coli isocitrate dehydrogenase. Eur. J. Biochem., **250**, 578-582 (1997)
[28] Palomo, J.; Gallardo, F.; Suárez, M.F.; Cánovas, F.M.: Purification and characterization of $NADP^+$-linked isocitrate dehydrogenase from scots pine. Plant Physiol., **118**, 617-626 (1998)
[29] Martinez-Rivas, J.M.; Garcia-Diaz, M.T.; Vega, J.M.: Purification and properties of cytosolic NADP-isocitrate dehydrogenase from Chlamydomonas reinhardtii. Plant Physiol. Biochem., **34**, 673-676 (1996)
[30] Steen, I.H.; Steine Madsen, M.; Birkeland, N-K.; Lien. T.: Purification and characterization of a monomeric isocitrate dehydrogenase from Desulfobacter vibrioformis and demonstration of the presence of a monomeric enzyme in other bacteria. FEMS Microbiol. Lett., **160**, 75-79 (1998)
[31] Steen, I.H.; Lien. T.; Birkeland, N-K.: Biochemical and phylogenetic characterization of isocitrate dehydrogenase from a hyperthermophilic archaeon, Archaeoglobus fulgidus. Arch. MIcrobiol., **168**, 412-420 (1997)
[32] Novotny, Jr., J.F.; Perry, J.J.: Characterization of a heat-stable NADP-dependent isocitrate dehydrogenase from the obligate thermophile Thermoleophilum minutum YS-4. Appl. Microbiol. Biotechnol., **35**, 461-465 (1991)
[33] Gálvez, S.; Bismuth, E.; Sarda, C.; Gadal, P.: Purification and characterization of chloroplastic NADP-isocitrate dehydrogenase from mixotrophic tobacco cells. Plant Physiol., **105**, 593-600 (1994)
[34] Fukunaga, N.; Imagawa, S.; Sahara, T.; Ishii, A.; Suzuki, M.: Purification and characterization of monomeric isocitrate dehydrogenase with $NADP^+$-specificity from Vibrio parhaemolyticus Y-4. J. Biochem., **112**, 849-855 (1992)
[35] Sechi, S.; Parmalee, D.; Roller, P.P.; Jennings, G.T.: Structural characterization of cytosolic NADP(+)-dependent isocitrate dehydrogenase from rat ovary. Enzyme Protein, **48**, 27-36 (1994)
[36] Audette, G.F.; Quail, J.W.; Hayakawa, K.; Bai, C.; Chen, R.; Delbaere, L.T.: Crystallization and preliminary X-ray diffraction studies of monomeric isocitrate dehydrogenase from Corynebacterium glutamicum. Acta Crystallogr.Sect.D, **55**, 1584-1585 (1999)

Phosphogluconate 2-dehydrogenase 1.1.1.43

1 Nomenclature

EC number
1.1.1.43

Systematic name
6-phospho-D-gluconate:NAD(P)$^+$ 2-oxidoreductase

Recommended name
phosphogluconate 2-dehydrogenase

Synonyms
2-keto-6-phosphogluconate reductase
6-PGADHase <5> [2]
6-phosphogluconate dehydrogenase (NAD)
6-phosphogluconic dehydrogenase
6PGAD <11> [4]
gluconate 6-phosphate dehydrogenase
phosphogluconate dehydrogenase

CAS registry number
9001-82-5

2 Source Organism

<1> *Pseudomonas fluorescens* (strain A 3.12 [1]) [1]
<2> *Leuconostoc mesenteroides* [1]
<3> *Aerobacter cloacae* [1]
<4> *Saccharomyces cerevisiae* (yeast [1]) [1]
<5> *Pseudomonas multivorans* (ATCC 17616, synonymous with Pseudomonas cepacia [2, 4]; type I [6]) [2, 4, 6]
<6> *Streptococcus faecalis* (type I [6]) [2, 4, 6]
<7> *Ovis aries* (sheep [3]) [3]
<8> *Rattus norvegicus* [3, 8]
<9> *Candida utilis* [3]
<10> *Neurospora crassa* [4]
<11> *Pseudomonas aeruginosa* [4]
<12> *Candida tropicalis* (yeast, strain pK 233 [5]) [5]
<13> *Pseudomonas sp.* (C [6]; W5 [6]) [6]
<14> *Candida boidinii* (methanol utilizing yeast [6, 9]) [6, 9]
<15> *Ankistrodesmus braunii* (C.C.A.P 202/7C [7]) [7]

<16> *Chlorella fusca* [7]
<17> *Chlamydomonas reinhardtii* (wild-type strain 2192, mutant strains nit A and nit B [7]) [7]
<18> *Zea mays* (maize, L. Pioneer HiBred International line B73 [9]) [9]
<19> *Ricinus communis* (castor bean [9]) [9]
<20> *Nicotiana tabacum* (tobacco [9]) [9]
<21> *Spinacia oleracea* (spinach [9]) [9]
<22> *Strongylocentrotus purpuratus* (sea urchin [10]) [10]
<23> *Lytechinus pictus* (sea urchin [10]) [10]
<24> *Xenopus sp.* [10]

3 Reaction and Specificity

Catalyzed reaction

6-phospho-D-gluconate + NAD(P)$^+$ = 6-phospho-2-dehydro-D-gluconate + NAD(P)H

Reaction type

oxidation
redox reaction
reduction

Natural substrates and products

S 6-phospho-D-gluconate + NAD(P)$^+$ <1-24> (<12> enzyme involved in hexose monophosphate pathway, correlated to the rate of fatty acid synthesis [5]; <15-18> enzyme belongs to the oxidative pentose phosphate pathway [7, 9]) (Reversibility: r <1-24> [1-10]) [1-10]
P 6-phospho-2-dehydro-D-gluconate + NAD(P)H

Substrates and products

S 2-ketogluconate + NAD(P)H <1> (Reversibility: r <1> [1]) [1]
P D-gluconate + NAD(P)$^+$
S 6-phospho-2-dehydro-D-gluconate + NAD(P)H <1-3> (Reversibility: r <1-3> [1]) [1]
P 6-phospho-D-gluconate + NAD(P)$^+$ <1> [1]
S 6-phospho-D-gluconate + NAD(P)$^+$ <7> (Reversibility: r <7> [3]) [3]
P 6-phospho-2-dehydro-D-gluconate + NAD(P)H
S D-glucose-6-phosphate + NAD$^+$ <5> (<5> enzyme active with NAD, 2% of the activity with 6-phosphogluconate when NAD is cofactor, no activity detected with NADP as cofactor [4]) (Reversibility: ? <5> [4]) [4]
P 2-dehydro-6-phospho-D-glucose + NADH
S D-glucose-6-phosphate + NADP$^+$ <5> (<5> NADP-specific enzyme, 4% of the activity with 6-phosphogluconate [4]) (Reversibility: ? <5> [4]) [4]
P 2-dehydro-6-phospho-D-glucose + NADPH
S Additional information <1, 5, 11, 12> (<1> fructose-6-phosphate and 2-keto-3-deoxy-6-phosphogluconate are not reduced [1]; <5> glucose, 2-deoxyglucose, gluconate, fructose-6-phosphate, glucose-1-phosphate,

fructose-1,6-diphosphate, ribulose-5-phosphate and 1,3-diphosphoglycerate are no substrates [4]; <11> appears to lack 6PGAD except during growth on glucose or gluconate [4]; <12> glucose does not serve as substrate [5]) [1, 4, 5]

P ?

Inhibitors

ATP <6, 13> (<6> NAD^+-specific enzyme [4]; <13> with NAD^+ as coenzyme, with $NADP^+$ as coenzyme activity is less affected [6]) [4, 6]
NADH <13> [6]
NADPH <5, 13, 18> (<5> inhibition is competitive with respect to $NADP^+$ [4]) [2, 4, 6, 9]
NaF <1> [1]
adenosine 5'-triphosphate <5> (<5> ATP inhibits the activity with NAD^+ [4]) [4]
fructose 1,6-bisphosphate <6, 12> (<6> $NADP^+$-specific enzyme [4]) [4, 5]
guanosine 5'-triphosphate <5> (<5> GTP inhibits the $NADP^+$ specific enzyme [4]) [4]

Cofactors/prosthetic groups

NAD^+ <5, 13> (<13> dual specificity for both NAD^+ and $NADP^+$ [6]) [2, 4, 6]
$NADP^+$ <5, 8, 12, 13> (<12> NAD^+ does not substitute $NADP^+$ [5]; <13> dual specificity for both NAD^+ and $NADP^+$ [6]) [2, 5, 6, 8]

Activating compounds

ADP <12> [5]
AMP <12> [5]
ATP <12> [5]
D-glucose 6-phosphate <12> [5]
phosphoenolpyruvate <12> [5]

Metals, ions

Mg^{2+} <13, 14, 18, 19> (<13> 2 mM required for maximum activity [6]; <14> 20 mM required for maximum activity [6]; <18> activity increases slightly with 7.5 mM $MgCl_2$ [9]; <19> stimulated by $MgCl_2$ [9]) [6, 9]
Mn^{2+} <18> (<18> activity increases slightly with 7.5 mM $MnCl_2$ [9]) [9]

Specific activity (U/mg)

0.119 <12> [5]
20.8 <18> [9]
24 <22, 23> [10]
34 <5> (<5> $NADP^+$ as cofactor [4]) [4]
39.7 <14> [6]
42.3 <5> (<5> NAD^+ as cofactor [4]) [4]
56.4 <13> [6]
98.6 <1> [1]
102 <5> (<5> coenzyme NAD^+ [2]) [2]
141 <5> (<5> coenzyme $NADP^+$ [2]) [2]

K_m-Value (mM)

0.013 <14> ($NADP^+$) [6]
0.014 <22> (6-phospho-D-gluconate, <22> homogenates [10]) [10]
0.02 <5> (6-phospho-D-gluconate, <5> isoform specific for $NADP^+$ and isoform active with NAD^+, with $NADP^+$ as cofactor [2]) [2, 4]
0.02 <13> ($NADP^+$) [6]
0.02 <5> ($NADP^+$, <5> enzyme active with NAD^+ and $NADP^+$-specific enzyme [4]) [4]
0.023 <18> ($NADP^+$, <18> 7.5 mM $MgCl_2$ [9]) [9]
0.028 <12> ($NADP^+$) [5]
0.04 <5> (6-phospho-D-gluconate, <5> isoform specific for $NADP^+$, with $NADP^+$ as cofactor, 5 mM ATP, glucose grown bacteria [2]) [2]
0.04 <5> (6-phospho-D-gluconate, <5> isoform specific for $NADP^+$, with $NADP^+$ as cofactor, glucose grown bacteria [2]) [2]
0.04 <13> (NAD^+) [6]
0.0422 <18> ($NADP^+$) [9]
0.094 <12> (6-phospho-D-gluconate) [5]
0.1 <13> (6-phospho-D-gluconate) [6]
0.107 <22> (6-phospho-D-gluconate, <22> permeabilized cells [10]) [10]
0.14 <18> (6-phospho-D-gluconate, <18> 7.5 mM $MgCl_2$ [9]) [9]
0.15 <18> (6-phospho-D-gluconate, <18> 1 mM $MnCl_2$ [9]) [9]
0.2 <5> (6-phospho-D-gluconate, <5> isoform active with NAD^+, $NADP^+$ as cofactor, glucose grown bacteria [2]) [2, 4]
0.2 <5> (NAD^+) [4]
0.25 <18> ($NADP^+$, <18> 0.01 mM NADPH [9]) [9]
0.3 <5> (6-phospho-D-gluconate, <5> enzyme active with NAD^+ or $NADP^+$ [4]) [4]
0.32 <18> (6-phospho-D-gluconate) [9]
0.38 <1> (2-keto-6-phospho-D-gluconate, <1> with NADPH as coenzyme [1]) [1]
0.39 <18> (6-phospho-D-gluconate, <18> 0.01 mM NADPH [9]) [9]
0.5 <13> (NAD^+) [6]
0.66 <1> (2-keto-6-phospho-D-gluconate, <1> with NADH as coenzyme [1]) [1]
0.7 <5> (6-phospho-D-gluconate, <5> isoform active with NAD as cofactor and isoform specific for NADP as cofactor, glucose grown bacteria [2]) [2, 4]
1 <5> (6-phospho-D-gluconate, <5> 0.005 M GTP [4]) [4]
2 <5> (6-phospho-D-gluconate, <5> isoform active with NAD^+ as cofactor, 5 mM ATP, glucose grown bacteria [2]) [2]

pH-Optimum

7.2-8 <1> (<1> coenzyme NADH [1]) [1]
7.4 <1> (<1> coenzyme NADPH [1]) [1]
7.8-8 <19> [9]
8 <12> [5]
8-8.5 <14> [6]
8-9 <13> [6]

8.5 <5> (<5> NADP-specific enzyme [4]) [4]
8.5-9.5 <5> (<5> enzyme active with NAD^+ [4]) [4]

4 Enzyme Structure

Molecular weight

52000 <22, 23> [10]
55000 <18> (<18> SDS-PAGE [9]) [9]
82000 <5> (<5> enzyme active with NAD^+, sucrose density gradient sedimentation [4]) [4]
84000 <5> (<5> enzyme active with NAD^+, gel filtration [4]) [4]
92000 <7> (<7> PAGE [3]) [3]
94000 <7> (<7> sedimention-equilibrium measurement by the meniscus-depletion method, initial protein concentration 0.2 mg/ml [3]) [3]
95000 <7> (<7> gel filtration [3]) [3]
98000 <7> (<7> sedimention-equilibrium measurement by the meniscus-depletion method, initial protein concentration 0.4 mg/ml [3]) [3]
100000 <9> [3]
102000 <8> [3]
110000 <14> [6]
110000-120000 <10> (<10> $NADP^+$-specific enzyme [4]) [4]
120000 <5> (<5> $NADP^+$-specific enzyme, gel filtration [4]) [4]
123000 <5> (<5> $NADP^+$-specific enzyme, sucrose gradient sedimentation [4]) [4]
129000 <7> (<7> short-column equilibrium sedimentation [3]) [3]
220000 <13> (<13> gel filtration [6]) [6]

Subunits

dimer <5, 7, 10, 18> (<5> 2 * 40000, enzyme active with NAD^+ [4]; <5> 2 * 60000, enzyme specific for $NADP^+$ [4]; <7> 2 * 47000, SDS-PAGE [3]; <10> 2 * 57000 [4]) [3, 4, 9]
tetramer <13> (<13> 4 * 55000, SDS-PAGE [6]) [6]

5 Isolation/Preparation/Mutation/Application

Source/tissue

egg <22, 23> (<22, 23> fertilized and unfertilized eggs [10]) [10]
embryo <24> [10]
endosperm <19> [9]
leaf <18, 21> [9]
liver <7, 8> [3, 8]
root <18> (<18> from seedlings [9]) [9]

Localization

cytoplasm <8, 18> [8, 9]
cytosol <18, 19> [9]

microsome <8> [8]
plastid <18> [9]
proplastid <19> [9]

Purification
<1> [1]
<4> [1]
<5> [4]
<7> [3]
<12> [5]
<13> [6]
<14> [6]
<18> [9]
<19> [9]

Crystallization
<7> [3]

6 Stability

General stability information
<12>, 20% glycerol stabilizes the enzyme [5]

Storage stability
<1>, 0°C freezing preserves activity for several days [1]
<5>, 0°C both 6-phosphogluconate dehydrogenase species obtained by treatment with ammonium sulfate retain 100% of their activity when stored for several months [4]

References

[1] Frampton, E.W.; Wood, W.A.: Carbohydrate oxidation by Pseudomonas fluorescens VI. Conversion of 2-keto-6-phosphogluconate to pyruvate. J. Biol. Chem., **236**, 2571-2577 (1961)
[2] Lessie, T.G.; Vander Wyk, J.C.: Multiple forms of Pseudomonas multivorans glucose-6-phosphate and 6-phosphogluconate dehydrogenases: Differences in size, pyridine nucleotide specificity, and suspectibility to inhibition by adenosine 5'-triphosphate. J. Bacteriol., **110**, 1107-1117 (1972)
[3] Silverberg, M.; Dalziel, K.: Crystalline 6-phosphogluconate dehydrogenase from sheep liver. Biochem. Soc. Trans., **1**, 1132-1134 (1973)
[4] Lee, Y.N.; Lessie, T.G.: Purification and characterization of the two 6-phosphogluconate dehydrogenase species from Pseudomonas multivorans. J. Bacteriol., **120**, 1043-1057 (1974)
[5] Tanaka, A.; Takenaka, Y.; Hirai, M.; Fukui,S.: Glucose-6-phosphate dehydrogenase and 6-phosphogluconate dehydrogenase of Candida tropicalis. Par-

tial purification and some properties. Agric. Biol. Chem., **40**, 1829-1835 (1976)
[6] Ben-Bassat, A.; Goldberg, I.: Purification and properties of glucose-6-phosphate dehydrogenase(NADP$^+$/NAD$^+$) and 6-phosphogluconate dehydrogenase (NADP$^+$/NAD$^+$) from methanol grown Pseudomonas C. Biochim. Biophys. Acta, **611**, 1-10 (1980)
[7] Hipkin, C.R.; Cannons, A.C.: Inorganic nitrogen assimilation and the regulation of glucose 6-phosphate dehydrogenase in unicellular algae. Plant Sci., **41**, 155-160 (1985)
[8] Bublitz, C.; Lawler, C.A.; Steavenson, S.: The topology of phosphogluconate dehydrogenases in rat liver microsomes. Arch. Biochem. Biophys., **259**, 22-28 (1987)
[9] Bailey-Serres, J.; Nguyen, M.T.: Purification and characterization of cytosolic 6-phosphogluconate dehydrogenase in isozymes from maize. Plant Physiol., **100**, 1580-1583 (1992)
[10] Swezey, R.R.; Epel, D.: The use of caged substrates to assess the activity of 6-phosphogluconate dehydrogenase in living sea urchin eggs. Exp. Cell Res., **201**, 366-372 (1992)

Phosphogluconate dehydrogenase (decarboxylating) 1.1.1.44

1 Nomenclature

EC number
1.1.1.44

Systematic name
6-phospho-D-gluconate:$NADP^+$ 2-oxidoreductase (decarboxylating)

Recommended name
phosphogluconate dehydrogenase (decarboxylating)

Synonyms
6-PGDHase <4> [29]
6-phospho-D-gluconate dehydrogenase
6-phosphogluconate dehydrogenase (decarboxylating)
6-phosphogluconic carboxylase
6-phosphogluconic dehydrogenase
D-gluconate-6-phosphate dehydrogenase <11> [22]
GNTZII [Swissprot]
phosphogluconic acid dehydrogenase

CAS registry number
9073-95-4

2 Source Organism

<1> *Candida utilis* (type I, type II [23, 33]) [1, 6, 8, 10-13, 15, 19, 23, 26-28, 29, 30, 31, 33, 37, 46, 48, 50-52, 58-61, 67, 69-72, 78-80]
<2> *Dictyostelium discoideum* (cellular slime mould, strain NC-4, haploid [2]) [2]
<3> *Rattus norvegicus* (rat [3, 4, 6, 10, 13, 15, 17, 23, 26, 27, 36, 40, 45, 54, 59, 63, 69, 72]; Sprague-Dawley [4]; Charles River albino [40]; Wistar strain [63, 72]) [3, 4, 6, 10, 13, 15, 17, 23, 26, 27, 36, 40, 45, 54, 59, 63, 69, 72]
<4> *Ovis aries* (sheep [4, 6-8, 10, 14, 15, 17, 21, 26, 27, 29, 32, 35, 48, 52, 58-60, 69, 70, 75, 76, 78, 79, 80]; lamb [49, 67, 71, 78]) [4, 6-8, 10, 14, 15, 17, 21, 26, 27, 29, 32, 35, 39, 48, 49, 52, 58, 60, 61, 67, 69, 71, 75, 76, 78, 79-81]
<5> *Mus musculus* (mouse [4]) [4]
<6> *Spinacia oleracea* (spinach, var. Long Standing Bloomsdale [5]) [5, 34, 68]
<7> *Neurospora crassa* [8, 9, 13, 20, 23, 28, 36, 59, 78]
<8> *Bacillus stearothermophilus* (NCA 1503 [11, 13, 24, 25, 51, 57]) [11, 13, 24, 25, 28, 47, 49, 51, 56, 57, 59]

<9> *Homo sapiens* (human [11, 13-16, 18, 23, 26-28, 58, 67, 69, 72, 74, 80]; Krebs ascites cells [23]) [11, 13-16, 18, 23, 26, 27, 58, 67, 69, 72, 74, 80]
<10> *Saccharomyces cerevisiae* (baker's yeast [42]) [12, 16, 20, 35, 40, 42, 53-55, 57, 81]
<11> *Streptococcus faecalis* (ATCC 27792, formerly designated as strain MR [22, 23]) [22, 23, 28, 59, 62]
<12> *Sus scrofa* (pig [12, 27, 36]) [12, 27, 36]
<13> *Escherichia coli* (MRE 600 [28]; JM109, M15 [79]; RW226/pLC33-5 [38]) [24-26, 28, 38, 51, 59, 79, 81]
<14> *Cuniculus sp.* (rabbit, Primiparous New Zealand white [26]) [26]
<15> *Ricinus communis* (castor bean, Baker 296 "Dwarf" [34, 54]) [34, 54, 68]
<16> *Candida boidinii* (methanol utilizing yeast [35, 81]; ATCC 32195 [35]) [35, 81]
<17> *Corylus avellana* (hazel nut [36]) [36]
<18> *Avena fatua* (wild oat [41]) [41]
<19> *Streptomyces griseus* [42]
<20> *Nicotiana tabacum* (W-38 [43, 44]) [43, 44, 54]
<21> *Acer pseudoplatanus* (sycamore cells [68]) [47, 68, 81]
<22> *Drosophila sp.* [15]
<23> *Synechococcus sp. 6307* (mesophilic cyanobacterium, formerly called Coccochloris peniocystis [56]) [56, 59]
<24> *Synechococcus sp. 6716* (thermophilic cyanobacterium, formerly called Synechococcus lividus [56]) [56, 59]
<25> *Phormidium sp.* (filamentous cyanobacterium [59]) [59]
<26> *Brevibacterium flavum* (No. 2247 ATCC 14067 [62]) [62, 81]
<27> *Dicentrarchus labrax* (bass [64, 78]) [64, 78]
<28> *Methylobacillus flagellatum* [65]
<29> *Arthrobacter globiformis* (VKM B-175 [66]) [66]
<30> *Catharanthus roseus* ((L.) G. Don = Vinca rosea L., strain B/S [68]) [68]
<31> *Beta vulgaris* (silver beet [73]) [73]
<32> *Trypanosoma brucei* (parasitic protozoan [76]) [76, 80]
<33> *Lactococcus lactis* [77]
<34> *Schizosaccharomyces pombe* (fission yeast, NCYC 132S2-2 [78]) [78]
<35> *Corynebacterium glutamicum* (ATCC 13032 [81]) [81]
<36> *Corynebacterium glutamicum ssp. flavum* [81]
<37> *Pseudomonas sp.* (C [81]; P. fluorescens [78]) [78, 81]
<38> *Gluconobacter suboxydans* [78, 81]
<39> *Phaseolus aureus* (black gram, mung bean [34, 36]) [34, 36]
<40> *Daucus carota* (carrot [34, 68]) [34, 68]
<41> *Leuconostoc mesenteroides* [69, 78]
<42> *Synechococcus sp. strain PCC* [78]
<43> *Salmonella typhimurium* (LT2 [78]) [78]
<44> *Gallus sp.* (chicken [72]) [72]
<45> *Synechocystis 6714* (heterotrophic strain, formerly called Aphanocapsa sp. [56]) [56]
<46> *Carassius carassius* (crucian carp [64]) [64]
<47> *Pisum sativum* (pea [68]) [68]

<48> *Glycine max* (soybean [68]) [68]
<49> *Haemophilus influenza* [69]
<50> *Vigna radiata* (black gram, mung bean [68]) [68]

3 Reaction and Specificity

Catalyzed reaction

6-phospho-D-gluconate + $NADP^+$ = D-ribulose 5-phosphate + CO_2 + NADPH (certain preparations reduce NAD^+ as well as $NADP^+$)

Reaction type

oxidation
redox reaction
reduction

Natural substrates and products

S 6-phospho-D-gluconate + $NADP^+$ <1-50> (<3, 4, 6, 7-10, 13, 15, 17, 18, 20, 23-27, 30, 35> enzyme specific for the oxidative pentose phosphate pathway [5-9, 16, 28, 29, 34, 36, 39, 40, 41, 43, 44, 49, 51, 52, 56, 59, 60, 62, 64, 68, 72, 74, 78, 81]; <14> mammary gland enzyme involved in NADPH production which is essential for lipogenesis [26]; <15> endosperm contains both cytosolic and plastic glycolytic and pentose phosphate pathway [34]; <34> pivotal enzyme links the gluconate route and the oxidative phase of the pentose phosphate pathway [78]) (Reversibility: r <1-27, 29-50> [1-81]; ir <28> [65]) [1-81]
P D-ribulose 5-phosphate + CO_2 + NADPH

Substrates and products

S 2-deoxy-3-oxo 6-phosphogluconate + ? <9, 32> (<4> no decarboxylation by the liver enzyme [80]) (Reversibility: ? <9, 32> [80]) [80]
P ?
S 2-deoxy-6-phosphogluconate + $NADP^+$ <1, 4> (Reversibility: r <1, 4> [67, 70, 80]) [67, 70, 80]
P 2-deoxy-3-oxo 6-phosphogluconate + NADPH <4> [67, 80]
S 2-deoxy-6-phosphogluconate + $NADP^+$ <1> (<1> oxidation, 1.5% activity of that of 6-phosphogluconate [12]) (Reversibility: r <1> [12, 19]) [12, 19]
P 3-oxo-2-deoxy-6-phosphogluconate + NADPH
S 3-oxo-2-deoxy-6-phosphogluconate + $NADP^+$ <1> (<1> decarboxylation [12]; <1> reaction requires presence of NADPH, but the reduced coenzyme does not play a redox role [19]) (Reversibility: r <1> [12]) [12, 19]
P 1-deoxyribulose 5-phosphate + CO_2 + NADPH
S 3-oxo-2-deoxy-6-phosphogluconate + NADPH <1> (<1> reduction [12]) (Reversibility: r <1> [12, 19]) [12, 19]
P 2-deoxy-6-phosphogluconate + $NADP^+$
S 6-phospho-D-gluconate + $NADP^+$ <1-27, 29-50> (Reversibility: r <1-27, 29-50> [1-81]; ir <28> [65]) [1-81]
P D-ribulose 5-phosphate + CO_2 + NADPH

S 6-phosphomannoate + $NADP^+$ <1, 4> (Reversibility: ? <1, 4> [70, 75]) [70, 75]
P ? + NADPH
S D-ribulose 5-phosphate + CO_2 + NADPH <4, 11> (<11> reverse reaction only to a very limited extent [23]) (Reversibility: r <4, 11> [17, 23]) [17, 23]
P 6-phospho-D-gluconate + $NADP^+$
S β-3-acetylpyridine adenine dinucleotide 2'-phosphate + 6-phosphogluconate <1> (Reversibility: ? <1> [71]) [71]
P ?
S β-nicotinamide 1-N^6-ethenoadenine dinucleotide 2'-phosphate + 6-phosphogluconate <1> (Reversibility: ? <1> [71]) [71]
P ?
S β-nicotinamide adenine dinucleotide 3'-phosphate + 6-phosphogluconate <1> (Reversibility: ? <1> [71]) [71]
P ?
S β-nicotinamide hypoxanthine dinucleotide 2'-phosphate + 6-phosphogluconate <1> (Reversibility: ? <1> [71]) [71]
P ?
S ribulose 5-phosphate + H_2O + NADPH <1> (<1> tritium exchange reaction between tritiated ribulose 5-phosphate and water, reaction requires presence of NADPH, but the reduced coenzyme does not play a redox role [19]) (Reversibility: ? <1> [19, 70]) [19, 70]
P ?
S Additional information <4, 7, 35> (<4> 6-phosphogalactonate is no substrate [75]; <7> neither gluconic acid nor glucose-6-phosphate can replace 6-phosphogluconate as substrate [20]; <35> glucose-6-phosphate is no substrate [81]) [20, 75, 81]
P ?

Inhibitors

(5-(dimethylamine)naphthalin-1-sulfonylchloride) <1> (<1> dansyl chloride, DNSCl, rapid inactivation at pH 7.5, 95% inhibition at pH 6.2 [30]) [30]
2'-AMP <4> [7, 17]
3'-AMP <4> [7, 17]
3-oxo-2-deoxy-6-phosphogluconate <32> (<32> competitive inhibitor [80]) [80]
5,5'-dithiobis(2-nitrobenzoic acid) <1, 4, 8, 11, 13, 25> (<8> Ellman's reagent [13, 51]; <1> DTNB [33]) [10, 13, 23, 28, 33, 50, 51, 59, 60]
6-methoxy-7-hydroxycoumarin <20> (<20> scopolin [43]) [43]
7-chloro-4-nitrobenzo-2-oxa-1,3-diazole <8> (<8> NBD-Cl [51]) [51]
ADP <3, 4, 25> [3, 7, 59]
AMP <3> [3]
ATP <4, 13, 16, 25, 28, 29, 35> [7, 17, 35, 38, 59, 65, 66, 81]
CDP <4> [7, 17]
CMP <4> [7, 17]
CO_2 <12> [27]

CTP <4> [7, 17]
Cd^{2+} <11, 25> [22, 59]
Co^{2+} <29> [66]
Cu^{2+} <11, 25> [22, 59]
D-2-deoxyglucose <1, 4> [12, 60]
D-5-phosphoarabonate <1> [70]
D-5-phosphoribonate <1> [70]
D-5-phosphoribosyl-1-diphosphate <26, 30> [62, 68]
D-6-phosphoallonate <1> [70]
D-6-phosphogluconate <1> (<1> competitive substrate inhibition at pH 7 [71]) [71]
D-6-sulfogluconate <1, 9> [58, 70]
D-ATP-ribose <1> [69]
D-erythrose 4-phosphate <26, 31, 35> [62, 73, 81]
D-fructose 1,6-diphosphate <4, 7, 11, 13, 16, 25, 26, 31, 35, 37, 38> [4, 7, 8, 14, 17, 21-23, 28, 35, 38, 59, 62, 73, 81]
D-fructose 1-phosphate <4> [7, 17]
D-fructose 2,6-bisphosphate <15> [54]
D-fructose 6-phosphate <4, 25> [7, 17, 59]
D-glucose 1,6-bisphosphate <3, 15, 20> [40, 43, 54]
D-glucose 6-phosphate <4> [7, 17]
D-glyceraldehyde 3-phosphate <26, 35> [62, 81]
D-phosphate <4, 7, 9> (<9> competitive inhibitor of the substrates [14]) [7, 14, 17, 20, 39]
D-ribulose 5-phosphate <1, 16, 25, 26, 31, 34, 35> [35, 59, 62, 70, 73, 78, 81]
D-xylulose 5-phosphate <26> [62]
Fe^{2+} <29> [66]
GDP <4> [7, 17]
GMP <4> [7, 17]
GTP <4> [7, 17]
Hg^{2+} <8, 25> (<25> $HgCl_2$ [59]) [13, 59]
IDP <4> [7, 17]
IMP <4> [7, 17]
ITP <4> [7, 17]
Mg^{2+} <12> (<12> acts as an inhibitor above 20 mM [27]) [27]
N-ethylmaleimide <25> [59]
NADH <29> [66]
$NADP^+$ <1, 4> [46, 61]
NADPH <3, 7-9, 12, 13, 16, 25-30, 32, 34, 35> (<1> noncompetitive inhibition versus 6-phosphogluconate [69]; <9> product inhibition [14]; <3, 9, 16, 34> competitive inhibition [35, 58, 72, 78]) [4, 8, 14, 20, 27, 28, 35, 58, 59, 62, 64-66, 68, 69, 72, 78, 80, 81]
Ni^{2+} <29> [66]
Rose Bengal <7, 8, 13, 25> (<7, 8, 13> photooxidation [8, 28]) [8, 28, 59]
TDP <4> [7, 17]
TMP <4> [7, 17]
TTP <4> [7, 17]

UDP <4> [7, 17]
UMP <4> [7, 17]
UTP <4, 16> [7, 17, 35]
Zn^{2+} <11, 25, 29> [22, 59, 66]
acetyl-CoA <26> [62]
adenosine <4> [7]
arsenate <1> [31]
butan-2,3-dione <1> [31]
caffeic acid <20> [43]
chlorogenic acid <20> [43]
chromate <1> [31]
citrate <4, 14> [7, 17, 26]
diethyldicarbonate <1> [50]
diphosphate <4, 14> [7, 17, 26]
esculetin <20> [43]
esculin <20> [43]
ferulic acid <20> [43]
hexachloroplatinate <1> [31]
iodoacetamide <7-9> [8, 11, 20, 51]
molybdate <1> [31]
nicotinamide <4> [7]
oxalate <4> [7]
oxaloacetate <4, 26, 35> [7, 17, 62, 81]
p-chloromercuribenzoate <8, 11, 13, 25> [13, 22, 28, 59]
p-coumaric acid <20> [43]
p-hydroxymercuribenzoate <4, 8, 11> [10, 23, 51, 60]
perchlorate <1> [31]
periodate <1> [31, 50]
permanganate <1> [31]
phosphoenolpyruvate <27> [64]
pyridoxal 5'-phosphate <1, 7> (<1, 7> photooxidation [8, 31]) [8, 31, 50]
scopoletin <20> [43]
sulfate <1> [31]
sulfate <4> [7]
tetranitromethane <1> (<1> irreversible loss of catalytic activity [1]) [1]
tungstate <1> [31]

Cofactors/prosthetic groups

3-acetylpyridine adenine dinucleotide phosphate <1> (<1> APADP [71]) [71]
3-acetylpyridine-$NADP^+$ <7> [8]
NAD^+ <1, 3, 8, 11, 28> (<1> uses NAD^+ very poorly as a substrate [71]; <3> reacts also with $NADP^+$ [3]; <8> some activity with NAD^+ [13]; <11> NAD^+ specific [62]) [3, 13, 62, 65, 71]
$NADP^+$ <1-4, 7, 8, 11, 13, 17, 25, 26, 28, 29, 34> (<1> NAD^+ being totally inactive [46]; <1> uses NAD^+ very poorly as a substrate [71]; <2> no reaction with NAD^+ [2]; <3> reacts also with NAD^+ [3]; <4> does not utilize NAD^+ [17]; <7> NAD^+ is completely ineffective as hydrogen acceptor [20];

<8> strictly specific for $NADP^+$, no activity observed with NAD^+ [28]; <11> specific for $NADP^+$, no activity observed when NAD^+ is substituted for $NADP^+$ [22]; <17> specific for $NADP^+$, unable to utilize NAD^+ [36]; <25> no activity observed when NAD^+ is substituted for $NADP^+$ [59]; <26> specific to $NADP^+$ [62]; <29> absolutely specific with respect to $NADP^+$, no reaction can be detected in the presence of NAD^+ [66]; <34> requires $NADP^+$ obligatorily for activity [78]) [2, 3, 7, 13, 17, 20, 22, 28, 36, 39, 46, 59, 62, 65, 66, 78]
deamino-$NADP^+$ <7> [8]
nicotinamide 1,N^6-ethenoadenine dinucleotide phosphate <1> (<1> ε-$NADP^+$ [71]) [71]
nicotinamide adenine dinucleotide 3'-phosphate <1> (<1> 3'-$NADP^+$ [71]) [71]
nicotinamide hypoxanthine dinucleotide phosphate <1> (<1> HX-$NADP^+$ [71]) [71]
thio-$NADP^+$ <1> [71]

Activating compounds

2,3-diphosphoglycerate <20> (<20> activates both isoenzymes up to 200% of their original activity [43]) [43]
6-phosphogluconate <4> (<4> bona fide allosteric activator [80]) [80]
ADP <26> [62]
AMP <25> (<25> 1 mM, relative activity 105% [59]) [59]
ATP <26> [62]
D-fructose 1-phosphate <26> [62]
D-fructose 6-phosphate <1> (<1> slight activation [70]) [70]
D-gluconate <11> (<11> induced by [62]) [62]
L-aspartate <26> [62]
L-histidine <26> [62]
L-tryptophan <26> [62]
glycylglycine <2> [2]

Metals, ions

Ca^{2+} <25, 29> (<25> activated by 1 mM $CaCl_2$, relative activity 146.6% [59]) [59, 66]
Co^{2+} <34> (<34> activated by divalent metal ions [78]) [78]
K^+ <26> (<26> stabilized by KCl [62]) [62]
KCl <8> (<8> activates the enzyme at high concentrations [13]) [13]
Mg^{2+} <1, 8, 12, 15, 27, 29> (<1, 8> requires Mg^{2+} for full activity [59]; <8> activates the enzyme at low concentrations 0.01-0.04 M [13]; <15> $MgCl_2$ stimulation for the oxidative decarboxylation of 6-phosphogluconate [34]; <27> activated by $MgCl_2$ [64]) [13, 27, 34, 51, 59, 64, 66]
Mn^{2+} <25, 29, 34> (<25> activated by 1 mM $MnCl_2$, relative activity 142.1% [59]; <34> activated by divalent metal ions [78]) [59, 66, 78]
NaCl <8> (<8> activates the enzyme at high concentrations [13]) [13]

Turnover number (min^{-1})

13 <9> (6-phosphogluconate) [18]

Specific activity (U/mg)

0.1207 <3> (<3> kidney enzyme, optimum conditions, pH 8.0, 37°C [72]) [72]
0.41 <12> [27]
0.4242 <3> (<3> liver enzyme, optimum conditions, pH 8.0, 37°C [72]) [72]
1.33 <4> [67]
1.89 <12> (<12> supernatant fraction [27]) [27]
4.7 <4> (<4> recombinant enzyme [79]) [79]
10 <9> [14, 18]
12 <1> (<1> total enzyme after gel filtration in presence of 6-phosphogluconate [37]) [37]
13.6 <15> [34]
14.47 <11> [22, 23]
15 <9> [18]
17.4 <4> [6]
18.7 <4> (<4> ultrafiltration [17]) [17]
19 <4> (<4> pure enzyme [17]) [17]
20 <14> [26]
21 <4> [17]
21.2 <4> [10]
22 <14> (<14> 150fold purification [26]) [26]
25.7 <29> [66]
27 <8> [51]
30 <7> [9]
30.8 <7> [8, 20]
32 <3, 13> [4, 28]
33.5 <1> (<1> total enzyme after gel filtration in absence of 6-phosphogluconate [37]) [37]
38 <15, 25> (<15> cytoplasmic isoenzyme [34]) [34, 59]
39.4 <34> [78]
39.7 <16> [35]
40 <31> [73]
41.8 <1> [19]
48 <1> [69]
72 <35> [81]
79.5 <21> [47]
80 <21> (<21> pH 7.5, 25°C [47]) [47]
87 <8> [13]
92.3 <4> [49]
143 <35> [81]
156 <26> [62]
175 <26> (<26> cell-free extract [62]) [62]
Additional information <30> (<30> plastid isoenzyme 4.8 nkat mg-1 protein, cytosolic isoenzyme 2.5 nkat mg-1 protein [68]) [68]

K_m-Value (mM)

0.00022 <4> (NADPH) [17]

0.00088 <27> ($NADP^+$) [64]
0.002 <1> (NADPH) [69]
0.0041 <15> ($NADP^+$, <15> cytosolic isoenzyme [34]) [34]
0.0054 <15> ($NADP^+$, <15> proplastid isoenzyme [34]) [34]
0.00571 <4> ($NADP^+$, <4> 10 mM Tris-acetate, 10 mM $MgCl_2$, pH 7.7 [6]) [6]
0.006 <40> (6-phosphogluconate) [68]
0.006 <18> ($NADP^+$, <18> source endosperm [41]) [41]
0.00631 <4> ($NADP^+$, <4> 10 mM Tris-acetate, 80 mM KCl, pH 7.7 [6]) [6]
0.00676 <4> ($NADP^+$, <4> 50 mM Tris-acetate, pH 7.7 [6]) [6]
0.0068 <1> (6-phosphogluconate, <1> pH 10 [71]) [71]
0.0068 <4> ($NADP^+$) [17, 23]
0.0069 <4> (6-phosphogluconate) [17, 23]
0.007 <50> (6-phosphogluconate) [68]
0.007 <4> ($NADP^+$) [5, 7]
0.007 <6> ($NADP^+$) [5, 7]
0.008 <1> ($NADP^+$) [69]
0.008 <30> ($NADP^+$, <30> cytosolic isoenzyme [68]) [68]
0.00861 <4> ($NADP^+$, <4> 50 mM Tris-acetate, 20 mM $MgCl_2$, pH 7.7 [6]) [6]
0.00893 <4> ($NADP^+$, <4> 10 mM Tris-acetate, pH 7.7 [6]) [6]
0.0096 <32> (2-deoxy-3-oxo-6-phosphogluconate) [80]
0.0096 <15> (6-phosphogluconate, <15> proplastid isoenzyme [34]) [34]
0.01 <7> (6-phosphogluconate) [8, 9, 20, 23]
0.01 <6> ($NADP^+$) [5]
0.01 <24> ($NADP^+$) [56, 59]
0.01 <7> ($NADP^+$, <7> wild-type, mutant col 10, heterocaryon col-10-wild-type [9]) [9]
0.011 <16> (6-phosphogluconate) [35]
0.011 <47> ($NADP^+$) [68]
0.011 <30> ($NADP^+$, <30> plastid isoenzyme [68]) [68]
0.011 <8> ($NADP^+$, <8> 60°C [13]) [13]
0.012 <15> (6-phosphogluconate, <15> cytosolic isoenzyme [34]) [34]
0.013 <16> ($NADP^+$) [35]
0.013 <3> ($NADP^+$) [4, 23]
0.013 <25> ($NADP^+$) [59]
0.0135 <12> (6-phosphogluconate) [27]
0.014 <8> (6-phosphogluconate, <8> 60°C or 65°C [13]) [13]
0.0149 <4> (6-phosphogluconate, <4> 50 mM Tris-acetate, 5-10 mM $MgCl_2$, pH 7.7 [6]) [6]
0.015 <4> (6-phosphogluconate) [6]
0.015 <30> (6-phosphogluconate, <30> cytosolic isoenzyme [68]) [68]
0.015 <11> ($NADP^+$) [22, 23]
0.015 <7> ($NADP^+$, <7> heterocaryon col 3-wild-type [9]) [9, 23]
0.015 <8> ($NADP^+$, <8> 65°C [13]) [13]
0.016 <4> (6-phosphogluconate) [7, 80]
0.016 <8> (6-phosphogluconate, <8> 30°C [13]) [13]
0.016 <17> ($NADP^+$) [36]
0.016 <4> ($NADP^+$) [6]

0.0161 <4> (6-phosphogluconate, <4> 50 mM Tris-acetate, pH 7.7 [6]) [6]
0.0167 <4> (6-phosphogluconate, <4> 10 mM Tris-acetate, 80 mM KCl, pH 7.7 [6]) [6]
0.017 <35> ($NADP^+$) [81]
0.017 <8> ($NADP^+$, <8> 30°C [13]) [13]
0.018 <30> (6-phosphogluconate, <30> plastid isoenzyme [68]) [68]
0.018 <2> ($NADP^+$) [2]
0.018 <7> ($NADP^+$, <7> heterocaryon col-3-col-10 [9]) [9]
0.02 <9> (6-phosphogluconate) [14, 18, 23, 58]
0.02 <8> (6-phosphogluconate, <8> 43°C [13]) [13]
0.02 <1> ($NADP^+$, <1> type I and II [23]) [12, 19, 23]
0.02 <4> (ribulose-5-phosphate) [17]
0.021 <8> ($NADP^+$, <8> 45°C [13]) [13]
0.0217 <4> ($NADP^+$, <4> 23.5°C [61]) [61]
0.022 <8> (6-phosphogluconate, <8> 45°C [13]) [13]
0.0227 <4> (6-phosphogluconate) [61]
0.023 <9> (6-phosphogluconate) [80]
0.023 <14> ($NADP^+$) [26]
0.023 <47> ($NADP^+$, <47> plastid isoenzyme [68]) [68]
0.024 <11> (6-phosphogluconate) [22, 23]
0.025 <23> ($NADP^+$) [56, 59]
0.025 <31> ($NADP^+$) [73]
0.025 <10> ($NADP^+$, <10> pH 8.0 [55]) [55]
0.025 <7> ($NADP^+$, <7> source of enzyme: peak I [8]) [8]
0.025 <8> ($NADP^+$, <8> 43°C [13]) [13]
0.02666 <27> (6-phosphogluconate) [64]
0.029 <4> ($NADP^+$) [61]
0.03 <9> ($NADP^+$) [14, 18, 23]
0.03 <4> ($NADP^+$) [17]
0.03 <7> ($NADP^+$, <7> source of enzyme: whole cells [8]; <7> mutant col-3 [9]) [8, 9, 20, 23]
0.031 <1> (6-phosphogluconate, <1> type II [23]) [23]
0.031 <10> ($NADP^+$, <10> pH 9.03 [55]) [55]
0.032 <18> ($NADP^+$, <18> source embryo [41]) [41]
0.033 <8> ($NADP^+$) [28]
0.033 <10> ($NADP^+$, <10> pH 5.93 [55]) [55]
0.033 <7> ($NADP^+$, <7> source of enzyme: peak II [8]) [8]
0.034 <35> (6-phosphogluconate) [81]
0.035 <32> (6-phosphogluconate) [80]
0.035 <29> ($NADP^+$) [66]
0.036 <25> (6-phosphogluconate) [59]
0.036 <26> ($NADP^+$) [62]
0.04 <2> (6-phosphogluconate) [2]
0.04 <24> (6-phosphogluconate) [56, 59]
0.04-0.05 <7> (6-phosphogluconate, <7> heterocaryon col-3-wild-type [9]) [9]
0.0413 <4> (6-phosphogluconate, <4> 10 mM Tris-acetate, pH 7.7 [6]) [6]

0.043 <4> (6-phosphogluconate) [67]
0.045 <26> (6-phosphogluconate) [62]
0.05 <8> (6-phosphogluconate) [28]
0.05 <4> (6-phosphogluconate, <4> 50 mM Tris-acetate, 20 mM $MgCl_2$, KCl, pH 7.7 [6]) [6]
0.05 <7> (6-phosphogluconate, <7> heterocaryon col-10-wild-type [9]) [9]
0.052 <1> (6-phosphogluconate, <1> type I [23]) [23]
0.054 <1> (6-phosphogluconate) [12, 19]
0.054 <14> (6-phosphogluconate) [26]
0.06 <7> (6-phosphogluconate, <7> col-3-col-10 [9]) [9]
0.065 <31> (6-phosphogluconate) [73]
0.068 <10> (6-phosphogluconate, <10> pH 5.93 [55]) [55]
0.071 <3> (6-phosphogluconate) [4, 23]
0.071 <1> (6-phosphogluconate, <1> pH 7 [71]) [71]
0.071 <18> (6-phosphogluconate, <18> source embryo [41]) [41]
0.08 <6> (6-phosphogluconate, <6> isoenzyme 2 [5]) [5]
0.082 <9> (2-deoxy-3-oxo-6-phosphogluconate) [80]
0.085 <1> (6-phosphgluconate, <1> pH 8 [71]) [71]
0.09 <23> (6-phosphogluconate) [56, 59]
0.09 <7> (6-phosphogluconate, <7> mutant col 3 [9]) [9]
0.098 <48> (6-phosphogluconate) [68]
0.1 <6> (6-phosphogluconate, <6> isoenzyme 1 [5]) [5]
0.109 <48> (6-phosphogluconate, <48> plastid isoenzyme [68]) [68]
0.11 <17> (6-phosphogluconate) [36]
0.11 <29> (6-phosphogluconate) [66]
0.11 <7> (6-phosphogluconate, <7> mutant col 10 [9]) [9]
0.153 <10> (6-phosphogluconate, <10> pH 8.0 [55]) [55]
0.157 <3> (6-phosphogluconate) [72]
0.204 <1> (6-phosphogluconate, <1> pH 9 [71]) [71]
0.22 <1> (6-phosphogluconate) [69, 70]
0.258 <3> ($NADP^+$) [72]
0.29 <4> (6-phosphogluconate) [17]
0.313 <4> (6-phosphogluconate, <4> 23.5°C [61]) [61]
0.402 <18> (6-phosphogluconate, <18> source endosperm [41]) [41]
0.55 <1> (2-deoxy-6-phosphogluconate) [13]
0.64 <1> (6-phosphgluconate, <1> pH 9.5 [71]) [71]
1 <1> (NAD^+, <1> 43°C [13]) [13]
1 <1> (ribulose-5-phosphate) [69]
1.4 <4> (2-deoxy-6-phosphogluconate) [75]
1.42 <4> (2-deoxy-3-oxo-6-phosphogluconate) [80]
1.42 <4> (2-deoxy-6-phosphogluconate) [67]
1.7 <15> ($MgCl_2$, <15> cytosolic isoenzyme [34]) [34]
2.4 <15> ($MgCl_2$, <15> proplastid isoenzyme [34]) [34]
3.13 <10> (6-phosphogluconate, <10> pH 9.03 [55]) [55]
3.33 <27> (Mg^{2+}) [64]
3.5 <1> (NAD^+, <1> 30°C [13]) [13]
9 <1> (NAD^+, <1> pH 8 [71]) [71]

16 <1> (NAD^+, <1> pH 7 [71]) [71]
17 <1> (NAD^+, <1> pH 9.5 [71]) [71]
18 <1> (NAD^+, <1> pH 9 [71]) [71]
34 <4> (CO_2) [17]
50 <1> (CO_2) [69]
114 <1> (NAD^+, <1> pH 10 [71]) [71]

pH-Optimum

6.8-7.2 <9, 32> (<9, 32> decarboxylation in absence of activators [80]) [80]
7-9 <20> (<20> both isoenzymes [44]) [44]
7.2 <23, 24, 45> [56]
7.5 <1, 34> [70, 78]
7.5-8 <8, 11, 26> [59, 62]
7.6-8 <11> [22]
7.7 <4> (<4> low substrate concentrations [6]) [6]
7.8-8 <15> (<15> cytosolic and proplastid isoenzymes [34]) [34]
7.8-9 <7> [8, 20]
8 <1, 3, 4, 8, 9, 25, 32> (<3> both, liver and kidney cortex enzyme [72]; <4, 9, 32> 6-phosphogluconate oxidation [80]) [13, 19, 51, 59, 72, 80]
8.3 <3, 9> (<3> Tris-Cl orbicine-acetate buffers [4]) [4, 14, 18]
8.5 <12> [27]
8.6-8.8 <2> (<2> both Tris and tricine buffers, more active in tricine buffers [2]) [2]
9 <6, 32> (<6> isoenzyme 1, chloroplasts [5]; <32> 6-phosphogluconate-activated decarboxylation [80]) [5, 80]
9.1 <4, 9> (<4, 9> reduction [80]) [6, 80]
9.3 <6> (<6> isoenzyme 2, cytoplasm [5]) [5]
10 <32> [80]

pH-Range

4-9.6 <25> [59]
5.5-9 <4> [6]
5.5-9.1 <4> [61]
6-9 <8> [51]
7.5-10 <12> [27]

Temperature optimum (°C)

25 <8> (<8> maximum activity at pH 8 also at 43°C [51]) [51]
40 <25> [59]
43 <8> (<8> maximum activity at pH 8 also at 25°C [51]) [51]
55 <3, 23> [56, 72]
55-60 <13> [28]
60 <8> [11]
65 <24> [56]
70-75 <8> [56]
75-80 <8> [28]

Temperature range (°C)

25-70 <8> (<8> retains full activity at 60°C [11]) [11]

50-60 <13> [51]
65-75 <8> [51]

4 Enzyme Structure

Molecular weight

69000-72000 <20> (<20> 2 isoenzymes, gel filtration [43, 44]) [43, 44]
79000 <9> (<9> gel filtration of haemolysates [15]) [15]
83000 <12> (<12> gel filtration, method of Ornstein and Davis [27]) [27]
90000 <4> (<4> gel filtration, Bio-Gel 200 [10]) [10]
92000 <4> (<4> gel filtration, PAGE [10]) [10]
94000 <4> (<4> gel filtration, in presence of $MgCl_2$ [10]; <4> gel filtration, sedimentation equlibrium [17]) [10, 17, 23, 29, 59]
95000 <4> (<4> gel filtration [10]) [10]
98000 <4> (<4> sedimentation equilibrium [10]) [10]
100000 <1, 8, 13> (<1, 13> gel filtration [28, 31]; <8> sedimentation equilibrium [11, 51]; <8> sucrose density gradient centrifugation [28, 51]) [11, 28, 31, 51, 57, 59]
101000 <1, 8> (<1> gel filtration [10]; <1> type I, centrifugation in sucrose density gradient [15, 23]; <8> sucrose density gradient centrifugation [13]) [10, 13, 15, 23]
102000 <3, 31> (<3> gel filtration [4, 10, 15]) [4, 10, 15, 23, 26, 73]
104000 <4, 9, 14, 22, 25> (<9> sedimentation equilibrium, gel filtration [15, 18]; <9, 14, 25> gel filtration [26, 58, 59]) [4, 15, 18, 23, 26, 58, 59]
106000-115000 <11> (<11> gel filtration [23]) [23, 28]
108000 <11> (<11> sedimentation equilibrium [22, 23]) [22, 23, 59]
110000 <6, 7, 16> (<6> both isoenzymes [5]; <16> gel filtration [35]) [5, 35, 59]
110000-120000 <7> (<7> gel filtration [8, 20]) [8, 20, 23, 28]
111000 <1> (<1> sucrose density gradient centrifugation [15]; <1> type II [23]) [15, 23]
112000 <3> [59]
129000 <4> (<4> sedimentation equilibrium, short column method [10]; <4> presence of impurities [17]) [4, 10, 15, 17]
158000 <34> (<34> gel filtration [78]) [78]
229000 <29> (<29> gel filtration [66]) [66]

Subunits

dimer <1, 3, 4, 7-9, 11, 13-16, 21, 22, 25, 31, 32> (<1> 2 * 50000, type I, SDS-PAGE [15, 23, 33, 75]; <3> 2 * 50000, SDS-PAGE [15]; <3> 2 * 52000, SDS-PAGE [4, 23]; <4> 2 * 47000, SDS-PAGE [10, 21, 23, 32]; <4> 2 * 50000 [39, 75]; <4> 2 * 51000, SDS-PAGE [52, 79]; <4> 2 * 52000 [75]; <7> 2 * 57000, SDS-PAGE [8, 20, 23]; <7> heteromultimeric molecule composed of 2 nonidentical subunits [9]; <8> 2 * 50000, SDS-PAGE [11, 51]; <8> 2 * 51000, SDS-PAGE [13]; <9> 2 * 52000, SDS-PAGE [15, 18, 23]; <9> 2 * 54000, maleic anhydride electrophoresis [15]; <11> SDS-PAGE [22]; <11> 2 * 52000, SDS-

PAGE [23]; <14> 2 * 52000, SDS-PAGE [26]; <15> 2 * 54000, cytosolic isoenzyme, SDS-PAGE [34]; <15> 2 * 55000, SDS-PAGE [34]; <16> 2 * 58000, SDS-PAGE [35]; <21> 1 * 49000 + 1 * 45000, SDS-PAGE [47]; <22> 2 * 50000, SDS-PAGE [15]; <25> 2 * 52000, SDS-PAGE [59]; <31> 2 * 52000, SDS-PAGE [73]) [1, 4, 8-11, 13, 15, 17, 18, 20-23, 26, 28, 29, 32-34, 35, 37, 39, 46, 47, 52, 57-60, 67, 72, 73, 75, 76, 79, 80]
tetramer <29, 34> (<29> 4 * 57000, SDS-PAGE [66]; <34> 4 * 38000, SDS-PAGE [78]) [66, 78]

5 Isolation/Preparation/Mutation/Application

Source/tissue

adipose tissue <3> [3]
caryopse <18> [41]
cotyledon <17> [36]
diaphragm <3> [40]
embryo <18> [41]
endosperm <15, 18, 21> [34, 41, 47, 54, 68]
erythrocyte <4, 9> [11, 13-16, 18, 26-28, 58, 67, 69, 72, 74, 80]
fat pad <3> [40]
fat pad <3> (<3> epididymal [3]) [3]
heart <9> [16, 80]
hepatocyte <3> [63]
kidney <3, 9, 12> (<3> cortex [72]) [12, 16, 72]
leaf <6, 15, 20, 31> [5, 34, 43, 54, 68, 73]
lens <9> [74]
liver <3-5, 9, 12, 27> [4, 6-8, 10, 13, 14-17, 21, 23, 26, 27, 29, 32, 35, 36, 39, 40, 45, 48, 49, 52, 53, 58-61, 64, 67, 69-72, 75, 78-80, 81]
lung <9> [16]
mammary gland <3, 14> [3, 26]
muscle <3> (<3> leg and skeletal [40]) [40]
mycelium <7> [20]
nodule <48> [68]
root <21, 40, 47> [34, 47, 68]
seed <15, 17, 18> [34, 36, 41, 54]
seedling <39> [34]
skeletal muscle <9> [16]
testis <3> [40]

Localization

amyloplast <30> [68]
chloroplast <6> [5]
cytoplasm <1, 3, 9, 12, 23, 24, 31, 45> [4, 16, 27, 45, 56, 69, 73]
cytosol <3, 6, 15, 17, 18, 21, 30, 31, 47> [5, 34, 36, 41, 47, 54, 68, 72, 73]
microsome <3> [45]
plastid <15, 21, 30, 31, 47> [47, 54, 68, 73]
proplastid <15> [34]

Purification

<1> [19, 23]
<3> [4, 10, 17, 23, 27, 72]
<4> (recombinant enzyme [79]) [4, 7, 10, 17, 21, 23, 27, 29, 39, 49, 52, 60, 61, 79, 80]
<7> (wild-type, mutant col-3, mutant col-10 [9]) [8, 9, 20, 23]
<8> [11, 13, 24, 49, 51]
<9> [14, 15, 18, 23, 27, 58, 80]
<10> [12, 40]
<11> [22, 23]
<12> [27]
<13> [24, 28, 38]
<14> [26]
<15> [34]
<16> [35, 81]
<18> [41]
<20> [43, 44]
<21> [47, 81]
<25> [59]
<26> [62]
<27> [64]
<28> [65]
<29> [66]
<31> [73]
<32> (purified from an overexpressing Escherichia coli strain [80]) [80]
<33> [77]
<34> [78]
<35> [81]

Crystallization

<1> [1, 10, 12, 19, 30, 31, 37, 46, 49, 50]
<4> [10, 11, 17, 21, 28, 29, 32, 39, 52, 53, 78, 80]
<8> [11, 51]
<32> [76, 80]
<33> [77]

Cloning

<4> (mRNA encoding the 51-kDa subunit of 6-phosphogluconate dehydrogenase from sheep liver reverse transcribed and amplified resulting cDNA subcloned into pQE-30 and overproduced in Escherichia coli M15 [79]; amino acid sequence of the sheep liver enzyme deduced from cDNA clones [75]) [75, 79]
<9> (6PGDH cDNA encodes a protein with 94% sequence identity with the sheep liver enzyme [80]) [80]
<42> (cloning of the strain PCC 7942 gene gnd encoding 6-phosphogluconate dehydrogenase [78]) [78]
<43> (cloning of the LT2 gnd gene [78]) [78]

Application

medicine <9, 32> (<9> with a deficiency of 6-phosphogluconate dehydrogenase some reticulocytosis occurs [14]; <32> 6PGDH might be a target for chemotherapy against parasitic infections as trypanosomiasis [80]) [14, 16, 80]
synthesis <8> (<8> thermostability may lead to some practical applications [13]) [13]

6 Stability

pH-Stability

6-9 <25> [59]
9.1 <4> (<4> unstable at higher pH values [6]) [6]

Temperature stability

45 <13> (<13> stable below [28]) [28]
48 <10> [57]
50 <7, 13, 25> (<7> stable for at least 120 min [8]; <7> 6-phosphogluconate increases heat stability [9]; <25> stable below, retains 50% original activity at 57°C [59]) [8, 9, 24, 59]
60 <8> (<8> stable at 60°C [13]; <8> retains more than 90% of its structure [28]; <8> retains full activity after heating for 15 min [51]; <8> stable at optimal growth temperature [56]) [13, 28, 51, 56]
67 <8> (<8> stable below [28]) [24, 28, 57]
70 <8> (<8> stable at relatively high temperatures [11]; <8> denaturation half-time of 80 min [13]; <8> stable up to [28]) [11, 13, 28]
80 <8> (<8> denaturation half-time of 3 min [13]) [13]

Organic solvent stability

acetone <8> (<8> enzyme shows 15-20% enhanced activity [57]) [57]
dimethylformamide <8> (<8> enzyme shows 15-20% enhanced activity [57]) [57]
dioxane <8> (<8> enzyme shows 15-20% enhanced activity [57]) [57]

General stability information

<8>, noteworthy stability to heat and protein denaturants [51]
<8>, remarkably stable to the action of trypsin, chymotrypsin and elastase [57]
<8>, stable in 8 M urea, nearly unaffected, only 10% loss of activity after exposure for 60 min [25, 28]
<12>, enzyme loses no activity on dialysis overnight against 10 mM EDTA [27]
<13>, denaturation by urea [28]
<15>, both isoenzymes stabilized by the $H_2PO_4^-$ [34]
<25>, stable in 1 mM urea [59]

Storage stability

<1>, 4°C, stable for several days, as an $(NH_4)_2SO_4$ suspension stable for several months without any loss of enzyme activity [13, 19]

<2>, 0°C, stable to freezing and thawing, incubated for 24 h retains at least 70% of its original activity [2]
<3>, 4°C, purified enzyme, stored in 20% glycerol is stable for months [4]
<7>, 5°C, purified enzyme is stable up to 3 months if care is taken to exclude molecular oxygen from the enzyme solution [20]
<8>, 4°C, retains full activity for several months when stored as an ammonium sulfate suspension or in acetate buffer, pH 6 in presence of a crystal of thymol to prevent bacterial growth [51]
<9>, 4°C, stable for 5 h, loses 20% of its activity after 24%, no activity after 3 days [16]
<11>, 5°C, purified enzyme is stable to storage in 50 mM Tris HCl buffer, pH 7.5, 10 mM dithiothreitol and 20% glycerol, no significant decrease in activity occurs over a 2 week period [22, 23]
<13>, 20°C, in phosphate buffer pH 6.8 in presence of the reducing agent β-mercaptoethanol enzyme remains fully active for 24 h [28]
<28>, 4°C, the isoenzyme with an isolectric point of 6.5 retains virtually all its activity after storing for 30 days, the isoenzyme with isoelectric point 6.3 loses about 90% activity after storage for 3 days [65]
<29>, 4°C, purified enzyme in 50% glycerol buffer remains active for at least 3 months [66]

References

[1] Rippa, M.; Picco, C.; Signorini, M.; Pontremoli, S.: Evidence for a tyrosine residue at the triphosphopyridine nucleotide-binding site of 6-phosphogluconate dehydrogenase. Arch. Biochem. Biophys., **147**, 487-492 (1971)

[2] Emundson, T.D.; Ashworth, J.M.: 6-Phosphogluconate dehydrogenase and the assay of uridine diphosphate glucose pyrophosphorylase in the cellular slime mould Dictyostelium discoideum. Biochem. J., **126**, 593-600 (1972)

[3] Beitner, R.; Naor, Z.: The effect of adenine nucleotides and dehydroepiandrosterone on the isoenzymes of $NADP^+$-and NAD^+-glucose-6-phosphate dehydrogenase and 6-phosphogluconate dehydrogenase from rat adipose tissue. Biochim. Biophys. Acta, **286**, 437-440 (1972)

[4] Procsal, D.; Holten, D.: Purification and properties of rat liver 6-phosphogluconate dehydrogenase. Activity at normal in vivo concentration. Biochemistry, **11**, 1310-1314 (1972)

[5] Schnarrenberger, C.; Oeser, A.; Tolbert, N.E.: Two isozymes each of glucose-6-phosphate dehydrogenase and 6-phosphogluconate dehydrogenase in spinach leaves. Arch. Biochem. Biophys., **154**, 438-448 (1973)

[6] Dyson, J.E.D.; D'Orazio, R.E.; Hanson, W.H.: : Sheep liver 6-phosphogluconate dehydrogenase: Isolation procedure and effect of pH, ionic strength, and metal ions on the kinetic parameters. Arch. Biochem. Biophys., **154**, 623-635 (1973)

[7] Dyson, J.E.D.; D'Orazio, R.E.: Sheep liver 6-phosphogluconate dehydrogenase. Inhibition by nucleoside phosphates and by other metabolic intermediates. J. Biol. Chem., **248**, 5428-5435 (1973)

[8] Scott, W.A.; Abramsky, T.: Neurospora 6-phosphogluconate dehydrogenase. I. Purification and characterization of the wild type enzyme. J. Biol. Chem., **248**, 3535-3544 (1973)

[9] Scott, W.A.; Abramsky, T.: Neurospora 6-phosphogluconate dehydrogenase. II. Properties and two purified mutant enzymes. J. Biol. Chem., **248**, 3542-3545 (1973)

[10] Silverberg, M.; Dalziel, K.: Crystalline 6-phosphogluconate dehydrogenase from sheep liver. Eur. J. Biochem., **38**, 229-238 (1973)

[11] Pearse, B.M.F.; Harris, J.I.: 6-Phosphogluconate dehydrogenase from Bacillus stearothermophilus. FEBS Lett., **38**, 49-52 (1973)

[12] Rippa, M.; Signorini, M.; Dallocchio, F.: A multiple role for the coenzyme in the mechanism of action of 6-phosphogluconate dehydrogenase. J. Biol. Chem., **248**, 4920-4925 (1973)

[13] Veronese, F.M.; Boccu, E.; Fontana, A.; Benassi, C.A.; Scoffone, E.: Isolation and some properties of 6-phosphogluconate dehydrogenase from Bacillus stearothermophilus. Biochim. Biophys. Acta, **334**, 31-44 (1974)

[14] Pearse, B.M.F.; Rosemeyer, M.A.: Human 6-phosphogluconate dehydrogenase. Purification of the erythrocyte enzyme and the influence of ions and NADPH on its activity. Eur. J. Biochem., **42**, 213-223 (1974)

[15] Pearse, B.M.F.; Rosemeyer, M.A.: The molecular weight and subunit structure of human erythrocyte 6-phosphogluconate dehydrogenase. Eur. J. Biochem., **42**, 225-235 (1974)

[16] King, J.: 6-Phosphogluconat-dehydrogenase. Methods Enzym. Anal., 3rd Ed. (Bergmeyer, H.U., ed.), **1**, 668-672 (1974)

[17] Silverberg, M.; Dalziel, K.: 6-Phospho-D-gluconate dehydrogenase from sheep liver. Methods Enzymol., **41**, 214-220 (1975)

[18] Pearse, B.M.F.; Rosemeyer, M.A.: 6-Phosphogluconate from human erythrocytes. Methods Enzymol., **41**, 220-226 (1975)

[19] Rippa, M.; Signorini, M.: 6-Phosphogluconate dehydrogenase from Candida utilis. Methods Enzymol., **41**, 237-240 (1975)

[20] Scott, W.A.; Abramsky, T.: 6-Phosphogluconate dehydrogenase from Neurospora crassa. Methods Immunol., **41**, 227-231 (1975)

[21] Silverberg, M.; Dalziel, K.: Fluorescence studies of coenzyme binding to 6-phosphogluconate dehydrogenase. Arch. Biochem. Biophys., **168**, 646-651 (1975)

[22] Bridges, R.B.; Wittenberger, C.L.: 6-Phosphogluconate dehydrogenase from Streptococcus faecalis. Methods Enzymol., **41**, 232-237 (1975)

[23] Bridges, R.B.; Palumbo, M.P.; Wittenberger, C.L.: Purification and properties of an NADP-specific 6-phosphogluconate dehydrogenase from Streptococcus faecalis. J. Biol. Chem., **250**, 6093-6100 (1975)

[24] Fontana, A.; Grandi, C.; Boccu, E.; Veronese, F.M.: Comparative thermostability of thermophilic and mesophilic 6-phosphogluconate dehydrogenase. Hoppe-Seyler's Z. Physiol. Chem., **356**, 1191-1193 (1975)

[25] Veronese, F.M.; Boccu, E.; Fontana, A.: Denaturation of thermophilic and mesophilic 6-phosphogluconate dehydrogenase by 8 M urea. Int. J. Pept. Protein Res., **7**, 341-343 (1975)

[26] Betts, S.A.; Mayer, R.J.: Purification and properties of 6-phosphogluconate dehydrogenase from rabbit mammary gland. Biochem. J., **151**, 263-270 (1975)
[27] Toews, M.L.; Kanji, M.I.; Carper, R.W.: 6-Phosphogluconate dehydrogenase. Purification and kinetics. J. Biol. Chem., **251**, 7127-7131 (1976)
[28] Veronese, F.M.; Boccu, E.; Fontana, A.: Isolation and properties of 6-phosphogluconate dehydrogenase from Escherichia coli. Some comparisons with the thermophilic enzyme from Bacillus stearothermophilus. Biochemistry, **15**, 4026-4033 (1976)
[29] Adams, M.J.; Helliwell, J.R.; Bugg, C.E.: Structure of 6-phosphogluconate dehydrogenase from sheep liver at 6 A resolution. J. Mol. Biol., **112**, 183-197 (1977)
[30] Dallocchio, F.; Signorini, M.; Rippa, M.: Evidence for the proximity of a cysteinyl and a tyrosyl residue in the active site of 6-phosphogluconate dehydrogenase. Arch. Biochem. Biophys., **185**, 57-60 (1978)
[31] Rippa, M.; Signorini, M.; Pernici, A.; Dallocchio, F.: Evidence for the proximity of two sulfhydryl groups at the active site of 6-phosphogluconate dehydrogenase. Arch. Biochem. Biophys., **186**, 406-410 (1978)
[32] Rippa, M.; Signorini, M.; Bellini, T.; Dallocchio, F.: The active site of 6-phosphogluconate dehydrogenase. A phosphate binding site and its surroundings. Arch. Biochem. Biophys., **189**, 516-523 (1978)
[33] Adams, M.J.; Archibald, I.G.; Helliwell, J.R.; Jenkins, S.E.: The crystal structure of 6-phosphogluconate dehydrogenase. Biochem. Soc. Trans., **6**, 1112-1115 (1978)
[34] Simcox, P.D.; Dennis, D.T.: 6-Phosphoglyconate dehydrogenase isoenzymes from the developing endosperm of Ricinus communis L.. Plant Physiol., **62**, 287-290 (1978)
[35] Kato, N.; Sahm, H.; Schuette, Horst; Wagner, F.: Purification and properties of glucose-6-phosphate dehydrogenase and 6-phosphogluconate dehydrogenase from a methanol-utilizing yeast, Candida boidinii. Biochim. Biophys. Acta, **566**, 1-11 (1979)
[36] Gosling, P.G.; Ross, J.D.: Characterization of glucose-6-phosphate dehydrogenase and 6-phosphogluconic acid dehydrogenase from hazel cotyledons. Phytochemistry, **18**, 1441-1445 (1979)
[37] Rippa, M.; Bellini, T.; Signorini, M.; Dallocchio, F.: The stabilization by a coenzyme analog of a conformational change induced by substrate in 6-phosphogluconate dehydrogenase. Arch. Biochem. Biophys., **196**, 619-623 (1979)
[38] Orozco de Silva, A.; Fraenkel, D.G.: The 6-phosphogluconate dehydrogenase reaction in Escherichia coli. J. Biol. Chem., **254**, 10237-10242 (1979)
[39] Abdallah, M.A.; Adams, M.J.; Archibald, I.G.; Bielmann, J.-F.; Helliwell, J.R.; Jenkins, S.E.: Binding of coenzyme and substrate and coenzyme analogues to 6-phosphogluconate dehydrogenase from sheep liver. Eur. J. Biochem., **98**, 121-130 (1979)
[40] Beitner, R.; Nordenberg, J.: Inhibition of 6-phosphogluconate dehydrogenase(decarboxylating) by glucose 1,6-bisphosphate. Biochim. Biophys. Acta, **583**, 266-269 (1979)

[41] Adkins, S.W.; Gosling, P.G.: Ross, J.D.: Glucose-6-phosphate dehydrogenase and 6-phosphogluconic acid dehydrogenase of wild oat seeds. Phytochemistry, **19**, 2523-2525 (1980)
[42] Graefe, U.; Bormann, E.J.; Truckenbrodt, G.: Control by phospho-adenosinediphospho-ribose of NADP-dependent isocitrate dehydrogenase and 6-phosphogluconate dehydrogenase in Streptomyces griseus. Z. Allg. Mikrobiol., **20**, 607-611 (1980)
[43] Al-Quadan, F.; Wender, S.H.; Smith, E.C.: The effect of some phenolic compounds on the activity of 6-phosphogluconate dehydrogenase from tobacco tissue cultures. Phytochemistry, **20**, 961-964 (1981)
[44] Al-Quadan, F.; Wender, S.H.; Smith, E.C.: Comparison of tryptic peptide maps of two isoenzymes of 6-phosphogluconate dehydrogenase from tobacco tissue cultures. Phytochemistry, **20**, 1201-1203 (1981)
[45] Bublitz, C.: Physical separation of cytoplasmic and microsomal 6-phosphogluconate dehydrogenase from rat liver. Biochem. Biophys. Res. Commun., **98**, 588-594 (1981)
[46] Dallocchio, F.; Matteuzzi, M.; Bellini, T.: Effect of the substrate on the binding of coenzyme and coenzyme analogues to 6-phosphogluconate dehydrogenase from Candida utilis. J. Biol. Chem., **256**, 10778-10780 (1981)
[47] Jessup, W.; Dean, P.D.G.: Purification of 6-phosphogluconate dehydrogenase from Acer pseudoplatanus L. using immobilised procion red He-3B. J. Chromatogr., **219**, 419-426 (1981)
[48] Wong, C.-H.; Pollak, A.; McCurry, S.D.; Sue, J.M.; Knowles, J.R.; Whitesides, G.M.: Synthesis of ribulose 1,5-bisphosphate: Routes from glucose-6-phosphate (via 6-phosphogluconate) and from adenosine monophosphate (via ribose 5-phosphate). Methods Enzymol., **89**, 108-121 (1982)
[49] Carne, A.: Purification by dye-ligand chromatography of an NADP-dependent enzyme, 6-phosphogluconate dehydrogenase from lamb's liver. Anal. Biochem., **121**, 227-229 (1982)
[50] Dallocchio, F.; Matteuzzi, M.; Bellini, T.: Non-enzymic protein phosphorylation. Biochem. J., **203**, 401-404 (1982)
[51] Veronese, F.M.; Boccu, E.; Fontana, A.: 6-Phosphogluconate dehydrogenase from Bacillus stearothermophilus. Methods Enzymol., **89**, 282-291 (1982)
[52] Adams, M.J.; Archibald, I.G.; Bugg, C.E.; Carne, A.; Gover, S.; Helliwell, J.R.; Pickersgill, R.W.; White, S.W.: The three dimensional structure of sheep liver 6-phosphogluconate dehydrogenase at 2.6 A resolution. EMBO J., **6**, 1009-1014 (1983)
[53] Adams, M.J.; Gover, S.; Pickersgill, R.W.; Helliwell, J.R.: Structure-function relationships of NAD^+ and $NADP^+$-dependent dehydrogenases with particular reference to the three-dimensional structure of 6-phosphogluconate dehydrogenase. Biochem. Soc. Trans., **11**, 429-435 (1983)
[54] Miernyk, J.A.; MacDougall, P.S.; Dennis, D.T.: In vitro inhibition of the plastid and cytosolic isozymes of 6-phosphogluconate dehydrogenase from developing endosperm of Ricinus communis by fructose 2,6-bisphosphate. Plant Physiol., **76**, 1093-1094 (1984)

[55] Rendina, A.R.; Hermes, J.D.; Cleland, W.W.: Use of multiple isotope effects to study the mechanism of 6-phosphogluconate dehydrogenase. Biochemistry, **23**, 6257-6262 (1984)
[56] Lubberding, H.J.; Bot, P.V.M.: The influence of temperature on glucose-6-phosphate dehydrogenase and 6-phosphogluconate dehydrogenase and the regulation of these enzymes in a mesophilic and a thermophilic cyanobacterium. Arch. Microbiol., **137**, 115-120 (1984)
[57] Veronese, F.M.; Boccu, E.; Schiavon, O.; Grandi, C.; Fontana, A.: General stability of thermophilic enzymes: Studies on 6-phosphogluconate dehydrogenase from Bacillus stearothermophilus and yeast. J. Appl. Biochem., **6**, 39-47 (1984)
[58] Dallocchio, F.; Matteuzzi, M.; Bellini, T.: Half-site reactivity in 6-phosphogluconate dehydrogenase from human erythrocytes. Biochem. J., **227**, 305-310 (1985)
[59] Sawa, Y.; Suzuki, K.; Ochiai, H.: Purification and characterization of 6-phosphogluconate dehydrogenase from Phormidium sp.. Agric. Biol. Chem., **49**, 2543-2549 (1985)
[60] Topham, C.M.; Dalziel, K.: Chemical modification of sheep-liver 6-phosphogluconate dehydrogenase by diethylpyrocarbonate. Eur. J. Biochem., **155**, 87-94 (1986)
[61] Topham, C.M.; Matthews, B.; Dalziel, K.: Kinetic studies of 6-phosphogluconate dehydrogenase from sheep liver. Eur. J. Biochem., **156**, 555-567 (1986)
[62] Sugimoto, S.-i.; Shiho, I.: Regulation of 6-phosphogluconate dehydrogenase in Brevibacterium flavum. Agric. Biol. Chem., **51**, 1257-1263 (1987)
[63] Hansen, R.J.; Jungermann, K.: Sex differences in the control of glucose-6-phosphate dehydrogenase and 6-phosphogluconate dehydrogenase. Biol. Chem. Hoppe-Seyler, **368**, 955-962 (1987)
[64] del mar Medina-Puerta, M.; Gallego-Iniesta, M.; Garido-Pertierra, A.: Kinetic analysis of 6-phosphogluconate dehydrogenase from bass liver: Effects of temperature and pH on its catalytic function. Arch. Biochem. Biophys., **262**, 130-141 (1988)
[65] Kletsova, L.V.; Kiryukhin, M.Y.; Chistoserdov, A.Y.; Tsygankov, Y.D.: Glucose-6-phosphate dehydrogenase and 6-phosphogluconate dehydrogenase from Methylobacillus flagellatum. Methods Enzymol., **188**, 335-339 (1990)
[66] Sokolov, A.P.; Trotsenko, Y.A.: Glucose-6-phosphate dehydrogenase and 6-phosphogluconate dehydrogenase from Arthrobacter globiformis. Methods Enzymol., **188**, 339-345 (1990)
[67] Hanau, S.; Dallocchio, F.; Rippa, M.: NADPH activates a decarboxylation reaction catalyzed by lamb liver 6-phosphogluconate dehydrogenase. Biochim. Biophys. Acta, **1122**, 273-277 (1992)
[68] Ishida, S.; Ashihara, H.: Isoenzymes of phosphogluconate dehydrogenase (decarboxylating) from suspension-cultured Cataranthus roseus cells. Phytochemistry, **33**, 1307-1311 (1993)
[69] Berdis, A.J.; Cook, P.F.: Overall kinetic mechanism of 6-phosphogluconate dehydrogenase from Candida utilis. Biochemistry, **32**, 2036-2040 (1993)

[70] Berdis, A.J.; Cook, P.F.: Chemical mechanism of 6-phosphogluconate dehydrogenase from Candida utilis from pH studies. Biochemistry, **32**, 2041-2046 (1993)

[71] Berdis, A.J.; Cook, P.F.: The 2'-phosphate of NADP is critical for optimum productive binding to 6-phosphogluconate dehydrogenase from Candida utilis. Arch. Biochem. Biophys., **305**, 551-558 (1993)

[72] Corpas, F.J.; Garcia-Salguero, L.; Barroso, J.B.; Aranda, F.; Lupianez, J.A.: Kinetic properties of hexose-monophosphate dehydrogenases. II. Isolation and partial purification of 6-phosphogluconate dehydrogenase from rat liver and kidney cortex. Mol. Cell. Biochem., **144**, 97-104 (1995)

[73] Signorini, M.; Bregoli, A.M.; Caselli, L.; Bergamini, C.M.: Purification and properties of 6-phosphogluconate dehydrogenase from beet leaves. Biochem. Mol. Biol. Int., **35**, 669-675 (1995)

[74] Ganea, E.; Harding, J.J.: Inhibition of 6-phosphogluconate dehydrogenase by carbamylation and protection by α-crystallin, a chaperone-like protein. Biochem. Biophys. Res. Commun., **222**, 626-631 (1996)

[75] Price, N.E.; Cook, P.F.: Kinetic and chemical mechanisms of the sheep liver 6-phosphogluconate dehydrogenase. Arch. Biochem. Biophys., **336**, 215-223 (1996)

[76] Phillips, C.; Dohnalek, J.; Barrett, M.P.; Adams, M.J.: A 2.8 A resolution structure of 6-phosphogluconate dehydrogenase from the protozoan parasite Trypanosoma brucei: Comparison with the sheep enzyme accounts for differences in activity with coenzyme and substrate analogues. J. Mol. Biol., **25**, 667-681 (1998)

[77] Tetaud, E.; Hall, D.R.; Gourley, D.G.; Leonard, G.A.; Arkison, S.; Barrett, M.P.; Hunter, W.N.: Crystallization and preliminary x-ray diffraction studies of 6-phosphogluconate dehydrogenase from Lactococcus lactis. Acta Crystallogr. Sect. D, **54**, 1422-1424 (1998)

[78] Tsai, C.S.; Chen, Q.: Purification and kinetic characterization of 6-phosphogluconate dehydrogenase from Schizosaccharomyces pombe. Biochem. Cell Biol., **76**, 637-644 (1998)

[79] Chooback, L.; Price, N.E.; Karsten, W.E.; Nelson, J.; Sundstrom, P.; Cook, P.F.: Cloning, expression, purification, and characterization of the 6-phosphogluconate dehydrogenase from sheep liver. Protein Expr. Purif., **13**, 251-258 (1998)

[80] Rippa, M.; Giovannini, P.P.; Barrett, M.P.; Dallocchio, F.; Hanau, S.: 6-Phosphogluconate dehydrogenase: The mechanism of action investigated by a comparison of the enzyme from different species. Biochim. Biophys. Acta, **1429**, 83-92 (1998)

[81] Moritz, B.; Striegel, K.; de Graaf, A.A.; Sahm, H.: Kinetic properties of the glucose-6-phosphate and 6-phosphogluconate dehydrogenases from Corynebacterium glutamicum and their application for predicting pentose phosphate pathway flux in vivo. Eur. J. Biochem., **267**, 3442-3452 (2000)

L-Gulonate 3-dehydrogenase 1.1.1.45

1 Nomenclature

EC number
1.1.1.45

Systematic name
L-gulonate:NAD^+ 3-oxidoreductase

Recommended name
L-gulonate 3-dehydrogenase

Synonyms
L-3-aldonate dehydrogenase
L-3-aldonic dehydrogenase
L-3-hydroxyacid dehydrogenase
L-β-hydroxy-acid-NAD-oxidoreductase
L-β-hydroxyacid dehydrogenase
L-gulonic acid dehydrogenase

CAS registry number
9028-51-7

2 Source Organism

<1> *Sus scrofa* (hog) [1]
<2> *Schwanniomyces occidentalis* [2]
<3> *Drosophila melanogaster* [3, 5, 6, 7]
<4> *Ovis aries* (sheep) [4]
<5> *Rattus norvegicus* (rat) [8]
<6> *Bos taurus* [8]

3 Reaction and Specificity

Catalyzed reaction
L-gulonate + NAD^+ = 3-dehydro-L-gulonate + NADH

Reaction type
oxidation
reduction

Natural substrates and products

S L-gulonic acid + NAD^+ <1-3, 5, 6> (<1, 2, 5, 6> glucuronate-xylulose cycle [1, 2, 8]) (Reversibility: r <1> [1]) [1-3, 5-8]
P 3-dehydro-L-gulonic acid + NADH <1> [1]

Substrates and products

S 2-keto-L-gulonate + NADH <1> (Reversibility: ir <1> [1]) [1]
P 2,3-diketo-L-gulonate + NAD^+
S D-galactonic acid + NAD^+ <1> [1]
P ?
S D-gluconic acid + NAD^+ <1, 2> [1, 2]
P ?
S D-lyxonic acid + NAD^+ <1, 2> [1, 2]
P ?
S D-mannonic acid + NAD^+ <1> [1]
P ?
S D-talonic acid + NAD^+ <1> [1]
P ?
S D-xylonic acid + NAD^+ <1> [1]
P ?
S L-3-hydroxybutanoate + NADH <1, 3, 4> (Reversibility: r <1, 3, 4> [1, 3, 4]) [1, 3, 4]
P acetoacetate + NAD^+
S L-altronic acid + NAD^+ <1> [1]
P ?
S L-erythronic acid + NAD^+ <1> [1]
P ?
S L-gulonic acid + NAD^+ <1-3, 5, 6> (<1, 2> stereospecificity for L-configuration of the β-carbon hydroxyl group [1, 2]) (Reversibility: r <1-3, 5, 6> [1-3, 5-8]) [1-3, 5-8]
P 3-dehydro-L-gulonic acid + NADH
S L-idonic acid + NAD^+ <1, 2> [1, 2]
P ?
S L-ribonic acid + NAD^+ <1> [1]
P ?
S L-threonic acid + NAD^+ <1> [1]
P ?
S Additional information <1> (<1> no substrates are: L-arabonic acid, D-glucuronate, D-mannuronate, l-iduronate, 6-phospho-D-gluconate, L-serine, L-threonine, β-D-hydroxy acids, acetoacetyl-CoA, acetoacetyl-S-pantetheine) [1]
P ?
S Additional information <3> (<3> no substrates are: DL-2-hydroxybutanoate, DL-4-hydroxybutanoate, 2-butanol, DL-3-hydroxypropionate, L-threonine) [3]
P ?

Inhibitors

NADH <3> (<3> competitive inhibitor [6]) [6]
acetoacetate <3> (<3> competitive inhibitor [6]) [6]
iodoacetate <2> [2]
p-chloromercuribenzoate <1, 2> (<1> in both directions [1]) [1, 2]
phenylmercurinitrate <2> [2]
Additional information <1> (<1> not inhibitory iodoacetate, ethylendiamine tetraacetate, 2,2'-bipyridine, 8-hydroxyquiniline, O-phenanthroline, diethyldithiocarbamate, thiamine pyrophosphate [1]) [1]

Cofactors/prosthetic groups

NAD^+ <1-6> [1-8]
NADH <1, 3-6> [1, 3-8]
$NADP^+$ <3> (<3> 10% activity of NAD^+ [6]) [6]

Activating compounds

cysteine <1> (<1> in mM-concentration, no activation of the reverse reaction [1]) [1]

Metals, ions

Additional information <1> (<1> no activation or inhibition Zn^{2+}, Cd^{2+}, Cu^{2+}, Fe^{3+}, Mn^{2+}, Ni^{2+}, Co^{3+} [1]) [1]

Turnover number (min^{-1})

44 <3> (DL-3-hydroxybutanoate) [6]
444 <3> (L-gulonate) [6]

Specific activity (U/mg)

12 <3> [3, 6]
21 <3> [7]
63.8 <3> [7]

K_m-Value (mM)

0.04 <1> (NAD^+, <1> L-gulonate, pH 8.5 [1]) [1]
0.09 <3> (NAD^+, <3> L-gulonate, pH 8.2 [6]) [6]
0.13 <3> (NAD^+, <3> L-gulonate, pH 8.2 [6]) [6]
0.25 <3> (L-gulonate, <3> NAD^+ [3]) [3]
0.45 <1> (β-keto-L-gulonate, <1> NADH, pH 6.3 [1]) [1]
0.85 <3> (L-gulonate, <3> NAD^+, pH 8.2 [6]) [6]
1.1 <2> (NAD^+, <2> L-gulonate, pH 7.5 and pH 8.5 [2]) [2]
5.1 <1> (L-gulonate, <1> NAD^+, pH 8.5 [1]) [1]
5.3 <1> (L-3-hydroxybutanoate) [1]
5.3 <1> (L-idonate, <1> NAD^+, pH 8.5 [1]) [1]
6.4 <1> (acetoacetate, <1> NADH, pH 6.3 [1]) [1]
7.3 <1> (2,3-diketo-L-gulonate, <1> NADH, pH 6.3 [1]) [1]
14.9 <3> (L-3-hydroxybutanoate, <3> NAD^+, pH 8.2 [6]) [6]
19 <2> (D-gluconate, <2> NAD^+, pH 8.5 [2]) [2]
22.8 <3> (L-3-hydroxybutanoate, <3> NAD^+ [3]) [3]
55 <2> (L-gulonate, <2> NAD^+, pH 8.5 [2]) [2]

67 <2> (L-gulonate, <2> NAD^+, pH 7.5 [2]) [2]
Additional information <1, 3> (<1, 3> kinetic studies [1, 3]) [1, 3]

pH-Optimum

6.3 <1> (<1> reduction of 3-dehydro-L-gulonic acid [1]) [1]
8.2 <3> [3, 6]
8.5 <1> [1]
9 <2> (<2> oxidation of L-gulonic acid [2]) [2]

pH-Range

4-8 <1> (<1> reduction of acetoacetate [1]) [1]
6.2-10 <1> (oxidation of L-3-hydroxybutanoate [1]) [1]
7-10 <3> (oxidation of 3-hydroxybutanoate [3]) [3]

4 Enzyme Structure

Molecular weight

60000-70000 <5, 6> (<5, 6> HPLC gel filtration [8]) [8]
63000 <3> (<3> sucrose density gradient centrifugation [3]) [3]
63500 <3> (<3> SDS-PAGE [6]) [6]
66000-70000 <3> (<3> 33000/unit in vitro, 35000/unit in vivo, SDS-PAGE [7]) [7]

Subunits

dimer <3> (<3> $\alpha\beta$, 1 * 23500 + 1 * 40000, SDS-PAGE [6]; <3> α_2, 2 * 33000-35000, SDS-PAGE [7]) [6, 7]

Posttranslational modification

no modification <3> [7]

5 Isolation/Preparation/Mutation/Application

Source/tissue

brain <3> [5]
fat body <3> [5]
heart <4> [4]
intestine <3> [5]
kidney <1, 4> [1, 4]
lens <5, 6> [8]
liver <4> [4]
malpighian tubule <3> [5]
rumen epithelium <4> [4]

Localization

cytosol <1, 4> [1, 4]

Purification

<1> [1]
<2> (67% yield [2]) [2]
<3> (12%-52% yield [7]) [3, 7]

Application

medicine <5, 6> (<5, 6> role in an alternative pathway of the production of xylitol in the lens that causes cataracts in mammalian organisms [8]) [8]

6 Stability

Temperature stability

Additional information <2> (<2> in solution instable, by addition of below mM-conc. of cyanide several days in the cold, minor effect through EDTA-addition [2]) [2]

General stability information

<3>, 3 M urea results in loss of activity of 83% in 7 min. [6]

Storage stability

<1>, -18°C, 0.05 M K_2HPO_4, several weeks, little loss of activity [1]

References

[1] Smiley, J.D.; Ashwell, G.: Purification and properties of β-L-hydroxy acid dehydrogenase. J. Biol. Chem., **236**, 357-364 (1961)
[2] Dworsky, P.; Hoffmann-Ostenhof, O.: L-3-Aldonic acid dehydrogenase from Schwanniomyces occidentalis. Acta Biochim. Pol., **9**, 269-277 (1964)
[3] Borack, L.I.; Sofer, W.: Drosophila β-L-hydroxy acid dehydrogenase. J. Biol. Chem., **246**, 5345-5350 (1970)
[4] Koundakjian, P.P.; Snoswell, A.M.: 3-Hydroxy acid dehydrogenases in Sheep tissues. Biochem. J., **127**, 449-452 (1972)
[5] Borack, L.I.: Organ distribution of Drosophila L-3-hydroxyacid: NAD oxidase. Experientia, **30**, 31 (1974)
[6] Cannistraro, V.J.; Borack, L.I.; Chase, T.: Subunit structure and kinetic properties of L-β-hydroxyacid dehydrogenase of Drosophila. Biochim. Biophys. Acta, **569**, 1-5 (1979)
[7] Menotti-Raymond, M.; Sullivan, D.T.: Further characterization of L-β-hydroxyacid dehydrogenase from Drosophila. Biochim. Biophys. Acta, **841**, 15-21 (1985)
[8] Goode, D.; Lewis, M.E.; Crabbe, J.C.: Accumulation of xylitol in the mammalian lens is related to glucuronate metabolism. FEBS Lett., **395**, 174-178 (1996)

L-Arabinose 1-dehydrogenase 1.1.1.46

1 Nomenclature

EC number
1.1.1.46

Systematic name
L-arabinose:NAD^+ 1-oxidoreductase

Recommended name
L-arabinose 1-dehydrogenase

Synonyms
arabinose dehydrogenase
dehydrogenase, L-arabinose

CAS registry number
9028-52-8

2 Source Organism

<1> *Rhizobium leguminosarum* (strain MNF 300 [1]) [1, 3]
<2> *Rhizobium sp.* (strain NGR 234 [1]; strain 32111 [3]) [1, 3]
<3> *Azospirillum brasilense* [2, 7]
<4> *Rhizobium meliloti* [3]
<5> *Rhizobium trifolii* [3]
<6> *Rhizobium phaseoli* [3]
<7> *Rhizobium japonicum* [3, 6]
<8> *Pseudomonas fluorescens* [4]
<9> *Pseudomonas saccharophila* [5]

3 Reaction and Specificity

Catalyzed reaction
L-arabinose + NAD^+ = L-arabinono-1,4-lactone + NADH

Reaction type
oxidation
redox reaction
reduction

Natural substrates and products

S L-arabinose + NAD(P)$^+$ <1-9> (Reversibility: r <1-9> [1-7]) [1-7]
P L-arabinono-1,4-lactone + NAD(P)H

Substrates and products

S L-arabinose + NAD(P)$^+$ <1-9> (<3> specific for substrates with L-arabino configuration at carbons 2,3 and 4 [2]) (Reversibility: r <1-9> [1-7]) [1-7]
P L-arabinono-1,4-lactone + NAD(P)H

Inhibitors

ATP <9> [5]
$CaCl_2$ <3> [2]
$CoCl_2$ <3> [2]
$FeSO_4$ <3> [2]
$MnCl_2$ <3> [2]

Cofactors/prosthetic groups

NAD$^+$ <1-7, 9> (<9> specific for [5]; <3> preferred [7]) [2, 3, 5-7]
NADP$^+$ <1-3, 7> (<3> greater affinity for NADP$^+$ than for NAD$^+$ [2]) [1, 2, 6, 7]

Metals, ions

Additional information <3> (<3> no activation by divalent cations [2]) [2]

Specific activity (U/mg)

0.006 <7> [3]
0.014 <4> [3]
0.034 <2> [3]
0.042 <1> [3]
0.043-0.101 <1> (<1> different values for different growth media [1]) [1]
0.068 <7> [6]
0.07 <3> [7]
0.102 <6> [3]
0.184 <5> [3]
4.12 <3> [2]

K_m-Value (mM)

0.075 <3> (L-arabinose, <3> coenzyme NADP$^+$ [2]) [2]
0.14 <3> (L-arabinose, <3> coenzyme NAD$^+$ [2]) [2]

pH-Optimum

8 <9> [5]
9.5 <3> [2]

4 Enzyme Structure

Molecular weight

175000 <3> (<3> gel filtration [2]) [2]

5 Isolation/Preparation/Mutation/Application

Purification

<3> (partial [2]) [2]
<9> (partial [5]) [5]

6 Stability

Temperature stability

55 <3> (<3> stable for 5 min [2]) [2]

References

[1] Dilworth, M.J.; Arwas, R.; McKay, I.A.; Saroso, S.; Glenn, A.R.: Pentose metabolism in Rhizobium leguminosarum MNF300 and in Cowpea Rhizobium NGR234. J. Gen. Microbiol., **132**, 2733-2742 (1986)
[2] Novick, N.J.; Tyler, M.E.: Partial purification and properties of an L-arabinose dehydrogenase from Azospirillium brasilense. Can. J. Microbiol., **29**, 242-246 (1983)
[3] Duncan, M.J.: L-Arabinose metabolism in Rhizobia. J. Gen. Microbiol., **113**, 177-179 (1979)
[4] Schimz, K.-L.; Kurz, G.: Regulation of L-arabinose metabolism in Pseudomonas fluorescens. Biochem. Soc. Trans., **3**, 1087-1089 (1975)
[5] Weimberg, R.; Doudoroff, M.: The oxidation of arabinose by Pseudomonas saccharophila. J. Biol. Chem., **217**, 607-624 (1955)
[6] Pedrosa, F.O.; Zancan, G.T.: L-Arabinose metabolism in Rhizobium japonicum. J. Bacteriol., **119**, 336-338 (1974)
[7] Novick, N.J.; Tyler, M.E.: L-Arabinose metabolism in Azospirillum brasiliense. J. Bacteriol., **149**, 364-367 (1982)

Glucose 1-dehydrogenase 1.1.1.47

1 Nomenclature

EC number

1.1.1.47

Systematic name

β-D-glucose dehydrogenase: $NAD(P)^+$ 1-oxidoreductase

Recommended name

glucose 1-dehydrogenase

Synonyms

D-glucose dehydrogenase (NAD(P))
GSP74 [Swissprot]
General stress protein 74 [Swissprot]
Hexose-6-phosphate dehydrogenase [Swissprot]
hexose phosphate dehydrogenase

CAS registry number

9028-53-9

2 Source Organism

<1> *Bacillus megaterium* [1, 7, 11, 14, 15, 18, 31, 32]
<2> *Halobacterium salinarum* [2]
<3> *Thermoproteus tenax* [3]
<4> *Haloferax mediterranei* (halophilic archae [4]) [4]
<5> *Thermoplasma acidophilum* [5, 8, 12]
<6> *Bos taurus* [6, 25]
<7> *Oncorhynchus mykiss* [6]
<8> *Bacillus subtilis* (strain NS168I-2 [19]) [9, 10, 19, 21, 30]
<9> *Sus scrofa* [13, 20, 22, 35]
<10> *Sulfolobus solfataricus* [16]
<11> *Nostoc sp.* (strain MAC [17]) [17, 33]
<12> *Gluconobacter suboxydans* [23, 28]
<13> *Corynebacterium 93-1* [24, 27]
<14> *Corynebacterium 150-1* [24, 26]
<15> *Bacillus sp.* (strain BG 1722, D-ribose producing mutant [29]) [29]
<16> *Schizothrix calciola* [33]
<17> *Lyngbya lagerheimii* [33]
<18> *Chlorogloea fritschii* [33]
<19> *Salmo gairdneri* [34]

3 Reaction and Specificity

Catalyzed reaction

β-D-glucose + NAD(P)$^+$ = D-glucono-1,5-lactone + NAD(P)H

Reaction type

oxidation (<2> ordered Bi Bi, formation of ternary complexes, NADP$^+$ adds first, NADPH formed dissociates last, for reverse reaction, NADPH adds first, NADP$^+$ leaves last [2]) [2]
redox reaction
reduction

Natural substrates and products

S β-D-glucose + NAD(P)$^+$ <1-19> (<3> 100fold higher affinity for NADP$^+$ [3]; <2> coenzyme NADP [2]) (Reversibility: r <2> [2]; ir <3> [3]; r <4> [4]) [1-35]
P D-glucono-1,5-lactone + NAD(P)H <1-19> [1-35]

Substrates and products

S 2-amino-2-deoxy-D-glucose + NAD(P)$^+$ <1, 10> (Reversibility: ? <1> [32]; ? <10> [16]) [16, 32]
P 2-amino-2-deoxy-D-glucono-1,5-lactone + NAD(P)H <1, 10> [16, 32]
S 2-deoxy-D-glucose + NAD(P)$^+$ <1, 8, 10, 15> (Reversibility: ? <1> [7]; ? <8> [21] ? <10> [16]; ? <15> [29]) [7, 16, 21, 29, 30, 32]
P 2-deoxy-D-glucono-1,5-lactone + NAD(P)H <1, 8, 10, 15> [7, 16, 21, 29, 30, 32]
S 2-deoxyglucose 6-phosphate + NAD(P)$^+$ <19> (Reversibility: ? <19> [34]) [34]
P 2-deoxyglucono-1,5-lactone 6-phosphate + NAD(P)H <19> [34]
S 4-O-(β-D-glucopyranosido)-β-D-glucopyranoside + NAD$^+$ <13, 14> (<13> 100 mM, 89% activity compared to glucose, 20 mM, 158% [27]) (Reversibility: ? <13, 14> [24]) [24, 27]
P 4-O-(β-D-glucopyranosido)-D-glucono-1,5-lactone + NADH <13, 14> [24, 27]
S 6-O-(β-D-glucopyranosido)-β-D-glucopyranoside + NAD$^+$ <13, 14> (<13> 100 mM, 86% activity compared to glucose, 20 mM, 158% [27]) (Reversibility: ? <13, 14> [24]) [24, 27]
P 6-O-(β-D-glucopyranosido)-D-glucono-1,5-lactone + NADH <13, 14> [24, 27]
S 6-deoxy-D-glucose + NAD(P)$^+$ <10> (Reversibility: ? <10> [16]) [16]
P 6-deoxy-D-glucono-1,5-lactone + NAD(P)H <10> [16]
S D-allose + NADP$^+$ <10> (Reversibility: ? <10> [16]) [16]
P D-allono-1,5-lactone + NADPH <10> [16]
S D-altrose + NADP$^+$ <10> (Reversibility: ? <10> [16]) [16]
P D-altrono-1,5-lactone + NADPH <10> [16]
S D-fructose + NAD(P)$^+$ <1, 4, 14> (<4> 3% relative activity compared with glucose, 1% with NAD as coenzyme [4]; <14> 5% activity compared to glucose [24]) (Reversibility: ? <4> [4]; ? <1> [7]; ? <14> [24]) [4, 7, 24]

P D-fructono-lactone + NAD(P)H <1, 4, 14> [4, 7, 24]

S D-fucose + NADP$^+$ <4> (<4> 18% relative activity compared with glucose, 3% with NAD$^+$ as coenzyme [4]) (Reversibility: ? <4> [4]) [4]

P D-fucono-1,5-lactone + NADPH <4> [4]

S D-galactose + NAD(P)$^+$ <1, 4, 5, 10, 12, 19> (<4> 13% relative activity compared with glucose, 1% with NAD$^+$ as coenzyme [4]; <5> 70% activity compared with glucose [12]; <10> coenzyme NADP$^+$ [16]; <12> 1% activity compared to glucose [23]) (Reversibility: ? <1, 4, 5, 10, 12, 19> [4, 7, 12, 16, 23, 34]) [4, 7, 12, 16, 23, 34]

P D-galactono-1,5-lactone + NAD(P)H <1, 4, 5, 10, 12, 19> [4, 7, 12, 16, 23, 34]

S D-glucose 6-phosphate + NAD(P)$^+$ <6, 7, 9, 12, 19> (<12> 1% activity compared to glucose [23]) (Reversibility: ? <6, 7> [6]; ? <9> [13]; ? <12> [23]; ? <19> [34]) [6, 13, 20, 23, 34]

P D-glucono-1,5-lactone 6-phosphate + NAD(P)H <6, 7, 9, 12, 19> [6, 13, 20, 23, 34]

S D-glucoseamine + NAD(P)$^+$ <1, 4, 8> (<4> 6% relative activity compared with glucose, 5% with NAD$^+$ as coenzyme [4]) (Reversibility: ? <4> [4]; ?<1> [7]; ? <8> [21]) [4, 7, 21]

P D-gluconamine-1,5-lactone + NAD(P)H <1, 4, 8> [4, 7, 21]

S D-gulose + NADP$^+$ <10> (Reversibility: ? <10> [16]) [16]

P D-gulono-1,5-lactone + NADPH <10> [16]

S D-iodose + NADP$^+$ <10> (Reversibility: ? <10> [16]) [16]

P D-iodono-1,5-lactone + NADPH <10> [16]

S D-mannose + NAD(P)$^+$ <1, 4, 10, 12> (<4> 7% relative activity compared with glucose, 7% with NAD$^+$ as coenzyme [4]; <10> coenzyme NADP$^+$ [16]; <12> 88% activity compared to glucose) (Reversibility: ? <4> [4]; ? <1> [7]; ? <10> [16]; ? <12> [23]) [4, 7, 16, 23, 28]

P D-mannono-1,5-lactone + NAD(P)H <1, 4, 10, 12> [4, 7, 16, 23, 28]

S D-ribose + NADP$^+$ <4, 10> (<4> 3% relative activity compared with glucose, 0% with NAD$^+$ as coenzyme [4]; <10> coenzyme NADP [16]) (Reversibility: ? <4> [4]; ? <10> [16]) [4, 16]

P D-ribono-1,4-lactone + NADPH <4, 10> [4, 16]

S β-D-glucose + NAD(P)$^+$ <1-19> (<3> 100fold higher affinity for NADP [3]; <2> coenzyme NADP$^+$ [2]; <12> coenzyme NADP$^+$, 4% of reaction in reverse direction [23]; <14> coenzyme NAD$^+$ [24]; <13> coenzyme NAD$^+$, 5.9% velocity for reverse reaction compared to forward reaction, pH 8.0 [27]; <16,17> coenzyme NADP$^+$ [33]) (Reversibility: r <2, 4, 12, 13> [2, 4, 23, 27]; ir <3> [3]; ? <1, 5-11, 14-19> [1, 4-22, 24, 25, 26, 28-35]) [1-35]

P D-glucono-1,5-lactone + NAD(P)H <1-19> [1-35]

S galactose-6-phosphate + NAD(P)$^+$ <19> (Reversibility: ? <19> [34]) [34]

P galactono-1,5-lactone-6-phosphate + NAD(P)H <19> [34]

S xylose + NAD(P)$^+$ <1, 3, 4, 10, 13, 14, 19> (<4> 88% relative activity compared with glucose, 79% with NAD$^+$ as coenzyme [4]; <14> 47% activity compared to glucose [24]; <13> 44% activity compared to glucose

[27]) (Reversibility: ? <1, 3, 4, 10, 13, 14, 19> [3, 4, 7, 16, 24, 27, 34]) [3, 4, 7, 16, 24, 27, 34]

P xylono-1,5-lactone + NAD(P)H <1, 3, 4, 10, 13, 14, 19> [3, 4, 7, 16, 24, 27, 34]

Inhibitors

$(NH_4)_2SO_4$ <13> (<13> 71% inhibition [27]) [27]
2-mercaptoethanol <10> (<10> 1 mM, room temperature, pH 9.0, rapid inactivation [16]) [16]
5,5'-dithiobis-(2-nitrobenzoic acid) <10> (<10> 1 mM, room temperature, pH 9.0, 50% inactivation [16]) [16]
Ag^+ <13> (<13> 99% inhibition [27]) [27]
Ca^{2+} <13> (<13> 14% inhibition [27]) [27]
Co^{2+} <13> (<13> 74% inhibition [27]) [27]
Cu^{2+} <9, 12, 13> (<13> 14% inhibition [27]) [23, 27, 28, 35]
D-glucono-1,5-lactone <2> (<2> forward reaction, oxidation, product inhibition, noncompetitive vs. *β*-D-glucose and $NADP^+$ [2]) [2]
EDTA <4, 10> (<4> 2 mM [4]; <10> 50 mM, 5% activity left [16]) [4, 16]
Fe^{2+} <9, 13> (<13> 41% inhibition [27]) [27, 35]
Fe^{3+} <13> (<13> 41% inhibition [27]) [27]
Hg^{2+} <12, 13> (<13> 100% inhibition [27]) [23, 27, 28]
KCl <13> (<13> 44% inhibition [27]) [27]
LiCl <13> (<13> 64% inhibition [27]) [27]
Mn^{2+} <13> (<13> 16% inhibition [27]) [27]
N-ethylmaleimide <10> (<10> 3 mM, room temperature, pH 9.0, 50% inactivation [16]) [16]
$NADP^+$ <2, 13> (<2> reverse reaction, reduction, product inhibition, competitive vs. NADPH and noncompetitive vs. D-glucono-1-5-lactone [2]; <13> 3 mM, 33% inhibition [27]) [2, 27]
NADPH <2, 13> (<2> forward reaction, oxidation, product inhibition, competitive vs. NADP [2]; <13> 0.1 mM, 19% inhibition [27]) [2, 27]
NaCl <13> (<13> 28% inhibition [27]) [27]
Ni^{2+} <12> [23, 28]
Zn^{2+} <9, 13> (<13> 8% inhibition [27]) [27, 35]
dithiothreitol <13> (<13> 10 mM, 28% inhibition [27]) [27]
p-chloromercuribenzoate <9, 12, 13> (<13> 100% inhibition [27]) [23, 27, 35]
thioredoxin <11> (<11> in the presence of 0.5 mM dithiothreitol [17]) [17]

Cofactors/prosthetic groups

$NAD(P)^+$ <1-19> [1-35]

Activating compounds

KCl <4> (<4> optimal activity at 1.3 M [4]) [4]
NAD^+ <3> [3]
$NADP^+$ <3> (<3> 2 mM $NADP^+$ transforms enzyme at 70°C within 10 min from inactive to active state [3]) [3]
NaCl <4> (<4> optimal activity at 1.3 M [4]) [4]

Metals, ions

Ca^{2+} <10> (<10> maximal activity at 20 mM [16]) [16]
Mg^{2+} <4, 10, 19> (<10> maximal activity at 20 mM [16]; <19> without Mg^{2+} low activity with coenzyme $NADP^+$ [34]) [4, 16, 34]
Mn^{2+} <4, 10> (<10> maximal activity at 20 mM [16]) [4, 16]
Ni^{2+} <4> [4]
Zn^{2+} <10> [16]

Turnover number (min^{-1})

90 <6> (D-glucose 6-phosphate, <7> coenzyme $NADP^+$ [6]) [6]
96 <6> (D-glucose 6-phosphate, <6> coenzyme NAD^+ [6]) [6]
666 <7> (D-glucose, <7> coenzyme NAD^+ [6]) [6]
762 <7> (D-glucose 6-phosphate, <7> coenzyme $NADP^+$ [6]) [6]
852 <7> (D-glucose 6-phosphate, <7> coenzyme NAD^+ [6]) [6]
882 <7> (D-glucose, <7> coenzyme $NADP^+$ [6]) [6]
1080 <1> (β-D-glucose, <1> E96K mutant, coenzyme $NADP^+$ [11]) [11]
1380 <1> (β-D-glucose, <1> E96K mutant, coenzyme NAD^+ [11]) [11]
2460 <6> (D-glucose, <6> coenzyme $NADP^+$ [6]) [6]
2754 <6> (D-glucose, <6> coenzyme NAD^+ [6]) [6]
5520 <1> (β-D-glucose, <1> E96G mutant, coenzyme $NADP^+$ [11]) [11]
5820 <1> (β-D-glucose, <1> E96A mutant, coenzyme NAD^+ [11]) [11]
5880 <1> (β-D-glucose, <1> E96A mutant, coenzyme $NADP^+$ [11]) [11]
6840 <1> (β-D-glucose, <1> E96G mutant, coenzyme NAD^+ [11]) [11]
7200 <1> (β-D-glucose, <1> coenzyme $NADP^+$, isoenzyme I [7]) [7]
8400 <1> (β-D-glucose, <1> coenzyme $NADP^+$, isoenzyme II [7]) [7]
11160 <1> (β-D-glucose, <1> Y253C mutant, coenzyme $NADP^+$ [11]) [11]
11400 <1> (β-D-glucose, <1> coenzyme NAD^+, isoenzymes I-II [7]) [7]
12000 <1> (β-D-glucose, <1> Y253C mutant, coenzyme NAD^+ [11]) [11]
15600 <1> (β-D-glucose, <1> wild type, coenzyme $NADP^+$ [11]) [11]
18000 <1> (β-D-glucose, <1> Q252L mutant, coenzyme $NADP^+$ [11]) [11]
18600 <1> (β-D-glucose, <1> Q252L mutant, coenzyme NAD^+ [11]) [11]
23400 <1> (β-D-glucose, <1> wild type, coenzyme NAD^+ [11]) [11]
25200 <1> (β-D-glucose, <1> coenzyme NAD^+, isoenzyme IV [7]) [7]
25800 <1> (β-D-glucose, <1> coenzyme NAD^+, isoenzyme III [7]) [7]
27720 <1> (β-D-glucose, <1> coenzyme NADP, isoenzyme IV [7]) [7]
48120 <1> (β-D-glucose, <1> coenzyme $NADP^+$, isoenzyme III [7]) [7]

Specific activity (U/mg)

0.27 <9> [35]
2.25 <6> [25]
3.1 <9> [22]
3.6 <19> [34]
5.98 <9> [13]
6.2 <7> [6]
19 <16> (<16> cell-free extracts, coenzyme $NADP^+$ [33]) [33]
19 <18> (<17> cell-free extracts, coenzyme NAD^+ [33]) [33]
20 <17> (<17> cell-free extracts, coenzyme $NADP^+$ [33]) [33]
25 <18> (<18> cell-free extracts, coenzyme $NADP^+$ [33]) [33]

26 <8> [30]
31 <11> (<11> 0.5 mM dithiothreitol, enzyme reduced with thioredoxin preparation from spinach [17]) [17]
32 <11> (<11> 0.5 mM dithiothreitol, enzyme reduced with thioredoxin preparation from Aspergilus nidulans [17]) [17]
33.6 <6> [6]
38 <11> (<11> 0.5 mM dithiothreitol, enzyme reduced with thioredoxin preparation from Anabaena sp. [17]) [17]
39 <11> (<11> 0.5 mM dithiothreitol, enzyme reduced with thioredoxin preparation from Nostoc sp. [17]) [17]
49.2 <14> [24]
64.9 <8> [19]
95 <11> (<11> 0.5 mM dithiothreitol [17]) [17]
105 <11> [17]
106 <13> [27]
140 <12> [23]
140 <12> [28]
202 <3> [3]
240 <8> (<8> cysteine residue introduced at position 44 [10]) [10]
320 <5> [12]
368 <8> [21]
375 <8> [9]
437 <10> [16]
480 <1> [18]
550 <1> [32]
550 <4> [4]

K_m-Value (mM)

0.0003 <9> (NAD^+) [35]
0.0024 <9> (β-D-glucose, <9> coenzyme NAD^+, 37°C, pH 7.5 [20]) [20]
0.0028 <9> (NAD^+, <9> pH 9.0 [13]) [13]
0.0029 <9> (D-glucose-6-phosphate, <9> coenzyme $NADP^+$, 37°C, pH 7.5 [20]) [20]
0.0032 <9> (β-D-glucose, <9> coenzyme NAD^+, 37°C, pH 10.00 [20]) [20]
0.0036 <9> ($NADP^+$, <9> substrate glucose-6-phosphate, 37°C, pH 7.5 [20]) [20]
0.0046 <9> ($NADP^+$, <9> , substrate β-D-glucose, 37°C, pH 10.0 [20]) [20]
0.0051 <9> (NAD^+, <9> substrate β-D-glucose, 37°C, pH 7.5 [20]) [20]
0.0051 <1> ($NADP^+$, <1> E96K mutant [11]) [11]
0.0072 <9> (D-glucose-6-phosphate, <9> coenzyme $NADP^+$, 37°C, pH 10.00 [20]) [20]
0.0072 <9> ($NADP^+$, <9> pH 10.0 [13]) [13]
0.0076 <1> ($NADP^+$, <1> E96G mutant [11]) [11]
0.0083 <9> (glucose-6-phosphate, <9> pH 10.0 [13]) [13]
0.0093 <1> ($NADP^+$, <1> E96A mutant [11]) [11]
0.011 <9> (β-D-glucose, <9> coenzyme $NADP^+$, 37°C, pH 7.5 [20]) [20]
0.013 <9> ($NADP^+$, <9> substrate β-D-glucose, 37°C, pH 7.5 [20]) [20]

0.018 <1> (NAD^+, <1> E96K mutant [11]) [11]
0.018 <1, 1> ($NADP^+$, <1> isoenzyme II [7]) [7]
0.02 <1> ($NADP^+$, <1> isoenzyme I [7]) [7]
0.0212 <6> (NAD^+, <6> substrate β-D-glucose, pH 9.5, 30°C [25]) [25]
0.024 <11> (NADP) [17]
0.024 <4> ($NADP^+$, <4> oxidation, forward reaction [4]) [4]
0.0255 <6> (D-glucose 6-phosphate, <6> coenzyme $NADP^+$ [25]) [25]
0.026 <1> ($NADP^+$, <1> Q252L mutant [11]) [11]
0.027 <1> ($NADP^+$, <1> wild type [11]) [11]
0.028 <9> ($NADP^+$, <9> , substrate β-D-glucose, 37°C, pH 10.0 [20]) [20]
0.03 <10> ($NADP^+$) [16]
0.032 <9> (β-D-glucose, <9> coenzyme $NADP^+$, 37°C, pH 10.00 [20]) [20]
0.037 <1> ($NADP^+$, <1> Y253C mutant [11]) [11]
0.061 <15> ($NADP^+$) [29]
0.062 <6> ($NADP^+$, <6> substrate D-glucose 6-phosphate, pH 9.5, 30°C [25]) [25]
0.062 <9> ($NADP^+$, <9> , substrate glucose-6-phosphate, 37°C, pH 10.0 [20]) [20]
0.081 <1> (NAD^+, <1> isoenzyme I [7]) [7]
0.082 <15> (NAD^+) [29]
0.088 <1> (NAD^+, <1> isoenzyme IV [7]) [7]
0.09 <19> (NAD^+) [34]
0.09 <19> ($NADP^+$) [34]
0.1 <19> (D-glucose 6-phosphate) [34]
0.11 <8> (NAD^+) [9]
0.113 <5> ($NADP^+$) [12]
0.14 <2> (β-D-glucose, <2> reduction, reverse reaction [2]) [2]
0.16 <1> (NAD^+, <1> isoenzyme III [7]) [7]
0.17 <1> (NAD^+, <1> E96G mutant [11]) [11]
0.19 <1> (NAD^+, <1> Q252L mutant [11]) [11]
0.19 <2> ($NADP^+$, <2> oxidation, forward reaction [2]) [2]
0.23 <8> (NAD^+) [30]
0.25 <1> (NAD^+, <1> Y253C mutant [11]) [11]
0.26 <8> ($NADP^+$) [30]
0.285 <11> ($NADP^+$, <11> enzyme reduced with Nostoc sp. thioredoxin [17]) [17]
0.3 <3> (β-D-glucose) [3]
0.31 <1> (NAD^+, <1> E96A mutant [11]) [11]
0.37 <1> (NAD^+, <1> wild type [11]) [11]
0.44 <13> (NAD^+) [24]
0.44 <10> (β-D-glucose, <10> coenzyme $NADP^+$ [16]) [16]
0.49 <2> (NADPH, <2> reduction, reverse reaction [2]) [2]
0.67 <14> (NAD^+) [24]
0.7 <3> ($NADP^+$) [3]
0.86 <3> (NAD^+) [3]
0.99 <1> (NAD^+, <1> isoenzyme II [7]) [7]
1 <12> ($NADP^+$) [23]

1.2 <10> (NAD^+) [16]
1.2 <4> (NAD^+, <4> oxidation, forward reaction [4]) [4]
1.9 <1> ($NADP^+$, <1> isoenzyme IV [7]) [7]
2.2 <10> (β-D-glucose, <10> coenzyme $NADP^+$ [16]) [16]
2.57 <2> (β-D-glucose, <2> oxidation, forward reaction [2]) [2]
2.7 <1> (β-D-glucose, <1> E96K mutant, coenzyme NAD^+ [11]) [11]
3 <19> (D-galactose 6-phosphate, <19> coenzyme NAD^+ [34]) [34]
3.2 <13> (6-O-(β-D-glucopyranosido)-β-D-glucopyranoside) [24]
3.38 <6> (β-D-glucose, <6> coenzyme NAD^+, pH 9.5, 30°C [25]) [25]
3.9 <4> (β-D-glucose, <4> oxidation, forward reaction [4]) [4]
4 <1> (β-D-glucose, <1> E96K mutant, coenzyme $NADP^+$ [11]) [11]
4.2 <8> (2-deoxyglucose) [21]
4.2 <15> (β-D-glucose) [29]
4.5 <1> (NAD^+) [32]
4.5 <1> (β-D-glucose, <1> isoenzyme IV, coenzyme $NADP^+$ [7]) [7]
4.6 <1> (β-D-glucose, <1> isoenzyme I, coenzyme NAD^+ [7]) [7]
4.6 <9> (β-D-glucose, <9> pH 9.0 [13]) [13]
4.8 <8> (β-D-glucose, <8> pH 6.5, K_M decreases with decreasing pH [21]) [21]
5 <19> (D-galactose 6-phosphate, <19> coenzyme $NADP^+$ [34]) [34]
5 <12> (β-D-glucose) [23]
5.6 <14> (6-O-(β-D-glucopyranosido)-β-D-glucopyranoside) [24]
7 <1> (NAD^+) [14]
7.5 <1> ($NADP^+$) [14]
7.9 <1> (β-D-glucose, <1> isoenzyme I, coenzyme $NADP^+$ [7]) [7]
8 <10> (β-D-glucose, <10> coenzyme NAD^+ [16]) [16]
8 <3> (xylose) [3]
8.7 <8> (β-D-glucose) [9]
8.7 <1> (β-D-glucose, <1> E96A mutant, coenzyme NAD^+ [11]) [11]
9 <1> (β-D-glucose, <1> Q252L mutant, coenzyme NAD^+ [11]) [11]
9.2 <1> (β-D-glucose, <1> isoenzyme II, coenzyme NAD^+ [7]) [7]
9.5 <1> (β-D-glucose, <1> wild type, coenzyme NAD^+ [11]) [11]
9.8 <14> (4-O-(β-D-glucopyranosido)-β-D-glucopyranoside) [24]
9.8 <1> (β-D-glucose, <1> isoenzyme IV, coenzyme NAD^+ [7]) [7]
10 <5> (β-D-glucose) [12]
10 <1> (β-D-glucose, <1> Q252L mutant, coenzyme $NADP^+$ [11]) [11]
11 <13> (4-O-(β-D-glucopyranosido)-β-D-glucopyranoside) [24]
11 <1> (β-D-glucose, <1> wild type, coenzyme $NADP^+$ [11]) [11]
12.4 <1> (β-D-glucose, <1> isoenzyme II, coenzyme $NADP^+$ [7]) [7]
13 <1> (β-D-glucose, <1> E96A mutant, coenzyme $NADP^+$ [11]) [11]
14 <11> (β-D-glucose) [17]
15 <11> (β-D-glucose, <11> enzyme reduced with Nostoc sp. thioredoxin [17]) [17]
16 <1> (β-D-glucose, <1> E96G mutant, coenzyme NAD^+ [11]) [11]
18 <1> (β-D-glucose, <1> E96G mutant, coenzyme $NADP^+$ [11]) [11]
18.9 <1> ($NADP^+$, <1> isoenzyme III [7]) [7]
20 <19> (β-D-glucose, <19> coenzyme NAD^+ [34]) [34]

22 <10> (β-D-galactose, <10> coenzyme $NADP^+$ [16]) [16]
25 <19> (2-deoxy-D-glucose 6-phosphate) [34]
30 <7> (NAD^+) [6]
35 <8> (D-glucosamine) [21]
40 <19> (β-D-glucose, <19> coenzyme $NADP^+$ [34]) [34]
42.9 <8> (β-D-glucose, <8> coenzyme $NADP^+$ [30]) [30]
44 <1> (β-D-glucose, <1> isoenzyme III, coenzyme NAD^+ [7]) [7]
47.22 <6> (β-D-glucose, <6> coenzyme $NADP^+$, pH 9.5, 30°C [25]) [25]
47.5 <1> (β-D-glucose) [32]
47.8 <8> (2-deoxy-β-D-glucose) [30]
48 <6> (NAD^+) [6]
51 <13> (β-D-glucose) [24]
55 <1> (β-D-glucose, <1> Y253C mutant, coenzyme NAD^+ [11]) [11]
55.9 <8> (β-D-glucose, <8> coenzyme NAD^+ [30]) [30]
59 <14> (β-D-glucose) [24]
63 <6> ($NADP^+$) [6]
68 <10> (β-D-xylose, <10> coenzyme NAD^+ [16]) [16]
70 <9> (β-D-glucose) [35]
70 <19> (xylose, <19> coenzyme NAD^+ [34]) [34]
85 <1> (β-D-glucose, <1> Y253C mutant, coenzyme $NADP^+$ [11]) [11]
100 <19> (xylose, <19> coenzyme $NADP^+$ [34]) [34]
105 <7> ($NADP^+$) [6]
125 <13> (D-xylose) [24]
135 <1> (β-D-glucose, <1> isoenzyme III, coenzyme $NADP^+$ [7]) [7]
148 <14> (D-xylose) [24]
1000 <19> (galactose, <19> coenzyme NAD^+ [34]) [34]
4000 <19> (galactose, <19> coenzyme $NADP^+$ [34]) [34]

pH-Optimum

8 <1> (<1> sharp decrease above [32]) [32]
8 <10> (<10> 70°C [16]) [16]
8 <8> (<8> rapid loss of activity above [30]) [30]
8.5-9 <12> [28]
8.5-9 <12> (<12> higher reaction rate in Tris-HCl compared to potassium phosphate [23]) [23]
9 <15> [29]
9.5 <6> (<6> 2fold activity compared to pH 9.0 and 10.0 [25]) [25]
9.8 <13> [24]
9.8 <13> (<13> 50 mM glycine-NaOH [27]) [27]
9.9 <9> (<9> sharp drop above pH 10.0, 0.1 M ionic strength [35]) [35]
10 <14> [24]
10 <19> [34]
10 <9> (<9> at 37°C, 50 mM glycine buffer [20]) [20]
10.7 <14> (<14> cell free extracts [26]) [26]

pH-Range

6.5-12.5 <19> [34]

7.5-10.5 <6> (<6> at pH 7.5 and 10.0 respectively 75% and 50% activity compared to optimal pH [25]) [25]

Temperature optimum (°C)
30-35 <13> [27]
30-35 <9> (<9> enzyme retains 7% activity at 60°C [35]) [35]
50 <12> [23, 28]
50 <15> (<15> imediate inactivation above [29]) [29]
77 <10> (<10> pH 8.0, 20% activity at 37°C [16]) [16]

4 Enzyme Structure

Molecular weight
54000 <14> (<14> gel filtration [24]) [24]
55000 <13> (<13> gel filtration [24]) [24, 27]
80000 <15> (<15> gel filtration [29]) [29]
84000 <3> (<3> in the presence of 1 mM $NADP^+$, without $NADP^+$ enzyme forms aggregates, gel filtration [3]) [3]
89000 <4> (<4> gel filtration [4]) [4]
115000 <8> (<8> gel filtration [30]) [30]
116000 <1> (<1> gel filtration [32]) [32]
124000 <10> (<10> gel filtration [16]) [16]
126000 <8> (<8> gel filtration [21]) [21]
130000 <10> (<10> sedimentation equilibrium [16]) [16]
153000 <12> (<12> gel filtration [23, 27]) [23, 27]
155000 <5> (<5> gel filtration [12]) [12]
180000 <19> (<19> gel filtration [34]) [34]
230000 <6> (<6> native PAGE [6]; <7> gel filtration [6]) [6]
235000 <6> (<6> gel filtration [25]) [25]
235000 <9> (<9> gel filtration [35]) [35]

Subunits
dimer <6, 7> (<6, 7> α,α, 2 * 90000, SDS-PAGE [6]) [6]
dimer <15> (<15> α_2, 2 * 45000, gel filtration of inactive monomeric enzyme [29]) [29]
homodimer <3> (<3> α,α, 2 * 41000, SDS-PAGE [3]) [3]
homodimer <4> (<4> α,α, 2 * 53000, SDS-PAGE [4]; 2 * 39000, CTAB-PAGE [4]) [4]
monomer <14> (<13> α, 1 * 48000, SDS-PAGE [24]) [24]
monomer <13> (<13> α, 1 * 51000, SDS-PAGE [24,27]) [24, 27]
tetramer <1> (<1> α_4, 4 * 30000, SDS-PAGE [32]) [32]
tetramer <10> (<10> α_4, 4 * 33000, SDS-PAGE [16]) [16]
tetramer <12> (<12> α_4, 4 * 40000, SDS-PAGE [23]) [23]
tetramer <5> (<5> α_4, 4 * 38000, SDS-PAGE [12]) [12]
tetramer <6> (<6> α_4, 4 * 59000, SDS-PAGE [25]) [25]
tetramer <8> (<8> α_4, 4 * 31500, SDS-PAGE [21]) [21]
tetramer <9> (<9> α_4, 4 * 58000, SDS-PAGE [20]) [20]

5 Isolation/Preparation/Mutation/Application

Source/tissue

liver <6, 7, 9> [6, 13, 20, 22, 25, 34, 35]

Localization

endoplasmic reticulum <6, 7, 9> (<6, 7> enzyme resides on the luminal side of liver microsomes [6]; <6> membrane bound enzyme [25]) [6, 13, 20, 22, 25]

forespore <8> [30]

Purification

<1> (hydroxyapatite, QAE-Sephadex [32]) [32]

<2> (ammonium sulfate, Sepharose 6B, DEAE-Sepharose, hydroxyapatite chromatography [2]) [2]

<3> (80°C, anion-exchange chromatography, hydrophobic chromatography, gel filtration with 1 mM $NADP^+$ [3]) [3]

<4> (Sepharose-4B, DEAE-cellulose, Blue-Sepharose, Red-Sepharose [4]) [4]

<5> (methanol extraction, Mono Q, Matrex Gel Red A, Mono P, gel filtration [12]) [12]

<6> (Triton X-100, acid treatment, ammonium sulfate, Sephacryl S-200, Sephacryl S-300, DEAE-cellulose, Sephacryl S-300 [25]) [25]

<8> (DEAE-Sephacel, gel filtration, hydroxyapatite, gelfiltration, hydroxyapatite, hydrophbic Sepharose, ω-aminohexyl-Sepharose, chromatofocusing [21]) [21]

<8> (ammonium sulfate, DEAE-Sepharose, Blue Fractogel [9]; expressed in Escherichia coli, introduced cystein residue at position 44, ammonium sulfate, phenyl-Sepharose, thiopropyl-Sepharose [10]) [9, 10]

<8> (gel filtration, DEAE-cellulose, hydroxylapatite, ω-aminohexyl-agarose [30]) [30]

<9> (Triton X-114, acid treatment, ammonium sulfate, Sephacryl S-200, Sephacryl S-300, SE-cellulose, DEAE-cellulose, purified enzyme contains 1.7% lipid [20]) [20]

<9> (ammonium sulfate, Sephacryl S-300, cation column, affinity column [13]) [13]

<9> (ammonium sulfate, acetic acid pH 5.5, Sephadex G200, G150, crystallization [35]) [35]

<10> (ammonium sulfate, SP-Sephadex, Phenyl-Sepharose, Affi-Gel Blue [16]) [16]

<11> (ammonium sulfate, hydroxyapatite, Sephacryl S-200 [17]) [17]

<12> (DEAE-Sephadex, Blue-Dextran Sepharose, DEAE-Sephadex, crystallization [23, 28]) [23, 28]

<13> (streptomycin treatment, ammonium sulfate, DEAE-cellulose, gel filtration, DE 52 cellulose, hydroxyl apatite, gel filtration [27]) [27]

<15> (DEAE-cellulose, partially purified enzyme, 10 mM NAD(P) protects from inactivation on column [29]) [29]

<19> (calcium-phosphate gel, ammonium sulfate, DEAE-cellulose [34]) [34]

<6, 7> (DEAE-Sepharose, 2',5'-ADP-Sepharose, Sephacryl [6]) [6]

Renaturation

<1> (deactivation at pH 9.0, reactivation at pH 6.5 [18]; 1 M NaCl inhibits unfolding of the enzyme by 8 M urea, optimal temperature for refolding 25°C, monomeric intermediate [31]) [18, 31]
<8> (shifting the pH from 6.5 to 9.0 leads to inactivation, reactivation to 100% activity by shifting pH back to 6.5, protein concentration higher than 0.1 mg/ml [9]) [9]
<15> (NAD^+, NADH, $NADP^+$, NADPH, AMP, ADP, ATP, dipicolinic acid, $(NH_4)_2SO_4$, Na_2SO_4, NaCl and KCl reactivate partially purified inactive enzyme, optimal reactivation at 30°C and pH 7.0 [29]) [29]

Crystallization

<1> (1.7 A resolution, four subunits are interrelated by three mutually perpendicular diad axes P,Q and R [1]) [1]
<5> (2.9 A resolution, monomer comprises a central nucleotide-binding domain [5]) [5]
<5> (hanging-drop vapour diffusion, 2% polyethylene glycol, pH 5.3, 0.4 mM $NADP^+$, 3.0 A resolution, space group P6122 [8]) [8]

Cloning

<1> (expressed in Escherichia coli, isoenzymes III and IV [7]; E96K mutant enzyme [11]) [7, 11]
<8> (expressed in Escherichia coli [9]) [9]

Engineering

E96A <1> (<1> obtained by random mutagenesis [11]) [11]
E96G <1> (<1> obtained by random mutagenesis [11]) [11]
E96K <1> (<1> obtained by random mutagenesis [11]) [11]
Q252L <1> (<1> obtained by random mutagenesis [11]) [11]
Y253C <1> (<1> obtained by random mutagenesis [11]) [11]

6 Stability

pH-Stability

4-10 <1> (<1> E96K, E96G and E96A mutant [11]) [1]
4-10 <1> (<1> isoenzymes I-IV, in the presence of 2 M NaCl [7]) [7]
6-6.5 <8> [21]
6-7.5 <1> (<1> isoenzymes IV and I, absence of 2 M NaCl [7]) [7]
6-7.5 <1> (<1> wild type, Q252L, Y253C mutant [11]) [11]
6-8 <13> [27]
6.5 <1> (<1> at pH 9.0 tetrameric enzyme dissociates into inactive subunits, inactivation proceeds by two distinct phases, 50% inactivation at the rapid first phase, 50% at slow second phase [14]) [14,18]
8 <8> (<8> rapid inactivation of enzyme at alkaline pH [30]) [30]

Temperature stability

4-28 <1> (<1> isoenzyme III, 80% activity at 28°C, absence of NaCl [7]) [7]
4-45 <1> (<1> isoenzyme IV, 80% activity at 45°C, absence of NaCl [7]) [7]

4-48 <1> (<1> 90% activity remains at 48°C, 50 mM phosphate, pH 6.5, 20 min [11]) [11]
4-50 <1> (<1> isoenzyme I, 95% activity at 50°C, absence of NaCl [7]) [7]
4-55 <1> (<1> Q252L mutant, activity decreases rapidly above 55°C, 50 mM phosphate, pH 6.5, 20 min [11]) [11]
4-55 <1> (<1> Q253L mutant, activity decreases rapidly above 55°C, 50 mM phosphate, pH 6.5, 20 min [11]) [11]
4-60 <1> (<1> E96G mutant, activity decreases rapidly above 60°C, 50 mM phosphate, pH 6.5, 20 min [11]) [11]
4-65 <1> (<1> E96K mutant, activity decreases rapidly above 65°C, 50 mM phosphate, pH 6.5, 20 min [11]) [11]
4-70 <1> (<1> E96A mutant, activity decreases rapidly above 70°C, 50 mM phosphate, pH 6.5, 20 min [11]) [11]
4-70 <1> (<1> isoenzymes I-IV, in the presence of NaCl [7]) [7]
10-40 <13> (<13> 50 mM phosphate buffer, pH 7.0, 10 min, 100% activity until 40°C, sharp drop above [27]) [27]
40 <19> (<19> 5 min, 60% activity at 45°C, complete inactivation at 50°C [34]) [34]
55 <8> (<8> 90% activity lost after 5 h in soluble native enzyme and cystein 44 mutant enzyme, immobilized cysteine 44 enzyme looses 20% activity [10]) [10]
55-75 <5> (<5> full activity after 9 h at 55°C, half-life at 75°C 3 h [12]) [12]
65 <4> (<4> half-life of inactivation, 56.8 hours, in the presence of 3 M NaCl, stability decreases significantly with lower salt concentrations [4]) [4]
70 <10> (<10> 50% activity lost after 40 days at 37°C and after 45 h at 70°C [16]) [16]
103 <3> (<3> half-life of inactivation, 10 min [3]) [3]

Organic solvent stability

acetone <5, 10> (<5> 50%, room temperature, 6 h, 100% activity [12]; <10> 50%, 24 h, room temperature, no loss of activity [16]) [12, 16]
ethanol <5, 10> (<5> 50%, room temperature, 6 h, 100% activity [12]; <10> 100%, slow decrease [16]) [12, 16]
methanol <5, 10> (<5> 50%, room temperature, 6 h, 100% activity [12]; <10> 50%, 24 h, room temperature, no loss of activity [16]) [12, 16]

General stability information

<1>, 2 M NaCl stabilizes isoenzymes [7]
<1>, enzyme is completely deactivated at pH 9.0 in the absence of NaCl, 3 M NaCl prevent enzyme from dissociation into subunits at pH 9.0, unstable at low ionic strength, in 67 mM phosphate buffer enzyme activity decreases to 80% within 5 hours at pH 6.5 and 0.01 mg/ml protein, reduction to 57% at 40 mM phosphate, high concentration of NAD inhibit dissociation [18, 31]
<5>, 4 M urea, room temperature, 6 h, 100% activity [12]
<8>, stability increases in the presence of glycerol [21]
<10>, 0.05% SDS, no loss of activity, 0.1%, slow decrease [16]
<10>, 2 M urea, 24 h, room temperature, no loss of activity, slow decrease at 4 M [16]

<10>, 4 M guanidinium chloride, rapid inactivation [16]
<10>, H4-furan, rapid inactivation [16]
<15>, crude enzyme looses 90% activity after 3 days at 5°C [29]

Storage stability

<1>, 4°C, 50 mM sodium phosphate pH 6.5, 2 M NaCl, 0% inactivation, room temperature, pH 6.5, 3 M NaCl, several months [7, 31]
<8>, -20°C, 20% glycerol, 3 months [30]
<8>, -20°C, 20% glycerol, 4 years, 0% inactivation [21]
<8>, 0°C, 67 mM KH_2PO_4, pH 6.5, 0.5 M NaCl, a few months, 0% inactivation [9]
<10>, 4°C, 20 mM $MgCl_2$, 20% ethylene glycol, several months, 0% inactivation [16]
<12>, 5°C, crystalline preparation, ammonium sulfate suspension, several months, 0% inactivation, without ammonium sulfate, 2 weeks, 0% inactivation [23]

References

[1] Yamamoto, K.; Kurisi, G.; Kusunoki, M.; Tabata, S.; Urabe, I.; Osaki, S.: Crystal structure of glucose dehydrogenase from Bacillus megaterium IWG3 at 1.7 A resolution. J. Biochem., **129**, 303-312 (2001)

[2] Bhaumik, S.R.; Sonawat, H.M.: Kinetic mechanism of glucose dehydrogenase from Halobacterium salinarum. Indian J. Biochem. Biophys., **36**, 143-149 (1999)

[3] Siebers, B.; Wendisch, V.F.; Hensel, R.: Carbohydrate metabolism in Thermoproteus tenax: in vivo utilization of the non-phosphorylative Entner-Doudoroff pathway and characterization of its first enzyme, glucose dehydrogenase. Arch. Microbiol., **168**, 120-127 (1997)

[4] Bonete, M.-J.; Pire, C.; Llorca, F.I.; Camacho, M.L.: Glucose dehydrogenase from the halophilic Archaeon Haloferax mediterranei: Enzyme purification, characterization and N-terminal sequence. FEBS Lett., **383**, 227-229 (1996)

[5] John, J.; Crennell, S.J.; Hough, D.W.; Danson, M.J.; Taylor, G.L.: The crystal structure of glucose dehydrogenase from Thermoplasma acidophilum. Structure, **15**, 385-393 (1994)

[6] Tranulis, M.A.; Christophersen, B.; Borrebaek, B.: Glucose dehydrogenase in beef (Bos taurus) and rainbow trout (Oncorhynchus mykiss) liver: a comparative study. Comp. Biochem. Physiol., **109**, 427-435 (1994)

[7] Nagao, T.; Mitamura, T.; Wang, X.H.; Negoro, S.; Yomo, T.; Urabe, I.; Okada, H.: Cloning, nucleotide sequences, and enzymatic properties of glucose dehydrogenase isoenzymes from Bacillus megaterium IAM1030. J. Bacteriol., **174**, 5013-5020 (1992)

[8] Bright, J.R.; Mackness, R.; Danson, M.J.; Hough, D.W.; Taylor, G.L.; Towner, P.: Crystallization and preliminary crystallographic study of glucose dehydrogenase from the archaebacterium Thermoplasma acidophilum. J. Mol. Biol., **22**, 143-144 (1991)

[9] Hilt, W.; Pfleiderer, G.; Fortnagel, P.: Glucose dehydrogenase from Bacillus subtilis expressed in Escherichia coli I: purification, characterization and comparison with glucose dehydrogenase from Bacillus megaterium. Biochim. Biophys. Acta, **1076**, 298-304 (1991)
[10] Persson, M.; Bülow, L.; Mosbach, K.: Purification and site-specific immobilization of genetically engineered glucose dehydrogenase on thiopropyl-sepharose. FEBS Lett., **270**, 41-44 (1990)
[11] Yamamot, K.; Nagao, T.; Makino, Y.; Urabe, I.; Okada, H.: Characterization of mutant glucose dehydrogenase with increasing stability. Ann. N. Y. Acad. Sci., **613**, 362-365 (1990)
[12] Smith, L.D.; Budgen, N.; Bungard, S.J.; Danson, M.J.; Hough, D.W.: Purification and characterization of glucose dehydrogenase from the thermoacidophilic archaebacterium Thermoplasma acidophilum. Biochem. J., **261**, 973-977 (1989)
[13] Carper, W.R.; Groutas, W.C.; Coffin, D.B.: Affinity chromatography of glucose dehydrogenase. Experientia, **44**, 29-32 (1988)
[14] Maurer, E.; Pfleiderer, G.: Reversible pH-induced dissociation of glucose dehydrogenase from Bacillus megaterium. II. Kinetics and mechanism. Z. Naturforsch. C, **42**, 907-915 (1987)
[15] Hönes, J.; Jany, K.-D.; Pfleiderer, G.; Wagner, A.F.V.: An integrated prediction of secondary, tertiary and quarternary structure of glucose dehydrogenase. FEBS Lett., **212**, 193-198 (1987)
[16] Giardina, P.; DeBasia, M.-G.; DeRosa, M.; Gambacort, A.; Buonocore, V.: Glucose dehydrogenase from the thermoacidophilic archaebacterium Sulfolobus solfataricus. Biochem. J., **239**, 517-522 (1986)
[17] Juhasz, A.; Csizmadia, V.; Borbely, G.; Udvardy, J.; Farkas, G.L.: The pyridine nucleotide-dependent D-glucose dehydrogenase of Nostoc sp. strain Mac, a cyanobacterium, is subject to thioredoxin modulation. FEBS Lett., **194**, 121-125 (1986)
[18] Maurer, E.; Pfleiderer, G.: Reversible pH-induced dissociation of glucose dehydrogenase from Bacillus megaterium. I. Conformational and functional changes. Biochim. Biophys. Acta, **827**, 381-388 (1985)
[19] Strauss, N.: Role of glucose dehydrogenase in germination of Bacillus subtilis spores. FEMS Microbiol. Lett., **20**, 379-384 (1983)
[20] Carper, W.R.; Campbell, D.P.; Morrical, S.W.; Thompson, R.E.: Characterization studies of glucose dehydrogenase. Experientia, **39**, 1295-1297 (1983)
[21] Ramaley, R.F.; Vasantha, N.: Glycerol protection and purification of Bacillus subtilis glucose dehydrogenase. J. Biol. Chem., **258**, 12558-12565 (1983)
[22] Thompson, R.E.; Morrical, S.W.; Campbell, D.P.; Carper, W.R.: Guanidinium- and temperature-induced conformational changes in glucose dehydrogenase. Biochim. Biophys. Acta, **745**, 279-284 (1983)
[23] Adachi, O.; Ameyama, M.: D-Glucose dehydrogenase from Gluconobacter suboxydans. Methods Enzymol., **89**, 159-163 (1982)
[24] Kobayashi, Y.; Ueyama, H.; Horikoshi, K.: Comparative studies of NAD^+-dependent maltose dehydrogenase and D-glucose dehydrogenase produced by two strains of alkalophilic Corynebacterium species. Agric. Biol. Chem., **46**, 2139-2142 (1982)

[25] Campbell, D.P.; Carper, W.R.; Thompson, R.E.: Bovine liver glucose dehydrogenase: isolation and characterization. Arch. Biochem. Biophys., **215**, 289-301 (1982)
[26] Kobayashi, Y.; Ueyama, H.; Horikoshi, K.: NAD-dependent maltose dehydrogenase and NAD-dependent D-glucose dehydrogenase of alkalophilic Corynebacterium sp. No. 150-1. Agric. Biol. Chem., **44**, 2837-2841 (1980)
[27] Kobayashi, Y.; Horikoshi, K.: Purification and properties of NAD-dependent D-glucose dehydrogenase produced by alkalophilic Corynebacterium sp. No. 93-1. Agric. Biol. Chem., **44**, 2261-2269 (1980)
[28] Adachi, O.; Matsushita, K.; Shinagawa, E.; Ameyama, M.: Crystallization and characterization of NADP-dependent D-glucose dehydrogenase from Gluconobacter suboxydans. Agric. Biol. Chem., **44**, 301-308 (1980)
[29] Yokota, a.; Sasajima, K.; Yoneda, M.: Reactivation of inactivated D-glucose dehydrogenase of a Bacillus species by pyridine and adenine nucleotides. Agric. Biol. Chem., **43**, 271-278 (1979)
[30] Fujita, Y.; Ramaley, R.; Freese, E.: Location and properties of glucose dehydrogenase in sporulating cells and spores of Bacillus subtilis. J. Bacteriol., **132**, 282-293 (1977)
[31] Pauly, H.E.; Pfleiderer, G.: Conformational and functional aspects of the reversible dissociation and denaturation of glucose dehydrogenase. Biochemistry, **16**, 4599-4604 (1977)
[32] Pauly, H.E.; Pfleiderer, G.: D-Glucose dehydrogenase from Bacillus megaterium M1286: purification, properties and structure. Hoppe-Seyler's Z. Physiol. Chem., **356**, 1613-1623 (1975)
[33] Pulich, W.M.; van Baalen, C.: Pyridine nucleotide-dependent glucose dehydrogenase activity in blue-green algae. J. Biol. Chem., **114**, 28-33 (1973)
[34] Shatton, J.B.; Halver, J.E.; Weinhouse, S.: Glucose (hexose 6-phosphate) dehydrogenase in liver of rainbow trout. J. Biol. Chem., **246**, 4878-4885 (1971)
[35] Thompson, R.E.; Carper, W.R.: Glucose dehydrogenase from pig liver. I. Isolation and purification. Biochim. Biophys. Acta, **198**, 397-406 (1970)

Galactose 1-dehydrogenase 1.1.1.48

1 Nomenclature

EC number
1.1.1.48

Systematic name
D-galactose:NAD$^+$ 1-oxidoreductase

Recommended name
galactose 1-dehydrogenase

Synonyms
D-galactose dehydrogenase
NAD-dependent D-galactose dehydrogenase
β-galactose dehydrogenase

CAS registry number
9028-54-0

2 Source Organism

<1> *Pseudomonas fluorescens* [1, 2, 3, 4, 5, 6, 8, 9, 10, 12, 13, 15, 16, 17]
<2> *Azotobacter vinelandii* [3]
<3> *Pseudomonas saccharophila* [3, 13, 14, 15, 16, 17, 18, 19, 20]
<4> *Rhizobium meliloti L5-30* [3, 7]
<5> *Rhizobium meliloti UR23* (galactose mutant [7]) [7]
<6> *Caulobacter crescentus CB13* (unable to grow on galactose as sole carbon source, growth occurs if dibutyryl cyclic AMP is added [11]) [11]
<7> *Caulobacter crescentus AE22* (mutant strain, grows on galactose in the absence of dibutyryl cyclic AMP [11]) [11]
<8> *Caulobacter crescentus AE23* (mutant strain, grows on galactose in the absence of dibutyryl cyclic AMP [11]) [11]
<9> *Pseudomonad sp. G6* (constitutive galactose dehydrogenase activity [21]) [21]

3 Reaction and Specificity

Catalyzed reaction
D-galactose + NAD$^+$ = D-galactono-1,4-lactone + NADH (<1, 3> depending on the pH, the immediately formed D-galactono-1,5-lactone rearranges intra-

molecularly to the corresponding 1,4-lactone or hydrolyzes to D-galctonate or both [9, 13, 16]; <1,3> B-type enzyme with regard to stereochemistry, transfers hydrogen from carbon-1 of D-galactose to si face of NAD, in the reverse reaction the pro-S hydrogen atom from carbon-4 of NAD(P)H is transferred to the re face of carbon-1 of D-galactono-1,5-lactone [12, 13])

Reaction type

redox reaction

Natural substrates and products

S D-galactose + NAD^+ <1-9> (Reversibility: ? <1-9> [1-21]) [1-21]
P D-galactono-1,5-lactone + NADH <1-9> [1-21]
S D-galactose + $NADP^+$ <1, 4> (Reversibility: ? <1, 4> [3, 7]) [3, 7]
P D-galactono-1,5-lactone + NADPH <1, 4> [3, 7]

Substrates and products

S 2-deoxy-D-galactose + NAD^+ <1, 3> (Reversibility: ? <1, 3> [17, 19]) [17, 19]
P 2-deoxy-galactono-1,5-lactone + NADH <1, 3> [17, 19]
S 2-deoxy-D-galactose + $NADP^+$ <1> (Reversibility: ? <1> [17]) [17]
P 2-deoxy-galactono-1,5-lactone + NADPH <1> [17]
S 3-deoxy-D-galactose + NAD^+ <3> (Reversibility: ? <3> [19]) [19]
P 3-deoxy-galactono-1,5-lactone + NADH <3> [19]
S 4-deoxy-D-galactose + NAD^+ <3> (Reversibility: ? <3> [19]) [19]
P 4-deoxy-galactono-1,5-lactone + NAD(P)H <3> [19]
S 6-deoxy-α-D-galactose + NAD^+ <1, 2, 3> (Reversibility: ? <1, 2, 3> [3, 17, 19]) [3, 17, 19]
P 6-deoxy-galactono-1,5-lactone + NADH <1, 2, 3> [3, 17, 19]
S D-galactose + NAD^+ <1-9> (Reversibility: ? <1-9> [1-21]) [1-21]
P D-galactono-1,5-lactone + NADH <1-9> [1-21]
S D-galactose + $NADP^+$ <1, 4> (Reversibility: ? <1, 4> [3, 7]) [3, 7]
P D-galactono-1,5-lactone + NADPH <1, 4> [3, 7]
S L-arabinose + NAD^+ <1, 2, 3, 4, 5> (<2> 1 mM, better substrate than D-galactose [3]) (Reversibility: ? <1, 2, 3, 4, 5> [3, 7, 17]) [3, 7, 17]
P L-arabinono-1,5-lactone + NADH <1, 2, 3, 4, 5> [3, 7, 17]
S succinate + $NAD(P)^+$ <4> (Reversibility: ? <4> [7]) [7]
P succinic acid anhydride + NAD(P)H <4> [7]

Inhibitors

5,5'-dithiobis(2-nitrobenzoic acid) <3> (<3> reacts with first 4 sulfhydryl groups of enzyme, inhibition is completely reversible upon addition of reducing agents, NAD^+ and NADH protect against inactivation [18]) [18]
p-chloromercuribenzoate <3> [18]

Turnover number (min^{-1})

21500 <3> (D-galactose) [19]
71000 <1> (D-galactose) [17]

Specific activity (U/mg)

0.0074 <5> (<5> substrate combination L-arabinose and D-galactose, coenzyme $NADP^+$, cell free extracts [7]) [7]
0.008 <4> (<4> substrate succinate, coenzyme $NADP^+$, cell-free extracts [7]) [7]
0.009 <8> (<8> crude extracts, cells grown on D-galactose, inducible enzyme [11]) [11]
0.012 <4> (<4> substrate succinate, coenzyme NAD^+, cell-free extracts [7]) [7]
0.023 <7> (<7> crude extracts, cells grown on D-galactose, inducible enzyme [11]) [11]
0.026 <6> (<6> crude extracts, cells grown on D-galactose, inducible enzyme [11]) [11]
0.031 <4> (<4> coenzyme $NADP^+$, cell-free extracts [7]) [7]
0.036 <4> (<4> coenzyme NAD^+, cell-free extracts [7]) [7]
0.06 <9> (<9> crude extracts, cells grown on sucrose [21]) [21]
0.071 <5> (<5> substrate combination L-arabinose and galactose, coenzyme NAD, cell-free extracts [7]) [7]
0.078 <4> (<4> substrate L-arabinose, coenzyme $NADP^+$, cell-free extracts [7]) [7]
0.079 <4> (<4> substrate L-arabinose, coenzyme NAD^+, cell-free extracts [7]) [7]
0.42 <1> (<1> cell extracts [10]) [10]
17 <1> (<1> β-galactosidase-galactose dehydrogenase fusion enzyme, 3 amino acids as linker [1]) [1]
27 <1> (<1> β-galactosidase-galactose dehydrogenase fusion enzyme, 13 amino acids as linker [1]) [1]
29 <1> (<1> β-galactosidase-galactose dehydrogenase fusion enzyme, 9 amino acids as linker [1]) [1]
34 <1> [1]
36 <4> [3]
76 <1> (<1> β-galactosidase-galactose dehydrogenase fusion enzyme [4]) [4]
76 <4> (<4> substrate L-arabinose [3]) [3]
202 <3> [3, 19]
260 <2> [3]
272 <4> (<4> coenzyme $NADP^+$ [3]) [3]
390 <2> (<2> substrate L-arabinose [3]) [3]
849 <1> [3, 9, 17]

K_m-Value (mM)

0.125 <2> (NAD^+) [3]
0.14 <3> (NAD^+) [3, 19]
0.15 <1> (NAD^+) [1]
0.185 <1> (D-galactose, <1> pH 7.5 [2]) [2]
0.205 <1, 1> (D-galactose, <1> pH 8.0 [2]) [2]
0.215 <1> (L-arabinose, <1> pH 7.5 [2]) [2]
0.23 <1> (D-galactose) [1]

0.24 <1> (NAD^+) [17]
0.24 <1> (NAD^+) [3]
0.25 <1> (NAD^+, <1> β-galactosidase-galactose dehydrogenase fusion enzymes [1]) [1]
0.255 <1> (D-galactose, <1> pH 8.5 [2]) [2]
0.26 <1> (D-galactose, <1> pH 7.5, β-galactosidase-galactose dehydrogenase fusion enzyme [2]) [2]
0.275 <1> (L-arabinose, <1> pH 8.0 [2]) [2]
0.32 <1> (D-galactose, <1> pH 8.0, β-galactosidase-galactose dehydrogenase fusion enzyme [2]) [2]
0.37 <1> (D-galactose, <1> β-galactosidase-galactose dehydrogenase fusion enzyme, 13 amino acids as linker [1]) [1]
0.37 <1> (L-arabinose, <1> pH 8.5 [2]) [2]
0.4 <1> (D-galactose, <1> β-galactosidase-galactose dehydrogenase fusion enzyme, 9 amino acids as linker [1]) [1]
0.43 <1> (D-galactose, <1> pH 8.5, β-galactosidase-galactose dehydrogenase fusion enzyme [2]) [2]
0.43 <1> (L-arabinose, <1> pH 7.5, β-galactosidase-galactose dehydrogenase fusion enzyme [2]) [2]
0.45 <1> (D-galactose, <1> β-galactosidase-galactose dehydrogenase fusion enzyme, 3 amino acids as linker [1]) [1]
0.5 <2> (D-galactose) [3]
0.5 <1> (L-arabinose, <1> pH 8.0, β-galactosidase-galactose dehydrogenase fusion enzyme [2]) [2]
0.68 <1> (L-arabinose, <1> pH 8.5, β-galactosidase-galactose dehydrogenase fusion enzyme [2]) [2]
0.7 <1> (D-galactose) [17]
0.7 <1> (D-galactose) [3]
1 <3> (D-galactose) [3, 19]
2.3 <1> ($NADP^+$) [17]

pH-Optimum

8-9 <3> [19]
9 <2> [3]
9.1-9.5 <1> [3, 17]
9.9 <1> [1]
10 <1> [5]

4 Enzyme Structure

Molecular weight

64000 <1> (<1> gel filtration [17]) [17]
74000 <2> (<2> gel filtration [3]) [3]
102000 <3> (<3> gel filtration [17]) [17, 19]

Subunits

dimer <1> (<1> α,α, 2 * 32000, SDS-PAGE [17]) [17]
tetramer <3> (<3> α_4, 4 * 25000, SDS-PAGE [19]) [19]
tetramer <3> (<3> α_4, 4 * 26000, 100 mM phosphate buffer, pH 7.4, 8 M urea, 100 mM 2-mercaptoethanol, sedimentation equilibrium [14]) [14, 17]

5 Isolation/Preparation/Mutation/Application

Localization

cytoplasm <1, 2, 3, 4> [3]

Purification

<1> (ammonium sulfate, CM-Sephadex, DEAE-Sephadex, hydroxyapatite, disc electrophoresis [17]) [17]
<1> (ammonium sulfate, DEAE-Sepharose, FPLC-Superose 12, β-galactosidase-galactose dehydrogenase fusion enzyme [4]) [4]
<2> (Mono Q column, partially purified enzyme [3]) [3]
<3> (ammonium sulfate, heat, phosphate precipitation, DEAE-cellulose, hydroxyapatite [19]) [19]

Cloning

<1> (β-galactosidase-galactose dehydrogenase fusion enzymes, expressed in Escherichia coli [1]) [1, 4]
<1> (expressed in Escherichia coli, Escherichia coli β-galactosidase promotor [10]) [10]
<1> (expressed in Escherichia coli, point mutations in natural promotor increase expression [6]) [6]

Application

analysis <1> (<1> determination of lactose using a biosensor made of immobilized β-galctosidase-galactose dehydrogenase fusion enzyme [2]; <1, 3> determination of galactose in various hepatic diseases and galactosaemia in clinical biochemistry [8,15]) [2, 8, 15]

6 Stability

pH-Stability

4.6-9.2 <1> [17]

Temperature stability

20-65 <3> (<3> 5 min, 100% activity at 57°C, rapid drop above 60°C [19]) [19]
50 <1> (<1> native enzyme, 70% activity after 20 min, 30% after 60 min, 20% after 90 min, β-galactosidase-galactose dehydrogenase fusion enzymes, 80% activity after 90 min [1]; <1> 50% activity after 60 min, 100 mM, pH 8.0, 1 mM $MgCl_2$, 1 mM dithiothreitol [5]) [1, 4, 5]

Oxidation stability

<3>, oxidation of sulfhydryl residues inactivates enzyme, reactivation with 100 mM 2-mercaptoethanol or 10 mM dithiothreitol, pH 7.0, 30°C, enzymatically active oligomers of native enzyme are formed upon storage without reducing agents [15, 18, 19, 20]

Storage stability

<1>, -20°C, 100 mM phosphate buffer, pH 6.5-7.6, 30-50% glycerol, 18 months, no loss of activity [9, 17]

<3>, 4°C, 3.5 M $(NH_4)_2SO_4$, 20 mM 2-mercaptoethanol, pH 7.0, several months, no loss of activity [19]

References

[1] Carlsson, H.; Ljung, S.; Bülow, L.: Physical and kinetic effects on introduction of various linker regions in β-galactosidase/galactose dehydrogenase fusion enzymes. Biochim. Biophys. Acta, **1293**, 154-160 (1996)

[2] Carlsson, H.; Ljungcrantz, P.; Lindbladh, C.; Persson, M.; Bülow, L.: Use of geneticallly prepared enzyme conjugates in lactose and galactose analyses. Anal. Biochem., **218**, 278-283 (1994)

[3] Wong, T.Y.; Yao, X.-T.: The deLey-Doudoroff pathway of galactose metabolism in Azotobacter vinelandii. Appl. Environ. Microbiol., **60**, 2065-2068 (1994)

[4] Lindbladh, C.; Persson, M.; Bülow, L.; Mosbach, K.: Characterization of a recombinant bifunctional enzyme, galactose dehydrogenase/bacterial luciferase, displaying an improved bioluminescence in a three-enzyme system. Eur. J. Biochem., **204**, 241-247 (1992)

[5] Ljungcrantz, P.; Bülow, L.; Mosbach, K.: Construction and characterization of a recombinant tripartite enzyme, galactose dehydrogenase/β-galactosidase/galactokinase. FEBS Lett., **275**, 91-94 (1990)

[6] Fiedler, S.; Buckel, P.: The D-galactose dehydrogenase gene from Pseudomonas fluorescens: characterization of mutations leading to increased expression in Escherichia coli. Appl. Microbiol. Biotechnol., **33**, 418-423 (1990)

[7] Arias, A.; Cervenansky, C.: Galactose metabolism in Rhizobium meliloti L5-30. J. Bacteriol., **167**, 1092-1094 (1986)

[8] Fjimura, Y.: Flourometric method with galactose dehydrogenase. Methods Enzymol., **6**, 288-296 (1984)

[9] Maier, E.; Kurz, G.: D-Galactose dehydrogenase from Pseudomonas fluorescens. Methods Enzymol., **89**, 176-181 (1982)

[10] Buckel, P.; Zehelein, E.: Expression of Pseudomonas fluorescens D-galactose dehydrogenase in E. coli. Gene, **16**, 149-159 (1981)

[11] Kurn, N.; Contreras, I.; Shapiro, L.: Galactose catabolism in Caulobacter crescentus. J. Bacteriol., **135**, 517-529 (1978)

[12] Brendel, K.; Bressler, R.; Alizade, M.A.: Stereochemistry of the hydrogen transfer to NAD catalyzed by D-galactose dehydrogenase from Pseudomonas fluorescens. Biochim. Biophys. Acta, **397**, 1-4 (1975)

[13] Ueberschär, K.-H.; Blachnitzky, E.-O.; Lehman, J.; Kurz, G.: Stereospecificity of hydrogen transfer catalyzed by D-galactose dehydrogenase from Pseudomonas Saccharophila and Pseudomonas fluorescens. Biochim. Biophys. Acta, **391**, 15-18 (1975)

[14] Wengenmayer, F.; Kurz, G.: Subunit structure of D-galactose dehydrogenase from Pseudomonas Saccharophila. Biochim. Biophys. Acta, **386**, 590-602 (1975)

[15] Kurz, G.; Wallenfels, K.: UV-Test mit Galactose Dehydrogenase. Methods Enzym. Anal., 3rd Ed. (Bergmeyer, H.U., ed.), **2**, 1324-1327 (1974)

[16] Ueberschär, K.-H.; Blachnitzky, E.-O.; Kurz, G.: Reaction mechanism of D-galactose dehydrogenases from Pseudomonas saccharophila and Pseudomonas fluorescens. Eur. J. Biochem., **48**, 389-405 (1974)

[17] Blachnitzky, E.-O.; Wengenmayer, F.; Kurz, G.: D-Galactose dehydrogenase from Pseudomonas fluorescens. Purification, properties and structure. Eur. J. Biochem., **47**, 235-250 (1974)

[18] Wengenmayer, F.; Ueberschär, K.-H.; Kurz, G.: Sulfhydryl group reactivity of D-galactose dehydrogenase from Pseudomonas saccharophila. Eur. J. Biochem., **43**, 49-58 (1974)

[19] Wengenmayer, f.; Ueberschär, K.-H.; Kurz, G.: D-Galactose dehydrogenase from Pseudomonas saccharophila. Purification, properties and structure. Eur. J. Biochem., **40**, 49-61 (1973)

[20] Wengenmayer, F.; Blachnitzky, E.-O.; Kurz, G.: Artefakt-Formen der D-Galaktose-Dehydrogenase. Hoppe-Seyler's Z. Physiol. Chem., **354**, 131-135 (1973)

[21] Hu, A.S.L.; Cline, A.L.: The regulation of some sugar dehydrogenases in a Pseudomonad. Biochim. Biophys. Acta, **93**, 237-245 (1964)

Glucose-6-phosphate 1-dehydrogenase 1.1.1.49

1 Nomenclature

EC number

1.1.1.49

Systematic name

D-glucose 6-phosphate:NADP$^+$ 1-oxidoreductase

Recommended name

glucose-6-phosphate 1-dehydrogenase

Synonyms

6-phosphoglucose dehydrogenase
D-glucose 6-phosphate dehydrogenase
Entner-Doudoroff enzyme
G6PD [Swissprot]
G6PD1 [Swissprot]
G6PD2 [Swissprot]
G6PD3 [Swissprot]
G6PD4 [Swissprot]
G6PD5 [Swissprot]
G6PD6 [Swissprot]
G6PDH1 [Swissprot]
G6PDH2 [Swissprot]
G6PDH3 [Swissprot]
G6PDH4 [Swissprot]
G6PDH5 [Swissprot]
G6PDH6 [Swissprot]
NADP-dependent glucose 6-phophate dehydrogenase
NADP-glucose-6-phosphate dehydrogenase
VEG11 [Swissprot]
Vegetative protein 11 [Swissprot]
Zwischenferment
glucose 6-phosphate dehydrogenase (NADP)
glucose-6-phosphate dehydrogenase

CAS registry number

9001-40-5

2 Source Organism

<1> *Candida utilis* [1, 2, 8]
<2> *Homo sapiens* (2 isoforms: A(+) and A(-), where A(-) is a mutant derivative of A(+) with V68M exchange [34]; 2 isoenzymes [42]) [9, 27, 29, 34, 41, 42]
<3> *Rattus norvegicus* [27, 35, 39]
<4> *Bos taurus* [4, 5, 36]
<5> *Pseudomonas fluorescens* [10, 18]
<6> *Neurospora crassa* [3]
<7> *Leuconostoc mesenteroides* [6]
<8> *Penicillium duponti* [7]
<9> *Anabaena sp.* [11]
<10> *Candida boidinii* [12]
<11> *Lactobacillus buchneri* [13]
<12> *Pseudomonas sp.* (C [14]) [14]
<13> *Mus musculus* [16, 27, 40]
<14> *Bacillus subtilis* [17]
<15> *Methylomonas sp.* (M 15 [19]) [19]
<16> *Aspergillus parasiticus* [20]
<17> *Zea mays* [21]
<18> *Candida maltosa* [22]
<19> *Saccharomyces cerevisiae* [15]
<20> *Pisum sativum* [23]
<21> *Agaricus bisporus* [24]
<22> *Plasmodium falciparum* [25]
<23> *Bacillus stearothermophilus* [26]
<24> *Canis familiaris* [27]
<25> *Azotobacter vinelandii* [28]
<26> *Aspergillus niger* [30]
<27> *Aspergillus nidulans* [30]
<28> *Schizosaccharomyces pombe* [31]
<29> *Acetobacter hansenii* (i.e. Acetobacter xylinum [32]) [32]
<30> *Cryptococcus neoformans* [33]
<31> *Trypanosoma brucei* [37]
<32> *Dictyostelium discoideum* [38]
<33> *Sus scrofa* (2 isoenzymes [42]) [42]

3 Reaction and Specificity

Catalyzed reaction

D-glucose 6-phosphate + $NADP^+$ = D-glucono-1,5-lactone 6-phosphate + NADPH (<4> physiological regulating function is discussed [36]; <13> reaction mechanism is important for regulation [40]; <2> rapid equilibrium ordered bi bi kinetic [41])

Reaction type

oxidation
reduction

Natural substrates and products

S D-glucose 6-phosphate + $NAD(P)^+$ <1-8, 10, 12-18, 20, 21, 23, 25, 28-33> (<25> sigmoidal relationship in cofactor binding [28]; <2> isoform A(-) shows sigmoidal enzyme-substrate interaction, isoform A(+) shows hyperbolic enzyme-substrate interaction [34]; <2> hyperbolic cofactor binding [41]) (Reversibility: r <1-8, 10, 12-18, 23, 25, 27, 29, 31> [1-10, 12, 14, 16-22, 26, 28, 32, 34-37]; ? <2, 13, 20, 21, 28, 30, 32, 33> [23, 24, 31, 33, 38-42]) [1-10, 12, 14, 16-24, 26, 28, 31-42]

P D-gluconate 6-phosphate + NAD(P)H <1-8, 10, 12-18, 20, 21, 23, 31, 32> (<1, 2, 4-8, 10, 12-18, 20, 21, 23, 31, 32> 6-phosphoglucono-δ-lactone [1-10, 12, 14, 16-24, 26, 37, 38]; <2> D-glucose-δ-lactone 6-phosphate [34]) [1-10, 12, 14, 16-24, 26, 34-38]

S Additional information <3, 29> (<29> physiological role of NAD^+- and $NADP^+$-specific enzymes are discussed [32]; <3> physiological regulation of $NADP^+$-NADPH-ratio in the brain is dicussed [39]) [32, 39]

Substrates and products

S 2-deoxyglucose-6-phosphate + $NADP^+$ <2> (Reversibility: ? <2> [41]) [41]

P 2-deoxygluconate 6-phosphate + NADPH

S 6-phosphonogluconate + NADPH <2, 33> (<2, 33> isoenzyme II [42]) (Reversibility: ? <2, 33> [42]) [42]

P ?

S D-galactose 6-phosphate + $NAD(P)^+$ <2, 29, 33> (<29> at 3.5 mM 5.5% of the activity with glucose 6-phosphate [32]) (Reversibility: r <29> [32]; ? <2, 33> [41, 42]) [32, 41, 42]

P D-galactonate 6-phosphate + NADH

S D-glucose + $NAD(P)^+$ <29, 30> (<30> activity only with dimethylsulfoxide added by 40% (v/v) [33]; <29> at 0.7 mM only 2% of the activity with glucose 6-phosphate for the oxidation, at 3.5 mM only 0.6% activity [32]) (Reversibility: ? <30> [33]; r <29> [32]) [32, 33]

P D-gluconate + NAD(P)H

S D-glucose 6-phosphate + $NAD(P)^+$ <1-8, 10, 12-18, 20, 21, 23, 25, 28-33> (<25> sigmoidal relationship in cofactor binding [28]; <2> isoform A(-) shows sigmoidal enzyme-substrate interaction, isoform A(+) shows hyperbolic enzyme-substrate interaction [34]; <2> hyperbolic cofactor binding [41]) (Reversibility: r <1-8, 10, 12-18, 23, 25, 27, 29, 31> [1-10, 12, 14, 16-22, 26, 28, 32, 34-37]; ? <2, 13, 20, 21, 28, 30, 33> [23, 24, 31, 33, 40-42]; ir <3, 32> [38, 39]) [1-10, 12, 14, 16-24, 26, 28, 31-42]

P D-gluconate 6-phosphate + NAD(P)H <1-8, 10, 12-18, 20, 21, 23, 31-33> (<1, 2, 4-8, 10, 12-18, 20, 21, 23, 31, 32> 6-phosphoglucono-δ-lactone [1-10,12,14,16-24,26,37,38]; <2> D-glucose-δ-lactone 6-phosphate [34]; <13> glucono-δ-lactone-6-phosphate [40]) [1-10, 12, 14, 16-24, 26, 34-42]

S Additional information <2, 26, 27, 33> (<26, 27> no activity with glucose, glucose 1-phosphate, 2-deoxyglucose 6-phosphate, fructose, fructose 6-phosphate, fructose 1,6-diphosphate, mannose, galactose 6-phosphate, NAD^+ [30]; <28> no activity with D-glucose, D-ribose, D-fructose-6-phosphate, D-ribose-5-phosphate [31]; <2, 33> both isoenzymes show no activity with glucose and NAD^+, isoenzyme I shows no activity with 6-phosphonogluconate [42]) [30, 31, 42]

P ?

Inhibitors

2'-phosphoadenosine 5'-diphosphoribose <30> [33]
2,3-diphosphoglycerate <2> [41]
ADP <3, 5, 17, 30> [10, 21, 33, 39]
AMP <5, 10, 30> (<30> only 2'-AMP [33]) [10, 12, 33]
ATP <1-3, 5, 7, 9, 12, 17, 18, 28, 29, 33> (<29> inhibition only at pH 5.4 [32]) [6, 8, 10, 11, 14, 21, 22, 31, 32, 39, 42]
Cd^{2+} <11, 16, 30> [13, 20, 33]
Co^{2+} <16> [20]
CoA <5, 7> [6, 10]
Cu^{2+} <11> [13]
EDTA <2, 33> (<2, 33> slightly [42]) [42]
GTP <5, 10, 28> [10, 12, 31]
Hg^{2+} <11> [13]
ITP <5> [10]
NADH <5, 12> [10, 14]
$NADP^+$ <2, 28> (<28> at high concentration, dead-end ternary complexes are formed [31]; <2> at concentrations above 0.3 mM [41]) [31, 41, 42]
NADPH <1-3, 5, 9, 10, 13, 20, 21, 28, 30, 32, 33> (<3> liver enzyme is more sensitive to inhibition than kidney enzyme [35]; <13> inhibition patterns of substrates and product [40]; <2> NADPH binding is important for physiological regulation of pentose phosphate pathway [41]) [8, 10-12, 23, 24, 31, 33, 35, 38, 40-42]
Zn^{2+} <2, 11, 16, 30, 33> [13, 20, 33, 42]
acetyl-CoA <5, 7> [6, 10]
blue dextran <11> [13]
estrogen <4> [5]
glucose <4> (<4> at concentration above 10 mM [36]) [36]
glyceraldehyde 3-phosphate <10> [12]
glyoxylate <18> [22]
isocitrate <18> [22]
p-chloromercuribenzoate <3, 4> (<3> can be reversed 100% by dithiothreitol, partially by glutathione [39]) [5, 39]
palmitoyl-CoA <2, 21, 25, 33> (<25> non-competitive [28]; <2, 33> leads to dissociation of active tetramers to inactive dimers [42]) [24, 28, 42]
phosphate <1> [8]
phosphoenolpyruvate <5> [10]
progesterone <4> [5]

pyridoxal 5'-phosphate <1> [1, 8]
ribulose 1,5-diphosphate <9> [11]

Cofactors/prosthetic groups

NAD^+ <4, 5, 7, 12, 14, 15, 23, 25, 29> (<29> preference for [32]) [5, 6, 10, 14, 17-19, 26, 28, 32]
$NADP^+$ <1-8, 10, 12-16, 18, 23, 25-28, 30-33> (<2, 26, 27, 30, 32, 33> absolute specificity [30, 33, 38, 42]; <2, 28> high preference for [31, 41]; <2> isoform A(-) has a higher affinity for $NADP^+$ [34]) [1-7, 10, 12, 14, 16-20, 22, 26, 28, 30-34, 36-42]
NADPH <1, 4-8, 10, 12-16, 18, 23> [1-7, 10, 12, 14, 16-20, 22, 26]
deamino-$NADP^+$ <2> [41]

Activating compounds

glucosamine 6-phosphate <25> [28]

Metals, ions

Ca^{2+} <2, 14, 33> (<2, 14, 33> activation [17, 42]) [17, 42]
Mg^{2+} <1, 2, 4, 9, 14, 23, 33> (<1, 9> required [8, 11]; <2, 4, 14, 23, 33> activation [5, 17, 26, 41, 42]) [5, 8, 11, 17, 26, 41, 42]
Mn^{2+} <14> (<14> activation [17]) [17]
Additional information <2, 33> (<2, 33> no metal ions necessary [42]) [42]

Turnover number (min^{-1})

2580 <2> (2-deoxyglucose 6-phosphate) [41]
5880 <2> (deamino-$NADP^+$) [41]
9420 <2> (D-galactose 6-phosphate) [41]
34230 <2> (D-glucose 6-phosphate) [41]
67000 <5> ($NADP^+$) [10]
83000 <5> (NAD^+) [10]
868500 <29> (D-glucose 6-phosphate, <29> cofactor $NADP^+$ [32]) [32]
1803000 <29> (D-glucose 6-phosphate, <29> cafactor NAD^+ [32]) [32]

Specific activity (U/mg)

0.007-0.009 <31> (<31> cellular extract [37]) [37]
0.077 <3> (<3> nephridic enzyme [35]) [35]
0.327 <3> (<3> hepatic enzyme [35]) [35]
14 <31> (<31> purified enzyme [37]) [37]
50 <30, 33> (<33> purified isoenzyme I [42]) [33, 42]
52 <2> (<2> purified isoenzyme II [42]) [42]
54.1 <32> (<32> purified enzyme [38]) [38]
180 <2> [29]
201 <33> (<33> purified isoenzyme II [42]) [42]
210 <2> (<2> purified isoenzyme I [42]) [42]
222.8 <2> (<2> recombinant from Escherichia coli [29]) [29]
377 <29> (<29> purified enzyme [32]) [32]
480 <3> (<3> purified enzyme [39]) [39]
745 <27> (<27> purified enzyme [30]) [30]
790 <26> (<26> purified enzyme [30]) [30]

Additional information <1, 2, 5, 16, 18, 20, 23, 28> (<28> steady state kinetics [31]; <2> isoform A(-) shows about ten times lower specific activity than isoform A(+) [34]) [1, 2, 10, 18, 22, 20, 23, 26, 31, 34]

K_m-Value (mM)

0.0003 <3> ($NADP^+$, <3> nephridic enzyme [35]) [35]
0.001 <3> ($NADP^+$, <3> hepatic enzyme [35]) [35]
0.0016 <30> ($NADP^+$) [33]
0.0021 <3> (glucose 6-phosphate, <3> nephridic enzyme [35]) [35]
0.0033 <3> (glucose 6-phosphate, <3> hepatic enzyme [35]) [35]
0.0053 <31> ($NADP^+$) [37]
0.0056 <4> ($NADP^+$) [5]
0.006-0.008 <2, 33> ($NADP^+$, <2,33> isoenzymes I and II, with glucose 6-phosphate [42]) [42]
0.011 <32> ($NADP^+$) [38]
0.012 <2, 6> ($NADP^+$, <2> recombinant from E. coli [29]) [3, 29]
0.013 <2> ($NADP^+$) [29]
0.014 <2> (NADPH, <2> recombinant from E. coli [29]) [29]
0.015 <2> (NADPH) [29]
0.017 <3> ($NADP^+$, <3> 0.0025 mM glucose 6-phosphate [39]) [39]
0.02 <2> ($NADP^+$, <2> with D-glucose 6-phosphate or D-galactose 6-phosphate [41]) [41]
0.02 <27> (NADPH) [30]
0.023 <33> (NADPH, <33> isoenzyme II, with 6-phosphonogluconate 0.1 mM [42]) [42]
0.024 <30> (glucose 6-phosphate) [33]
0.025 <2> ($NADP^+$, <2> isoenzyme II, with 6-phosphonogluconate 0.1 mM [42]) [42]
0.026 <26> ($NADP^+$) [30]
0.03 <2, 27> ($NADP^+$, <2> with 2-deoxyglucose 6-phosphate, K_M decreases to 10 mM with increasing substrate concentration from 0.3 to 1.5 mM [41]) [30, 41]
0.031 <2> (D-galactose 6-phosphate, <2> isoenzyme II [42]) [42]
0.037 <6> (glucose 6-phosphate) [3]
0.04 <2, 32> (glucose 6 phosphate) [38, 41]
0.042 <4> (glucose 6-phosphate, <4> + $NADP^+$ [5]) [5]
0.046-0.047 <2, 33> (D-galactose 6-phosphate, <2, 33> isoenzyme I [42]) [42]
0.047 <4> (NAD^+) [5]
0.051 <33> (D-galactose 6-phosphate, <33> isoenzyme II [42]) [42]
0.056 <33> (glucose 6-phosphate, <33> isoenzyme II [42]) [42]
0.061 <2> (glucose 6-phosphate, <2> isoenzyme I [42]) [42]
0.067 <1> ($NADP^+$) [1]
0.069 <2> (glucose 6-phosphate, <2> recombinant from E. coli [29]) [29]
0.071 <29> (glucose 6-phosphate, <29> pH 5.4, NAD^+ [32]) [32]
0.072 <2> (glucose 6-phosphate) [29]
0.089 <29> (glucose 6-phosphate, <29> pH 5.4, $NADP^+$ [32]) [32]
0.092 <27> (glucose 6-phosphate) [30]

0.104 <29> (NAD^+, <29> pH 5.4 [32]) [32]
0.12 <7> (NAD^+) [6]
0.138 <31> (glucose 6-phosphate) [37]
0.153 <26> (glucose 6-phosphate) [30]
0.167 <33> (glucose 6-phosphate, <33> isoenzyme I [42]) [42]
0.18 <2> (glucose 6-phosphate, <2> isoenzyme II [42]) [42]
0.23 <1> (glucose 6-phosphate) [1]
0.279 <33> (6-phosphonogluconate, <33> isoenzyme II [42]) [42]
0.34 <29> ($NADP^+$, <29> pH 5.4 [32]) [32]
0.864 <2> (6-phosphonogluconate, <2> isoenzyme II [42]) [42]
Additional information <2, 7-15, 17, 20, 23> [6, 7, 9, 11-14, 16, 17, 19, 21, 23, 26]

pH-Optimum

7 <11> [13]
7-9 <26, 27> [30]
7.4 <9, 16, 21> [11, 20, 24]
7.4-8.2 <6> [3]
7.5-9 <22> [25]
7.6-10 <32> [38]
7.8 <30> [33]
8-8.2 <18> [22]
8-9 <4, 8, 10> [5, 7, 12]
8.3 <2> (<2> isoform A(+) [34]) [34]
8.5 <23> [26]
8.5-9.5 <12> [14]
8.9 <5> [10]
9 <2, 31> (<2> isoform A(-) [34]) [34, 37]
9.1 <1, 15> [1, 19]
9.2 <1, 14> [2, 17]
9.4 <3> [35]

pH-Range

6.5-8.5 <30> [33]
6.5-10 <3, 32> [35, 38]
7.4-8.2 <6> [3]

Temperature optimum (°C)

25 <1, 2, 4, 6-8, 14> (<1, 2, 4, 6-8, 14> assay at [1-4, 6-9, 17]) [1-4, 6-9, 17]
30 <5, 12, 15, 16, 21> (<5, 12, 15, 16, 21> assay at [10, 14, 18-20, 24]) [10, 14, 18-20, 24]
37 <4> [5]
40 <11> [13]
45 <15> [19]
55 <3, 10> [12, 35]

4 Enzyme Structure

Molecular weight

103000-110000 <1, 6, 7, 15> (<1> gel filtration [1, 2, 8]; <6> gel electrophoresis [3]; <7> sedimentation equilibrium method [6]; <15> sedimentation equilibrium method [19]) [1-3, 6, 8, 19]
112000 <28> (<28> gel filtration [31]) [31]
114000 <17> (<17> dimer dimerization to MW 209000 Da in presence of $NADP^+$, gel filtration [21]) [21]
117000 <13> (<13> gel filtration [40]) [40]
118000-120000 Da <9, 10, 14> (<9> three principal forms with MW of 120000 Da, 240000 Da and 345000 Da, gel electrophoresis [11]; <10> gel filtration [12]; <14> sucrose density gradient centrifugation [17]) [11, 12, 17]
121000 <3> (<3> gel filtration [39]) [39]
126000 <8> (<8> gel filtration [7]) [7]
138000 <21> (<21> gel filtration [24]) [24]
150000 <12> (<12> gel filtration [14]) [14]
158000 <31> (<31> gel filtration [37]) [37]
160000 <26> (<26> gel filtration [30]) [30]
180000 <16, 27> (<16, 27> gel filtration [20, 30]) [20, 30]
200000 <23> (<23> gel filtration [26]) [26]
206000 <6> (<6> gel electrophoresis [3]) [3]
210000 <2> [9]
220000 <2, 5, 33> (<5> Entner-Doudoroff enzyme, gel chromatography [10,18]; <2, 33> isoenzymes I and II, gel filtration [42]) [10, 18, 42]
238700 <4> (<4> ultracentrifugation [5]) [5]
240000 <9, 11, 13, 20> (<9> three principal forms with MW of 120000 Da, 240000 Da and 345000 Da, gel electrophoresis [11]; <11> gel filtration [13]; <20> gel filtration [23]) [11, 13, 16, 23]
243000 <29> (<29> gel filtration [32]) [32]
265000 <5> (<5> Zwischenferment, gel chromatography [10, 18]) [10, 18]
345000 <9> (<9> three principal forms with MW of 120000 Da, 240000 Da and 345000 Da, gel electrophoresis [11]) [11]

Subunits

? <2, 26, 27, 30> (<2> ? * 59000, SDS-PAGE, native and recombinant from E. coli [29]; <26> ? * 55000 + ? * 53000, SDS-PAGE [30] <27> ? * 60000 + ? * 57000, SDS-PAGE [30]; <30> ? * 50000 SDS-PAGE [33]) [29, 30, 33]
dimer <2, 3, 6, 7, 10, 14, 15, 17, 21, 28, 31> (<6, 28> 2 * 57000, SDS-PAGE [3, 31]; <7> 2 * 54800, SDS gel electrophoresis [6]; <10> 2 * 61000, SDS gel electrophoresis [12]; <14> 2 * 58000, SDS gel electrophoresis [17]; <15> 2 * 55000, SDS-PAGE [19]; <17> 2 * 54000, SDS-PAGE [21]; <21> 2 * 55300, SDS-PAGE [24]; <3, 31> 2 * 62000, SDS-PAGE [37, 39]) [3, 6, 12, 17, 19, 21, 24, 31, 37, 39, 41]
tetramer <2, 5, 6, 11, 13, 16, 20, 23, 25, 29, 33> (<11, 13, 20, 29> 4 * 60000, SDS-PAGE [13, 16, 23, 32]; <2, 6, 33> 4 * 57000, SDS-PAGE [3, 42]; <2> 4 * 53000, SDS-PAGE [9]; <5> 4 * 55000, Entner-Doudoroff enzyme [18]; <5> 4 *

65000, Zwischenferment [18]; <16> 4 * 48000, SDS-PAGE [20]; <23, 25> 4 * 52000, SDS-PAGE [26, 28]; <13> 4 * 31000, SDS-PAGE [40]) [3, 9, 13, 16, 18, 20, 23, 26, 28, 32, 40, 42]
Additional information <2> (<2> presence of SH-reagents, high enzyme and Mg^{2+} or Mn^{2+} concentrations, low pH and ionic strength favour oligomeric, active forms of the enzyme, while the opposite conditions favour inactive monomers [41]) [41]

5 Isolation/Preparation/Mutation/Application

Source/tissue

adrenal cortex <4> [5]
brain <2, 3, 33> [39, 42]
commercial preparation <19> [15]
erythrocyte <2, 3, 13, 24> [9, 27, 29, 34]
kidney <3, 13> [16, 35]
leaf <17, 20> [21, 23]
liver <1-3, 13, 24> [2, 27, 35, 40]
mammary gland <4> [4]
mycelium <6, 16> [3, 20]
placenta <2> [41]
procyclic form <31> [37]

Localization

chloroplast <17> [21]
cytoplasm <17, 20> [21, 23]
cytosol <2, 20, 31, 33> [23, 37, 42]
glycosome <31> [37]
mitochondrion <2, 33> [42]

Purification

<1> [1, 2, 8]
<2> (native and recombinant from Escherichia coli [29]; both isoforms A(-) and A(+) [34]; isoenzymes I and II [42]) [9, 29, 34, 42]
<3> (partial from liver and kidney [35]) [35, 39]
<4> [4, 5]
<5> [10, 18]
<6> [3]
<7> [6]
<8> [7]
<9> [11]
<10> [12]
<12> [14]
<13> [16, 40]
<14> [17]
<15> [19]
<16> [20]

<17> [21]
<18> [22]
<20> [23]
<21> [24]
<22> [25]
<23> [26]
<25> [28]
<28> [31]
<29> [32]
<30> [33]
<32> [38]
<33> (isoenzymes I and II [42]) [42]
<26, 27> [30]

Renaturation
<19> (inactivated by potassium myristate [15]) [15]

Crystallization
<1, 4, 7> [1, 2, 4-6, 8]

Cloning
<2> (expression in Escherichia coli [29]) [29]

6 Stability

pH-Stability
3-10 <11> (<11> stable [13]) [13]
5.4-7 <8> (<8> highest stability [7]) [7]
6-10.5 <5, 10> (<5, 10> stable at 5°C for 48 h [12, 18]) [12, 18]
6.5-10 <5> (<5> Entner-Doudoroff enzyme stable [10, 18]) [10, 18]
7.5-9 <15> (<15> stable [19]) [19]

Temperature stability
5 <10> (<10> 48 h [12]) [12]
25 <20> (<20> half-life: 15 min [23]) [23]
30 <23> (<23> 1 month [26]) [26]
55 <11> (<11> completely inactivated in 15 min [13]) [13]

Organic solvent stability
dimethylsulfoxide <30> (<30> substrate glucose and 40% DMSO (v/v): 10% remaining enzyme activity in HEPES buffer, 20% remaining enzyme activity in phosphate buffer [33]) [33]

General stability information
<2>, isoform A(-) is less stable during purification than A(+) due to V68M mutation that makes it suceptible to oxidation of sulfhydryl group [34]
<3>, $NADP^+$ stabilizes, while the enzyme is rapidly inactivated in presence of substrate glucose 6-phosphate [39]
<4>, EDTA stabilizes [4]

<4>, thioglycolate stabilizes [4]
<29>, stabilized by glycerol, $MgSO_4$, 2-mercaptoethanol, pH 5.4 [32]

Storage stability

<2>, -10°C, 10 mM Tris-HCl, pH 7.4, 5 mM 2-mercaptoethanol, 2 mM EDTA, 40% glycerol (v/v), weeks [41]
<7>, -20°C, 2 months [6]
<8>, -20°C or -40°C, impure state, more than 1 week [7]
<10>, 5°C, pH 6.0-10.5, stable for 48 h [12]
<23>, 30°C, Bacillus stearothermophilus enzyme maintains more than 80% of activity after 1 month [26]
<32>, 4°C, purified 650fold, several weeks [38]
<2, 5>, 4°C, several months [9, 10, 18]

References

[1] Engel, H.J.; Domschke, W.; Alberti, M.; Domagk, G.F.: Protein structure and enzymatic activity. II. Purification and properties of a crystalline glucose-6-phosphate dehydrogenase from Candida utilis. Biochim. Biophys. Acta, **191**, 509-516 (1969)
[2] Chilla, R.; Doering, K.M.; Domagk, F.; Rippa, M.: A simplified procedure for the isolation of a highly active crystalline glucose-6-phosphate dehydrogenase from Candida utilis. Arch. Biochem. Biophys., **159**, 235-239 (1973)
[3] Scott, W.A.: Glucose-6-phosphate dehydrogenase from Neurospora crassa. Methods Enzymol., **41**, 177-182 (1975)
[4] Julian, G.R.; Reithel, F.J.: Glucose-6-phosphate dehydrogenase from bovine mammary gland. Methods Enzymol., **41**, 183-188 (1975)
[5] McKerns, K.W.: Glucose-6-phosphate dehydrogenase from cow adrenal cortex-1. Methods Enzymol., **41**, 188-196 (1975)
[6] Olive, C.; Levy, H.R.: Glucose-6-phosphate dehydrogenase from Leuconostoc mesenteroides. Methods Enzymol., **41**, 196-201 (1975)
[7] Shepherd, M.G.: Glucose-6-phosphate dehydrogenase from Penicillium duponti. Methods Enzymol., **41**, 201-205 (1975)
[8] Domagk, G.F.; Chilla, R.: Glucose-6-phosphate dehydrogenase from Candida utilia. Methods Enzymol., **41**, 205-208 (1975)
[9] Cohen, P.; Rosemeyer, M.A.: Glucose-6-phosphate dehydrogenase from human erythrocytes. Methods Enzymol., **41**, 208-214 (1975)
[10] Lessmann, D.; Schimz, K.-L.; Kurz, G.: D-Glucose-6-phosphate dehydrogenase (Entner-Doudoroff enzyme) from Pseudomonas fluorescens. Purification, properties and regulation. Eur. J. Biochem., **59**, 545-559 (1975)
[11] Schaeffer, F.; Stanier, R.Y.: Glucose-6-phosphate dehydrogenase of Anabaena sp. Kinetic and molecular properties. Arch. Microbiol., **116**, 9-19 (1978)
[12] Kato, N.; Sahm, H.; Schütte, H.; Wagner, F.: Purification and properties of glucose-6-phosphate dehydrogenase and 6-phosphogluconate dehydrogenase from a methanol-utilizing yeast, Candida boidinii. Biochim. Biophys. Acta, **566**, 1-11 (1979)

[13] Kawai, K.; Eguchi, Y.: Properties of Lactobacillus glucose 6-phosphate dehydrogenase. J. Ferment. Technol., **57**, 369-371 (1979)

[14] Ben-Bassat, A.; Goldberg, I.: Purification and properties of glucose-6-phosphate dehydrogenase ($NADP^+/NAD^+$) and 6-phosphogluconate dehydrogenase ($NADP^+/NAD^+$) from methanol-grown Pseudomonas C. Biochim. Biophys. Acta, **611**, 1-10 (1980)

[15] Tortora, R.; Burlini, N.; Hanozet, G.M.; Guerritore, A.: Reactivation of yeast glucose-6-phosphate dehydrogenase denaturated by saturated fatty acids. Experientia, **38**, 427-429 (1982)

[16] Lee, C.-Y.: Glucose-6-phosphate dehydrogenase from mouse. Methods Enzymol., **89**, 252-257 (1982)

[17] Ujita, S.; Kimura, K.: Glucose-6-phosphate dehydrogenase, vegetative and spore Bacillus subtilis. Methods Enzymol., **89**, 258-261 (1982)

[18] Maurer, P.; Lessmann, D.; Kurz, G.: D-Glucose-6-phosphate dehydrogenases from Pseudomonas fluorescens. Methods Enzymol., **89**, 261-270 (1982)

[19] Steinbach, R.A.; Schütte, H.; Sahm, H.: Glucose-6-phosphate dehydrogenase from Methylomonas M15. Methods Enzymol., **89**, 271-275 (1982)

[20] Niehaus, W.G.; Dilts, R.P.: Purification and characterization of glucose-6-phosphate dehydrogenase from Aspergillus parasiticus. Arch. Biochem. Biophys., **228**, 113-119 (1984)

[21] Valenti, V.; Stanghellini, M.A.; Pupillo, P.: Glucose 6-phosphate dehydrogenase isozymes of maize leaves. Plant Physiol., **75**, 521-526 (1984)

[22] Röber, B.; Stolle, J.; Reuter, G.: Eigenschaften der Glucose-6-phosphat-Dehydrogenase aus der Hefe Candida maltosa II. Z. Allg. Mikrobiol., **24**, 629-636 (1984)

[23] Fickenscher, K.; Scheibe, R.: Purification and properties of the cytoplasmic glucose-6-phosphate dehydrogenase from pea leaves. Arch. Biochem. Biophys., **247**, 393-402 (1986)

[24] Hammond, J.B.W.: Glucose-6-phosphate dehydrogenase from Agaricus bisporus: purification and properties. J. Gen. Microbiol., **131**, 321-328 (1985)

[25] Yoshida, A.; Roth, E.F.: Glucose-6-phosphate dehydrogenase of malaria parasite Plasmodium falciparum. Blood, **69**, 1528-1530 (1987)

[26] Okuno, H.; Nagata, K.; Nakajima, H.: Purification and properties of glucose-6-phosphate dehydrogenase from Bacillus stearothermophilus. J. Appl. Biochem., **7**, 192-201 (1985)

[27] Schraven, E.; Gruber, C.: Organ- und speziesspezifische Eigenschaften von Glucose-6-phosphatdehydrogenase und deren Beeinflussung durch Molsidomin. Arzneim. Forsch., **38**, 36-39 (1988)

[28] Anderson, B.M.; Wise, D.J.; Anderson, C.D.: Azotobacter vinelandii glucose 6-phosphate dehydrogenase properties of NAD- and NADP-linked reactions. Biochim. Biophys. Acta, **1340**, 268-276 (1997)

[29] Bautista, J.M.; Mason, P.J.; Luzzatto, L.: Purification and properties of human glucose-6-phosphate dehydrogenase made in E. coli. Biochim. Biophys. Acta, **1119**, 74-80 (1992)

[30] Wennekes, L.M.J.; Goosen, T.; Van den Broek, P.J.M.; Van den Broek, H.W.J.: Purification and characterization of glucose-6-phosphate dehydrogenase

from Aspergillus niger and Aspergillus nidulans. J. Gen. Microbiol., **139**, 2793-2800 (1993)
[31] Tsai, C.S.; Chen, Q.: Purification and kinetic characterization of hexokinase and glucose-6-phosphate dehydrogenase from Schizosaccharomyces pombe. Biochem. Cell Biol., **76**, 107-113 (1998)
[32] Ragunathan, S.; Levy, H.R.: Purification and characterization of the NAD-preferring glucose 6-phosphate dehydrogenase from Acetobacter hansenii (Acetobacter xylinum). Arch. Biochem. Biophys., **310**, 360-366 (1994)
[33] Niehaus, W.G.; Mallett, T.C.: Purification and characterization of glucose-6-phosphate dehydrogenase from Cryptococcus neoformans: identification as "nothing dehydrogenase". Arch. Biochem. Biophys., **313**, 304-309 (1994)
[34] Adediran, S.A.: Kinetic and thermodynamic properties of two electrophoretically similar genetic variants of human erythrocyte glucose-6-phosphate dehydrogenase. Biochimie, **78**, 165-170 (1996)
[35] Corpas, F.J.; Garcia-Salguero, L.; Peragon, J.; Lupianez, J.A.: Kinetic properties of hexose-monophosphate dehydrogenases. I. Isolation and partial purification of glucose-6-phosphate dehydrogenase from rat liver and kidney cortex. Life Sci., **56**, 179-189 (1995)
[36] Zhang, Z.; Apse, K.; Pang, J.; Stanton, R.C.: High glucose inhibits glucose-6-phosphate dehydrogenase via cAMP in aortic endothelial cells. J. Biol. Chem., **275**, 40042-40047 (2000)
[37] Heise, N.; Opperdoes, F.R.: Purification, localisation and characterization of glucose-6-phosphate dehydrogenase of Trypanosoma brucei. Mol. Biochem. Parasitol., **99**, 21-32 (1999)
[38] Reimers, J.M.; Huang, Q.; Albe, K.R.; Wright, B.E.: Purification and kinetic characterization of glucose-6-phosphate dehydrogenase from Dictyostelium discoideum. Exp. Mycol., **17**, 1-6 (1993)
[39] Askar, M.A.; Sumathy, K.; Baquer, N.Z.: Regulation and properties of purified glucose-6-phosphate dehydrogenase from rat brain. Indian J. Biochem. Biophys., **33**, 512-518 (1996)
[40] Velasco, P.; Barcia, R.; Ibarguren, I.; Sieiro, A.M.; Ramos-Martinez, J.I.: Purification, characterization and kinetic mechanism of glucose-6-phosphate dehydrogenase from mouse liver. Int. J. Biochem., **26**, 195-200 (1994)
[41] Ozer, N.; Aksoy, Y.; Ogüs, I.H.: Kinetic properties of human placental glucose-6-phosphate dehydrogenase. Int. J. Biochem. Cell Biol., **33**, 221-226 (2001)
[42] Cho, S.-W.; Joshi, J.G.: Characterization of glucose-6-phosphate dehydrogenase isozymes from human and pig brain. Neurosci., **38**, 819-828 (1990)

3α-Hydroxysteroid dehydrogenase (B-specific) 1.1.1.50

1 Nomenclature

EC number
1.1.1.50

Systematic name
3α-hydroxysteroid:NAD(P)$^+$ oxidoreductase (B-specific)

Recommended name
3α-hydroxysteroid dehydrogenase (B-specific)

Synonyms
3-α-HSD [Swissprot]
3α-hydroxysteroid oxidoreductase <3> (<3> NADH- and NADPH-linked [37]) [37]
3α-hydroxysteroid-5β-oxidoreductase activity <16> [39]
5α-dihydroprogesterone 3α-hydroxysteroid oxidoreductase <3> (<3> NADH- and NADPH-linked [38, 40]) [36-38, 40]
HSD28 [Swissprot]
HSD29 [Swissprot]
NADPH:5α-dihydroprogesterone 3α-hydroxysteroid oxidoreductase <3> [23]
hydroxyprostaglandin dehydrogenase
sterognost 3α dehydrogenase, 3α-hydroxy steroid
Additional information (see also EC 1.1.1.213 for enzymes catalyzing the same reaction, but which are A-specific with respect to hydrogen transfer to/from NAD$^+$/NADH. Enzymes without information of stereospecificity are summerized under EC 1.1.1.50 but may in fact belong to EC 1.1.1.213; <3, 9, 11> enzymes belong to family of short-chain dehydrogenases [19, 20, 25]; <9> member of aldo-keto-reductase superfamily [26, 29, 30])

CAS registry number
9028-56-2

2 Source Organism

<1> *Pseudomonas putida* [1, 2, 27]
<2> *Pseudomonas testosteroni* [3, 12, 13, 15]
<3> *Rattus norvegicus* (NADPH- and NADH-linked 5α-dihydroprogesterone 3α-hydroxysteroid oxidoreductase activity [36-38, 40]) [4, 5, 7, 9-11, 14, 19, 23, 33, 35-38, 40]
<4> *Mus musculus* [6]

<5> *Canis familiaris* [8]
<6> *Eubacterium lentum* [16, 18]
<7> *Clostridium irregularis* [17]
<8> *Gallus gallus* [22, 34]
<9> *Homo sapiens* (1-4 different enzyme isoforms [20]; type 2 isoform [26]) [19, 20, 24, 26, 29, 30]
<10> *Eubacterium sp.* (strain VPI 12708 [21]) [21]
<11> *Comamonas testosteroni* [25]
<12> *Bos taurus* [28]
<13> *Pseudomonas sp.* [31]
<14> *hamster* [32]
<15> *Digitalis lanata* (floxglove, NADH-linked 5β-dihydroprogesterone 3α-hydroxysteroid oxidoreductase activity [41]) [41, 42]
<16> *Digitalis purpurea* [39]

3 Reaction and Specificity

Catalyzed reaction

androsterone + NAD(P)$^+$ = 5α-androstane-3,17-dione + NAD(P)H (<9> enzyme isoforms bind substrate in different ways [26]; <14> reaction mechanism: ordered bi bi with random substrate binding and ordered release of product, NADPH is released last, pro-R-hydrogen atom of NAD(P)H is transferred to carbonyl substrate [32])

Reaction type

oxidation
redox reaction
reduction

Natural substrates and products

S 3-hydroxysteroid + NAD(P)$^+$ <13, 14> (Reversibility: ? <13> [31]; r <14> [32]) [31, 32]
P 3-ketosteroid + NAD(P)H <13, 14> [31, 32]
S 5α-dihydrotestosterone + NAD(P)H <3, 9> (<3, 9> discussion of physiological role [7, 19]; <3> NADH-linked, particulate enzyme prefers oxidative reaction, while the NADPH-linked, cytosolic enzyme prefers the reductive reaction [36]) (Reversibility: r <3, 9> [4, 7, 19, 30, 36]) [4, 7, 19, 30, 36]
P 5α-androstane-3α,17β-diol + NAD(P)$^+$ <3, 9> [19, 30, 36]
S 5α-pregnan-3,20-dione + NADPH <3> (Reversibility: r <3> [36-38]) [36-38]
P 3α-hydroxy-5α-pregnan-20-one + NADP$^+$ <3> [36-38]
S 5α-pregnan-3-20-dione + NAD(P)H <3, 16> (<3> 5α-dihydroprogesterone 3α-hydroxysteroid oxidoreductase activity, utilizing NADH as cofactor when of particulate origin and NADPH when of cytosolic origin [36-38, 40]; <3> activity higher during lactation in female rats [38]; <16> 5α-dihydroprogesterone 3α-hydroxysteroid oxidoreductase activity, NADPH

and NADH possible as cofactors [39]) (Reversibility: r <3> [36-38, 40]; ? <16> [39]) [36-40]

P 5α-pregnane-3α-ol-20-one + NAD(P)$^+$ <3, 16> [36-40]

S 5β-pregnan-3,20-dione + NAD(P)H <15, 16> (<15, 16> 3α-hydroxysteroid-5β-oxidoreductase activity [39, 41, 42]) (Reversibility: r <16> [39]; ? <15> [41, 42]) [39, 41, 42]

P 3α-hydroxy-5β-pregnan-20-one + NAD(P)$^+$ <15, 16> [39, 41, 42]

S Additional information <1, 3, 9, 12> (<1, 3> belongs to the group of Y'bile acid binders, bile acid metabolizing enzymes [27, 35]; <9> type 3 isoform in liver virtually identical with bile acid-binding protein [26]; <12> broad substrate range [28]) [26-28, 35]

P ?

Substrates and products

S 17β-hydroxy-dihydrotestosterone + NAD$^+$ <9> (Reversibility: r <9> [20]) [20]

P 5α-androstan-3-α,17β-dione + NADH <9> [20]

S 3-oxocholic acid + NADPH <3> (<3> reverse reaction rate negligible [35]) (Reversibility: ir <3> [35]) [35]

P 3α-hydroxycholic acid + NADP$^+$ <3> [35]

S 3α,12α-dihydroxy-5β-cholanoyl glycine + NAD$^+$ <6> (Reversibility: ? <6> [18]) [18]

P 12α-hydroxy-3-oxo-5β-cholanoyl glycine + NADH

S 3α,17α,21-trihydroxy-5β-pregnan-11,20-dione + NADP$^+$ <4> (Reversibility: r <4> [6]) [6]

P 17α,21-dihydroxy-5β-pregnan-3,11,20-trione + NADPH <4> [6]

S 3α,21-dihydroxy-5β-pregnan-20-one + NADP$^+$ <4> (Reversibility: r <4> [6]) [6]

P 21-hydroxy-5β-pregnan-3,20-dione + NADPH

S 3α,7α,12α-trihydroxy-5β-cholanoyl glycine + NAD$^+$ <1, 6> (Reversibility: ? <1, 6> [18, 27]) [18, 27]

P 7α,12α-dihydroxy-5β-cholan-3-one-24-oyl glycine + NADH

S 3α,7α-dihydroxy-5β-cholanoyl glycine + NAD$^+$ <6> (Reversibility: ? <6> [18]) [18]

P 7α-hydroxy-3-oxo-5β-cholanoyl glycine + NADH

S 3α-adiol + NAD(P)$^+$ <3, 9> (Reversibility: r <3, 9> [19]) [19]

P dihydrotestosterone + NAD(P)H <3, 9> [19]

S 3α-hydroxy-12-oxo-9,11-cholenic acid + NADP$^+$ <1> (Reversibility: ? <1> [1]) [1]

P 3,12-dioxo-9,11-cholenic acid + NADPH

S 3α-hydroxy-5α-pregnan-20-one + NADP$^+$ <3> (Reversibility: ? <3> [9]) [9]

P 5α-pregnan-3,20-dione + NADPH

S 3α-hydroxy-5β-androstan-17-one + NADP$^+$ <4, 6, 12> (Reversibility: r <4, 6, 12> [6, 18, 28]) [6, 18, 28]

P 5β-androstan-3,17-dione + NADPH <12> [28]

S 3α-hydroxy-5β-pregnan-20-one + $NADP^+$ <4, 15, 16> (<15> inhibition by Mg^{2+} only in oxidative direction [41]) (Reversibility: r <4, 15, 16> [6, 39, 41]) [6, 39, 41]
P 5β-pregnan-3,20-dione + NADPH
S 3α-tetrahydroprogesterone + NADH <9> (Reversibility: r <9> [20]) [20]
P 5α-dihydroprogesterone + NAD^+ <9> [20]
S 5α-androstan-3α,17β-diol + $NAD(P)^+$ <3, 4, 6, 9> (Reversibility: r <3, 4, 6, 9> [6, 18, 20, 26]) [6, 18, 20, 23, 26]
P 17β-hydroxy-5α-androstan-3-one + NADPH <4, 9> [6, 20]
S 5α-androstan-3α,17β-diol + NAD(P)H <9> (Reversibility: r <9> [20, 26]) [20, 26]
P 5α-dihydrotestosterone + $NAD(P)^+$ <9> [20]
S 5α-dihydrocortisol + NADPH <9> (Reversibility: ? <9> [24]) [24]
P 3α,5α-tetrahydrocortisol + $NADP^+$ <9> [24]
S 5α-dihydrotestosterone + NAD(P)H <1, 3, 5, 9, 11> (<1, 3, 5> i.e. 5α-androstan-17β-ol-3-one [1, 4, 7, 8, 10, 11]) (Reversibility: r <1, 3, 5, 9, 11> [1, 4, 7, 8, 10, 11, 19, 20, 25, 26]; ? <9> [24]) [1, 4, 7, 8, 10, 11, 19, 20, 24-26]
P 3β-adiol + $NAD(P)^+$ <3, 9> [19, 20]
S 5α-dihydrotestosterone + NAD(P)H <1, 3, 5, 9, 11> (<1, 3, 5> i.e. 5α-androstan-17β-ol-3-one [1, 4, 7, 8, 10, 11]) (Reversibility: r <1, 3, 5, 9, 11> [1, 4, 7, 8, 10, 11, 19, 20, 25, 26, 30]; ? <9> [24]) [1, 4, 7, 8, 10, 11, 19, 20, 24-26, 30]
P 5α-androstan-3α,17β-diol + $NAD(P)^+$ <1, 3, 5, 9> [1, 7, 8, 10, 11, 19, 30]
S 5α-pregnan-3,20-dione + NAD(P)H <3, 9> (<3> 5α-dihydroprogesterone 3α-hydroxysteroid oxidoreductase activity, utilizing NADH as cofactor when of particulate origin and NADPH when of cytosolic origin [36,37]) (Reversibility: r <3> [23]; ? <3, 9> [29, 36-38]) [23, 29, 36-38]
P 3α-hydroxy-5α-pregnan-20-one + $NAD(P)^+$ <3, 9> [23, 29, 36-38]
S 5α-pregnan-3,20-dione + NADPH <3> (Reversibility: ? <3> [36-38]) [36-38]
P 3α-hydroxy-5α-pregnan-20-one + $NADP^+$ <3> [36-38]
S 5β-androstan-3,17-dione + $NADP^+$ <12> (Reversibility: r <12> [28]) [28]
P 5β-androstan-3α,17α-diol + NADPH <12> [28]
S 5β-androstan-3-α,17β-diol + $NADP^+$ <4, 6, 9> (Reversibility: r <4, 6, 9> [6, 18, 26]) [6, 18, 26]
P 17β-hydroxy-5β-androstan-3-one + NADPH
S 5β-androstan-3α-ol-17-one + NADPH <12> (Reversibility: r <12> [28]) [28]
P 5β-androstan-3α,17α-diol + $NADP^+$ <12> [28]
S 5β-androstanedione + NAD(P)H <8, 9, 11> (Reversibility: r <9, 11> [25, 26]; ? <8> [22, 34]) [22, 25, 26, 34]
P 5β-androstan-17-one-3-ol + $NAD(P)^+$
S 5β-dihydrocortisol + NADPH <9> (Reversibility: ? <9> [24]) [24]
P 3α,5β-tetrahydrocortisol + $NADP^+$ <9> [24]

S 5β-dihydrotestosterone + NAD(P)H <3, 9> (<3> NADH-linked, particulate enzyme prefers oxidative reaction, while the NADPH-linked, cytosolic enzyme prefers the reductive reaction [36]) (Reversibility: r <3> [36]; ? <3, 9> [7, 11, 19, 24]) [7, 11, 19, 24, 36]
P 5β-androstan-3α,17β-diol + NAD(P)$^+$ <3> [11, 36]
S 5β-pregnan-3,20-dione + NAD(P)H <15, 16> (<15, 16> 3α-hydroxysteroid-5β-oxidoreductase activity [39, 41, 42]) (Reversibility: r <15, 16> [39, 41]; ? <15> [42]) [39, 41, 42]
P 3α-hydroxy-5β-pregnan-20-one + NAD(P)$^+$ <15, 16> [39, 41, 42]
S 6-ketolithocholate + NAD$^+$ <1> (Reversibility: r <1> [27]) [27]
P 5β-cholan-3,6-dione-24-oic acid + NADH
S 9,10-phenanthrenequinone + NADPH <9, 12> (Reversibility: ? <9, 12> [26, 28]) [26, 28]
P ?
S androsterone + NAD(P)$^+$ <1-4, 6, 9, 11, 12, 14> (<1-4,6> i.e. 3α-hydroxy-5α-androstan-17-one [1, 5, 6, 9, 12, 13, 18]) (Reversibility: r <1-4, 6, 9, 11, 12, 14> [1, 5, 6, 9, 12, 13, 18-20, 25, 26, 28, 32]; ? <1> [27]) [1, 5, 6, 9, 12, 13, 18, 19, 20, 25-28, 32]
P 5α-androstan-3,17-dione + NAD(P)H <2, 9> [13, 20, 28]
S benzoylpyridine + NADPH <12> (Reversibility: ? <12> [28]) [28]
P ?
S camphorquinone + NADPH <12> (Reversibility: ? <12> [28]) [28]
P ?
S chenodeoxycholic acid + NAD(P)$^+$ <1> (<1> i.e. 3α,7α-dihydroxy-5β-cholan-24-oic acid [1]) (Reversibility: r <1> [1, 27]) [1, 27]
P 7α-hydroxy-3-oxo-5β-cholan-24-oic acid + NAD(P)H
S cholic acid + NAD(P)$^+$ <1, 10, 11> (<1> i.e. 3α,7α,12α-trihydroxy-5β-cholan-24-oic acid [1]) (Reversibility: ? <1, 10> [1, 21, 27]; r <11> [25]) [1, 21, 25, 27]
P 7α,12α-dihydroxy-3-oxo-5β-cholan-24-oic acid + NAD(P)H <10> [21]
S deoxycholic acid + NAD(P)$^+$ <1, 3> (<1, 3> i.e. 3α,12α-dihydroxy-5β-cholan-24-oic acid [1, 5]) (Reversibility: ? <1, 3> [1, 5, 27]) [1, 5, 27]
P 12α-hydroxy-3-oxo-5β-cholan-24-oic acid + NAD(P)H
S estradiol + NAD(P)H <3, 9> (Reversibility: ?<3, 9> [19]) [19]
P estrone + NAD(P)$^+$ <3, 9> [19]
S etiocholan-3α-ol-17-one + NADP$^+$ <1> (Reversibility: ? <1> [1]) [1]
P etiocholan-3,17-dione + NADPH
S fusidic acid + NAD$^+$ <11> (Reversibility: r <11> [25]) [25]
P ?
S glycolithocholic acid + NADP$^+$ <3, 4> (i.e. 3α-hydroxy-5β-cholanoyl glycine [5,6]) (Reversibility: ? <3, 4> [5, 6]) [5, 6]
P 3-oxo-glycocholanoic acid + NADPH (i.e. 3-oxo-5β-cholanoyl glycine)
S hydroxybutyrate + NADP$^+$ <12> (Reversibility: ? <12> [28]) [28]
P oxobutanoate + NADPH
S lithocholic acid + NADP$^+$ <1, 3> (<1, 3> i.e. 3α-hydroxy-5β-cholan-24-oic acid [1,5]) (Reversibility: ? <1, 3> [1, 5]) [1, 5]
P 3-oxo-5β-cholan-24-oic acid + NADPH

S menadione + NADPH <12> (Reversibility: ? <12> [28]) [28]
P ?
S metapyrone + NADH <11> (Reversibility: ? <11> [25]) [25]
P ?
S *p*-nitrobenzaldehyde + NAD(P)H <9, 11, 14> (Reversibility: r <11, 14> [25, 26, 32]) [25, 26, 32]
P *p*-nitrobenzylalcohol + $NAD(P)^+$ <11, 14> [25, 32]
S *p*-nitrobenzylalcohol + NAD^+ <11> (Reversibility: r <11> [25]) [25]
P *p*-nitrobenzaldehyde + NADH <11> [25]
S progesterone + NADH <11> (Reversibility: r <11> [25]) [25]
P ?
S taurocholic acid + $NAD(P)^+$ <1> (<1> i.e. 3α,7α,12α-trihydroxy-5β-cholanoyl taurine [1]) (Reversibility: ? <1> [1, 27]) [1, 27]
P 2[[7α,12α-dihydroxy-3,24-dioxo-5β-cholan-24-yl]amino]ethane sulfonic acid + NADPH
S testosterone + NAD(P)H <3, 9, 11> (Reversibility: ? <3, 9> [19]; r <11> [25, 26]) [19, 25, 26]
P androstenedione + $NAD(P)^+$ <9> [19]
S tetrahydrodeoxycorticosterone + $NADP^+$ <14> (Reversibility: ? <14> [32]) [32]
P dihydrodeoxycorticosterone + NADPH
S ursodeoxycholate + NAD^+ <1> (Reversibility: r <1> [27]) [27]
P 7β-hydroxy-3-oxo-5β-cholan-24-oic acid + NADH
S Additional information <1, 3, 4, 9, 12> (<1, 4> absolutely specific for 3α-hydroxy group of C19, C21 and C24 steroids [1, 6]; <1> acts faster on A/B cis hydroxysteroids than on A/B trans hydroxysteroids [1]; <3> substrate specificity of various isozymes [5]; <4, 12> reduction of carbonyl compounds e.g. quinones, aliphatic-chain diketones, ketones [6, 28]; <9> a saturated 4,5 double bound is essential for enzyme activity, 3-ketosteroids are not used as substrates [24]; <9> can not catalyze conversion of 3α-androstanediol to 5α-dihydrotestosterone [26]; <12> can not catalyze the dehydrogenation of 3β-hydroxysteroids, epiandrosterone and 17β-hydroxysteroids, i.e. testosterone and 17β-estradiol [28]) [1, 5, 6, 24, 26, 28]
P ?

Inhibitors

1-methylpyrrole acetic acid <3> [4]
11β,17α,21-trihydroxy-4-pregnen-3,20-dione <4> [6]
11β,21-dihydroxy-4-pregnen-3,20-dione <4> [6]
17,21-dihydroxy-5α-pregnan-3,20-dione <5> [8]
17α,21-dihydroxy-1,4-pregnadien-3,11,20-trione <4> [6]
17α,21-dihydroxy-4-pregnen-3,11,20-trione <4> [6]
17α-hydroxy-4-androsten-3-one <4> [6]
17α-hydroxy-4-pregnen-3,20-dione <4> [6]
17α-methyl-5α-dihydrotestosterone <5> [8]
17α-methyltestosterone <12> (<12> competitive inhibition [28]) [28]
17β-hydroxy-4-androsten-3-one <4> [6]

21-hydroxy-5α-pregnan-3,20-dione <5> [8]
2α bromo 17β hydroxy-5α-androstan-3-one <4, 5> [6, 8]
2α-bromo-5α-androstan-3,17-dione <5> [8]
2α-cyano-17β-methoxy-4-androsten 3 one <5> [8]
2α-cyano-17β-methoxy-5α-androstan-3-one <5> [8]
2α-cyano-5α-dihydrotestosterone <5> [8]
2α-fluoro-5α-androstan-3,17-dione <5> [8]
2α-iodo-5α-androstan-3,17-dione <5> [8]
4-androsten-3,17-dione <4> [6]
4-aza-4-methyl-5α-pregnane-3,20-dione <3> (<3> only slight inhibition [40]) [40]
4-pregnen-3,20-dione <4> [6]
5,5'-dithiobis(2-nitrobenzoic acid) <12> [28]
5α-androstan-3,16-dione <5> [8]
5α-androstan-3,17-dione <5> [8]
5α-androstan-3-one <5> [8]
5α-dihydrotestosterone hemisuccinate <5> [8]
5α-pregnan-3,20-dione <5> [8]
5β-dihydrotestosterone <5> [8]
5β-pregnan-3,20-dione <5> [8]
6α-methylprednisolone <4> [6]
Hg^{2+} <4> [6]
Mg^{2+} <15> (<15> strong inhibition of oxidation of 3α-hydroxysteroids [41]) [41]
acetaminophen <3> [4]
arylpropionic acid <3> [4]
bethamethasone <4> [6]
cholic acid <3> (<3> complete inhibition [35]) [35]
Cibacron Blue <14> (<14> competitive with $NADP^+$ [32]) [32]
deoxycorticosterone <4> [6]
dexamethasone <4, 9> [6, 24]
dexamethasone 21-acetate <9> [24]
dexamethasone 21-hemisuccinate <9> [24]
dienstrol <4> [6]
ethacrynate <12> [28]
ethynylestradiol <12> (<12> competitive inhibition [28]) [28]
fenamates <3> [4]
fluoxetine <9> (<9> allosteric interaction [29]) [29]
heavy metal ions <1> [1]
hexestrol <4, 14> (<14> competitive with $NADP^+$ [32]) [6, 32]
indomethacin <3, 4, 12> (not, <3> [14]) [4, 6, 28]
medroxyprogesterone acetate <3, 4> [4, 6]
megestrol <12> (<12> competitive inhibition [28]) [28]
norethindrone <3> [4]
p-chloromercuribenzoate <1, 3> [1, 5]
prednisolone <4> [6]
progesterone <3> [4]

prostaglandin A_1 <4> [6]
prostaglandin D_2 <4> [6]
prostaglandin E_2 <4> [6]
prostaglandin $F_{2\alpha}$ <4> [6]
salicylates <3> [4]
stilbestrol <4> [6]
sulfhydryl oxidizing reagents <12> [28]
testosterone <9> (<9> at concentration 0.001 mM 75% inhibition [24]) [24]
triamcinolone <4> [6]
zomepirac <4> [6]
Additional information <2-4> (<2, 4> product inhibition [6, 12]; <3> steroids [7]; <3> not inhibitory: phenobarbital, pyrazole [5]) [5-7, 12]

Cofactors/prosthetic groups

NAD^+ <1, 9, 10, 14, 16> (<9> preference for [20]; <1> specific, enzyme does not use $NADP^+$ [27]) [20, 21, 27, 32, 39]
NADH <3, 9, 11, 14, 16> (<9> preference for [20]; <3> plasma membrane NADH-linked 5α-dihydroprogesterone 3α-hydroxysteroid oxidoreductase activity [36-38, 40]; <16> cytosolic 5β-dihydroprogesterone 3α-hydroxysteroid oxidoreductase activity [39]) [20, 25, 32, 36-40]
$NADP^+$ <9, 10, 12, 14> [21, 26, 28, 32]
NADPH <3, 8, 9, 12, 14-16> (<3, 8, 9> preference for [22-24, 35]; <3> cytosolic NADPH-linked 5α-dihydroprogesterone 3α-hydroxysteroid oxidoreductase activity [36-38, 40]; <16> NADPH-linked microsomal 5β-dihydroprogesterone 3α-hydroxysteroid oxidoreductase shows very low activity only with NADPH [39]) [22-24, 26, 28, 32, 35-41]
Additional information <9> (<9> cytosolic and microsomal enzymes have different cofactor preferences: cytosolic enzymes prefer $NADP^+$/NADPH, microsomal prefer NAD^+/NADH [20]) [20]

Activating compounds

Additional information <1> (<1> see also EC 1.1.1.213 for enzymes catalyzing the same reaction but with A-specificity concerning the cofactor, all enzymes without information of stereospecificity for cofactors are included in EC 1.1.1.50, but some of those may in fact belong to EC 1.1.1.213 [1]) [1]

Turnover number (min^{-1})

0.47 <9> (androsterone) [26]
1 <9> (dihydrotestosterone) [19]
2 <3> (androsterone) [19]
2.39 <9> (5α-dihydrotestosterone, <9> detection fluorometrically [26]) [26]
4.54 <9> (testosterone) [26]
5 <3> (3α-adiol) [19]
5.79 <9> (5α-dihydrotestosterone, <9> detection spectrophotometrically [26]) [26]
6.87 <9> (5α-dihydrotestosterone, <9> detection radiometrically [26]) [26]
8 <9> (androsterone) [19]
11 <9> (3α-adiol) [19]

12.2 <9> (3α-androstanediol, <9> detection spectrophotometrically [26]) [26]
19.2 <9> (3α-androstanediol, <9> detection radiometrically [26]) [26]
26.64 156 <3> (androsterone) [5]
30.4 <9> (3α-androstanediol, <9> detection fluorometrically [26]) [26]
64.7 <9> (4-nitrobenzaldehyde) [26]
114 <9> (androstanedione, <9> detection spectrophotometrically [26]) [26]
1136 <9> (9,10-phenanthrenequinone) [26]
Additional information <3> (<3> overview, values for various substrates and 6 different isozymes [5]) [5]

Specific activity (U/mg)

0.00024 <8> (<8> purified enzyme, substrate 5β-androstan-3,17-dione [22]) [22]
0.0005 <9> (<9> substrate androsterone with cofactor NADP$^+$, detection radiometrically [26]) [26]
0.0013 <9> (<9> substrate 5α-dihydrotestosterone with cofactor NADP$^+$, detection radiometrically [26]) [26]
0.0025 <9> (<9> substrate 5α-dihydrotestosterone with cofactor NADPH, detection spectrophotometrically [26]) [26]
0.0042 <9> (<9> substrate 5α-dihydrotestosterone with cofactor NADPH, detection radiometrically [26]) [26]
0.0048 <9> (<9> substrate 3α-androstanediol with cofactor NADP$^+$, detection spectrophotometrically [26]) [26]
0.005 <3> [19]
0.0085 <9> (<9> substrate 3α-androstanediol with cofactor NADP$^+$, detection radiometrically [26]) [26]
0.011 <9> [19]
0.0111 <9> (<9> substrate androstanedione with cofactor NADPH, detection spectrophotometrically [26]) [26]
0.0134 <9> (<9> substrate 3α-androstanediol with cofactor NADP$^+$, detection fluorometrically [26]) [26]
0.0135 <9> [20]
0.031 <9> (<9> substrate dihydrotestosterone [20]) [20]
0.04 <3> (<3> 5α-pregnan-3,20-dione [23]) [23]
0.13 <9> (<9> substrate 4-nitrobenzaldehyde with cofactor NADPH, detection spectrophotometrically [26]) [26]
0.15 <12> (<12> purified enzyme [28]) [28]
0.323 <3> (<3> 3α-hydroxy-5α-pregnan-20-one [23]) [23]
0.358 <3> [10]
0.538 <9> (<9> substrate 9,10-phenanthrenequinone with cofactor NADPH, detection spectrophotometrically [26]) [26]
0.7 <3> [14]
1.56 <15> (<15> substrate 5β-dihydroprogesterone, young leaves [41]) [41]
2.2 <3> [4]
2.4 <14> (<14> oxidation of androsterone [32]) [32]
2.5 <11> (<11> fusidic acid, oxidation [25]) [25]
6.1 <4> [6]

6.5 <14> (<14> reduction of 5α-androstane-3,17-dione [32]) [32]
23.6 <1> (<1> purified enzyme, substrate cholate with NAD^+ [27]) [27]
27.4 <11> (<11> cholic acid, oxidation [25]) [25]
63 <1> [2]
64.5 <11> (<11> 5α-dihydrotestosterone, reduction [25]) [25]
85.7 <11> (<11> androstanedione, reduction [25]) [25]
119 <11> (<11> androsteone, oxidation [25]) [25]
869.8 <15> (<15> substrate 5β-dihydroprogesterone, green shoot culture [41]) [41]
2999-4199 <15> (<15> green leaves [42]) [42]
Additional information <3> [37]

K_m-Value (mM)

0.000082 <3> (5α-pregnan-3,20-dione, <3> reduction [23]) [23]
0.0001 <3> (3α-adiol) [19]
0.0001 <3> (3α-adiol) [19]
0.0001 <9> (5α-dihydrotestosterone) [19]
0.0001 <3> (androsterone) [19]
0.0001 <9> (androsterone) [19]
0.00016 <8> (5β-androstanedione) [34]
0.00035-0.00049 <4> (5β-androstan-3,17-dione, <4> 5β-pregnan-3,20-dione, 5α-androstan-3,17-dione [6]) [6]
0.00062-0.00065 <5> (5α-dihydrotestosterone) [8]
0.0007-0.00075 <3, 4> (3-oxo-5β-cholan-24-oic acid, <4> [6]; <3> 5β-dihydrotestosterone, rat testis [7]) [6, 7]
0.00071 <3> (NADPH) [23]
0.00097 <4> (NADPH) [6]
0.0012-0.0017 <3, 4> (17β-hydroxy-5β-androstan-3-one, <4> 17α,21-dihydroxy-5β-pregnan-3,11,20-trione, acenaphthequinone [6]; <3> 5β-dihydrotestosterone, rat epididymis [7]) [6, 7]
0.0014 <3> (3α-hydroxy-5α-pregnan-20-one, <3> oxidation [23]) [23]
0.0016 <8> (5β-androstan-3,17-dione) [22]
0.0016 <3> ($NADP^+$) [23]
0.002-0.0024 <4, 6> (17β-hydroxy-5α-androstan-3-one, <4> [6]; <6> 3α,7α-dihydroxy-5β-cholanoyl glycine [18]) [6, 18]
0.0028 <3> (testosterone) [19]
0.0033 <3> (testosterone) [19]
0.004 <9> (androstanedione, <9> NADPH, detection spectrophotometrically [26]) [26]
0.005 <9> (3α-tetrahydroprogesterone) [20]
0.0052-0.0059 <3, 4> (glycolithocholic acid, <4> [6]; <3> 3α-hydroxy-5α-pregnan-20-one [9]) [6, 9]
0.006 <9> (5α-dihydrotestosterone, <9> $NADP^+$, detection radiometrically [26]) [26]
0.0061 <11> (fusidic acid) [25]
0.007-0.0083 <2-4, 6> (5α-dihydrotestosterone, <3> [4]; <4> 3α-hydroxy-5β-pregnan-17-one [6]; <3> 3α-hydroxy-5α-androstan-17-one [9]; <2> andros-

tanedione [12]; <6> 3α,12α-dihydroxy-5β-cholanoyl glycine [18]) [4, 6, 9, 12, 18]
0.0095 <4> (androsterone) [6]
0.01-0.025 <1, 4, 6> (cholic acid, <1> [1]; <4> 3α-hydroxy-5β-androstan-17-one, 3α,21-dihydroxy-5β-pregnan-20-one, 5β-androstan-3α,17β-diol, 5α-androstan-3α,17β-diol, $NADP^+$ [6]; <6> NAD^+, 3α,7α,12α-trihydroxy-5β-cholanoyl glycine [18]) [1, 6, 18]
0.011 <1> (NAD^+) [27]
0.012 <9> (5α-dihydrotestosterone) [20]
0.012 <9> (testosterone, <9> $NADP^+$, detection radiometrically [26]) [26]
0.0136 <1> (deoxycholate) [27]
0.015 <1> (cholate) [27]
0.017 <9> (3α-androstanediol, <9> $NADP^+$, detection spectrophotometrically [26]) [26]
0.018 <9> (5α-dihydrotestosterone, <9> NADPH, detection radiometrically [26]) [26]
0.018 <9> (5β-dihydrocortisol) [24]
0.0184 <11> (cholic acid) [25]
0.019 <9> (3α-androstanediol, <9> $NADP^+$, detection radio- or fluorometrically [26]) [26]
0.019 <9> (5α-dihydrotestosterone, <9> NADPH, detection spectrophotometrically [26]) [26]
0.021 <9> (9,10-phenanthrenequinone, <9> NADPH, detection spectrophotometrically [26]) [26]
0.0223 <11> (5α-dihydrotestosterone) [25]
0.024 <9> (androsterone) [20]
0.03-0.077 <4> (camphorquinone, <4> 4-nitrobenzaldehyde, 2,3-heptanedione [6]) [6]
0.0311 <11> (androsterone) [25]
0.0422 <11> (androstanedione) [25]
0.048 <9> (androsterone, <9> $NADP^+$, detection radiometrically [26]) [26]
0.087 <9> (4-nitrobenzaldehyde, <9> NADPH, detection spectrophotometrically [26]) [26]
0.2-0.8 <4> (4-nitroacetophenone, <4> 2,3-hexanedione, 1-phenyl-1,2-propanedione, 2,3-pentanedione, pyridine-4-aldehyde, 2,3-butanedione [6]) [6]
0.235 <16> (5β-pregnan-3α-ol-20-one, <16> cytosolic NADH-linked 5β-dihydroprogesterone 3α-hydroxysteroid oxidoreductase [39]) [39]
0.4 <16> (NADH, <16> cytosolic NADH-linked 5β-dihydroprogesterone 3α-hydroxysteroid oxidoreductase [39]) [39]
0.64 <13> (NAD^+) [31]
0.87 <1> (NAD^+) [1]
0.91 <16> (NADPH, <16> cytosolic NADH-linked 5β-dihydroprogesterone 3α-hydroxysteroid oxidoreductase [39]) [39]
1.33 <16> (NADPH, <16> microsomal NADPH-linked 5β-dihydroprogesterone 3α-hydroxysteroid oxidoreductase [39]) [39]
Additional information <3> (<3> values for various substrates and different isozymes [5]) [5]

pH-Optimum

6.5 <4> (<4> reduction with NADPH [6]) [6]
7 <3, 4> (<4> reduction of 3-oxo-steroid [6]; <3> α-HSD I [35]) [6, 35]
7.9 <1> (<1> reduction of etiocholan-3,17-dione [1]) [1]
8.5 <3> (<3> oxidation with NAD^+ [5]) [5]
9.5 <4> (<4> oxidation of 3α-hydroxysteroids [6]) [6]
10 <3, 13> (<3> oxidation with $NADP^+$ [5]; <13> oxidation of 3α-hydroxysteroids [31]) [5, 31]
10-11 <1> (<1> oxidation of cholate [27]) [27]
11.4 <6> [18]

pH-Range

6.8-7.4 <9> [24]
8.5-11 <13> [31]

Temperature optimum (°C)

37 <13> [31]
42 <15> (<15> assay at [41]) [41]
45-47.5 <3> [10]
50 <2, 8> [15, 22, 34]

Temperature range (°C)

16-36 <4> (<4> increase of activity over range [6]) [6]
28-46 <13> [31]

4 Enzyme Structure

Molecular weight

25000 <1> (<1> non-denaturing gel electrophoresis and gel filtration [27]) [27]
27000 <10> [21]
32500-34000 <3, 4> (<3, 4> gel filtration [4, 5, 6, 10]) [4-6, 10]
33000 <3> (<3> gel filtration [35]) [35]
34000 <3> [33]
35000 <12> (<12> gel filtration [28]) [28]
36000 <3> (<3> non-dissociating electrophoresis [23]) [23]
37000 <9> (<9> gel filtration [24]) [24]
45000 <1> (<1> sedimentation equilibrium centrifugation [1]) [1]
47000 <13> (<13> gel filtration [31]) [31]
47100 <2> (<2> gel filtration, concentration dependent monomer-dimer association [3]) [3, 15]
49400 <11> (<11> gel filtration [25]) [25]
66000 <8> (<8> gel filtration [22]) [22]
100000 <2> (<2> gel filtration, concentration dependent monomer-dimer association [3]) [3]
205000 <6> (<6> gel filtration [16]) [16]

Subunits

? <8> (<8> ? * 66000, SDS-PAGE [22,34]) [22, 34]
dimer <2, 11> (<2> 2 * 47100, concentration dependent monomer-dimer association [3]; <11> 2 * 26400, SDS-PAGE and sequence analysis [25]) [3, 25]
monomer <1-4, 9, 10, 12> (<3> 1 * 31000, SDS-PAGE [4]; <3, 4> 1 * 34000, SDS-PAGE [6, 33]; <1> 1 * 45000, sedimentation equilibrium, presence of 2-mercaptoethanol [1]; <2> 1 * 47100, concentration dependent monomer-dimer association [3]; <10> 1 * 27000, SDS-PAGE [21]; <3> 1 * 36000, SDS-PAGE [23]; <9> 1 * 37000, SDS-PAGE [24]; <1> 1 * 25000, SDS-PAGE [27]; <12> 1 * 35000, SDS-PAGE [28]) [1, 3, 4, 6, 21, 23, 24, 27, 28, 33]

5 Isolation/Preparation/Mutation/Application

Source/tissue

brain <3> (<3> distribution in brain regions of cytosolic and particulate 5α-dihydroprogesterone 3α-hydroxysteroid oxidoreductase activity [36,38]; <3> NADH-linked 5α-dihydroprogesterone 3α-hydroxysteroid oxidoreductase activity is decreased by 40-50% in lactating rats [38]) [4, 36, 38]
cell suspension culture <15> (<15> low activity [41]) [41]
colon <9> [26]
embryo <15> (<15> somatic, dark-grown [42]; <15> preembryonic mass [42]) [42]
epididymis <3> [7]
heart <9> [26]
hypophysis <3> (<3> female [23]; <3> anterior [38]; <3> cytosolic and particulate 5α-dihydroprogesterone 3α-hydroxysteroid oxidoreductase activity, increase of both enzymes during lactation [38]) [23, 38]
hypothalamus <3> (<3> cytosolic and particulate 5α-dihydroprogesterone 3α-hydroxysteroid oxidoreductase activity [38]) [37, 38, 40]
kidney <9> [26]
leaf <15> (<15> low activity in young leaves [41]; <15> illuminated or light grown leaves [42]) [41, 42]
liver <3, 4, 8, 9, 14> [5, 6, 11, 14, 19, 22, 24, 26, 28, 32-35]
lung <9> [26]
ovary <3, 9> [9, 26]
prostate gland <3, 5, 9> [8, 10, 19, 26, 30]
shoot culture <15, 16> [39, 41]
skeletal muscle <9> [26]
small intestine <9> [26]
testis <3> [7]
thymus <9> [26]
Additional information <9> (<9> variations in different tissues concerning number of enzyme isoforms [20]; <9> type 2 isoform is expressed in several tissues [26]) [20, 26]

Localization

cytoplasmic membrane <3> (<3> NADH-linked 5α-dihydroprogesterone 3α-hydroxysteroid oxidoreductase activity [36-38, 40]; <3> integral membrane protein [36]) [36-38, 40]

cytosol <1, 3-5, 8, 9, 12-16> (<3> NADPH-linked 5α-dihydroprogesterone 3α-hydroxysteroid oxidoreductase activity [36-38, 40]; <16> NADH-linked 5β-dihydroprogesterone 3α-hydroxysteroid oxidoreductase activity, 97% in the cytosol [39]; <15> NADPH as cofactor [41]) [4-10, 22-24, 27, 28, 31-41]

microsome <3, 5, 9, 16> (<16> low activity of NADPH-linked 5β-dihydroprogesterone 3α-hydroxysteroid oxidoreductase activity [39]) [8, 11, 14, 20, 39, 40]

Purification

<1> [2, 27]

<3> (multiple forms [5]; partial [10, 11, 14]; three forms I, II, III [35]; micropurification procedure [23]) [4, 5, 10, 11, 14, 23, 35]

<4> [6]

<8> [22, 34]

<9> (recombinant from Escherichia coli [26]) [20, 24, 26]

<10> (partial [21]) [21]

<11> (recombinant His-tagged from Escherichia coli [25]) [25]

<12> [28]

<13> [31]

<14> [32]

Crystallization

<1> (association to dimer during crystallization [2]) [1, 2, 27]

<9> (α/β-barrel, $NADP^+$ is bound in an extended conformation, type III isoform [29]; type III isoform [30]) [29, 30]

Cloning

<3> (subclones reveal enzyme function in inflammation due to sequence homology to hydroxyprostaglandin dehydrogenase, prostaglandin $F_{2\alpha}$ synthase [33]; expression in Escherichia coli [19]) [19, 33]

<9> (expression in Escherichia coli [19]; expression of type II isoform in Escherichia coli [26]; expression in baculovirus system with Sf9 cells [20]) [19, 20, 26]

<10> (overexpression in Escherichia coli [21]) [21]

<11> (overexpression as His-tagged fusion protein in Escherichia coli [25]) [25]

Application

medicine <13> (<13> enzyme as part of commercial enzyme test kit for bile acid content in serum as an index for hepatic diseases [31]) [31]

6 Stability

pH-Stability

5-10.5 <1> (<1> stable for 30 min at 30°C [1]) [1]
6-10 <1> [27]

Temperature stability

40 <1> (<1> 15 min stable in presence of mercaptoethanol, 10% loss of activity in absence of mercaptoethanol [1]) [1]
45 <1> (<1> heating for 10 min, rapid heat denaturation over 50°C [27]) [27]
45 <14> (<14> 10 min with $NAD(P)^+$ [32]) [32]
50 <3, 6> (<3> denaturation above [10]; <6> 30-60 min stable [18]) [10, 18]

General stability information

<1>, mercaptoethanol stabilizes [1]
<1>, stable in solution for a long period [27]
<4>, glycerol: 30%, partially stabilizes [6]
<8>, not stable at 4°C [22]
<9>, stable in phosphate 50 mM plus 10-20% glycerol at 4°C, while freezing and thawing results in a great loss of enzyme activity [24]

Storage stability

<1>, -80°C, lyophilized or ammonium sulfate precipitated, more than 1 year [27]
<1>, 0-3°C, lyophilized, at least 6 months [1]
<1>, 4°C, diluted in Tris-HCl, pH 8.0, with ß-mercaptoethanol, 20 days [27]
<3>, -20°C, 25% v/v glycerol, at least 1 year or at 4°C, 1 week, stability depends on isozyme [5]
<3>, -80°C, 20 mM potassium phosphate buffer, pH 7.0, 0.1 M EDTA, 1 mM mercaptoethanol, 20% glycerol, 3 months, 10-15% loss of activity [4]
<3>, 4°C, 14 days [11]
<4>, -20°C, purified enzyme, at least 2 weeks, crude extract loses 80% activity at 4°C in 3 days [6]
<6>, 4°C, 2 days [18]
<8>, -20°C to -80°C, plus glycerol and 2-mercaptoethanol [22]

References

[1] Uwajima, T.; Terada, O.: Properties of crystalline 3α-hydroxysteroid dehydrogenase from Pseudomonas putida. Agric. Biol. Chem., **43**, 1521-1528 (1979)

[2] Uwajima, T.; Takayama, K.; Terada, O.: Production, purification and crystallization of a 3α-hydroxysteroid dehydrogenase of Pseudomonas putida. Agric. Biol. Chem., **42**, 1577-1583 (1978)

[3] Skalhegg, B.A.: On the 3 α-hydroxysteroid dehydrogenase from Pseudomonas testosteroni. Purification and properties. Eur. J. Biochem., **46**, 117-125 (1974)

[4] Penning, T.M.; Sharp, R.B.; Krieger, N.R.: Purification and properties of 3α-hydroxysteroid dehydrogenase from rat brain cytosol. Inhibition by nonsteroidal anti-inflammatory drugs and progestins. J. Biol. Chem., **260**, 15266-15272 (1985)

[5] Ikeda, M.; Hattori, H.; Ikeda, N.; Hayakawa, S.; Ohmori, S.: Purification and characterization of the multiple forms of 3α-hydroxysteroid dehydrogenase in rat liver cytosol. Hoppe-Seyler's Z. Physiol. Chem., **365**, 377-391 (1984)

[6] Hara, A.; Inoue, Y.; Nakagawa, M.; Naganeo, F.; Sawada, H.: Purification and characterization of $NADP^+$-dependent 3α-hydroxysteroid dehydrogenase from mouse liver cytosol. J. Biochem., **103**, 1027-1034 (1988)

[7] Hastings, C.; Hansson, V.: Kinetic properties of the soluble 3α-hydroxysteroid dehydrogenase from rat testis and epididymis. J. Steroid Biochem., **14**, 705-711 (1981)

[8] Jacobi, G.H.; Moore, R.J.; Wilson, J.D.: Characterization of the 3α-hydroxysteroid dehydrogenase of dog prostate. J. Steroid Biochem., **8**, 719-723 (1977)

[9] Nimrod, A.; Lamprecht, S.A.; Lindner, H.R.: Separation and characterization of two 3-hydroxysteroid oxidoreductases from the soluble fraction of rat ovarian cells. J. Steroid Biochem., **6**, 1205-1209 (1975)

[10] Inano, H.; Hayashi, S.; Tamaoki, B.-i.: Prostate 3α-hydroxysteroid dehydrogenase: its partial purification and properties. J. Steroid Biochem., **8**, 41-46 (1977)

[11] Golf, S.W.; Graef, V.; Nowotny, E.: Solubilization and purification of a 3α-hydroxysteroid dehydrogenase in rat liver microsomes (authors transl). Hoppe-Seyler's Z. Physiol. Chem., **357**, 35-40 (1976)

[12] Skalhegg, B.A.: 3α-hydroxysteroid dehydrogenase from Pseudomonas testosteroni: kinetic properties with NAD and its thionicotinamide analogue. Eur. J. Biochem., **50**, 603-609 (1975)

[13] Jarabak, J.; Talalay, P.: Stereospecificity of hydrogen transfer by pyridine nucleotide-linked hydroxysteroid dehydrogenase. J. Biol. Chem., **235**, 2147-2151 (1960)

[14] Boutin, J.A.; Shikata, M.; Talalay, P.: Rat liver microsomal 3α-hydroxysteroid dehydrogenase and dihydrodiol dehydrogenase: solubilization, separation and partial purification. Biochem. Biophys. Res. Commun., **135**, 795-801 (1986)

[15] Squire, P.G.; Dehlin, S.; Porath, J.: Physical and chemical characterization of hydroxysteroid dehydrogenases from Pseudomonas testosteroni. Biochim. Biophys. Acta, **89**, 409-421 (1964)

[16] MacDonald, I.A.; Jellett, J.F.; Mahony, D.E.; Holdeman, L.V.: Bile salt 3α- and 12α-hydroxysteroid dehydrogenases from Eubacterium lentum and related organisms. Appl. Environ. Microbiol., **37**, 992-1000 (1979)

[17] Mahony, D.E.; Meier, C.E.; MacDonald, I.A.; Holdeman, L.V.: Bile salt degradation by nonfermentative clostridia. Appl. Environ. Microbiol., **34**, 419-423 (1977)

[18] MacDonald, D.E.; Mahony, D.E.; Jellet, J.F.; Meier, C.E.: NAD-dependent 3α- and 12α-hydroxysteroid dehydrogenase activities from Eubacterium lentum ATCC no. 25559. Biochim. Biophys. Acta, **489**, 466-476 (1977)

[19] Biswas, M.G.; Russel, D.W.: Expression cloning and characterization of oxidative 17β- and 3α-hydroxysteroid dehydrogenase from rat and human prostate. J. Biol. Chem., **272**, 15959-15966 (1997)

[20] Chetyrkin, S.V.; Belyaeva, O.V.; Gough, W.H.; Kedishvili, N.Y.: Characterization of a novel type of human microsomal 3α-hydroxysteroid dehydrogenase. J. Biol. Chem., **276**, 22278-22286 (2001)

[21] Mallonee, D.H.; Lijewski, M.A.; Hylemon, P.B.: Exression in Escherichia coli and characterization of a bile acid-inducible 3α-hydroxysteroid dehydrogenase from Eubacterium sp. strain VPI 12708. Curr. Microbiol., **30**, 259-263 (1995)

[22] Morikawa,T.; Tamaoki, B.-I.: Purification and characterization of 3α-hydroxysteroid dehydrogenase from chicken hepatic cytosol. J. Steroid Biochem. Mol. Biol., **37**, 569-574 (1990)

[23] Campbell, J.S.; Karavolas, H.J.: Purification of the NADPH:5α-dihydroxyprogesterone 3α-hydroxysteroid oxidoreductase from female rat pituitary cytosol. J. Steroid Biochem. Mol. Biol., **37**, 215-222 (1990)

[24] Iyer, R.B.; Binstock, J.M.; Schwartz, I.S.; Gordon, G.G.; Weinstein, B.I.; Southern, A.L.: Purification and properties of human hepatic 3α-hydroxysteroid dehydrogenase. J. Steroid Biochem. Mol. Biol., **43**, 343-349 (1992)

[25] Maser, E.; Möbus, E.; Xiong, G.: Functional expression, purification, and characterization of 3α-hydroxysteroid dehydrogenase/carbonyl reductase from Comamonas testosteroni. Biochem. Biophys. Res. Commun., **272**, 622-628 (2000)

[26] Lin, H.-K.; Jez, J.M.; Schlegel, B.P.; Peehl, D.M.; Pachter, J.A.; Penning, T.M.: Expression and characterization of recombinant type 2 3α-hydroxysteroid dehydrogenase (HSD) from human prostate: demonstration of bifunctional 3α/17β-HSD activity and cellular distribution. Mol. Endocrinol., **11**, 1971-1984 (1997)

[27] Adachi, O.; Toyama, H.; Matsushita, K.: Purification and properties of 3α-hydroxysteroid dehydrogenase from Pseudomonas putida. Biosci. Biotechnol. Biochem., **61**, 131-137 (1997)

[28] Nanjo, H.; Terada, T.; Umemura, T.; Nishinaka, T.; Mizoguchi, T.; Nishihara, T.: Characterization of bovine liver cytosolic 3α-hydroxysteroid dehydrogenase and its aldo-keto reductase activity. Int. J. Biochem., **24**, 815-820 (1992)

[29] Jin, Y.; Stayrook, S.E.; Albert, J.H.; Palackal, N.T.; Penning, T.M.; Lewis, M.: Crystal structure of human type III 3α-hydroxysteroid dehydrogenase/bile acid binding protein complexed with NADP(+) and ursodeoxycholate. Biochemistry, **40**, 10161-10168 (2001)

[30] Zhu, D.W.; Cantin, L.; Nahoum, V.; Rehse, P.; Luu-The, V.; Labrie, F.; Breton, R.; Lin S.X.: Crystallization and preliminary X-ray crystallographic analysis of the human type 3 3α-hydroxysteroid dehydrogenase at 1.8 A resolution. Acta Crystallogr. Sect. D, **57**, 589-591 (2001)

[31] Gao, X.; Liu, J.; Li, G.: Purification and properties of 3α-hydroxysteroid dehydrogenase and 3-keto-5β-steroid-Δ^4-dehydrogenase. Ann. N. Y. Acad. Sci., **864**, 383-392 (1998)

[32] Sawada, H.; Hara, A.; Ohmura, M.; Nakayama, T.; Deyashiki, Y.: Kinetic and stereochemical characterization of hamster liver 3α-hydroxysteroid dehydrogenase and 3α(17β)-hydroxysteroid dehydrogenase. J. Biochem., **109**, 770-775 (1991)

[33] Pawlowski, J.; Huizinga, M.; Penning, T.M.: Isolation and partial characterization of a full length cDNA clone for 3α-hydroxysteroid dehydrogenase: a potential target enzyme for nonsteroidal anti.inflammatory drugs. Agents Actions, **34**, 289-293 (1991)

[34] Morikawa, T.; Tamaoki, B.: Dehydrogenase from chicken hepatic cytosol. J. Steroid Biochem. Mol. Biol., **37**, 569-574 (1990)

[35] Stolz, A.; Takikawa, H.; Sugiyama, Y.; Kuhlenkamp, J.; Kaplowitz, N.: 3α-hydroxysteroid dehydrogenase activity of the Y'bile acid binders in rat liver cytosol. Identification, kinetics, and physiologic significance. J. Clin. Invest., **79**, 427-434 (1987)

[36] Li, X.; Bertics, P.J.; Karavolas, H.J.: Regional distribution of cytosolic and particulate 5α-dihydroprogesterone 3α-hydroxysteroid oxidoreductases in female rat brain. J. Steroid Biochem. Mol. Biol., **60**, 311-318 (1997)

[37] Krause, J.E.; Karavolas, H.J.: Subcellular location of hypothalamic progesterone metabolizing enzymes and evidence for distinct NADH- and NADPH-linked 3α-hydroxysteroid oxidoreductase activities. J. Steroid Biochem., **13**, 271-280 (1980)

[38] Karavolas, H.J.; Hodges, D.R.: Changes in pituitary, hypothalamic and brain progestin-metabolizing enzyme activities during lactation. J. Steroid Biochem. Mol. Biol., **44**, 299-303 (1993)

[39] Seitz, H.U.; Gaertner, D.E.: Enzymes in cardenolide-accumulating shoot cultures of Digitalis purpurea. Plant Cell Tissue Organ Cult., **38**, 337-344 (1994)

[40] Bertics, P.J.; Edman, C.F.; Karavolas, H.J.: Potent inhibition of the hypothalamic progesterone 5α-reductase by a 5α-dihydroprogesterone analog. J. Biol. Chem., **259**, 107-111 (1984)

[41] Stuhlemmer, U.; Kreis, W.: Cardenolide formation and activity of pregnane-modifying enzymes in cell suspension cultures, shoot cultures and leaves of Digitalis lanata. Plant Physiol. Biochem., **34**, 85-91 (1996)

[42] Lindemann, P.; Luckner, M.: Biosynthesis of pregnane derivatives in somatic embryos of Digitalis lanata. Phytochemistry, **46**, 507-513 (1997)

Printing: Saladruck, Berlin
Binding: Stein+Lehmann, Berlin